S0-DUU-619

INTERNATIONAL CENTRE FOR MECHANICAL SCIENCES

COURSES AND LECTURES - No. 275

ROCK FRACTURE MECHANICS

EDITED BY

H.P. ROSSMANITH

TECHNICAL UNIVERSITY OF VIENNA

SPRINGER - VERLAG

WIEN - NEW YORK

Printed in Italy

ISBN 3-211-81747-6 Springer Verlag Wien-New York
ISBN 0-387-81747-6 Springer Verlag New York-Wien

In Memoriam
Prof.Dr. H. Parkus
The Untiring Devoted Promoter
Of Applied Mechanics.

CONTENTS

PREFACE

I. BASICS OF ROCK FRACTURE MECHANICS 1
(R.A.Schmidt - H.P.Rossmanith)

1. Introduction 1
2. Linear Elastic Fracture Mechanics (LEFM) 3
   2.1. The Stress Intensity Factor 5
   2.2. Fracture Criteria 10
   2.3. Fracture Toughness 11
   2.4. Crack Tip Zone of Micro-Cracking in Rock 12
   2.5. Subcritical Crack Growth 17
3. Development of Fracture Toughness Testing 19
   3.1. CT-Specimen and 3PB-Specimen Testing 21
   3.2. Anisotropy 25
   3.3. Effects of Hydrostatic Compression 25
4. Closure 26
5. References 27

II. ANALYSIS OF CRACKS RELATED TO ROCK FRAGMENTATION 31
(F.Ouchterlony)

1. Introduction to Blasting Configurations 31
2. Idealized Crack Systems 32
3. Complex Representation and Conformal Mapping 37
   3.1. Method 37
   3.2. Results 40
   3.3. Formal Approach to Uniform Growth 52
4. Path-Independent Integrals 57
   4.1. Method 57
   4.2. Results and Applications 61
5. References 66

III. FRACTURE TOUGHNESS TESTING OF ROCK 69
(*F. Ouchterlony*)

1. Review of Toughness Testing 69
1.1. On Specimen Geometries 70
1.2. Specific Work of Fracture 70
1.3. Griffith's Balance of Energy Rates 72
1.4. Fracture Toughness 74
1.4.1. Validity of Metals Testing Criteria 75
1.4.2. Other Aspects 79
1.5. J-Integral Resistance 82
1.5.1. Applicability of J_{Ic} Test Practice for Metals to Rock 83
1.5.2. J_{Ic}-Measurements on Rock 85
1.6. Anisotropy Effects 90
1.6.1. Material Description 91
1.6.2. Crack Growth Resistance Values 92
1.7. Conclusions 96

2. Development of Core Bend Specimens 98
2.1. Single-Edge-Crack-Round-Bar in Bending (SECRBB) 98
2.1.1. Experimental Procedure 101
2.1.2. Results 105
2.1.3. Further Fracture Mechanics Formulas 107
2.2. Chevron-Edge-Notch-Round-Bar in Bending (CENRBB) 110

3. Crack Resistance Measurements on Core Specimens 114
3.1. Simple R-Curve Approach to SECRBB Testing 115
3.1.1. Prediction Formulas 115
3.1.2. Energy Rate Crack Resistance Data 118
3.1.3. Conclusions 121
3.2. Direct R-Curve Measurements on SECRBB Specimens 121
3.2.1. R-Curves from Complete Failure Curves 122
3.2.2. R-Curves from Sub-Critical Failure Cycles 131
3.2.3. Conclusions 141
3.3. Conclusions from Core Toughness Data 141

4. References 145

IV. NUMERICAL MODELLING OF FRACTURE PROPAGATION 151
(*A.R. Ingraffea*)

1. Introduction 151

2. The Nature of Fracture Propagation in Rock 155

3. Stress Intensity Factor Computation 159
3.1. A Historical Overview 160
3.2. Computation by Finite Element Method 165
3.3. Computation by Boundary Element Method 170

4. Theories of Mixed-Mode Fracture 174
4.1. The $\sigma_{\theta,max}$-Theory 176
4.2. The $S(\theta)_{min}$-Theory 179
4.3. Comparison of Mixed-Mode Fracture Theories 181
4.4. Predicting Crack Increment Length 183

5. Fracture Propagation Programs 189
5.1. Numerical Methods 189
5.2. User-Computer Interface 190
5.3. Automatic Remeshing 191
5.4. Example Solutions 192

6. Fracture Propagation Modelling - The Future 199

7. References 204

V. Dynamic Photoelasticity and Holography Applied to Crack and Wave Propagation 209

(H.P.Rossmanith - W.L.Fourney)

1. Photoelasticity 210

2. High-Speed Photography and Requirements of an Optimum Photographic System in Dynamic Photoelasticity 213
2.1. Cranz-Schardin Camera 215

3. Dynamic Holography 217

4. Applications 220
4.1. Wave Propagation 220
4.2. Fracture Mechanics 222
4.3. Crack-Wave Interaction 222

5. References 227

VI. Elastic Wave Propagation 229

(H.P.Rossmanith)

1. Waves in Unbounded Media 229
1.1. Plane Waves 229
1.2. Spherical and Cylindrical Waves 232
1.2.1. Spherical Waves 233
1.2.2. Cylindrical Waves 233
1.3. Superposition of Elastic Waves 233

2. Boundary Effects 234
2.1. Cohesive Joints 234
2.2. Loose Joints 238
2.3. Plate Waves 240

2.4. Surface Waves 241
2.4.1. Glancing Angle Diffraction 241
2.4.2. Rayleigh-Waves 241
2.4.3. P-Wave Propagation in Layers 241
2.5. Nonplanar Wave Fronts at Boundaries and Interfaces 243
3. Dynamic Layer Detachment and Spallation 249
References 251

VII. ANALYSIS OF DYNAMIC PHOTOELASTIC FRINGE PATTERNS 253
(H.P.Rossmanith)

1. Identification of Isochromatic Fringes 254
2. Data Analysis 256
2.1. Boundary Conditions 256
2.2. Wave Form Conditions 257
3. Fracture Analysis Using Photoelastic Data 358
3.1. Mixed-Mode Fracture Problem 259
3.2. A Multiparameter Approach 262
4. Crack Speed versus Stress Intensity Factor Characterization of Dynamic Fracture 268
5. References 269

VIII. DYNAMIC CRACK ANALYSIS AND THE INTERACTION BETWEEN CRACKS AND WAVES 271
(H.P.Rossmanith)

1. The Moving Crack 271
1.1. Discontinuous Change of Crack Speed 279
2. Crack-Wave Interaction 282
2.1. Cracks Subjected to Stress Wave Loading 282
2.1.1. Diffraction by a Stationary Crack 284
2.1.2. Crack Extension Following Wave Diffraction 286
2.2. Photoelastic Investigation of Crack-Wave Interaction 290
3. Interface Cracks and Joints 295
4. References 299

IX. FRACTURE CONTROL BLASTING 301
(*W.L.Fourney*)

1. Introduction 301
   1.1. Effect of Notches in a Borehole 302
   1.2. Control of Crack Initiation 303
2. Photoelastic Studies of Fracture Control 310
3. References 318

X. FRAGMENTATION STUDIES WITH SMALL FLAWS 321
(*W.L.Fourney*)

1. Introduction 321
2. Fragmentation of a Homogeneous Model 323
3. Effects of Small Flaws 332
4. Summary 338
5. References 340

XI. FRAGMENTATION STUDIES WITH LARGE FLAWS 341
(*W.L.Fourney*)

1. Introduction 341
2. Joint Initiated Fracture 344
3. Time Delays Between Boreholes 349
4. References 352

XII. GAS WELL STIMULATION STUDIES 353
(*W.L.Fourney*)

1. Introduction 353
2. Growth of Fractures from a Wellbore and Gas Flow into them 354
   2.1. Effect of Loading Rate 361
   2.2. Stem Induced Fracture 361
3. References 369

XIII. GROUND VIBRATION STUDIES 371
(*W.L.Fourney*)

1. Introduction 371
2. Theory 373
3. Results 375
4. Conclusion 380
5. References 381

XIV. MODELLING AND DEVELOPMENT OF HYDRAULIC FRACTURING TECHNOLOGY 383
(*M.P.Cleary*)

Summary 383
Introduction 384
1. Potential and Status of Fracturing Technology 391
1.1. Mechanisms of Fracture Creation in Rock 391
1.2. General Equations Governing Hydraulic Fracturing 394
1.3. Models of Fracture Processes, Existing Potential Technology 400
1.3.1. Modelling 400
1.3.2. Field Technology 401
2. First-Order Models and Design of Hydraulic Fractures 405
2.1. Mathematical Models 405
2.1.1. Equations Governing Lumped P3DH-Type Models 405
2.1.2. Algebraic Solutions of Lumped Model Equations 411
2.1.3. Numerical Solutions of Lumped Model Equations 416
2.2. Summary of P3DH-Model Equations and Results 416
2.2.1. Reduction of P3DH-Model to Ordinary Differential Equations 422
2.2.2. Self-Similar Approximations for Storage in Lateral Flow 425
2.2.3. Hybridisation of Self-Similar and O.D.E. Models 428
2.3. Hydrafrac Designs Based on Lumped Model Solutions 430
3. Detailed Theoretical Modelling Hydrafrac 439
3.1. Complete Simulation for a Representative Cross Section 440
3.2. Development of a Reference Circular Hydrafrac Model 440
3.3. Modelling of Multiple Fractures, Interaction with Reservoir Conditions of Stress, Pore Pressure and Material Variations 443

3.3. Modelling of Multiple Fractures, Interaction with Reservoir Conditions of Stress, Pore Pressure and Material Variations 443
3.4. Fracture Impedance Mechanisms, Branching and Slipping 445
3.5. Fully Three-Dimensional Simulation of Fracturing 445

4. Laboratory Simulation of Fracturing 448
4.1. Interface Separation Apparatus, DISLASH 449
4.2. Apparatus for Full 3-D-Fracture Growth and Interaction 454
4.3. Development of Data Acquisition and Control Systems 459
4.4. Monitoring of Fracture Growth in the Laboratory 460
4.5. Some Studies on Fluid Rheology 462

5. Laboratory Testing of Material Response 463
5.1. High Temperature Triaxial Test System 464
5.2. Permeability, PPIC, and High Pressure Triaxial Test Systems 466

List of References 467

AUTHOR INDEX 477

SUBJECT INDEX 481

PREFACE

The scientific branch of rock fracture mechanics serves the purpose of treating fracture problems in rock mechanics. With the known traditional fossil fuel reserves of the world being rapidly exhausted and the mushrooming of urban settlements in mining engineering areas ever increasing, attention has turned to alternative sources and also alternative and advanced technologies to be able to cope with mankind's ever growing demand and need of energy and minerals. Faster, safer and more efficient procedures for production of above and/or underground fractures and excavations with suitable extends and shapes are required. An understanding of the fracture mechanisms of rock is an essential prerequisite for designing mining excavations and civil engineering structures, for developing advanced rock-breaking processes and for establishing programs to prevent hazardous situations such as rock-bursts. The whole process of rock breakage is a complicated interaction of stress waves and crack propagation governed by material and environmental aspects. Knowledge of the basic principles of rock fracture and testing in parallel with numerical and experimental model studies allow for the development of optimal and controlled rock breakage procedures to consequently reduce the cost of mining operation.

The content of this course covers the fundamentals of rock fracture mechanics and rock materials testing with a firm stronghold on the application of rock fracture to modern problems in mining engineering and related fields.

Chapter I (R.A.Schmidt and H.P.Rossmanith) contains the basic principles of fracture mechanics as applied to rock. The stress analysis of radial crack systems associated with rock blasting is treated in Chapter II (F. Ouchterlony). A detailed review of fracture toughness testing of rock and

and specimen development is presented in Chapter III (F.Ouchterlony). Numerical modelling of crack propagation in rock and concrete is the subject of Chapter IV (A.R.Ingraffea). The application of dynamic photomechanics techniques to fracture and wave propagation phenomena in rock is the subject of nine chapters. Chapters V to VIII (H.P.Rossmanith) contain introductory material on dynamic photoelasticity, wave propagation, fringe data evaluation, dynamic fracture, and wave-crack interaction. Experimental model studies and their in situ application to fracture control, oil shale fragmentation and gas well stimulation is considered in Chapters IX to XIII (W.L.Fourney). The final Chapter XIV (M.P.Cleary) is devoted to research and application of hydraulic fracturing of rock-type materials.

On behalf of the lecturers I take great pleasure in expressing my sincere gratitude to the late Prof.Dr.H.Parkus and to the Secretary General of CISM, Prof.Dr.G.Bianchi, for having been given the challenging opportunity to organize and present this series of lectures to an interdisciplinary audience. Thanks to all members of CISM for their great hospitality which made our stay at the Centre so pleasant.

Finally, I am indepted to the Fonds zur Förderung der wissenschaftlichen Forschung in Austria for kindly supporting a research program on "Rock Fracture Mechanics" at the Technical University of Vienna, Austria, which gave the incentive to these CISM-lectures.

Vienna, June 1983 H.P.Rossmanith

BASICS OF ROCK FRACTURE MECHANICS

R.A. S c h m i d t
TerraTek International
Applied Geomechanics Division
Salt Lake City, Utah, USA

H.P. R o s s m a n i t h
Institute of Mechanics
Technical University Vienna, Austria

1. *Introduction*

An understanding of the mechanics and mechanisms of rock fracture is a key element in solving a great many engineering problems that involve geotechnical structures. A "geotechnical structure" may be simply a rock mass containing a fossile fuel such as coal, oil, or gas or a mineral with valuable elements such as copper, iron, or aluminum. The rock mass becomes a structure when man enters the picture by drilling a hole, boring a tunnel, or digging a longwall, room and pillar, or open pit mine. The methods used in extracting these fossil fuels and minerals or in simply making tunnels invariably involve rock fracture of one form or another. These take such forms as drilling, blasting, boring, and well stimulation by hydraulic fracturing and explosive fracturing.

With the exception of a few early investigations, rock fracture mechanics is a rather recent field of study. As an example of the rapid increase of interest in this field one notes that prior to 1977, the U.S.Symposium on Rock Mechanics typically had at most one or two papers per year dealing with fracture mechanics of rock, while in 1977 there were thirteen papers in two separate sessions devoted to the subject at the symposium.

A distinction should be made between what are commonly called failure mechanics and fracture mechanics. Rock failure mechanics generally refers to the global process of failure in a continuum sense by which a rock or rock mass undergoes permanent damage that affects its ability to sustain a load. Examples include rapid failure of a rock cylinder loaded slowly in compression in a laboratory load frame (press) or a sudden rock burst or roof collapse in a mine. On the other hand fracture mechanics, sometimes called crack mechanics, refers to the discrete propagation of an individual crack or cracks. This crack propagation is often cataclysmic in nature but can also occur slowly. Examples include the fracturing of a rock beam loaded in bending in the laboratory and hydraulic fracturing in which a single large fracture is made to propagate from an oil or gas well by the application of fluid pressure to increase its productivity. These and other examples will be discussed in more detail in later sections.

In order to understand many of the principles applied to the specific study of rock fracture mechanics it is important to understand first the basic principles of fracture mechanics in general. Fracture mechanics, or more specifically linear elastic fracture mechanics (LEFM), has become well developed in the past 25 years as engineers sought to understand the brittle failure of structures made of high stength metal alloys. The work of scientists and engineers has led to improved alloys, better design criteria, better inspection procedures, and standardized test methods for determining the important material properties.

A word of caution, however: LEFM principles were not developed with rock materials and geological structures in mind. While certain basic theories will apply, large differences in basic material response and engineering application between rock and metallic materials must be considered when adopting these principles, practice, and test methods.

For example, brittle rock often fails in tension with the inelastic behavior taking the form of microcracking at the crack tip; whereas even the most brittle metal alloys are known to possess an inelastic zone of plasticity formed by shear stresses at the crack tip. Also, one notes that standardized test methods for LEFM are based on conservative design criteria, since it is the prevention of failure that is usually desired in man-made structures. However, it is the creation and propagation of fractures that is desired in many rock fracture applications such as drilling and well stimulation. This difference may not impact the physics and mechanics of the fracture process itself, but the engineering application should dictate how the important parameters are measured and utilized.

2. Linear Elastic Fracture Mechanics (LEFM)

Unexpected and sudden brittle failures of structures have plagued engineers for some time. In the mid and late 1800's failure of railroad wheels, axles, and rails occurred almost daily in Great Britain resulting in an average 200 deaths per year. During World War II, 145 Liberty ships broke in half, some while simply sitting at the dock, due to the existence of flaws (cracks) and stress concentrations. In more recent times frequent structural failures have occurred in aircraft and aerospace applications. Weight-critical design, accurate stress analyses that have allowed for a reduction of safety factors, and development of high strength alloys have all resulted in structures that are sensitive to flaws. Many failures of aircraft such as the British Comet, American F-111, and C5A, as well as failures in rocket motor casings of the Polaris missile occurred at stresses below the design service loads. As a result linear elastic fracture mechanics was born and has grown rapidly since the late 1950's /1/.

Basically, the structural engineer and the mining engineer aim for different goals in their desire to put fracture to work. In contract to the structural engineers primary task to either reduce the probability of fracture by appropriate application of fracture mechanics methodology principles in design and fabrication of structures, the mining engineers attitude toward fracture culminates in optimization of fracture processes in individual areas such as optimal fragmentation, oil shale retort removal, hydro-fracturing, gas well stimulation, tunneling, etc.

The fundamental questions that arise in both the structural and mining engineering view point to fracture are 'how can one predict the failure load of these flawed structures?', 'what combination of load and flaw geometry parameters leads to failure?', and 'what material parameters is the fracture process governed by?'.

Many failure criteria and theories such as the well known Coulomb criterion /2,3/ exist that predict failure conditions for rock. However, these theories often deal directly with fracture processes. As a result, they cannot be expected to deal with questions of crack propagation such as (1) the length and width of a hydraulic fracture created in an oil or gas well, (2) conditions resulting in crack advance of a hydraulic fracture in geothermal applications, or (3) number and length of fractures caused by explosive or tailored-pulse loading of a borehole.

In the past 25 years, many investigations involving crack propagation in brittle metal alloys have employed the well-founded disciplin of linear elastic fracture mechanics, LEFM, with great success. Although this theory is based on linear elasticity and is directly related with the Griffith theory /4/, plastic flow and other nonlinear behavior can occur on a small scale without affecting its predictive success. Purely brittle behavior is not required and only when the size of the zone of nonlinear behavior at a crack tip cannot be considered small when compared to the crack length, does recourse to other fracture theories such as the J-integral /5/ become necessary.

LEFM applications to rock fracture are rather recent, occurring primarily within the last 10 years; the vast majority of investigations have actually been only within the last 5 years. A brief review of several fracture mechanis approaches to rock behavior is provided by Ref./6/. Interest in a fracture mechanics approach to fock fracture is now rather high and a recent review by Ouchterlony /7/ of fracture toughness testing of rock listed some 85 references. A new Subcommittee of the American Society for Testing and Materials entitled 'Fracture Testing of Brittle Non-Metallic Materials (E24.07)' has been created to set standards for fracture testing of ceramics, rock, and concrete.

2.1. *The Stress Intensity Factor*

The answers to the questions formulated above lie in a parameter known as the stress intensity factor, K. This parameter appears from a straightforward analysis of stresses at a crack-tip. A similar parameter known as the strain energy release rate, G, is derived from a simple energy approach that treats crack growth as an instability phenomenon. It is reassuring to know that crack criteria based on either parameter are essentially equivalent since it can be shown that K and G are directly related.

To analyze the field of stresses at the tip of a crack in a linear elastic body one must first define a crack. The simplest model defines a crack as line across and/or along which the displacement field exhibits a discontinuity. The opposing sides of this line, the crack faces, may or may not be stress free.

A crack can deform in three basic modes as shown in Fig.1. In mode-1, the opening mode, the crack surface displacements are perpendicular to the plane of the crack. This is the most commonly encountered mode in application

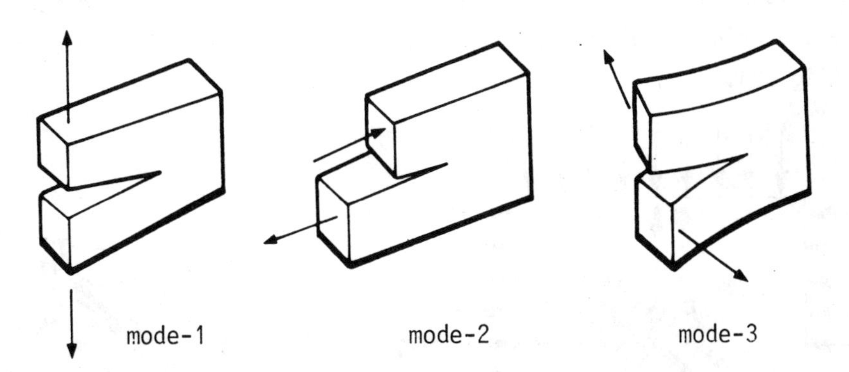

Fig.1 Three principal modes of crack loading

and it is the easiest to produce experimentally on laboratory specimens. In mode-2, the sliding mode, the crack surface displacements occur in the plane of the crack and perpendicular to the leading edge of the crack. In mode-3, the tearing mode, the crack surface displacements are also in the plane of the crack but parallel to the leading edge. Mode-3 is generally the easiest to analyze mathematically since there are no stresses or displacments produced in the plane perpendicular to the leading edge. In the discussions that follow only in-plane modes 1 and 2 will be given consideration.

LEFM /8/ is based on the stress intensity factor, K, which quantifies the intensity of the stress singularity at a crack tip. Fracture mechanics states that a crack will advance when its stress intensity reaches a critical value, K_c, assuming that the crack tip is in a state of plane strain. This value of K_c, known as plane strain fracture toughness, has been shown to be a measurable material constant for a vast number of metal alloys, glasses, polymers, and even some organic materials such as wood, paper and rubber. Many hundred of publications on rapid fracture, fatigue crack propagation, stress corrosion cracking, and crack branching have shown the stress intensity factor to be the controlling fracture parameter. Extensive ASTM standards /9/ have been defined for the measurement of K_c for metallic materials.

Consider a crack in a body loaded as shown in Fig.2a. Theory of elasticity - in particular the Muskhelishvili-Kolosov and Westergaard methods of stress analysis - provide a straightforward solution to this general problem /8,10,11/. The external (over burden pressure) loads and the internal pressure and fricitional forces can be reduced to a crack line loading system (Fig.2b).

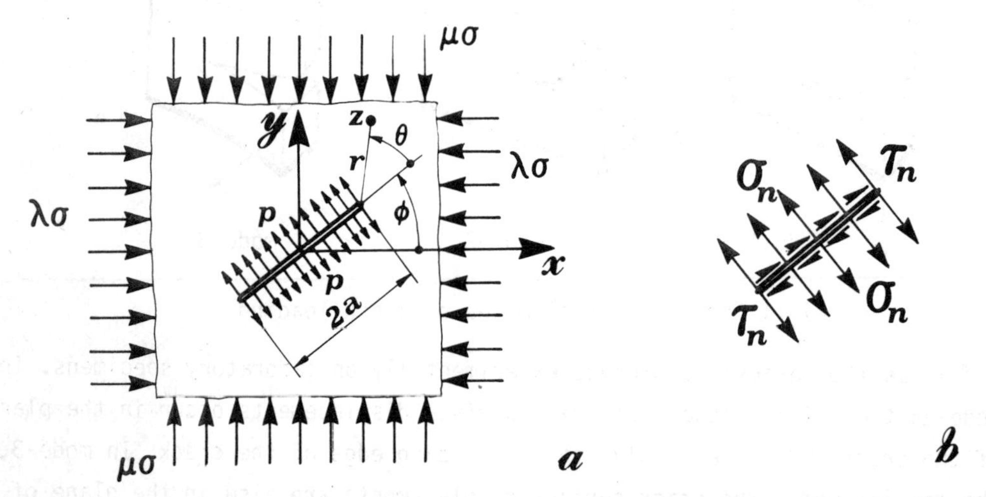

Fig.2 Crack under geomechanics loading conditions: a) natural loading configuration; b) crack line reduced loading system.

The general solution for the close-tip stress field components for mixed-mode loading (mode-1 and mode-2 operative) is given by

$$\begin{aligned}
\sigma_x &= \frac{K_1}{\sqrt{2\pi r}} \cos\frac{\theta}{2} \left(1-\sin\frac{\theta}{2}\sin\frac{3\theta}{2}\right) - \frac{K_2}{\sqrt{2\pi r}} \sin\frac{\theta}{2} \left(2+\cos\frac{\theta}{2}\cos\frac{3\theta}{2}\right) \\
\sigma_y &= \frac{K_1}{\sqrt{2\pi r}} \cos\frac{\theta}{2} \left(1+\sin\frac{\theta}{2}\sin\frac{3\theta}{2}\right) + \frac{K_2}{\sqrt{2\pi r}} \sin\frac{\theta}{2} \cos\frac{\theta}{2}\cos\frac{3\theta}{2} \\
\tau_{xy} &= \frac{K_1}{\sqrt{2\pi r}} \sin\frac{\theta}{2} \cos\frac{\theta}{2}\cos\frac{3\theta}{2} + \frac{K_2}{\sqrt{2\pi r}} \cos\frac{\theta}{2} \left(1-\sin\frac{\theta}{2}\sin\frac{3\theta}{2}\right)
\end{aligned} \qquad (1)$$

with the stress intensity factors K_1 and K_2 determined by

$$K_1 = \sigma_n \sqrt{\pi a} \qquad (2)$$

$$K_2 = \tau_{eff} \sqrt{\pi a} \qquad (3)$$

where

$$\sigma_n = \sigma(\mu \cos^2\phi + \lambda \sin^2\phi) - p \quad . \qquad (4)$$

If frictional forces are operative on the fracture surfaces for $\sigma_n < 0$,

$$\tau_{eff} = \begin{cases} \tau_{nt} & (\sigma_n > 0) \\ \tau_{nt} + (\rho\sigma_n - \tau_c)\,\mathrm{sgn}(\tau_{nt}) & (\sigma_n < 0) \end{cases} \qquad (5)$$

with

$$\tau_{nt} = \frac{1}{2}\sigma(\mu-\lambda)\sin 2\phi \quad , \qquad (6)$$

where p denotes the internal pressure, ρ is the coefficient of friction, and τ_c may be regarded as an inherent shear strength of the contacting fracture surfaces. Values of ρ and τ_c for some rock-type materials are given in Table I.

Table I: Values of ρ and τ_c for some rock-type materials (after Ref./3/)

Rock	ρ	τ_c(kPa)
Granite	0.64	310
Gabbro	0.66	380
Trachyte	0.68	415
Sandstone	0.51	275
Marble	0.75	1100

The complete solution for stresses throughout the generalized body include terms of higher order in r, but the stresses of eqs(1) are the only terms singular in r and are thus dominat at the crack tip. A detailed investigation shows that the stresses of eqs(1) are responsible for crack initiation; the higher order regular stress fields control crack growth and crack path stability.

Examination of eqs(1) shows that the singular stress field at the crack tip in any linear elastic body, no matter how large the loads or how long the crack, is always the same except for an undetermined constant, K. In other words, determination of the value of one parameter, K, for a particular problem determines the entire singular near-crack tip stress field. K, then, defines the 'intensity' of the stress field and depends on crack geometry and load configuration.

Take for example, a pressurized tunnel crack of length 2a in an infinite block loaded by uniform compressive stress, σ, at infinity (Fig.3a). Then, eqs(2) and (4) yield the stress intensity factor

$$K_1 = \sigma_n \sqrt{\pi a} = (p-\sigma) \sqrt{\pi a} \quad . \tag{7}$$

For a penny-shaped crack of radius a subjected to uniaxial external compression, σ,and internal pressure p (Fig.3b) the stress intensity factor for the open crack is given by

$$K_1 = \frac{2}{\pi} (p-\sigma) \sqrt{\pi a} \quad . \tag{8}$$

Many solutions for stress intensity can be derived from other solutions using the principle of superposition. Since linear elastic behavior is assumed, two solutions can be combined or superimposed to determine a third solution. This technique has actually been applied to derive equ.(7). Incidentally, changing the value of the homogeneous normal stress component acting parallel to the crack line will not affect the value of K.

The crack problems depicted in Fig.3 are just two of a great many crack geometries that have been solved for values of K. A compilation of stress intensity factors for various configurations can be found e.g. in Ref./12/. However, most of these solutions are based on situations that arise in man-made structures and several configurations that occur frequently in geologic

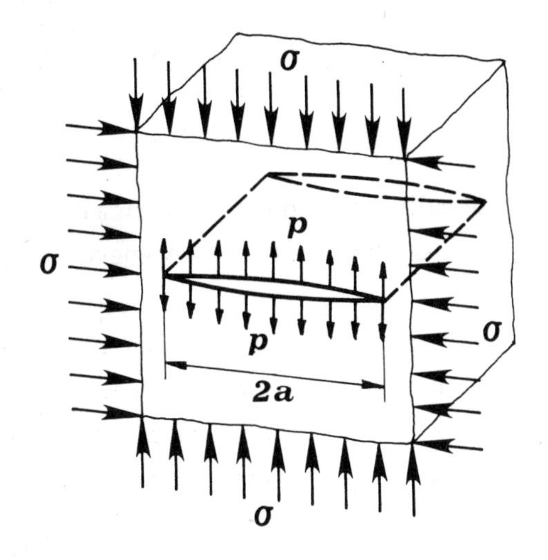

Fig.3a The pressurized tunnel crack

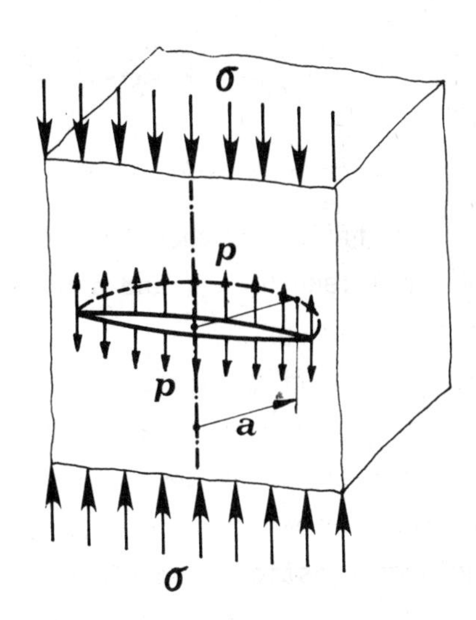

Fig.3b The pressurized circular crack

structures are not treated. For example, many rock fracture problems include finite boundaries, body loads, three-dimensional loading configurations, and pressurized cracks that don't often occur in man-made structures. Also, rock fracture problems often need solutions for displacements as well as stress intensity and these are often neglected.

Analytical methods such as that of Westergaard and the use of superposition to obtain solutions to problems are useful and compilations such as Ref./12/ can be helpful. But invariably a specific application or test specimen geometry will occur for which there is no readily available solution. Numerical techniques using finite and boundary element methods in modern computers have become so powerful and easy to use that one can rarely justify the time to solve these problems any other way. Most of these numerical methods now employ special crack tip elements that accomodate the high stress gradients that occur near crack tips. These methods will be discussed more fully in a following chapter by A.R.Ingraffea.

Crack surface displacements are often of interest. For the crack shown in Fig.3a these displacements, v, can be determined to be

$$v_{y=0} = 2\sqrt{a^2-x^2}\ \frac{1-\nu^2}{E}\ (p-\sigma)\quad . \tag{9}$$

2.2. *Fracture Criteria*

Griffith was the first to derive a formula relating the critical values of crack length, a_c, and applied stress, σ_c, on the basis of an energy balance concept.

The Griffith criterion /13/ states that crack growth will occur when the strain energy released by a virtual extension of the crack exceeds the energy absorbed by the material in that crack advance. The energy absorbed in crack propagation is assumed to be constant for a given material and includes plasticity and free surface energy.

Returning to the Griffith' crack of Fig.3a, it can be shown that the strain energy per unit thickness, U, due to the presence of the crack is

$$U = \pi\sigma^2\, a^2/E' \tag{10}$$

with $E'=E$ for plane stress and $E'=E/(1-\nu^2)$ for plane strain. The strain energy released when the crack grows an amount da is

$$-\,dU/da = 2\pi\sigma^2 a/E' \quad . \tag{11}$$

The strain energy release rate at one crack tip, G, is half that value or

$$G = \pi\sigma^2 a/E' \quad . \tag{12}$$

Comparing equ.(12) with equ.(2) we find that for mixed-mode problems K_1 and K_2 and G are related as follows

$$G = (K_1^2 + K_2^2)/E' \quad . \tag{13}$$

The Griffith' criterion states that crack growth will occur when G reaches a critical value, G_c, and fracture instability is achieved when

$$(G_c)_{mat} = (G)_{analyt.} \tag{14}$$

Equ.(13) shows that this criterion is essentially identical with that of crack growth when K reaches the critical value of K_c,

$$(K_c)_{mat} = (K)_{analyt.} \tag{15}$$

More sophisticated fracture criteria have been developed for anisotropic materials and combined-mode loading conditions /14/.

2.3. *Fracture Toughness*

The critical parameter K_c is known as the fracture toughness. When measured under certain conditions this parameter has been shown to be a true material constant for a great many materials.

Returning to the underground hydrafrac problem of Fig.3a or Fig.3b, the Griffith fracture criterion states that a certain combination of excess pressure $p_e=p-\sigma$ and crack length (or diameter) 2a is required to keep the fracture open or even increase its dimensions. Of course, the longer the crack in a given rock structure subjected to constant overburden (confining) pressure, σ, the lower the fluid pressure, p, required. The higher the fracture toughness, K_c, the higher the fluid pressure, p, required to open and initiate the crack.

Values of K_c and in particular its lower limit value, K_{Ic}, for plane strain, have been determined for a large number of materials following standard test methods and specifications. Many specimen configurations can be used and standard test methods have been established for fracture toughness testing of metallic materials. Whereas the development of such standard methods for toughness testing of rock materials has started only recentlydue to the relatively recent interest in fracture toughness of rock and the extreme variety in rock types and applications. Some representative values of fracture toughness and tensile strength are listed in Table II.

Table II: Typical values of fracture toughness and tensile strength

Material	Tensile Strength		Fracture Toughness	
	(MPa)	(ksi)	($MPa\sqrt{m}$)	($ksi\sqrt{in}$)
Indiana Limestone	5.4	0.78	1.0	0.9
Westerly Granite	13.7	1.98	2.6	2.4
Maraging 300 Steel	1900	280	57	52
4340 Steel	1400	200	68-90	62-82
7075-T6 Aluminum	540	78	18-29	16-26
Epoxy	70	10	0.7	0.6

Note that units of fracture toughness can be expressed as stress times square root of length; or, expressed in basic units: 1 MPa$\sqrt{m}$ = 1 MNm$^{-3/2}$ (1 psi$\sqrt{in}$ = 1000 lb-in$^{-3/2}$).

A failure criterion based on fracture toughness accounts for the degradation in load-carrying ability of a flawed structure. Conversely, if one knows the design load of a structure and the fracture toughness of the material from which it is constructed, then one can determine how large a crack that structure can tolerate without failure; or, how large a fracture fluid pressure is required to initiate fracture extension. In structural engineering applications this approach is commonly used to set flaw detection requirements for structures such as aircraft and pressure vessels.

While this failure criterion seems to be an obvious improvement over a simple tensile strength criterion which dos not account for structural flaws, the drawback comes in the difficulty in detecting and measuring flaws before a catastrophic failre occurs. With respect to underground mining engineering applications flaw size detection still represents an unresolved problem. Fracture mechanics seems to be applied all too often in post mortems when the cause of failure is being determined rather than in the design of the component to prevent failure, or, in mining engineering terms, in the design of optimally controlled fracture development. Stress analysis techniques have become so powerful with modern finite element techniques that determining stress is simple; the difficulty is in determining the size, shape, location, and orientation of cracks - and this holds especially for cracks in underground rock strata.

2.4. *Crack Tip Zone of Micro-Cracking in Rock*

By inspection one finds from eqs(1) that the linear-elastic crack tip stress field has stresses becoming infinite at the crack tip for all non-zero values of K. Infinite stress is an unacceptable physical concept, and the "error" is not in the mathematics, but our simplified model is not an accurate description of the physical state. Crack tips are never perfectly sharp and nonlinear behavior is known to take place in the immediate vicinity of the crack tip in even the most brittle materials. The inelastic behavior at the high stress levels near a crack tip is usually plasticity in metallic materials which gives rise to a plastic zone with finite stresses. In rock this inelastic behavior is often manifested by micro-cracking and a zone of micro-cracking forms at the crack tip /15-17,38/.

The development of micro-crack zone models for rock materials bases on the plastic zone models as developed for metallic materials. If the zone of inelastic behavior at the crack tip is small compared to the other dimensions in the problem, the crack tip stress field is still largely defined by the elastic parameter, K, and crack growth will occur when K reaches K_c. This condition of small inelastic behavior is known as small-scale yielding. The implication here is that a crack tip process uone occurs in rock, but that it is somewhat different from the plastic zone in metals. A description of this crack tip process zone for rock may provide an explanation of the fracture toughness effects and is needed to help provide guidelines for fracture testing. In order to develop a model for this crack tip process zone it may be helpful to review the steps that led to the description of the plastic zone that occurs at a crack tip in metals.

A rigorous solution for the size and shape of this plastic zone is not available, but an approximate description is obtained by determining the boundary from the elastic solution within which the stresses are greater than the yield stress. This is typically done /18/ by using a von Mises yield criterion

$$(\sigma_1 - \sigma_2)^2 + (\sigma_2 - \sigma_3)^2 + (\sigma_3 - \sigma_1)^2 = 2\sigma_{ys}^2 \quad , \tag{16}$$

where σ_{ys} is the uniaxial yield stress obtained from a standard tensile test.

Through a transformation of stresses for mode-1 (K_2=0), eqs(1) can be rewritten in terms of principal stresses as

$$\begin{aligned} \sigma_1 &= \frac{K_1}{\sqrt{2\pi r}} \cos\frac{\theta}{2} \left(1 + \sin\frac{\theta}{2} \right) \\ \sigma_2 &= \frac{K_2}{\sqrt{2\pi r}} \cos\frac{\theta}{2} \left(1 - \sin\frac{\theta}{2} \right) \\ \sigma_3 &= \nu(\sigma_1 + \sigma_2) \quad \text{(plane strain)} \quad , \end{aligned} \tag{17}$$

where ν is Poisson's ratio and conditions of plane strain have been assumed.

Substituting eqs(17) into equ.(16), and solving for r one obtains the shape of the plastic zone

$$r(\theta) = \frac{1}{4\pi} (K/\sigma_{ys})^2 \{\frac{3}{2} \sin^2\theta + (1-2\nu)^2(1+\cos\theta)\} \quad , \tag{18}$$

which is plotted in non-dimensional form in Fig.4 assuming a Poisson's ratio of 1/3 for the plane strain case. Recall that these boundaries merely define the region where the elastic stress solution predicts that yielding will occur. If the stresses inside this zone are limited to the yield stress, there must be an increase in stress outside the zone to carry the load. This results in an increase in the actual plastic zone size. It is estimated that the actual plastic zone has linear dimensions that are approximately double the lengths predicted from equ.(18) /8,18/.

Despite the fact that actual shapes of plastic zones differ from theoretical predictions the foregoing description is widely accepted for engineering purposes.

The variation of apparent fracture toughness of metals with crack length (Fig.5) can now be understood. The toughness values for small crack lengths differ from K_c because the relative size of the plastic zone is not small enough for linear elastic conditions to prevail. Years of testing many metallic alloys have resulted in an empirical determination, that measured toughness will be suffiently close to K_{Ic} when all in-plane specimen are at least 25 times larger than the plastic zone. The ASTM criterion for a valid test for K_{Ic} is

$$\text{crack length} \geqq 2.5 \, (K_{Ic}/\sigma_{ys})^2 \tag{19}$$

The apparent fracture toughness of rock has been shown to depend on crack length /19,20/(Fig.5), which is similar to the behavior of many metallic materials /21,22/. This dependence in metals is attributed to the size of the crack tip plastic zone relative to the dimensions of the specimen. For metals the effect of this plastic zone is also reflected in a thickness effect, but the fracture toughness of rock (and concrete) does not appear to be affected by thickness at all /19,23/. These findings indicate that a crack tip process zone exists for rock, but its desription must differ from the plastic zone model.

The stress state at a crack tip is not conducive to plastic flow in most

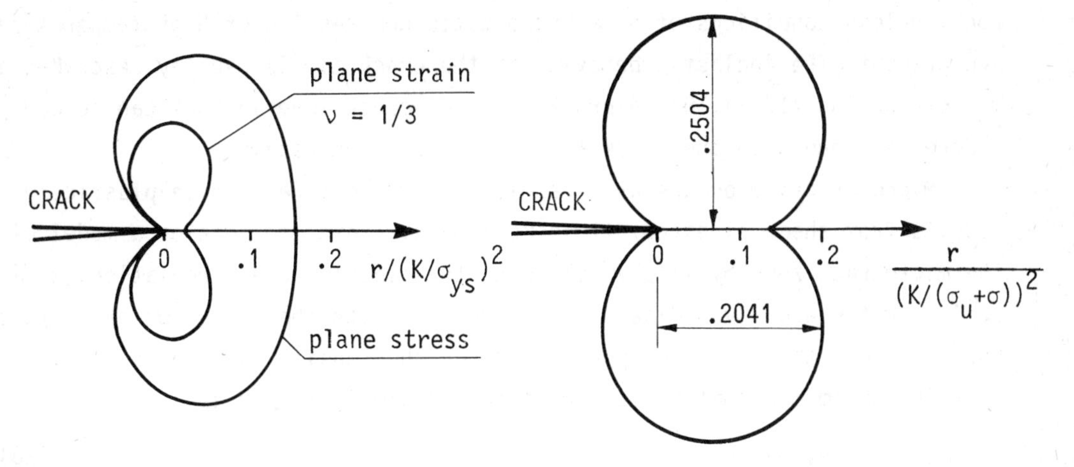

Fig.4 Plastic zone contours for Von Mises yield condition

Fig.6 Microcrack zone contour using maximum normal stress criterion

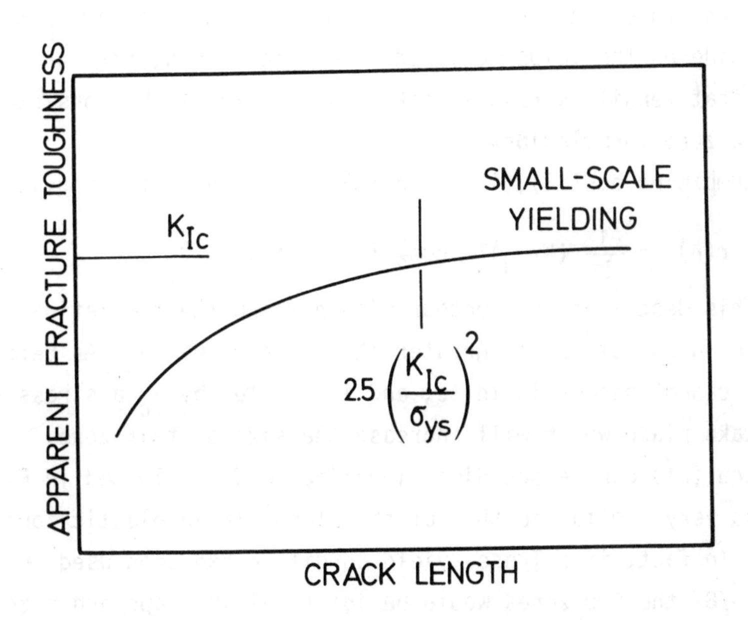

Fig.5 Effect of crack length on apparent fracture toughness generalized for metals

rock, unless conditions of high hydrostatic compression or high temperatures are present. The inelastic behavior at the crack tip is usually described as micro-cracking /17,24,38/. A crack tip microcrack zone in rock can be considered analogous to the crack tip plastic zone in metals.

Micro-cracking occurs as a result of tensile stress while plasticity results from shear stresses. It seems logical, then, to develop a model of the microcrack zone by simply following the plastic zone formulation, while replacing the usual von Mises yield condition with the criterion of maximal normal stress that is more appropriate for the tensile failure in rock.

The maximum normal stress criterion is simply stated as

$$\sigma_1 = \sigma_u \tag{20}$$

where σ_u is the ultimate tensile strength. The most appropriate choice for a value of σ_u for a given rock is unclear yet, since tensile tests are usually performed on samples that are considerably larger than the expected microcrack zone size. But, for lack of a better choice, σ_u will, for now, be considered the value measured in an unconfined, direct-pull, tension test. (Note that tensile stress is taken as posiitve to be consistent with the plastic zone formulation).

Combining eqs(16),(17), and (29) and solving for r gives

$$r(\theta) = \frac{1}{2\pi}(K/\sigma_u)^2 \{ \cos\frac{\theta}{2} (1 + \sin\frac{\theta}{2})\}^2 . \tag{21}$$

This describes the contour within which the theoretical (linear elastic) maximum normal stress is greater than some value, σ_u. As before, the actual stress cannot become infinite, and if limited by σ_u a stress redistribution must take place which will increase the size of this zone.

Equ.(21) can be non-dimensionalized and is plotted in Fig.6. The zone size is very similar to that of the plane strain plastic zone plotted in Fig.4. In fact, if a Tresca yield condition had been used instead of von Mises' /8/ the two zones would be identical in shape and size. The similar size probably accounts for the fact that the present fracture toughness data for rock, although limited, appear to follow a limiting criterion similar to that for metals /25/, equ.(19) with σ_{ys} replaced by σ_u.

The most obvious difference between the microcrack zone and the plastic zone models is that the microcrack zone, equ.(21), remains unchanged whether the out-of-plane stress, σ_3, is for plane stress, plane strain, or any state inbetween. In other words, the zone shape and size as displayed in Fig.6 holds whether the crack tip location is on a free surface (plane stress) or deep within the material (plane strain). This could well be the reason that rock and concrete, unlike metals, exhibit no effect of thickness on fracture toughness. Another difference between the plastic zone model and microcrack zone model can be seen if hydrostatic compression is applied. Many applications involving rock fracture take place where significant confining stresses are present. Consider, e.g. the underground tunnel crack problem shown in Fig.3a where hydrostatic compression of magnitude σ is superimposed. If the resulting equations are substituted into the von Mises yield criterion, equ. (16), the same plastic zone contours are achieved as before, equ.(18). That is, hydrostatic compression has no effect on the plastic zone size. However, upon substitution into the maximum normal stress criterion, equ.(20), the resulting microcrack zone description becomes again equ.(21) with the stress σ_u replaced by the stress $\sigma_u + \sigma$. This decrease in microcrack zone size with hydrostatic compression may account for the increase in fracture toughness with confining pressure observed for Indiana limestone /26/. The reduced process zone must somehow elevate the fracture toughness. This is similar to the increase in apparent fracture toughness with crack length observed for this material (Fig.5), since an increase in crack length is comparable to a decrease in the size of the microcrack zone.

This microcrack model has been shown useful in describing behavior and general trends in fracture toughness testing of rock and concrete. However, no direct confirmation of this model has been achieved as yet.

2.5. *Subcritical Crack Growth*

Under certain conditions crack growth can occur at stress intensity levels below K_{Ic}. The two principal types of subcritical crack growth are fatigue (mechanically influenced subcritical crack growth) and stress corrosion cracking (chemically influenced subcritical crack growth).

Fatigue crack growth can occur when cyclic loads (e.g. pressure fluctuations) are applied to a structure causing stable crack advance. Fatigue is at least partially responsible for most failures of engineering structures.

The growth rate of a fatigue crack has been shown to depend on the cyclic range of the stress intensity factor, K /8,27/. Referring to eqs(7) and (8) one obtains for cyclic pressure fluctuation, p

$$\Delta K = \Delta(p-\sigma)\sqrt{\pi a}\,Y = \Delta p\sqrt{\pi a}\,Y \qquad (22)$$

with $Y=1$ for the tunnel crack and $Y=2/\pi$ for a circular crack, respectively.

If K_{max} and K_{min} denote the maximum and minimum stress intensities reached in a given cycle then $\Delta K = K_{max}-K_{min}$. A typical plot of ΔK versus crack growth rate, da/dn, is seen in Fig.7. While fatigue crack growth is of great importance in metallic materials it is probably of little interest in most rock fracture applications, although rocks are known to exhibit fatigue crack propagation /28/.

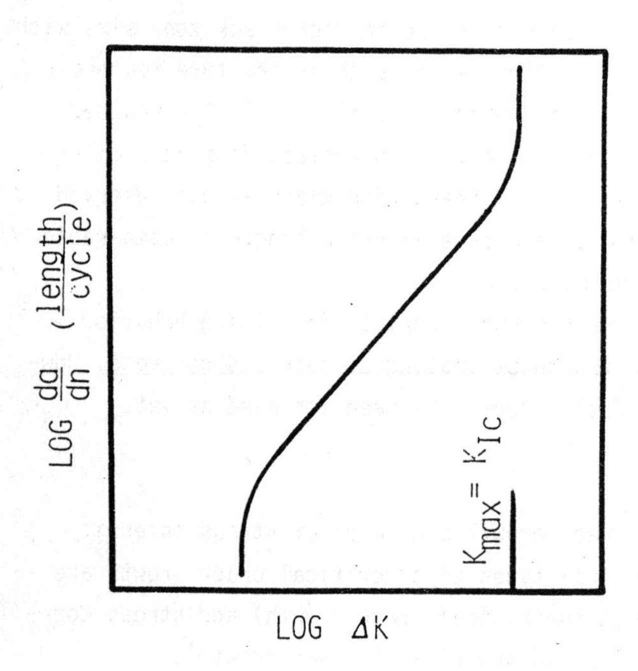

Fig.7 Fatigue crack growth rate versus stress intensity range

Stress corrosion cracking, more correctly known as environmental cracking, takes place under steady load conditions in combination with an environment that interacts with the highly stressed material at the crack tip. Some steels, for example, undergo stable crack growth when simple destilled water comes in contact with a crack under load. A 4340 steel structure may have a degradation in long-term load-carrying capacity by as much as a factor of five or more when subjected to an environment of sea water. The material properties of rock are also known to be affected by environmental conditions (thermal fluctuations and chemical agression) although little work has been performed as yet to investigate actual stress corrosion cracking in these materials. This is an important area for future research.

3. Development of Fracture Toughness Testing

Although treated in a later Chapter, the basic steps of development of fracture toughness testing will be outlined here.

The following discussion of fracture toughness testing of rock focusses on methods adapted from the ASTM Standards for metallic materials. These methods were modified and extended to suit the particular behavior of rock. This focus is admittedly biased by the first author's experiences with fracture toughness testing.

Several specimen configurations have been utilized in fracture toughness testing of rock. Two of the most popular are the three-point-bend and the single-edge-crack geometries depicted in Fig.8. Basically, one obtains a value of K_{Ic} from a precracked specimen such as this by loading it to failure and then calculating toughness from the formula appropriate for that specimen.

However, this procedure is oversimplified. The determined value of K_{Ic} may not be valid unless the specimen dimensions are sufficiently large compared to the size of the crack tip process zone. Unfortunately, there is little difinitive evidence available on the size of that zone and no standards are as yet available for a direct determination of adequate specimen size. As a result the careful experimenter, whenever possible, must determine several candidate values of fracture toughness, labelled K_Q, from several specimen sizes and make a judgement as to the validity of the values.

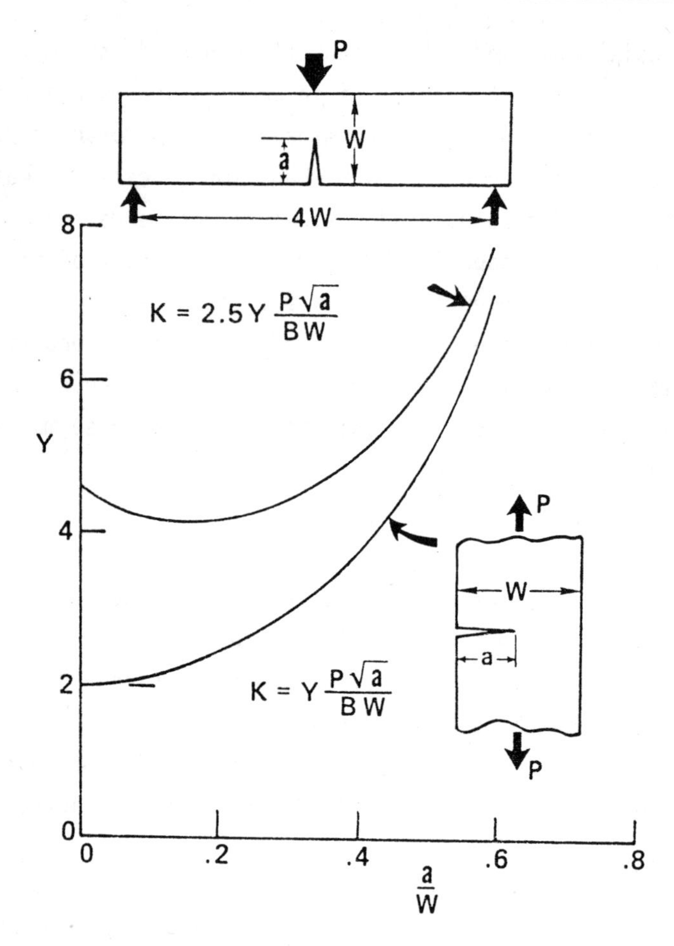

Fig.8 Three-point-bend and single-edge-crack specimen configurations and K-calibration curve.

In addition, the determination of crack length is often difficult. Even when a direct observation of crack length on the specimen surface is possible, this length may not be representative of the crack length through the specimen thickness. Machined notches that simulate cracks should be avoided when possible due to the uncertainty of how well a natural crack is being simulated by the notch. Indirect determinations of crack length by means of specimen compliance are possible /25,29/. Some specimen configurations such as the thick-walled cylinder /29/ and the short rod /30/ offer the advantage that the expected crack length at final unstable rupture is predetermined by the specimen's geometry. The short rod specimen is particularly suited for many rock fracture tests in that (1) its size and shape are convenient to obtain from core material, (2) loading can be performed simply, and (3) only maximum load is needed to determine a value of fracture toughness.

Several previous investigations on rock materials have examined the effect of specimen size on apparent fracture toughness, K_Q /20,25/. Indiana limestone, for example, was shown to have a K_Q versus crack length behavior that is similar to aluminum alloys (Fig.5). The ASTM requirement (ASTM-Test for Plane-Strain Fracture Toughness of Metallic Materials)(E 399-76) for crack length, a, was shown to apply reasonably well for this rock when in equ.(19) the yield stress σ_{ys} was replaced by σ_{ult} /25/. However, much more fracture data are required on various rock types using a wide range in specimen dimensions before standards can be reliably set for fracture toughness testing of rock in general. The primary purpose of this investigation, then, is as a parameter sensitivity study to measure the fracture toughness of Westerly granite for a variety of thicknesses and in-plane dimensions using two specimen configurations.

However, it should be noted that the stress-strain behavior of this rock in tension is known to be rather nonlinear and, as a result, the use of a linear elastic parameter, K, may be questionable. Since the use of the J-integral fracture theory does not require linear-elastic behavior, additional measurements were made on these tests to determine the critical J-integral, J_{Ic}. Effects on specimen size on J_{Ic} have been noticed and correlated with the K_{Ic}-measurements. In addition, results from direct-pull tension tests are used for the J_{Ic}-K_{Ic} correlation, for crack length determination by compliance, and for evaluating the size effects on K_Q.

3.1. CT-Specimen and 3PB-Specimen Testing

In a typical toughness test program the first author /19/ conducted fracture tests on 20 compact tension specimens and 6 three-point-bend specimens of various sizes. The dimensions and tolerances of these specimens were in accordance with ASTM recommendations including the use of chevron-notches to aid in establishing a well-behaved crack front (ASTM Method E-399). All specimens ranged in width from 25 to 400 mm and thickness from

12 to 100 mm . Notched beam specimens ranged in width from 25 to 100 mm and thickness from 13 to 50 mm. The rock used in this investigation, Westerly granite, comes from a quarry in Westerly, R.I., and is particularly well known to rock mechanics investigators. It is relatively fine-grained (average grain diameter of 0.75 mm) and is highly homogeneous on the macroscopic scale. A detailed description of this rock can be found elsewhere /31,32/.

All tension tests and fracture tests were performed in a 1.0 MN servo-controlled MTS load frame. Data for all tests were digitized, reduced, and plotted in real time by means of a PDP11 computer, MTS interface, graphics display terminal, and appropriate software.

Tension specimens were pulled to failure at a constant stress rate of 220 kPa/s. Bending stresses were minimized by the combination of careful machining described previously and by including high strength alloy link chain in the load train as recommended in ASTM Test for Direct Tensile Strength of Intact Rock Core Specimens (D 2936-71). Since granite is a poor conductor of heat compared to metals, strain gauge voltage was held to less than 3V to minimize heating effects.

Fracture tests of compact tension specimens were performed using a clevis design following ASTM recommendations (ASTM Method E-399) but reduced in size somewhat due to the lower load requirements /33,34/. Specimens were precracked in fatigue and loaded to failure by using the crack mouth displacement (CMD) as the control parameter. Since CMD at fracture ranged from 0.5 mm to as low as 0.03 mm, the standard clip-in displacement gauges (ASTM Method E-399) were replaced with a linear-variable differential transformer, LVDT, displacement transducer having a linear range of 0.25 mm. Load-line displacement measurements were made on the compact-tension-specimen tests for J-integral determinations. The measurement technique employed here is similar to one tested by Hills et al /35/ for J_{Ic}-measurements of metals at elevated temperatures.

Three-point-bend tests were performed in a similar manner except that no attempt was made to determine J_{Ic} or to measure the displacement of the load point. Loading was performed in three-point-bending using a fixture that conforms to ASTM recommendations. Fatigue pre-cracking and the final loadings were performed under CMD control using the LVDT as for the compact tension specimens.

Tensile Test Results: A typical pair of stress-strain curves from one of the tension tests is displayed in Fig.9 along with a summary of results from all tests. An unloading-reloading curve is also included. The stress-axial strain curve of this "brittle" material shows a large degree of non-linear and inelastic response. The unloading cycle demonstrates that most of the nonlinear behavior is indeed inelastic but this inelastic behavior is generally recognized as growth of microcracks rather than plasticity per se.

Fracture Toughness Results: Values of apparent fracture toughness, K_Q, were calculated from load versus crack mouth displacement (CMD) records, such as Fig.10. The compliance was established by fitting a straight line by the method of least squares through all the digitized data that fell between 50 and 90 percent of the maximum load. This compliance and the compliance calibration were used to establish the crack length. The load at "failure" was determined by the ASTM method using a 5 percent secant line constructed from the compliance line extrapolated to zero load. Crack length and failure load were then used to calculate K_Q from the ASTM formulas for compact tension and three-point-bend specimens (ASTM Method E-399).

Results from 19 compact tension tests are summarized in Fig.11. K_Q generally increases slightly with crack length /35/ and appears to level off for long cracks. For all practical purposes the curve can probably be considered to have reached a constant value, K_{Ic}, for cracks longer than 2.5 $(K_Q/\sigma_{ult})^2$ /3/. Data from the six bend tests (not shown here) appear to be consistent with the compact tension data and helps to establish K_{Ic} as a material property for this rock.

Critical J-integral data were determined by the J-resistance technique /33/ with the Merkle-Corten correction factors /36/ for the compact tension specimen geometry. Using guidelines recommended by ASTM Subcommittee E24.08 on Elastic-Plastic Fracture Mechanics Terminology, the J-integral data were reduced using the Merkle-Corten representation. Results were in full agreement with the results shown in Fig.11.

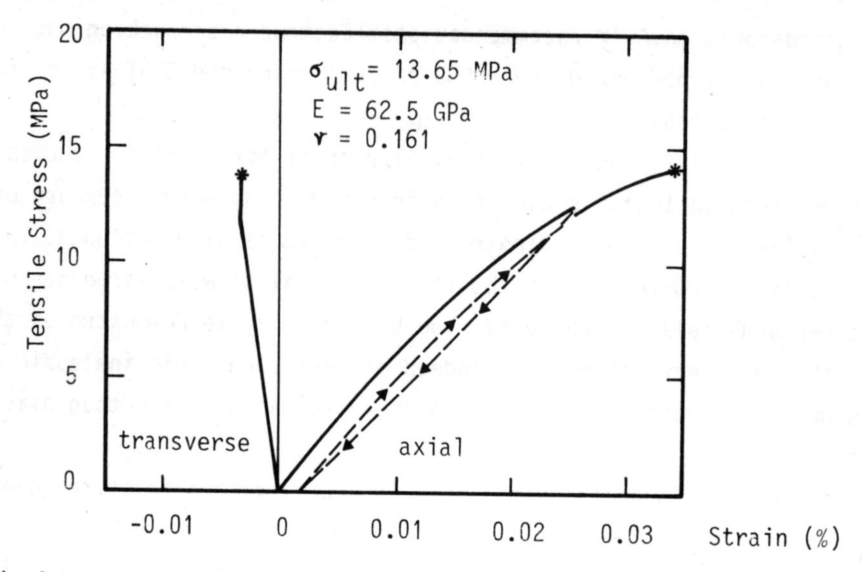

Fig.9 Typical stress-strain curve for direct-pull tension test

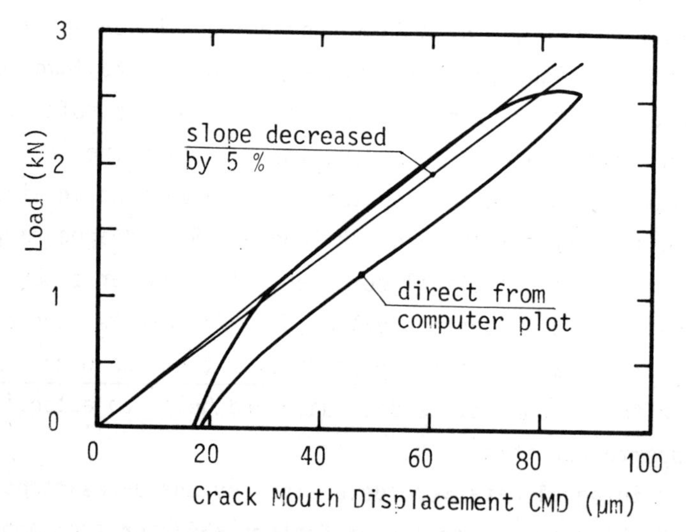

Fig.10 Load versus CMD for load to failure of compact tension specimen

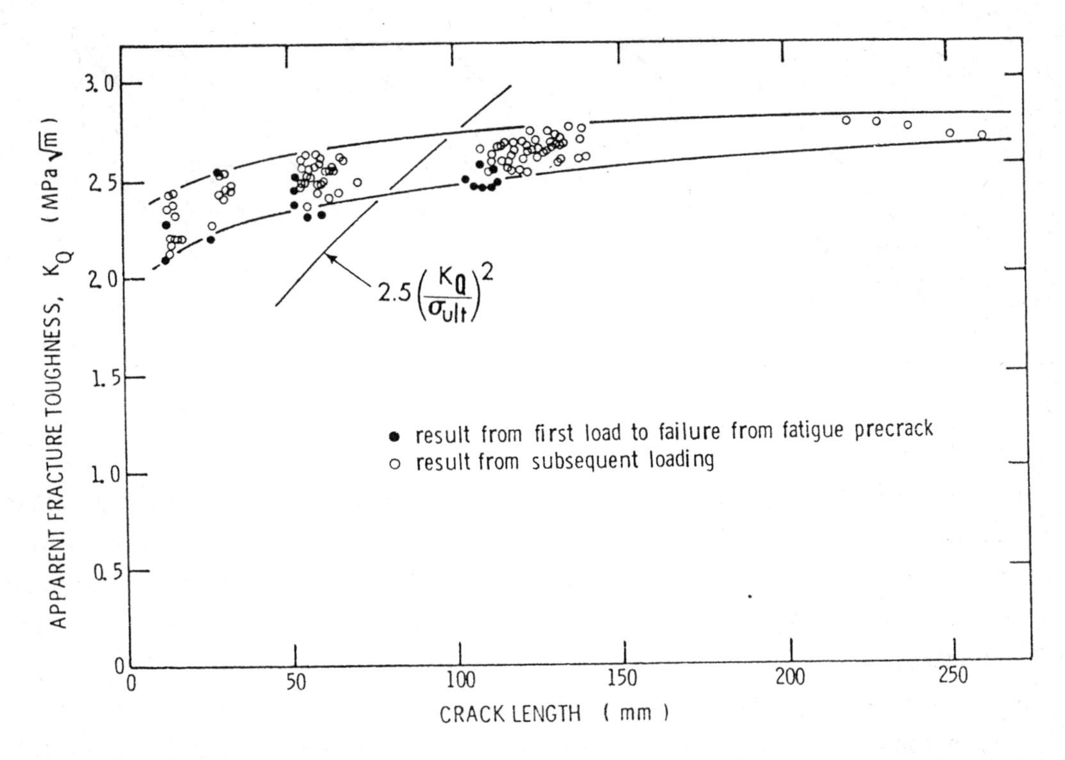

Fig.11 Apparent fracture toughness of Westerly granite from compact tension specimens

3.2. Anisotropy

A large number of rock types are anisotropic. Most sedimentary rock, for example, have a bedding as a result of the manner in which the sediments were deposited. In some, but not all, cases the rock will have mechanical properties that are directionally dependent with respect to these layers. Whenever fracture toughness tests are being performed on a rock having a known bedding plane, the specific crack orientation with respect to the bedding should be noted. Three principal orientations are recognized and are depicted in Fig.12.

3.3. Effects of Hydrostatic Compression

Applications involving rock fracture, such as mining and well stimulation, often occur at some depth beneath the earth's surface. In these cases

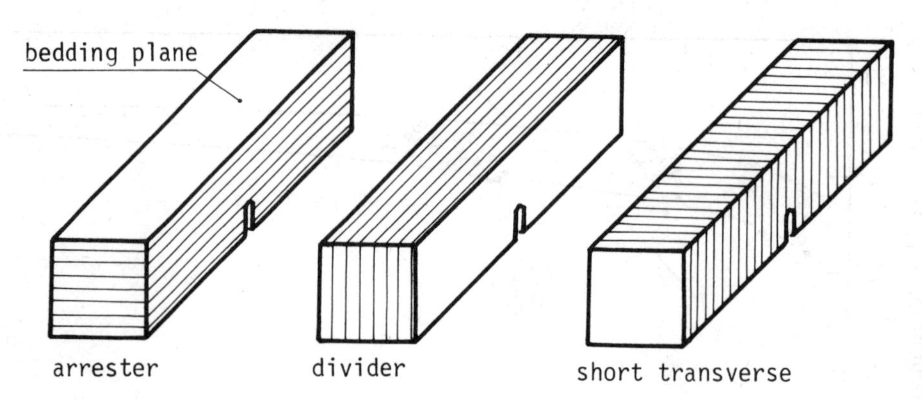

Fig.12 Principal crack orientations for layered rock

the state of stress includes the in situ stresses which are usually triaxial compression due to the overburden and lateral constraints. The strengths and elastic moduli of rock are known to vary with the hydrostatic component of this triaxial stress and thus one might expect a similar effect on fracture toughness.

Fracture toughness measurements under conditions of hydrostatic compression can be considerably more involved than confined tests /26/. Care should also be taken that the experimenter understands and accounts for the applied loading conditions /37/ when reducing data and evaluating results. Fracture toughness of Indiana limestone was shown to increase by a factor of 4 with an increase in hydrostatic compression from 0 to 62 MPa. Future research needs to address these effects of hydrostatic compression and improvements in test procedures are badly needed.

-. Closure

Mother Nature has provided the rock mechanics engineer with the formidable task of dealing with a material having an infinite variation in composition and properties. The application of fracture mechanics principles to rock fracture problems can, however, bring order and straightforward engineering to this seemingly impossible task. Parameters such as hydrostatic stress have effects on rock properties and many rock types exhibit nonlinear behavior. These facts add to the engineer's problems when compared to working with controlled man-made materials such as steel. However, reasonable engineering solutions can be achieved in many rock fracture applications if the proper variables are accounted for and if sound principles of fracture mechanics are properly applied.

References

/1/ Irwin, G.R. and A.A.Wells: A continuum-mechanics view of crack propagation, Metallurgical Review, 10, 223-270(1965)

/2/ Coulomb, C.A.: Sur une application des regles de maximis et minimis a quelques problemes de statique relatifs a l'architecture. Acad.Roy. des Sciences Memoires de Math. et de Phys. par divers savans,7, 343-382(1773)

/3/ Jaeger, J.C. and N.G.W.Cook: Fundamentals of Rock Mechanics, Chapman and Hall Ltd, London (1969)

/4/ Griffith, A.A.: The theory of rupture. Proc. of First Int'l Congr. on Appl.Mech., 55 (1924)

/5/ Begley, J.A. and J.D.Landes: The J-integral as a failure criterion. Fracture Toughness, ASTM STP 514, ASTM, p.1, (1972)

/6/ Rudnicki, J.W.: Fracture mechanics applied to the earth's crust. Ann. Rev. Earth Planet.Sci.8, 489-525(1980)

/7/ Ouchterlony, F.: Review of fracture toughness testing of rock. Swedish Detonic Research Foundation Report, DS 1980:15 (1980)

/8/ Broek, D.: Elementary Engineering Fracture Mechanics. Noordhoff Int. Publ., Leyden, Netherlands (1974)

/9/ Tentative method of test for plane strain fracture toughness of metallic materials (ASTM designation: E-399-72T), Annual Book of Standards, Part 31, Amer.Soc.for Testing and Materials, Philadelphia, Pa (1972)

/10/ Westergaard, H.M.: Bearing pressures and cracks. Transactions of ASME, J.Appl.Mech.6,49-53(1939)

/11/ Paris, P.C. and G.C.Sih: Stress analysis of cracks. Fracture Toughness Testing and its Applications. ASTM STP 381, Amer.Soc. for Testing and Materials,p.30 (1965)

/12/ Tada, H., P.C.Paris and G.R.Irwin: The Stress Analysis of Cracks Handbook, Del Research Corporation, Hellertown, Pa (1973)

/13/ Griffith, A.A.: The phenomena of rupture and flow in solids. Phil. Trans. of the Roy.Soc.of London, Series A-192, p.163(1920/21)

/14/ Cherepanov, G.P.: Mechanics of brittle fracture, McGraw-Hill (1979)

/15/ Schmidt, R.A.: A microcrack model and its significance to hydraulic fracturing and fracture toughness testing. Proc.21st U.S.Symp.Rock Mechanics, Rolla, MI (1980)

/16/ Hoagland, R.G., G.T.Hahn and A.R.Rosenfield: Influence of microstructure on fracture propagation in rock. Rock Mechanics, 5, 77-107(1973)

/17/ Hardy, M.P.: Fracture Mechanics Applied to Rock. PhD Thesis, Univ. of Minnesota (1973)

/18/ McClintock, F.A. and G.R.Irwin: Plasticity aspects of fracture mechanics. Fracture Toughness Testing and its Applications, ASTM STP 381, p.84(1965)

/19/ Schmidt, R.A. and T.J.Lutz: K_{IC} and J_{IC} of Westerly granite - effects of thickness and in-plane dimensions. ASTM STP 678, ASTM, p.166(1979)

/20/ Ingraffea, A.R. and R.A.Schmidt: Experimental verification of a fracture mechanics model for tensile strength prediction of Indiana limestone. Proc.19th U.S.Symp. Rock Mechanics, Stateline, NV, p.247(1978)

/21/ Nelson, F.G., P.E.Schilling and S.G.Kaufman: The effect of specimen size on the results of plane strain fracture toughness tests. Eng. Fract.Mechanics,4,p.33(1972)

/22/ Jones, M.H. and W.F.Brown Jr.: The influence of crack length and thicknessin plane-strain fracture toughness tests. Review of Developments in Plane-Strain Fracture Toughness Testing, ASTM STP 463, ASTM, p.63 (1970)

/23/ Mindess, S. and J.S.Nadeau: Cement and Concrete Research,6,529-534 (1976)

/24/ Hoagland, R.G., G.T.Hahn and A.R.Rosenfield: Influence of microstructure on fracture propagation in rock. Rock Mechanics,5,77-106(1973)

/25/ Schmidt, R.A.: Fracture toughness testing of limestone. Exp.Mech.16, p.161(1976)

/26/ Schmidt, R.A. and C.W.Huddle: Effect of confining pressure on fracture toughness of Indiana limestone. Int.J.Rock Mech.Min.Sci.& Geomech. Abstr.14,289-293(1977)

/27/ Paris, P.C.: Crack growth due to variations in load. PhD Thesis, Lehigh University, Pa(1962)

/28/ Schmidt, R.A.: Fracture toughness testing of limestone. Exp.Mech.16, p.161(1976)

/29/ Johnson, J.N. et al.: Analysis of fracture for hollow cylindrical and spherical rock specimens subject to internal pressure with application to underground nuclear containment. TerraTek Report on Contract No.DNA 001-73-C-0153, Salt Lake City, Utah (1973)

/30/ Barker, L.M.: A simplified method for measuring plane strain fracture toughness. Eng.Fract.Mech.9,p.361.(1977)

/31/ Birch, F.: J.Geophys.Research,65,4,p.1083 (1960)

/32/ Brace, W.F.: J.Geophys.Research,70,2,p.391 (1965)

/33/ Brown, Jr. W.F. and J.E.Srawley: Plane strain crack toughness testing of high strength metallic materials. ASTM STP 410, ASTM, 1-65 (1966)

/34/ See Ref./9/

/35/ Mills, W.J., L.A.James and J.A.Williams. J. of Testing and Evaluation, 5,6,446-451 (1977)

/36/ Merkle, J.G. and H.T.Corten: J.Pressure Vessel Technology,96, Series J,4,286-292 (1974)

/37/ Schmidt, R.A. and S.E.Benzley: Stress intensity factors of edge crack specimens under hydrostatic compression with application to measuring fracture toughness of rock. Int.J.Fracture,12,p.320 (1976)

/38/ Rossmanith, H.P.: Modelling of fracture process zones and singularity dominated zones. Eng.Fract.Mech.(to appear 1982).

ANALYSIS OF CRACKS RELATED TO ROCK FRAGMENTATION

Finn Ouchterlony
Swedish Detonic Research Foundation (SveDeFo)
S-12611 Stockholm, Sweden

1. INTRODUCTION TO BLASTING CONFIGURATIONS

In the late 1960's, fracture mechanics was seldom used in the analysis of rock fragmentation problems. Examplifying rock mechanics textbooks with Jaeger and Cook [1], they do present Griffith's theory of fracture and the term fracture mechanics in passing but not any applications to fragmentation.

The situation in rock fragmentation research was not much better then but during the 1970's an awareness of linear elastic fracture mechanics (LEFM) developed, see for example Hardy [2], Swan [3], or Wang and Clark [4], which is the proceedings of the 18th US Symposium of Rock Mechanics in 1977. Today LEFM is being taught at the under-graduate level and non-linear fracture mechanics methods pervade research, as will become evident during this course.

My first lectures will summarize some earlier work [5-11], mainly from 1972-1978, where I used LEFM analysis on crack configurations related to rock fragmentation to elucidate some basic aspects of the process. Thus they are not state of art but should still help the basic understanding.

Rock is in no way an ideal material, but it has often been regarded as a linear-elastic, homogeneous, and even isotropic material to bring out general conclusions about rock fragmentation. This may be reasonably true for dense competent rocks such as granite and marble. Rock is considerably more brittle than metals such as steel and aluminum, usually by at least a factor of ten in fracture toughness. Due to the low tensile strength however, the micro-crack zone at a crack tip is of the order 0.01 m for granite and marble. Thus the smallest crack size admissible in LEFM analysis is of the order 0.10 m, hardly small enough for rock cutting but sufficiently small in blasting with typical crack lengths of the order 1 m.

A drastic idealization of the crack geometries is usually necessary. I used analytical methods to study two-dimensional crack systems in plane strain elasticity. The numerical codes available today are more flexible and powerful but the analysis of three-dimensional crack systems is not trivial, especially if dynamics are included. Such aspects will be covered thorougly by the other lectures. Yet the quasi-static approach has brought out essential parts of the behavior of crack systems related to blasting.

2. IDEALIZED CRACK SYSTEMS

When an explosive charge detonates in a bore hole, the pressure is propagated into the rock as a shock wave in which all stresses are compressive. Near the hole their levels may exceed the strength of the rock material. The radial expansion behind the wave front relaxes the stresses and the tangential stress soon tends to become tensile. Hereby an initiation of radial cracks becomes possible and usually occurs. See figure 2.1 for model scale blasting.

The circular hole with radial cracks of alternating lengths in figure 2.2 resembles closely the situation at this point. The star crack

Figure 2.1: Crack system in PMMA after model scale blast.

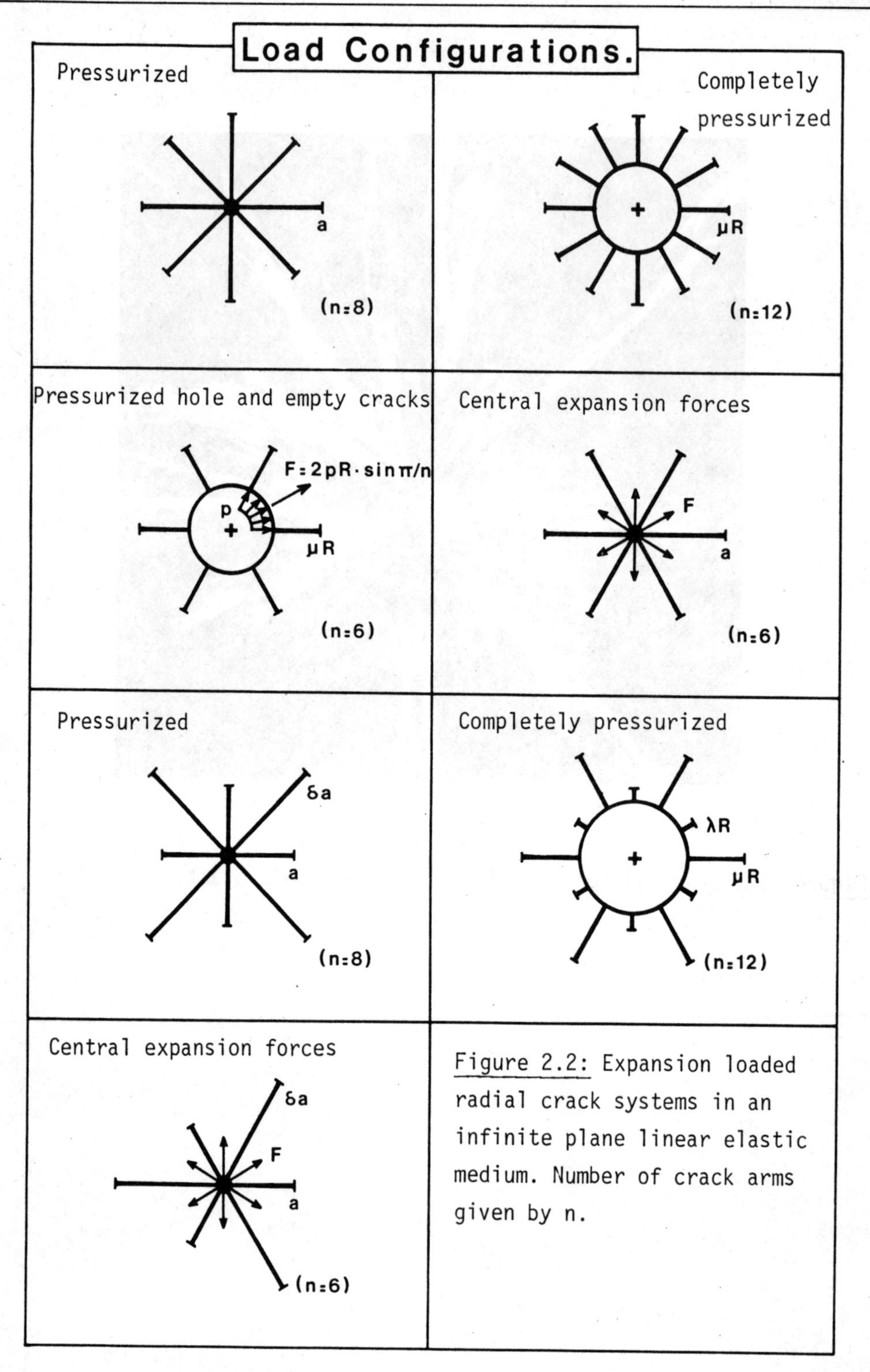

Figure 2.2: Expansion loaded radial crack systems in an infinite plane linear elastic medium. Number of crack arms given by n.

with arms of alternating lengths resembles the situation at a later stage when the cracks have grown large compared to the hole. The influence of the bench face or other fracture planes and cracks isn't included.

The volume expansion of the gaseous reaction products in the bore hole and in the cracks is essential to the blasting process, thus the volume of the crack system is an essential supplement to the stress intensity factors in a fracture analysis. The important question of gas penetration into the cracks is modelled by two extremes. One is the completely pressurized crack system with a constant internal pressure. The other is a pressurized hole with empty cracks, or, in the limit of long cracks, a star crack opened by resultant expansion forces. These statically self-equilibrating loads are all called expansion loads.

Figure 2.2 shows most of the expansion loaded configurations analyzed by Ouchterlony.[5-10] Because of the symmetry these crack systems will most likely continue growing radially. Crack branching is not discussed. They should also hold some relevance to hydraulic fracturing of bore holes and expansion splitting of boulders or blocks.

There are two basic ways of fragmenting rock with wedge shaped tools. One is splitting action with thrust force and tool motion perpendicular to the rock surface. The other is cutting action with the tool motion essentially parallel to the surface but where the applied force also has a thrust component. A combination is also possible. Rock splitting arises for example in impact breaking where fairly large cracks radiate out from the tool tip. Examples of rock cutting are rotary drag bit drilling and full face tunnel boring where an ideal chip is formed by a crack which propagates below the cutting plane before curving upwards to the surface.

All these fragmentation processes, where a few cracks dominate, are shown in figure 2.3. A plane infinite wedge with a radial crack can be made to resemble the early stages of the splitting and cutting situations. This configuration was also analyzed[8-11], but is given less attention here.

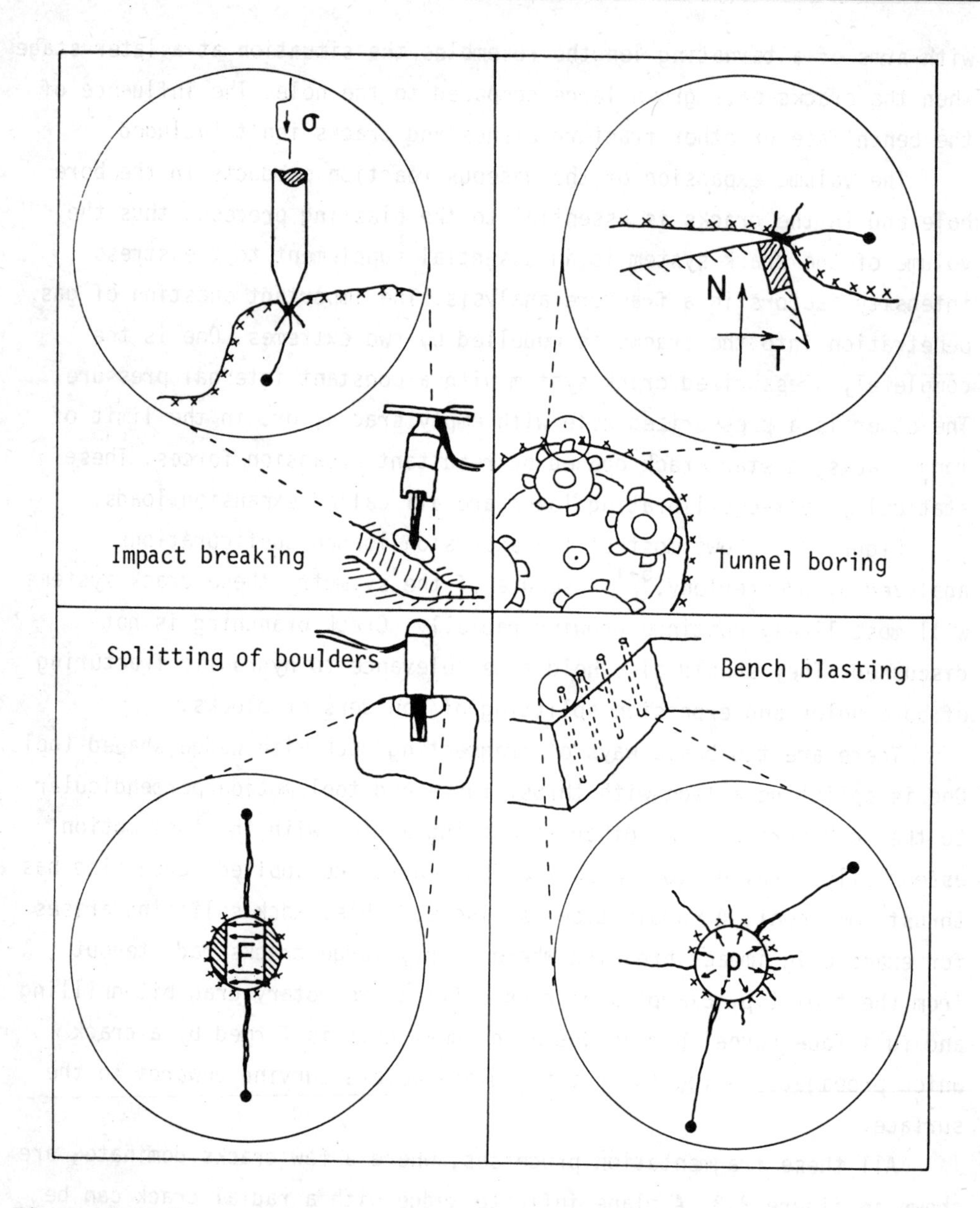

Figure 2.3: Some examples of rock fragmentation processes where the fragmentation may be caused by a few dominating cracks. The active areas at the crack tips are shown.

3. COMPLEX REPRESENTATION AND CONFORMAL MAPPING

3.1 Method

My use of complex representation for the analysis of plane problems in isotropic elastomechanics [5] follows Muskhelishvili [12] and Persson [13]. In it an Airy's stress function Φ is introduced that is the solution to the biharmonic equation $\Delta\Delta\Phi = 0$. Thus it represents Φ by two complex stress functions f(z) and g(z) through

$$\Phi(x,y) = Re\{\bar{z}f(z)+\int_0^z g(z')dz'\} \text{ and } z = x+iy \qquad (3.1)$$

such that $\Delta\Delta\Phi = 0$ is satisfied. f and g are determined by the boundary conditions and in turn uniquely determine the stress state in the body and the deformations too.

The stresses and displacements are given by

$$\left.\begin{aligned} \sigma_x+\sigma_y &= 4Re\{f'(z)\}, \\ \sigma_y-\sigma_x+2i\tau_{xy} &= 2[\bar{z}f''(z)+g'(z)], \text{ and} \\ 2G(u_x+iv_v) &= \kappa f(z)-\overline{zf'(z)}-\overline{g(z)}, \end{aligned}\right\} \qquad (3.2)$$

where G is the shear modulus, ν the Poisson's ratio, and where $\kappa = (3-\nu)/(1+\nu)$ in plane stress and $\kappa = 3-4\nu$ in plane strain. The two fundamental stress and displacement boundary value problems take on a similar form. With stresses known on the contour S

$$f(z)+\overline{zf'(z)}+\overline{g(z)} = i\int^S (X_n+iY_n)ds + \text{constant}, \qquad (3.3)$$

where X_n and Y_n are the boundary stress components and the constant may be chosen as to eliminate rigid body displacements.

In order to bring the boundary value problem on a more amenable form, an auxiliary complex variable ζ is introduced and the area around the radial crack system in the physical z-plane is mapped onto the area around the unit circle $|\zeta| = 1$ in the image plane with the inverse of the analytical function

$$z = \omega(\zeta),$$

see figure 3.1. A change of variable yields the transformed boundary condition

$$f(\sigma)+\omega(\sigma)\overline{f'(\sigma)}/\overline{\omega'(\sigma)}+\overline{g(\sigma)}= F(\sigma) \text{on } |\zeta|= 1, \tag{3.5}$$

the unit circle where $\zeta = \sigma = \exp(i\theta)$. An integration of this expression with the proper Cauchy kernel around the unit circle yields $f(\zeta)$ and $g(\zeta)$ on closed form if $\omega(\zeta)$ is a rational function.

The stress intensity factors are given by

$$K_I-iK_{II} = 2\sqrt{\pi}f'(\sigma_j)/[\omega''(\sigma_j)e^{i\phi_j}]^{1/2}, \tag{3.6}$$

where ϕ_j denotes the angle between crack tip and real axis in the z-plane and σ_j denotes the crack tip position in the image plane. The crack surface displacements become

$$2G(u_x+iv_y) = (\kappa+1)f(\sigma)-F(\sigma) \tag{3.7}$$

and hence the increase in crack volume ΔV_c [m³/m]

$$\Delta V_c = \frac{\pi}{G}\cdot Re\{\frac{1}{2\pi i}\oint[(\kappa+1)\overline{f(\sigma)}-\overline{F(\sigma)}]\omega'(\sigma)\frac{\sigma d\sigma}{\sigma-\zeta}\}_{\zeta = 0}. \tag{3.8}$$

Here only the constant term in the Laurent expansion of the source function

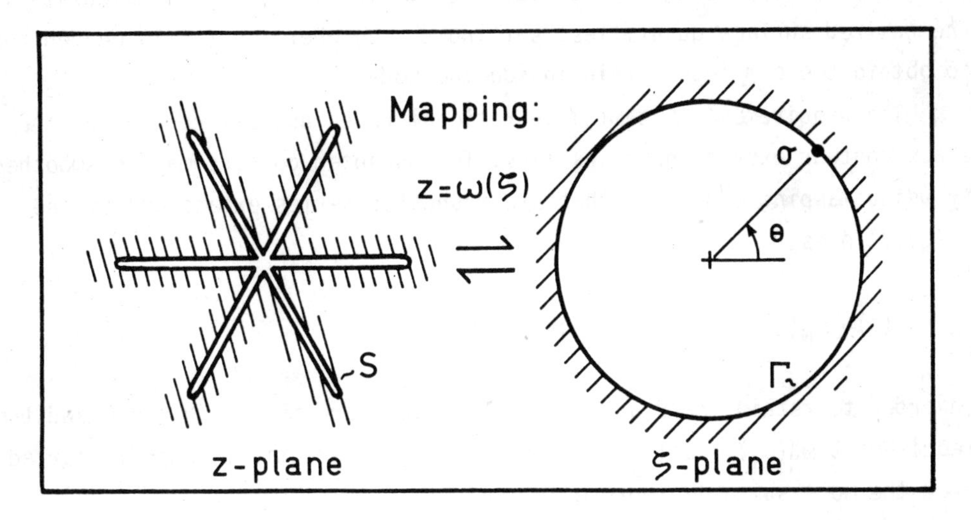

Figure 3.1: Mapping of area around the star crack in the physical z-plane onto area around the unit circle in the image plane via complex representation.

will give a contribution. A method of calculating gas volume from stress intensity factors is also presented.[5] Only $f(\zeta)$ is needed to calculate the desired surface quantities, setting $\zeta = \sigma$, whereas $g(\zeta)$ is needed too to obtain the complete field inside the body.

The requirement that $\omega(\zeta)$ be rational rules out all corners on the crack contour except the crack tips. The unwanted corners may be smoothed by using mapping functions that are truncated series expansions of the original ones,

$$\omega(\zeta) \simeq \omega_N(\zeta) = A\zeta\sum_{k=o}^{N} c_k\zeta^{-k} \text{ for } |\zeta| \geq 1. \tag{3.9}$$

In order to retain the crack tips it is necessary that $\omega_N'(\sigma_j) = 0$ and the requirement $\omega_N''(\sigma_j) = \omega''(\sigma_j)$ sharpens their reproduction. Hence I started from the more suitable representation

$$\omega_N'(\zeta) = A\cdot\prod_{j=1}^{n}(1-\sigma_j/\zeta)\cdot\sum_{k=o}^{N-n} c_k'\zeta^{-k} \text{ for } |\zeta| \geq 1 \tag{3.10}$$

and chose the last few terms to make $\omega_N''(\sigma_j) = \omega''(\sigma_j)$ for all the $j = 1$ to n crack tips. The coefficients c_k' were relatively simple to determine, many times involving orthogonal polynomials. The load configurations of figure 2.2 were analyzed this way. The details of the analysis may be found in Ouchterlony [5]. An English version exists in manuscript form.

In general the approximate crack contours reproduce the original ones well, see figures 3.2. The crack surfaces join smoothly at the tips as prescribed but overlap further in because of the forced retaining of the crack tips in a finite approximation. Despite this overlap the numerical accuracy of the stress intensity seems to be very good.[5,10]

3.2 Results

A physical interpretation of the most important results [5] is given in my paper, Ouchterlony.[6] A comparison of the configurations in figure 2.2

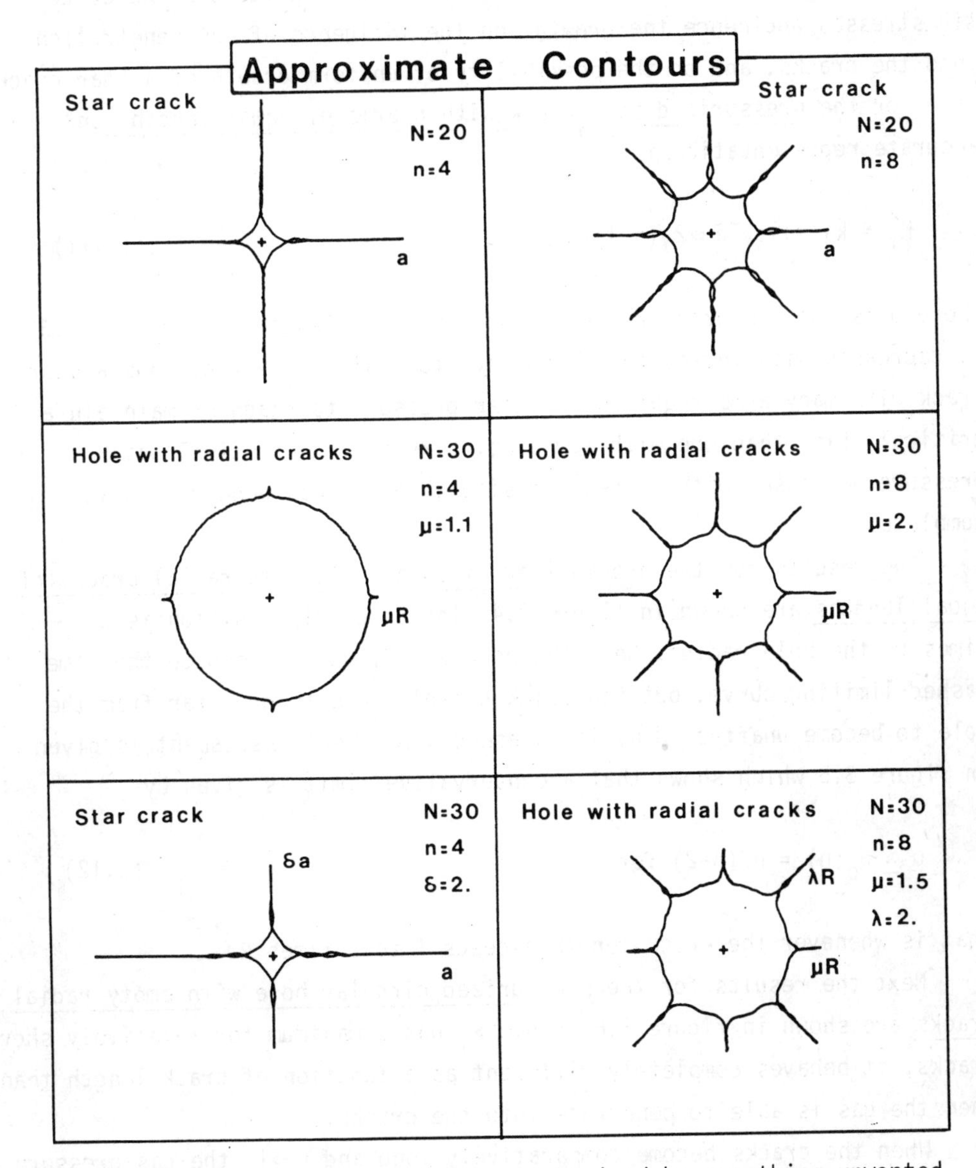

Figure 3.2: Approximate crack contours obtained by smoothing unwanted corners. N denotes the number of terms retained in the approximate mapping function.

gives information on how the presence of the hole influences the crack tip stresses and hence the growth, on the influence of gas penetration into the cracks, and on the possibility of uniform growth of a star crack.

For the pressurized star crack with n arms of equal lengths an accurate representation is [5,7]

$$K_I = k_1(n) \cdot p\sqrt{\pi a} \simeq 2\sqrt{(n-1)}/n \cdot p\sqrt{\pi a}, \quad (3.11)$$

where p is the gas pressure and a the crack arm length. See figure 3.3. K_I decreases with increasing n and asymptotically $k_1 \sim 2/\sqrt{n}$. Thus a star crack with many arms requires a higher pressure to reach or maintain a critical state than one with few arms. Likewise the critical gas pressure decreases with increasing size when the gas penetration is complete.

The results for the pressurized circular hole with radial cracks of equal lengths are shown in figure 3.4. The crack tip position is μ times R, the hole radius. Near the hole all K_I curves tend to the same dashed limiting curve, but the crack doesn't have to grow far from the hole to become unaffected by its presence. A direct assessment is given in figure 3.5 which shows that a conservative limit is given by

$$\mu \geq \mu_0(n) = n/(n-2) \text{ for } n > 2, \quad (3.12)$$

that is whenever the crack length exceeds 1 to 2 times R.

Next the results for the pressurized circular hole with empty radial cracks are shown in figure 3.6. Since K_I has a maximum for relatively short cracks, it behaves completely different as a function of crack length than when the gas is able to penetrate into the cracks.

When the cracks become comparatively long and $\mu \gg 1$, the gas pressure action on the bore hole sector between two adjacent cracks may be replaced by its force resultant $F = 2pR \cdot \sin(\pi/n)$. The result is the star crack with arms of equal lengths opened by expansion forces. K_I for this configuration is shown in figure 3.7. An exact representation is [7-9]

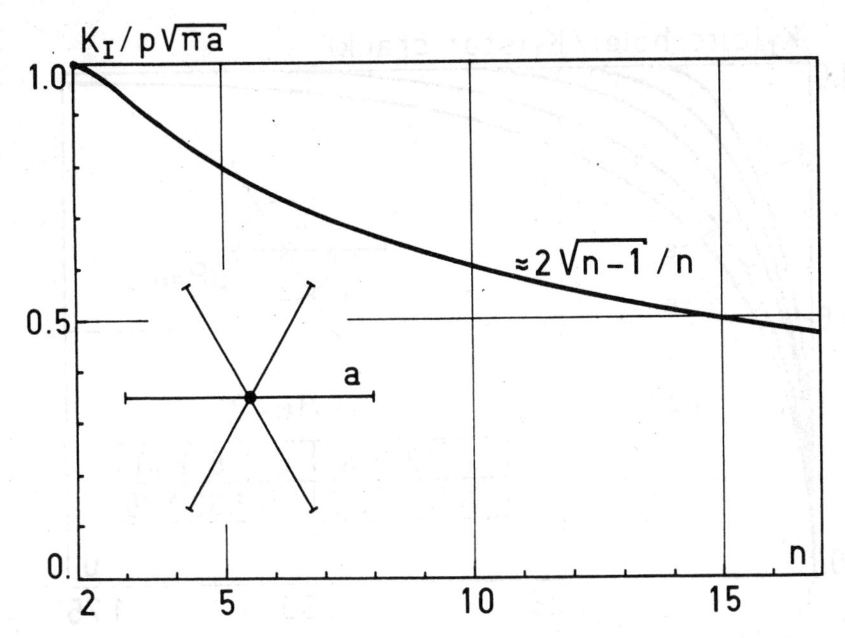

Figure 3.3: Normalized stress intensity factor for pressurized star crack with arms of equal lengths as function of n.

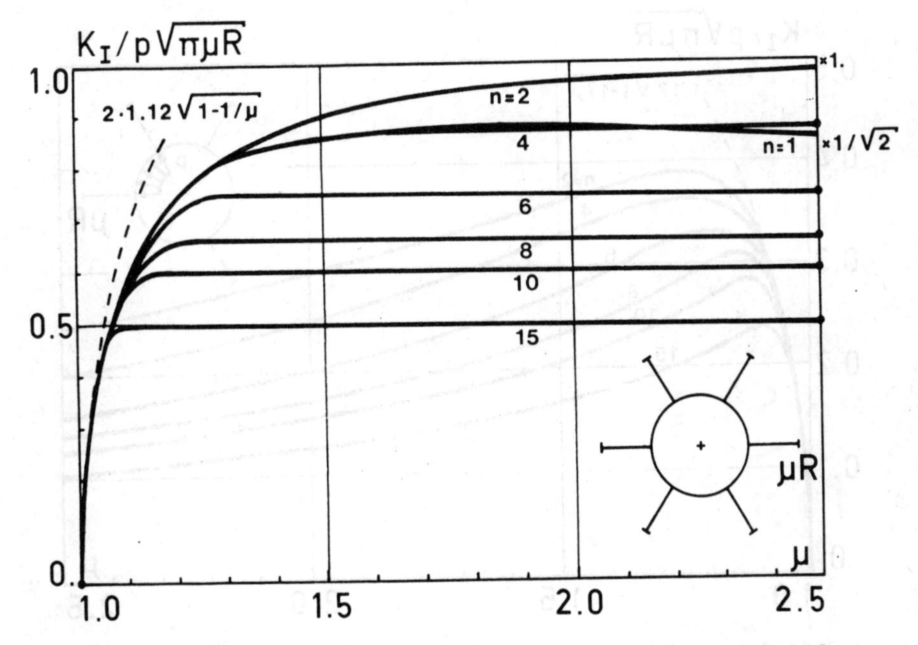

Figure 3.4: Stress intensity factors for pressurized circular hole with radial crack arms of equal lengths as function of crack length for different crack numbers.

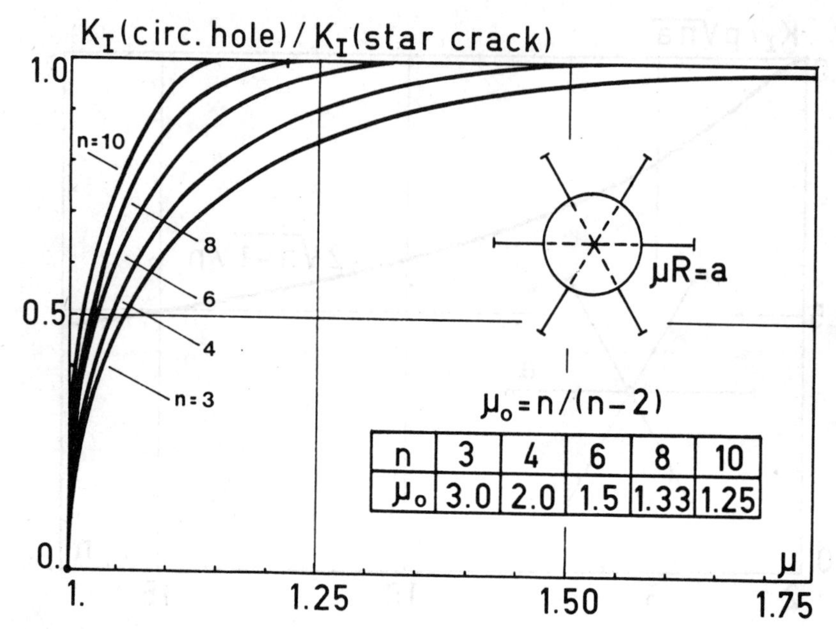

n	3	4	6	8	10
μ_o	3.0	2.0	1.5	1.33	1.25

Figure 3.5: Ratio of stress intensity factors for pressurized contours, circular hole versus star crack, as function of crack length.

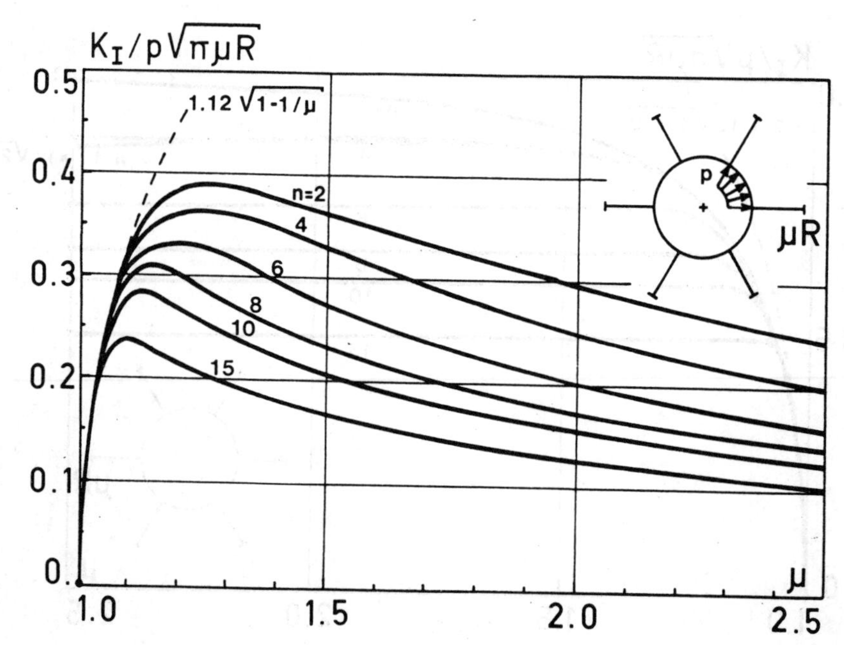

Figure 3.6: Stress intensity factors for pressurized circular hole with empty radial cracks as a function of crack length for different crack numbers.

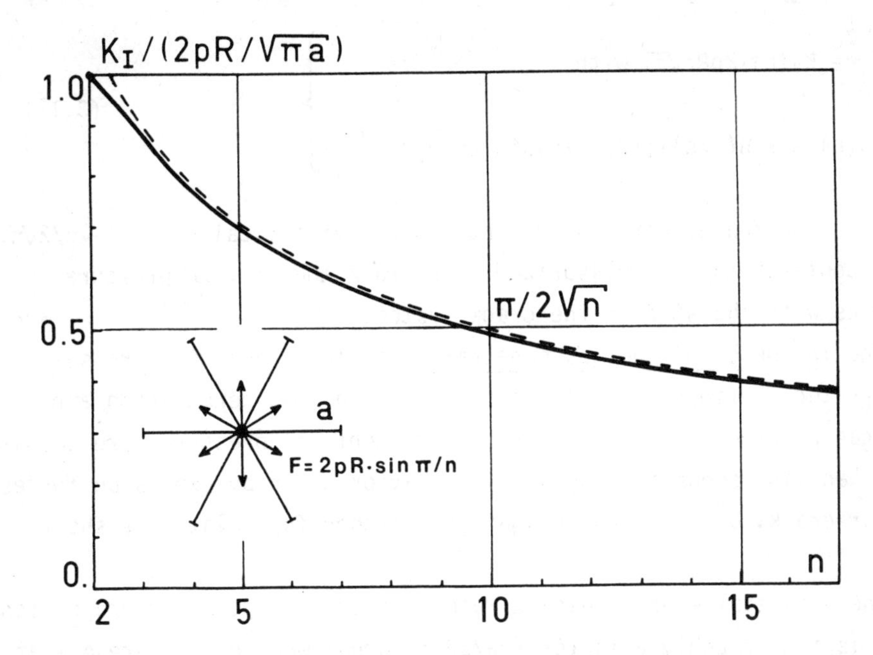

Figure 3.7: Stress intensity factor for star crack opened by expansion forces as a function of crack number.

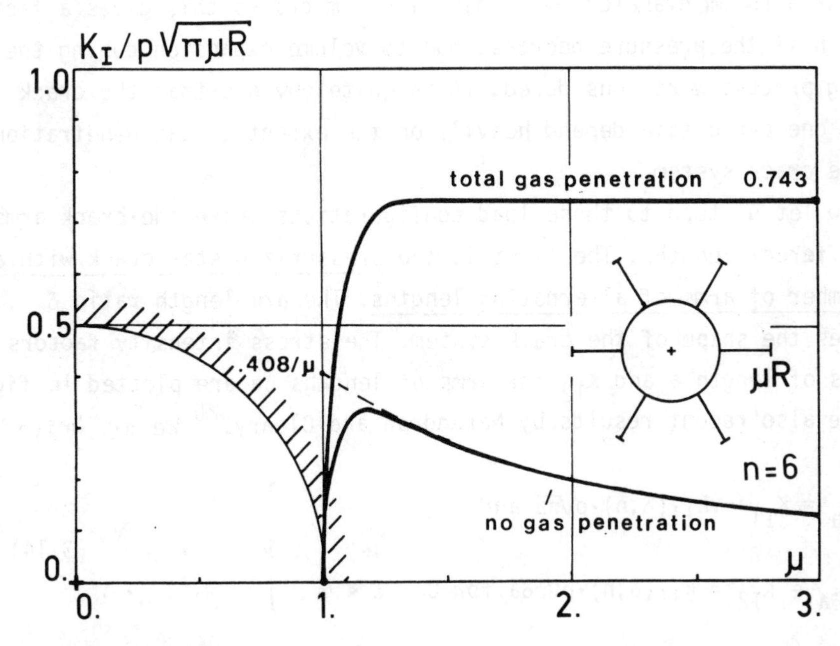

Figure 3.8: Comparison of stress intensity factor values for total and no gas penetration into six radial cracks surrounding pressurized circular hole.

$$\left.\begin{aligned} K_I &= k_2(n)\cdot 2pR/\sqrt{\pi a} \text{ with} \\ k_2(n) &= \sin(\pi/n)\cdot\{\pi/[2\pi/n+\sin(2\pi/n)]\}^{1/2}. \end{aligned}\right\} \quad (3.13)$$

Also $k_2(n)$ decreases with increasing n and asymptotically $k_2(n) \sim \pi/2\sqrt{n}$. But in contrast to the pressurized star crack the critical pressure increases with the size of the crack system.

The influence of the gas penetration on the crack tip stresses is demonstrated in figure 3.8. The two extremes no gas penetration and total gas penetration into the cracks at identical pressures are compared. The broken line shows the asymptotic behavior given by expansion forces in the star crack, $0.408/\mu$, which again is reached for relatively short cracks.

The potential energy release rate at a crack tip, being proportional to K_I^2, is thus roughly a factor $(\pi a/2R)^2$ larger when gas is present at the crack tip than when the same gas pressure is restricted to the bore hole. For a 25 mm diameter bore hole with 1 m cracks this gives a factor 10^4. Even if the pressure decrease due to volume expansion during the blasting process were considered, it is quite obvious that the crack lengths one can obtain depend heavily on the extent of gas penetration into the crack system.

Now let us turn to those load configurations where the crack arms have different lengths. The first is the pressurized star crack with an even number of arms of alternating lengths. The arm length ratio δ describes the shape of the crack system. The stress intensity factors K_{I1} for arms of length a and K_{I2} for arms of lengths δa are plotted in figure 3.9, see also recent results by Narendran and Cleary.[20] We may write

$$\left.\begin{aligned} K_{Ia} &= K_{I1} = k_{11}(\delta,n)\cdot p\sqrt{\pi a} \text{ and} \\ K_{I\delta a} &= K_{I2} = k_{12}(\delta,n)\cdot p\sqrt{\pi\delta a} \text{ for } 0 < \delta < \infty . \end{aligned}\right\} \quad (3.14)$$

For reasons of symmetry it follows that

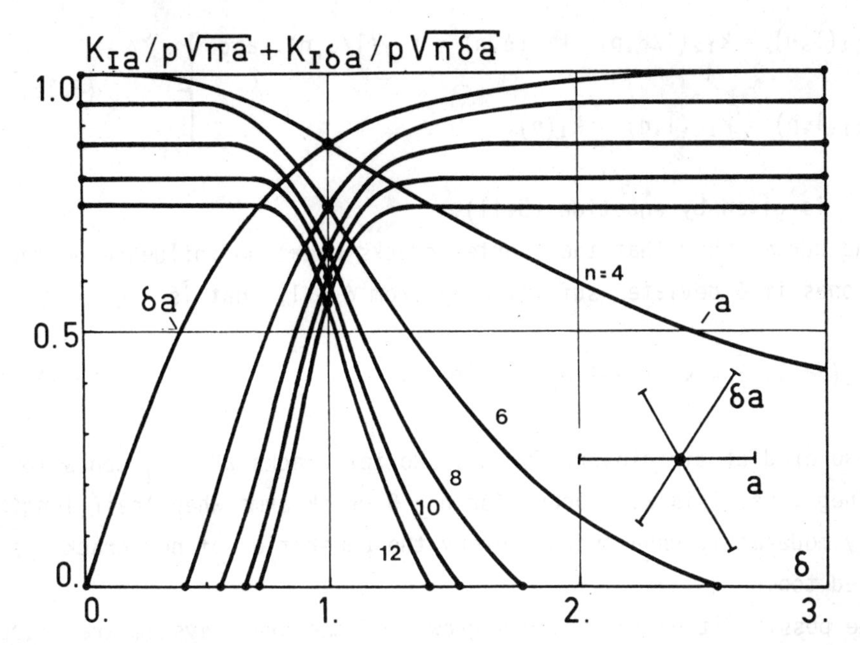

Figure 3.9: Normalized stress intensity factors for pressurized star crack with arms of alternating lengths as a function of the crack length ratio for different crack numbers.

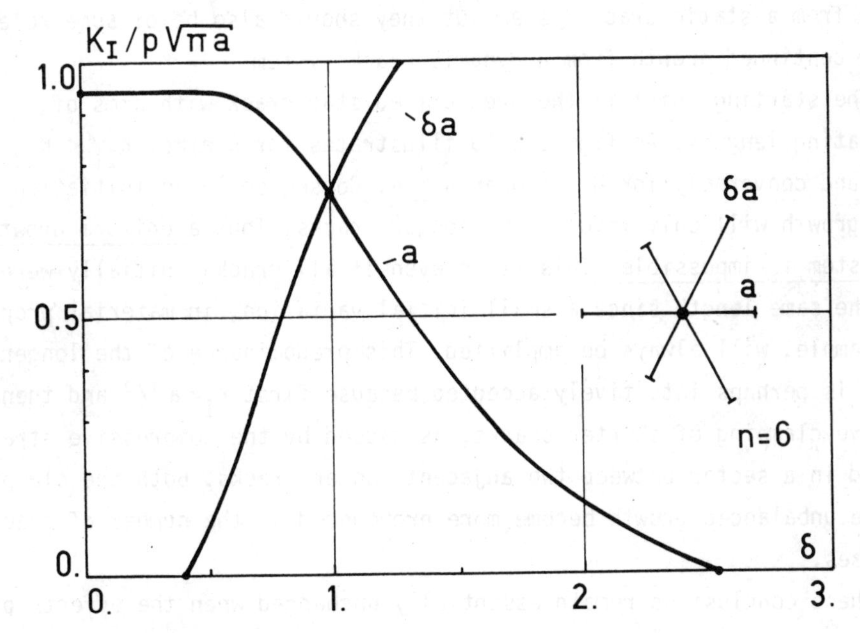

Figure 3.10: Direct comparison of stress intensity factors for the different length crack arms of a pressurized star crack with six arms of alternating lengths.

$$\left.\begin{aligned} k_{11}(\delta,n) &= k_{12}(1/\delta,n),\ k_{12}(\delta,n) = k_{11}(1/\delta,n), \text{ and} \\ k_{11}(1,n) &= k_{12}(1,n) = k_1(n), \end{aligned}\right\} \quad (3.15)$$

where k_1 is given by equation (3.11).

The curves show that the shorter cracks exert no influence on the longer ones if δ deviates sufficiently from $\delta = 1$, that is

$$k_{11}(\delta \ll 1,n) = k_{12}(\delta \gg 1,n) = k_1(n/2). \quad (3.16)$$

They also predict tip closure for the shorter cracks when K_I tends to become negative. This will occur for $n \geq 6$ crack arms when their lengths are only moderately unbalanced. Hereby the generation of new cracks is supressed too.

The possibilities for uniform growth of the crack system are discussed next, using the stress intensity factors as synonymous of the energy flow into the crack tips. Strictly seen the results concern an initiation of growth from a static crack system but they should also be of some relevance to the continued growth from a dynamic crack system.

The starting point is the pressurized star crack with arms of alternating lengths. As figure 3.10 illustrates for 6 arms, $K_{I1} < K_{I2}$ when $\delta > 1$ and conversely for $\delta < 1$ when $n \geq 4$. Consequently an initiation of crack growth will only involve the longer cracks. Thus a uniform growth of the system is impossible. This is so even if all cracks initially were to have the same length since a small initial variation, in material properties for example, will always be amplified. This predominance of the longer cracks is perhaps intuitively accepted because first $K_I \propto a^{1/2}$ and then the relative clamping of shorter cracks, as caused by the compressive stresses induced in a sector between two adjacent longer cracks. Both the clamping and the unbalanced growth become more pronounced as the number of cracks increases.

These conclusions remain essentially unchanged when the effects of the hole are considered as in figure 3.11 for $n = 6$ cracks. Here $\mu \cdot R$ and $\lambda \cdot R$

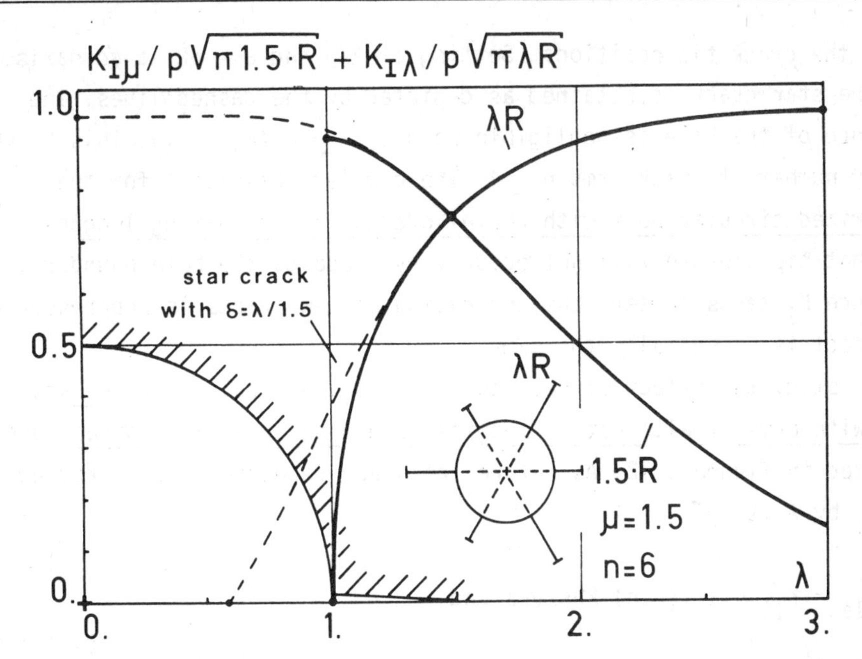

Figure 3.11: Comparison of stress intensity factors for pressurized star crack and circular hole with radial cracks, both with six arms of alternating lengths.

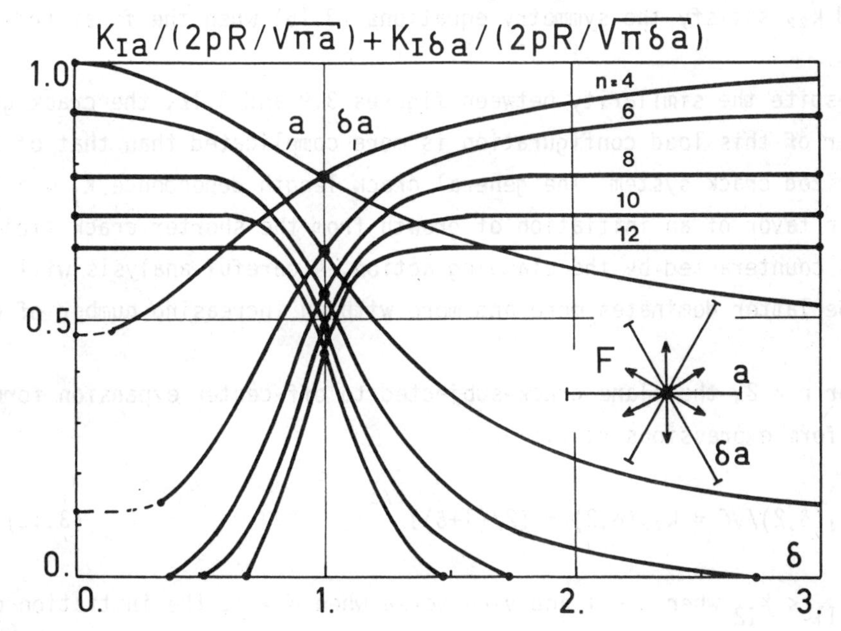

Figure 3.12: Normalized stress intensity factors for star crack with arms of alternating lengths opened by expansion forces, as a function of the crack length ratio for different n.

denote the crack tip positions. Setting $\delta = \lambda/\mu$ and $a = \mu R$, a comparison with the star crack is obtained as depicted by the dashed lines. The influence of the hole is negligible when $\mu, \lambda \geq \mu_0(6) = 1.5$. This is true for any number of crack arms $n \geq 4$. The complete results [5] for the pressurized circular hole with radial cracks of alternating lengths show that tip closure will not occur very close to the hole boundary. But since K_I tends to zero the generation of new cracks is suppressed and the effect is essentially the same.

To study the effect of gas penetration, the results for the star crack with arms of alternating lengths opened by expansion forces, are presented in figure 3.12. As before the shape dependence is extracted from K_I by writing

$$\left.\begin{aligned} K_{Ia} &= K_{I1} = k_{21}(\delta,n)\cdot 2pR/\sqrt{\pi a} \text{ and} \\ K_{I\delta a} &= K_{I2} = k_{22}(\delta,n)\cdot 2pR/\sqrt{\pi\delta a} \text{ for } 0 < \delta < \infty . \end{aligned}\right\} \tag{3.17}$$

k_{21} and k_{22} satisfy the symmetry equations (3.15) when the first index $1 \to 2$.

Despite the similarity between figures 3.9 and 3.12, the crack growth behavior of this load configuration is more complicated than that of the pressurized crack system. The general crack length dependence $K_I \propto a^{-1/2}$ works in favor of an initiation of growth from the shorter crack tips but this is counteracted by the clamping action. A careful analysis will show that the latter dominates more and more with an increasing number of crack arms.

For $n = 2$, the plane crack subjected to off-center expansion forces, closed form expressions result

$$k_{21}(\delta,2)/\sqrt{\delta} = k_{22}(\delta,2) = [2/(1+\delta)]^{1/2} . \tag{3.18}$$

Since $K_{I1} < K_{I2}$ when $\delta < 1$ and vice versa when $\delta > 1$, the initiation of crack growth will always start from the tip of the shorter crack. Eventually

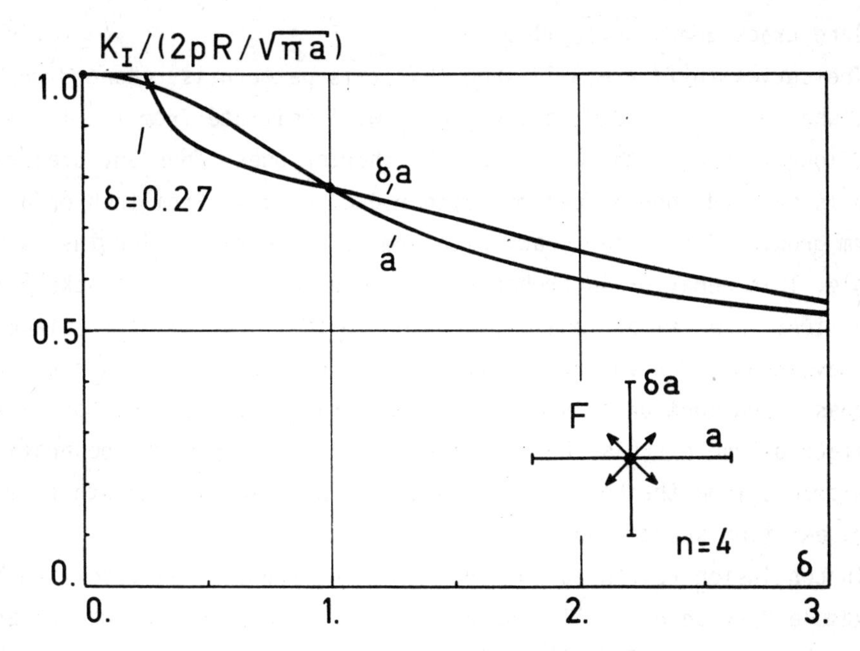

Figure 3.13: Direct comparison of stress intensity factors for the different length crack arms of a star crack with four arms of alternating lengths opened by expansion forces.

the plane crack grows uniformly.

The curves in figure 3.13 show that this pattern is changed when $n = 4$. When $0.27 < \delta < 3.7$, crack growth will initiate from the tips of the longer cracks otherwise from the shorter ones. Thus any preferential growth is stopped when δ becomes either 0.27 or 3.7. From then on a uniform growth of the star crack with arms of alternating lengths is possible. This behavior is probably also relevant for $n = 6$ cracks but with a wider interval of imbalance. For $n \geq 8$ the interval of imbalance is fully developed and a uniform growth is no longer possible. Just as for the pressurized contour this also involves crack tip closure. Most likely the effect of the circular hole will again be to suppress the generation of new cracks from the hole and to check any preferential growth from already existing short cracks.

In conclusion return to the rock blasting process, as described by for example Persson et al.[14], where many small cracks may be generated at the hole boundary. The discussion above shows that a radial crack system has an inherent tendency towards non-uniform growth in that cracks will successively fall behind the propagation front and their further growth be suppressed. This mechanism is largely insensitive to the gas penetration into the crack system. It provides a qualitative understanding of why only a few of the original cracks will grow to a substantial length. Persson et al.[14] state that the number of dominating cracks is often around five. Some sources give a lower number, others a higher number.

3.3 Formal Approach to Uniform Growth

The discussion of uniform growth of the star crack in section 3.2 was rather qualitative. It can be put on a more formal basis by using the potential energy of the cracked body in a stability criterion. Since the rate of decrease of this energy

$$-dU = pdV-dE = \sum_{i=1}^{n} G_i da_i , \tag{3.19}$$

equals the sum of the local energy release rates at potential growth, one can simply use the rate of change of this sum as a criterion on stability. E is the internal energy of the cracked body. All quantities are based on unit material thickness, J/m etc.

For a reversible quasi-static growth of the crack tips

$$\sum_{i=1}^{n} G_i da_i = G_c \sum_{i=1}^{n} da_i , \tag{3.20}$$

implying that $G_i = G_c$, the critical energy release rate, if the crack growth increment $da_i > 0$. A static crack tip and tip closure are also possible. Thus

a growing crack tip $da_i > 0$ with $G_i = G_c$,
a static crack tip $da_i = 0$ with $0 \leq G_i, \leq G_c$ and
crack tip closure $da_i < 0$ with $G_i = 0$
are admissible changes in the system.

If $-dU$ grows with increasing crack length, excess energy is available and the growth tends to be unstable. The reverse situation tends to make the crack growth stable. Thus the stability criterion related to a quasi-static growth situation becomes

$$\left.\begin{array}{lll} \text{instability when} & -d^2U \simeq \Sigma dG_i \cdot da_i > 0, & \\ \text{critical state when} & \Sigma dG_i \cdot da_i = 0, \text{ and} & \\ \text{stability when} & \Sigma dG_i \cdot da_i < 0 & \end{array}\right\} \tag{3.21}$$

Assuming a mode I situation in plane strain such that $G_i = (1-\nu^2)K_i{}^2/E$ there results with $G_i(a_j,p)$ that

$$-d^2U = \frac{2(1-\nu^2)}{E}\left\{\sum_{i=1}^{n}\sum_{j=1}^{n} K_i \frac{\partial K_i}{\partial a_j} da_j da_i + \sum_{i=1}^{n} K_i^2 da_i \cdot \frac{dp}{p}\right\}, \tag{3.22}$$

where a change of indici $i \rightleftharpoons j$ because of symmetry yields

$$K_i \cdot \partial K_i/\partial a_j = K_j \cdot \partial K_j/\partial a_i . \tag{3.23}$$

A decreasing gas pressure obviously tends to stabilize the potential growth of the crack system but since an inherent geometrical property is discussed we omit the gas pressure term here.

Specializing equation (3.22) to a star crack with arms of alternating lengths such that odd indici refer to arms of length a and even indici to arms of lengths δa there results

$$-d^2U = \frac{n^2(1-\nu^2)}{2E}\left\{K_{I1}\frac{\partial K_{I1}}{\partial a}\cdot(da_1)^2 + 2K_{I1}\frac{\partial K_{I1}}{\partial(\delta a)}\cdot da_1 da_2 + K_{I2}\frac{\partial K_{I2}}{\partial(\delta a)}\cdot(da_2)^2\right\} \tag{3.24}$$

with the use of equation (3.23).

Let us consider a situation of potential growth such that

$$da_1 = \alpha_1 \cdot da \text{ and } da_2 = \alpha_2 \cdot da, \tag{3.25}$$

where $\alpha_1, \alpha_2 \geq 0$ are constants. It covers static crack tips too but not tip closure. Hence $\alpha_2 = \delta\alpha_1$ describes uniform growth of a star crack and either of α_1 and $\alpha_2 = 0$ describes non-uniform growth. The crack system will prefer the form of growth which results in the largest increase of $-dU$ over a given crack extension. Under the restriction

$$\left.\begin{aligned} &\Sigma da_i = \frac{n}{2}(\alpha_1+\alpha_2)da = \text{constant} \\ &[-d^2U]_{\alpha_1=1+\delta,\alpha_2=0} - [-d^2U]_{\alpha_1=1,\alpha_2=\delta} > 0 \end{aligned}\right\} \tag{3.26}$$

thus defines non-uniform growth when $\delta \leq 1$. The case $\delta \geq 1$ is obtained by setting $\alpha_1 = 0$ and $\alpha_2 = 1 + \delta$ in the first term.

For the pressurized star crack, equation (3.23) yields

$$k_{11} \cdot k'_{11} + \delta^2 k_{12} \cdot k'_{12} = 0. \tag{3.27}$$

Here ' denotes differentiation with respect to δ. The only balanced situation for $n \geq 4$ arises when the crack arms are equally long, $\delta = 1$. At the point of incipient growth

$$\left.\begin{aligned} &K_{I1} = K_{I2} = K_{Ic}, \\ &k_{11}(1,n) = k_{12}(1,n) \text{ and } k'_{11}(1,n) = -k'_{12}(1,n). \end{aligned}\right\} \tag{3.28}$$

Substituting into equation (3.24) we obtain

$$-d^2U = \frac{n^2(1-\nu^2)}{2Eak_{11}} K_{Ic}^2 \cdot [\tfrac{1}{2}(\alpha_1^2+\alpha_2^2)k_{11} - (\alpha_1-\alpha_2)^2 k'_{11}] \cdot (da)^2, \tag{3.29}$$

valid when $\delta = 1$. Figure 3.9 shows that $k_{11}(1,n) > 0$ and that $k'_{11}(1,n) < 0$ for $n \geq 4$. Thus $-d^2U > 0$ and the crack growth tends to be unstable in any case.

The evaluation of (3.26) for $\delta = 1$ thus amounts to whether or not

$$\kappa_1(n) = k_{11}(1,n) - 4k'_{11}(1,n) > 0. \tag{3.30}$$

This is always the case. Thus non-uniform growth will always occur for a pressurized star crack with $n \geq 4$ arms. Since $k_{11}(\delta,2) = [(1+\delta)/2]^{1/2}$ there results with $k_{11}(1,2) = 1$ and $k'_{11}(1,2) = 1/4$ that $\kappa_1(2) = 0$. Thus uniform and non-uniform growth are equally likely for the plane pressurized crack, as expected because $K_{I1} \equiv K_{I2}$ for any δ.

For the star crack opened by expansion forces, equation (3.23) yields

$$k_{21} \cdot k'_{21} + k_{22} \cdot k'_{22} = 0. \tag{3.31}$$

Again start with a balanced situation $K_{I1} = K_{I2}$ such that

$$k_{21}(\delta,n) = k_{22}(\delta,n)/\sqrt{\delta} \tag{3.32}$$

For n = 2 and $n \geq 8$ the only solution seems to be $\delta = 1$.
For n = 4 and 6 there appears to be other solutions too, see figure 3.13 which yields $\delta = 0.27$ and 3.7 when n = 4. Equation (3.24) reduces to

$$-d^2U = \frac{n^2(1-\nu^2)}{2Eak_{21}}K_{Ic}^2\cdot[-\frac{1}{2}(\alpha_1^2+\alpha_2^2)k_{21}-\frac{1}{\delta}(\delta\alpha_1-\alpha_2)^2k'_{21}]\cdot(da)^2. \tag{3.33}$$

This equation shows that, since

$$[-d^2U]_{\alpha_1=1,\alpha_2=\delta} \propto -\frac{1}{2}(\delta^2+1)k_{21}(\delta,n) < 0 \tag{3.34}$$

for all open crack tips, a uniform growth would be stable if it could occur.

The possible preference for non-uniform growth amounts to whether or not

$$\kappa_2(\delta,n) = -\delta[k_{21}(\delta,n)+(\delta+1)^2k'_{21}(\delta,n)] > 0. \tag{3.35}$$

For a star crack with arms of equal lengths, $\delta = 1$, equation (3.18) shows that $k_{21}(1,2) = 1$ and $k'_{21}(1,2) = 1/4$. Hence $\kappa_2(1,2) < 0$ and a uniform growth is preferred. When n = 4 figure 3.12 shows that $k_{21}(1,4) \approx 0.78$ and $k'_{21}(1,4) \approx -0.29$. Thus $\kappa_2(1,4) > 0$ and $\kappa_2(1,n)$ increases with increasing n. Accordingly a non-uniform growth is preferred whenever $n \geq 4$. Equation (3.33) with inserted values $k_{21}(1,6) \approx 0.64$ and $k_{22}(1,6) \approx -0.61$ etc. gives

$$[-d^2U]_{\delta_1=2,\alpha_1=0} \propto -2k_{21}(1,n)-4k'_{21}(,n) > 0 \tag{3.36}$$

for $n \geq 6$. Thus the non-uniform growth will then be unstable despite the general crack length dependence $K_I \propto a^{-1/2}$ for expansion forces.

The cases of a star crack subjected to expansion forces with n = 4 or 6 balanced crack arms of alternating lengths remain to be analyzed. For n = 4 the balance occurs when $\delta = 0.27$. Then $k_{21}(0.27,4) \approx 0.98$, $k_{21}'(0.27,4) \approx -0.17$, and $\kappa_2(0.27,4) \approx -0.19$. Since $\kappa_2 < 0$ a uniform stable growth is the preference. This is also true for n = 6 arms but in practice δ is so small that the longer crack arms dominate the picture in both cases.

These formal results underscore the conclusions of section 3.2. They demonstrate clearly the tendency for non-uniform growth which is inherent in a radial crack system regardless of the gas penetration. Furthermore they show that such growth is usually unstable. A uniform stable growth appears to be possible with four and six crack arms when gas isn't present in the cracks. The longer cracks are however considerably longer in these cases, roughly by a factor $1/0.27 \approx 3.7$ or more. This would put the number of inherently dominating cracks to 3-4 or even less. This quantitative result should however be used with caution because both the pressure time history and the dynamics of the blasting process are important factors which have been left out in this analysis.

4. PATH - INDEPENDENT INTEGRALS

4.1 Method

The star cracks analyzed consist of radial crack arms with a common origin and no initial volume. So does a cracked wedge where two infinite radial lines form the sides. These two configurations are self-similar in the sense that a uniform expansion doesn't change the configuration.

A very powerful method of LEFM analysis of such cracks is the use of certain path-independent integrals given by Knowles and Stenberg [15], Eshelby [16], and Carlsson.[17] These depend on the expansional invariance of the surrounding elastic field. This approach was used by Ouchterlony [9-11] to determine closed form stress intensity factors.

The defining formulas [17] are in tensorial notation

$$\left.\begin{aligned} \underset{\sim}{J_x} &= \oint_C (W x_k n_k - T_i x_k u_{i,k})\,ds, \text{ and} \\ \underset{\sim}{\bar{J}_x} &= \oint_C (\bar{W} x_k n_k - T_i u_i - x_k \sigma_{ij,k} u_i n_j)\,ds. \end{aligned}\right\} \tag{4.1}$$

Here C is a closed curve in a plane of linearly elastic and isotropic material. W is the energy density and $\bar{W}$ its complementary form, both given by $\sigma_{ij}\varepsilon_{ij}/2$. σ_{ij} is the stress tensor and ε_{ij} the strain tensor. n_i denotes the unit outward normal to the curve, $T_i \doteq \sigma_{ij}n_j$ is the stress vector acting on the outside of C, u_i the displacement vector and ds an incremental line element along C.

These expressions simplify considerably when C coincides with the coordinate lines of a centered polar coordinate system (r,θ) since n_r or $n_\theta = 0$ along them. The energy terms vanish along radial lines since $x_k n_k = 0$ there. The other terms vanish along free boundaries such as the face of a wedge or a crack surface. Thus it is convenient to let C consist of segments of radial lines and centered circular arcs wherever possible. The only other type of segment necessary to close C is the crack tip loop in figure 4.1.

It may be shown that for a segment AB of C then [9]

$$\underset{\approx}{J_x^{AB}} + \underset{\approx}{\bar{J}_x^{AB}} = [r(\sigma_\theta v_\theta + \tau_{r\theta} u_r)]_A^B. \tag{4.2}$$

See figure 4.2. Thus

$$\underset{\approx}{J_x^{AB}} = - \underset{\approx}{\bar{J}_x^{AB}} \tag{4.3}$$

whenever the end points of AB lie on radial lines that are either free boundaries, shear free lines of symmetry, or fixed boundaries. Thus with a suitable choice of segments the two integrals can be interchanged easily.

Denoting the segments by C_i or C_∞ if both end-points are far from the origin it follows that $C \hat{=} \Sigma_i C_i + C_\infty$. The conservation law inherent in

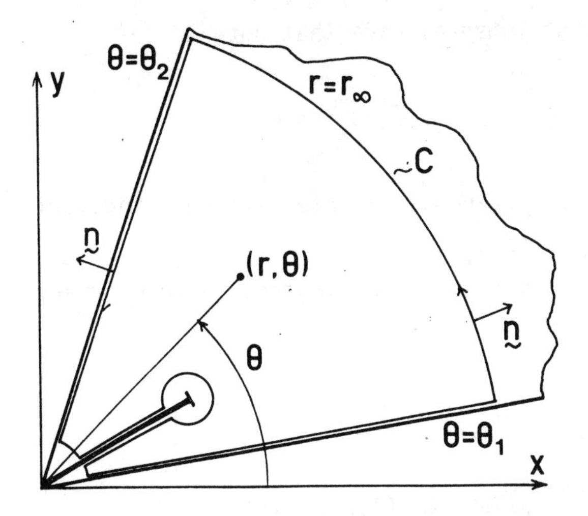

Figure 4.1: Cracked wedge in polar coordinate system with suggested integration contour C for path-independent integrals.

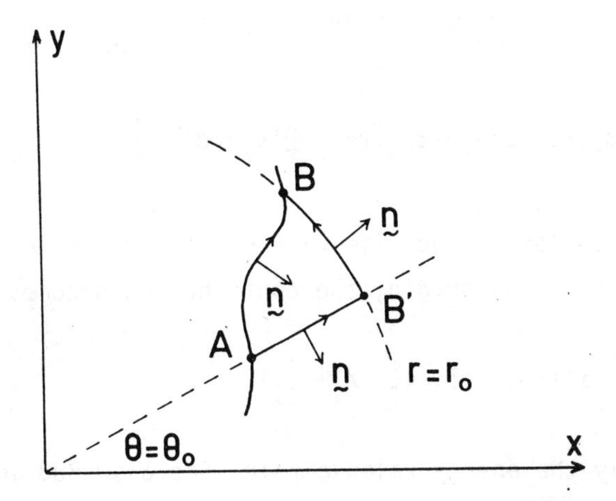

Figure 4.2: Arbitrary path AB and equivalent path AB'B consisting of radial line AB' and centered circular arc B'B.

the path-independent integrals are that both

$$J_{\underset{\sim}{x}} = \Sigma_i J^i_{\underset{\sim}{x}} + J^\infty_{\underset{\sim}{x}} \text{ and } \bar{J}_{\underset{\sim}{x}} = \Sigma_i \bar{J}^i_{\underset{\sim}{x}} + \bar{J}^\infty_{\underset{\sim}{x}} = 0, \tag{4.4}$$

when C encompasses homogeneous material without singularities. The superscript i refers to C_i etc.

The contributions from radial segments with $\underset{\sim}{n}$ oriented as on AB' in figure 4.2 become

$$\left.\begin{aligned} J^i_{\underset{\sim}{x}} &= \int_{C_i} (\sigma_\theta v_{\theta,r} + \tau_{r\theta} u_{r,r}) r\,dr, \text{ and} \\ \bar{J}^i_{\underset{\sim}{x}} &= \int_{C_i} [(\sigma_\theta + r\sigma_{\theta,r}) v_\theta + (\tau_{r\theta} + r\tau_{r\theta,r}) u_r] dr. \end{aligned}\right\} \tag{4.5}$$

Both vanish if C_i belongs to any of the previously mentioned categories since for example $\sigma_\theta = 0$ on a radial line implies that $\sigma_{\theta,r} = 0$ there too. The contributions from centered arcs like BB' in figure 4.2 become

$$\left.\begin{aligned} J^i_{\underset{\sim}{x}} &= \int_{C_i} (W - \sigma_r u_{r,r} - \tau_{r\theta} v_{\theta,r}) r^2 d\theta \text{ and} \\ \bar{J}^i_{\underset{\sim}{x}} &= \int_{C_i} [Wr - (\sigma_r + r\sigma_{r,r}) u_r - (\tau_{r\theta} + r\tau_{r\theta,r}) v_\theta] r\,d\theta. \end{aligned}\right\} \tag{4.6}$$

For a crack tip loop taken clockwise with $\underset{\sim}{n}$ pointing to the tip, situated at the distance a from the origin, the contribution becomes

$$J^i_{\underset{\sim}{x}} = -\bar{J}^i_{\underset{\sim}{x}} = -a(1+\nu)(\kappa+1)/4E \cdot (K_I^2 + K_{II}^2), \tag{4.7}$$

that is essentially the energy release rate. See also Budiansky and Rice.[18]

C_∞ is always a centered arc. Equations (4.6) imply that if the stresses decay faster than 1/r, like they do if the load is balanced, then the contribution from C_∞ will vanish. If an unbalanced loading is applied to the origin or the crack faces then $\sigma_{ij} \propto 1/r$ far away and stress terms like $\sigma_r + r\sigma_{r,r}$ vanish such that

$$\underset{\sim}{J_x^{\infty}} = \int_{C_\infty} Wr^2 d\theta. \tag{4.8}$$

The same expression holds for an arc near a point load or a dislocation where the 1/r term dominates the stress field. For such loadings closed form solutions may result.

4.2 Results and Applications

Four applications of equations (4.4) to (4.7) that yield closed form solutions are presented here. First consider the pressurized star crack in figure 4.3. There results for the i:th arm that[9]

$$\left.\begin{aligned} \underset{\sim}{J_x^{i1}} + \underset{\sim}{J_x^{i3}} &= -p\int_0^{a_i} (v_{\theta i}^+ - v_{\theta i}^-)\,dr = -pV_i, \text{ and} \\ \underset{\sim}{J_x^{i2}} &= a_i(1+\nu)(\kappa+1)(K_{Ii}^2 + K_{IIi}^2)/4E. \end{aligned}\right\} \tag{4.9}$$

The superscripts + and - refer to upper and lower crack faces respectively and V_i denotes the crack arm volume per unit thickness. Hence equation (4.4) yields that the total gas volume

$$V = (1+\nu)(\kappa+1)/4pE\cdot\sum_i a_i(K_{Ii}^2 + K_{IIi}^2). \tag{4.10}$$

Related expressions were given earlier.[5]

Next consider the star crack opened by expansions forces which is depicted in figure 4.4. For it[9]

$$\underset{\sim}{J_x^2} = \underset{\sim}{J_x^6} = -(1+\nu)(\kappa+1)/(2\alpha+\sin 2\alpha)\cdot F^2/8E. \tag{4.11}$$

J_x^4 is given by equation (4.7) with $K_{II} = 0$, the other contributions vanish. Thus there results that

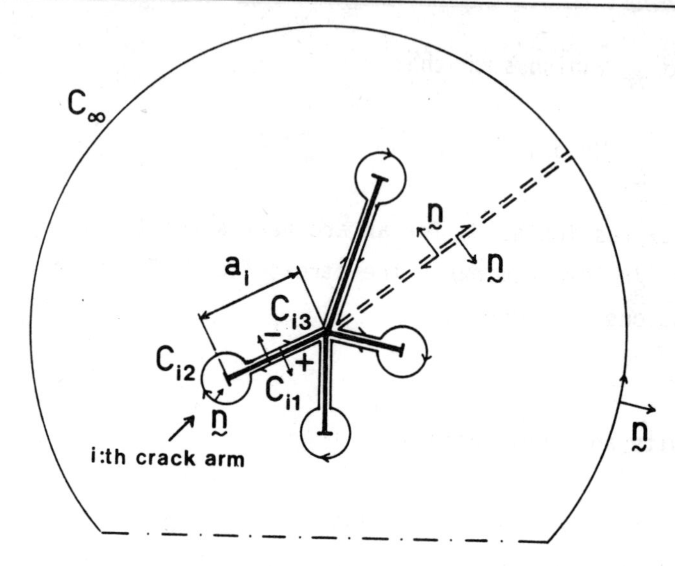

Figure 4.3: Pressurized star crack of arbitrary shape with matching integration contour.

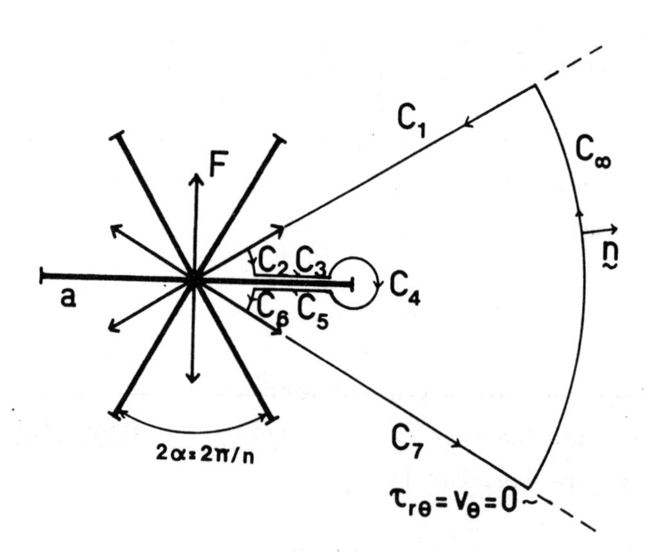

Figure 4.4: Star crack with arms of equal lengths opened by expansion forces and matching integration contour.

$$K_I = \frac{F}{\sqrt{\pi a}} \cdot [\pi/(2\alpha+\sin 2\alpha)]^{1/2}, \tag{4.12}$$

which with $\alpha = \pi/n$ and $F = 2pR \cdot \sin(\pi/n)$ reproduces equation (3.13).

For a wedge with a symmetric crack at the end, opened by a symmetric pair of unbalanced forces as in figure 4.5 there results[9]

$$\left.\begin{aligned} \tilde{J}_x^{\infty} &= (1+\nu)(\kappa+1)\cos^2\phi/(2\alpha+\sin 2\alpha) \cdot F^2/E \text{ and} \\ \tilde{J}_x^{2} = \tilde{J}_x^{6} &= -(1+\nu)(\kappa+1)F^2/4E \cdot \left[\frac{\cos^2(\phi-\alpha/2)}{\alpha+\sin\alpha} + \frac{\sin^2(\phi-\alpha/2)}{\alpha-\sin\alpha}\right] \end{aligned}\right\} \tag{4.13}$$

With $\tilde{J}_x^4$ again given by equation (4.7) with $K_{II} = 0$ and the other contributions vanishing, K_I becomes

$$K_I = \frac{F}{\sqrt{\pi a}} \cdot (\sin\phi - \cos\phi \frac{2\sin^2\alpha}{2\alpha+\sin 2\alpha}) \cdot \left[\pi \frac{2\alpha+\sin 2\alpha}{\alpha^2-\sin^2\alpha}\right]^{1/2}. \tag{4.14}$$

This result was also derived earlier.[8] Equation (4.14) implies that the thrust component $F \cdot \cos\phi$ tends to close the crack tip as opposed to the splitting component $F \cdot \sin\phi$. Closure will occur when

$$\phi < \phi_{min}(\alpha) = \operatorname{atan}[2\sin^2\alpha/(2\alpha+\sin 2\alpha)]. \tag{4.15}$$

For a cracked half-plane $\alpha = \pi/2$ and $\phi_{min} \approx 32.5^0$.

Regarding rock splitting as the wedging of a symmetrically cracked half-plane this means that the tool angle is critical. It must be slender enough to compensate both for friction and thrust force closure before any splitting action occurs. If the opening displacement caused by the tool is locked at some point, due to frictional or other effects, then the result of unloading will be an increase in crack driving force to start with. This implies that whether or not there is any crack growth during loading, the crack may grow upon unloading.

The final example is the wedge with a symmetric crack at the end, opened by transverse displacements of constant magnitude. It is shown in

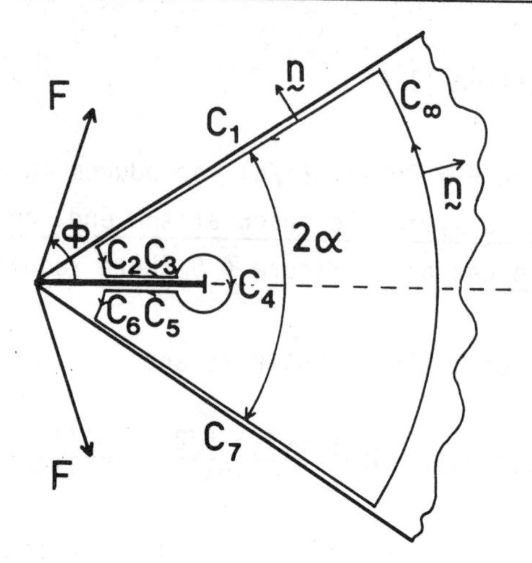

Figure 4.5: Symmmetrically cracked wedge subjected to a symmetric pair of unbalanced forces and matching integration contour.

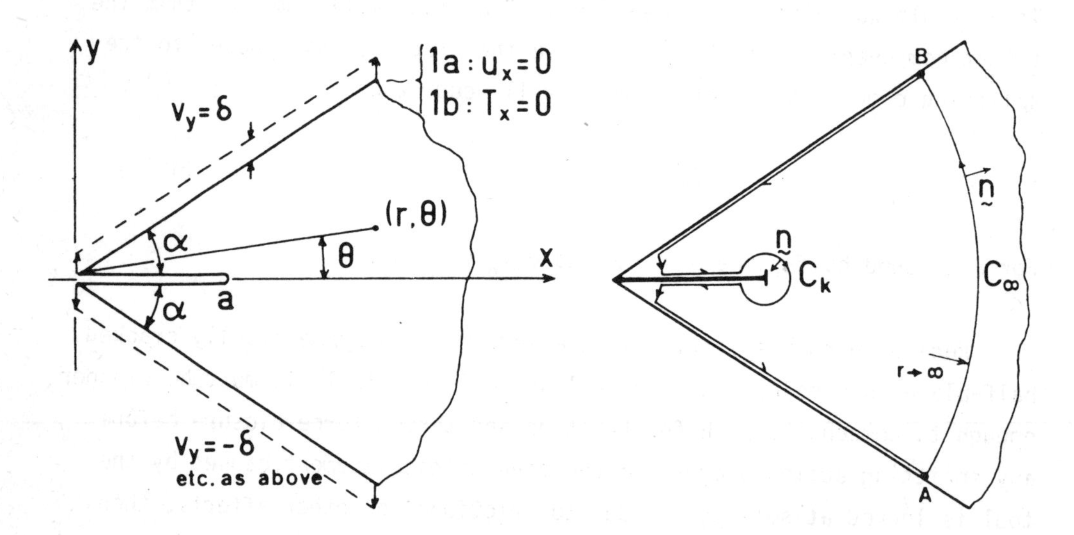

Figure 4.6: Symmetrically cracked wedge opened by transverse displacements of constant magnitude. Supplementary boundary conditions and matching integration contour also shown.

figure 4.6. The analysis [11] is more involved than before. The results are again simple in form however. For case 1a, the displacement boundary conditions $|v_y| = \delta$ and $u_x = 0$ along the wedge faces, there results

$$J_{\underset{\sim}{x}}^{\infty} = 2(\kappa+1)G\delta^2/(2\kappa\alpha-\sin 2\alpha). \tag{4.16}$$

with J_x^i for the crack tip given by equation (4.7) and the other contributions vanishing the result is

$$K_I = \frac{4G\delta}{\sqrt{\pi a}}\cdot[\pi/(2\kappa\alpha-\sin 2\alpha)]^{1/2}. \tag{4.17}$$

Here $G = E/2(1+\nu)$ is the shear modulus of the material.

For case 1b, the displacement - stress boundary conditions $|v_y| = \delta$ and $T_x = \sigma_x n_x = 0$ along the wedge faces, there results

$$J_{\underset{\sim}{x}}^{\infty} = 2G\delta^2/(\kappa+1)\cdot[(2\alpha+\sin 2\alpha)/\alpha^2]. \tag{4.18}$$

Again the crack tip loop gives the only remaining contribution and

$$K_I = \frac{4G\delta}{\sqrt{\pi a}}\frac{1}{\kappa+1}\cdot[\pi(2\alpha+\sin 2\alpha)/\alpha^2]^{1/2}. \tag{4.19}$$

The limiting cases $\alpha \rightarrow 0$ correspond to the well known cracked infinite strip configuration. The cases $\alpha = \pi$ can be construed as rough or friction free tools driving cracks ahead of them [11] but this is not pursued here. The cases $\alpha = \pi/2$ have simple interpretations too.

The path-independent integrals used here concentrate on fracture fundamentals and generally express the total energy release rate at self-similar growth. They won't distinguish between the different crack tips of an unsymmetric system nor usually between the modes at any one tip, not unless the crack geometry is symmetric.[9]

Other path-independent integrals that have been used together with finite element methods master these separation problems and anisotropy as well, see Wang et al.[19] for example. For two dimensional elastic problems the surface integral method seems to be an efficient general alternative, see Narendran and Cleary.[20]

The finite element programs in use today are capable of handling both mixed mode and three dimensional crack problems. Here K formulas that are based on the crack surface displacements behind the crack tip seem to dominate, see Tseng[21] and Saouma and Ingraffea[22] for example.

5. REFERENCES

1. Jaeger, J.C. and Cook, N.G.W., *Fundamentals of Rock Mechanics*, 1st ed, Methuen, London, 1969. 3rd ed 1979.

2. Hardy, M.P., Fracture mechanics applied to rock, PhD Thesis, Univ Minnesota, Minneapolis MN, 1973.

3. Swan, G., The mechanics and dynamics of certain rock fracture modes, PhD Thesis, Univ London and Imperial College of Science and Techn, London, 1974.

4. Wang, F-D. and Clark, G.B., Energy resources and excavation technology, *Proc 18th US Symp Rock Mechs*, Colorado School of Mines, Golden CO, 1977.

5. Ouchterlony, F., Analysis of the stress state around some expansion loaded crack systems in an infinite plane medium, Swedish Detonic Research Foundation (SveDeFo) report DS 1972:11, Stockholm, Sweden, 1972. In Swedish.

6. Ouchterlony, F., Fracture mechanics applied to rock blasting, in *Proc 3rd Int Congress of the ISRM*, vol II-B, Denver CO, 1974, 1377.

7. Ouchterlony, F., Stress intensity factors for the expansion loaded star crack, *Engng Fract Mechs*, 8, 1976, 447.

8. Ouchterlony, F., Symmetric cracking of a wedge by concentrated loads, *Int J Engng Sciences*, 15, 1977, 109.

9. Ouchterlony, F., Some stress intensity factors for self-similar cracks derived from path-independent integrals, *J Elasticity*, 8, 1978, 259.

10. Ouchterlony, F., Fracture analysis of cracks related to rock fragmentation, Dr Sci Thesis, Dept Strength of Mtrls and Solid Mechs, Royal Inst Techn, Stockholm, Sweden, 1978.

11. Ouchterlony, F., Symmetric cracking of a wedge by transverse displacements, *J Elasticity*, 10, 1980, 215.

12. Muskhelishvili, N.I., *Some Basic Problems of the Mathematical Theory of Elasticity*, 2nd ed, Noordhoff, Groningen, The Netherlands, 1963.

13. Persson, A., Plane states in elastomechanics described by complex functions, Solid Mechs Dept publ no 17, Chalmers Inst Techn, Gothenburg, Sweden, 1966.

14. Persson, P.A., Lundborg, N., and Johansson, C.H., The basic mechanisms in rock blasting, in *Proc 2nd Congr ISRM*, paper 5-3, Beograd, Yugoslavia, 1970.

15. Knowles, J.K. and Sternberg, E., On a class of conservation laws in linearized and finite elastostatics, *Arch Rat Mechs Analysis*, 44, 1972, 187.

16. Eshelby, J.D., The calculation of energy release rates, in *Prospects of Fracture Mechanics*, Sih,G.C. et al. eds, Noordhoff, Leyden, The Netherlands, 1975, 69.

17. Carlsson, J., Path independent integrals in fracture mechanics and their relation to variational principles, in *Prospects of Fracture Mechanics*, Sih, G.C. et al. eds , Noordhoff, Leyden, The Netherlands, 1975, 139.

18. Budiansky, B. and Rice, J.R., Conservation laws and energy release rates, *J Appl Mechs*, 40, 1973, 201.

19. Wang, S.S., Yau, J.F., and Corten, H.T., A mixed-mode analysis of rectilinear anisotropic solids using conservation laws of elasticity, *Int J Fracture*, 16, 1980, 247.

20. Narendran, V.M. and Cleary, M.P., Elastostatic interaction of multiple arbitrarily shaped cracks in plane inhomogeneous regions, Report REL-82-6, Dept Mech Engng, MIT, Cambridge MA, 1982.

21. Tseng, A.A., A comparison of three-dimensional finite element solutions for the compact specimen, *Int J Fracture*, 17, 1981, R125.

22. Saouma, V.E. and Ingraffea, A.R., Fracture mechanics analysis of discrete cracking, in *Advanced Mechanics of Reinforced Concrete*, Delft Univ Press, Delft, The Netherlands, 1981, 413.

FRACTURE TOUGHNESS TESTING OF ROCK

Finn Ouchterlony

Swedish Detonic Research Foundation
S-12611 Stockholm, Sweden

1. REVIEW OF TOUGHNESS TESTING

Results from crack growth resistance measurements on rock are presented in this section. They are given a fracture mechanics interpretation and the influence of different test parameters is studied. The review is condensed from previous work by Ouchterlony.[1,2,59]

A cornerstone of linear elastic fracture mechanics (LEFM) testing of metals is the standard E 399 of the American Society for Testing and Materials (ASTM) which gives the standard test method for plane-strain fracture toughness (K_{Ic}) of metallic materials. Since no similar standard exists for rock, E 399 is an indispensible reference in this review. One for rock is underway though through ASTM and its subcommittee E 24.07. It will hopefully be based on core specimens since minimal specimen preparation efforts are mandatory.

Apart from K_{Ic}, a number of other crack resistance measures are reviewed including the J-integral. Their mutual compatibility is given special consideration. The review is a basis for the development of core based specimens and their testing which is presented in sections 2 and 3.

1.1 On Specimen Geometries

Test specimens are usually based on the assumption that a two-dimensional stress and strain field exists inside them, at least the conditions along the crack front are assumed constant. In practice all specimens violate this more or less, see for example Tseng.[3] Thus all experimental results are through the thickness averages.

Two frequently used specimen types are shown in figure 1.1. The single edge notch beam (SENB) with proportions fixed to S/w = 4 and B/w = 0.5 is the first standard specimen recommended in E 399, here abbreviated 3PB. A core based variety is presented in section 2. The compact tension (CT) specimen as well as a disc shaped version are also standardized in E 399. The influence of specimen type is discussed throughout this review.

Common to all is that geometry and loading set up a stress field which is symmetric with respect to the crack, such that the crack will tend to propagate in its own plane. This is the technically important case, called mode I. Few results for other modes exist, examples are Awaji and Sato [4] and Ingraffea [5], and they are left out.

1.2 Specific Work of Fracture

Earlier, the most common crack resistance measure for rock was the effective surface energy γ_{eff}, which alludes to Griffith but avoids the energy rate balance. Instead use the term <u>specific work of fracture</u>

$$\bar{R} = 2\gamma_{eff} = W_f/A = \int_0^\infty F \cdot d\delta_F/A, \qquad (1.1)$$

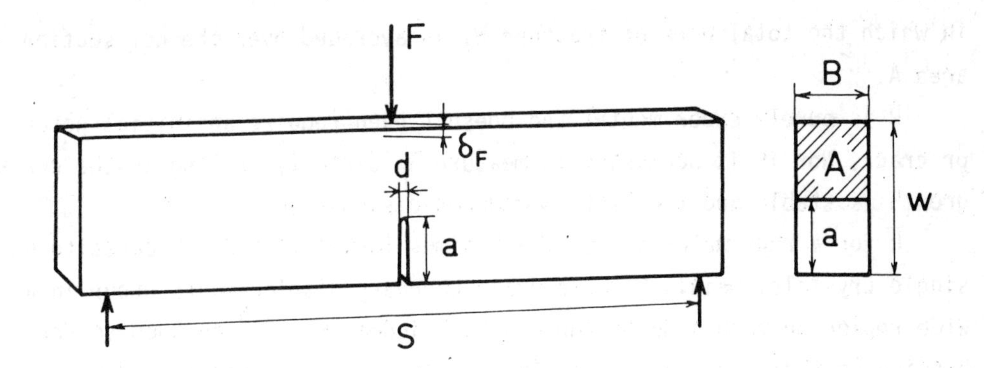

Basic notation:

- w = specimen depth
- B = specimen thickness
- S = support span
- a = crack length or notch depth
- A = net section area
- d = notch width
- F = load on specimen
- δ_F = (ideal) load point displacement
- a_0 = initial notch depth

Figure 1.1a: Single edge notch beam (SENB) under three point bending with basic notation.

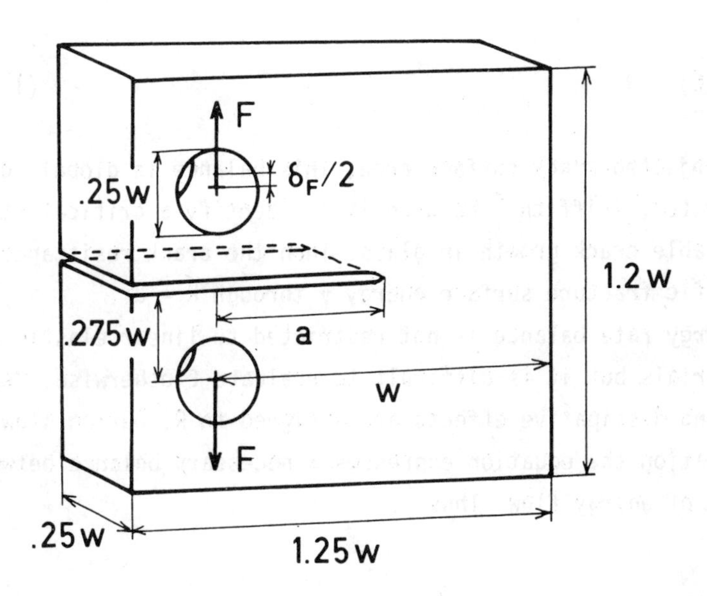

Figure 1.1b: Compact tension (CT) specimen with dimensions according to E 399.

in which the total work of fracture W_f is averaged over the net section area A.

$\overline{R}$ is purely experimental and doesn't even require an initial notch or crack. Nor is it necessary to measure δ_F directly as long as the crack growth is stable and the test system non-dissipative.

$\overline{R}$ for a rock material is 10-100 times larger than $\overline{R}$ for constituent single crystals, mainly because grain boundary displacements occur in a wide region on each side of the crack, Friedman et al.[6] An $\overline{R}$-curve for bending is shown in figure 1.2. The plateau level is interpreted as a material property, Forootan-Rad and Moavenzadeh.[7]

1.3 Griffith's Balance of Energy Rates

This is an equilibrium requirement which expresses a balance between the virtual release rate of available energy, that is work W minus specimen strain energy E, and the fracture energy consumption rate R through

$$\frac{d}{dA_c}(W-E) = R. \tag{1.2}$$

A_c is the projected crack surface area. This balance is global, comprising the whole system, Griffith.[8] He used it to identify a critical state at the onset of unstable crack growth in glass. Then the crack resistance is given by the specific fracture surface energy γ through $R = 2\gamma$.

The energy rate balance is not restricted to linear elastic and brittle materials but it is difficult to evaluate E otherwise. Usually non-linear and dissipative effects are assigned to R. During slow stable crack propagation the equation expresses a necessary balance between actual rates of energy flow. Thus

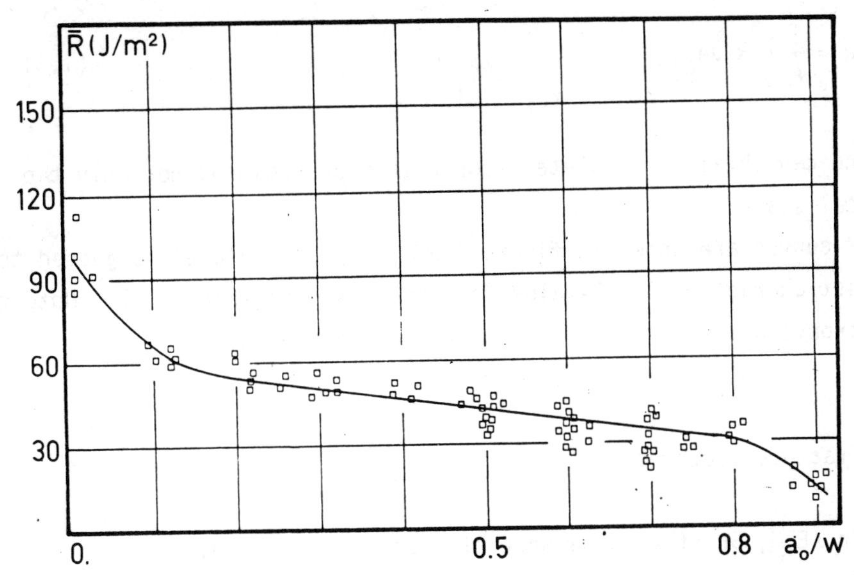

Figure 1.2: Specific work of fracture $\bar{R}$ for SENB of Chelmsford granite.[7]

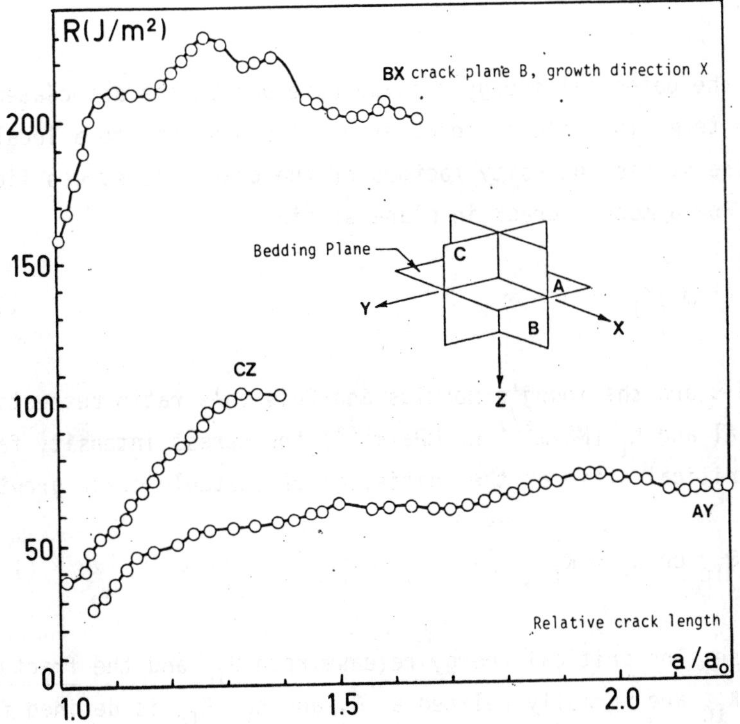

Figure 1.3: Fracture energy consumption rate R for Indiana limestone.[9]

$$\bar{R} = \frac{1}{A}\int_0^A R \cdot dA_c, \tag{1.3}$$

and one can obtain a complete R-curve from a specimen, not only the critical state.

R-curves are shown in figure 1.3. The plateau level is judged to be the most characteristic R-value for the rock, Hoagland et al.[9] Note the anisotropy. R doesn't seem to be very rate dependent.

1.4 Fracture Toughness

In LEFM, Griffith's balance of energy rates becomes

$$-\frac{\partial U}{\partial A_c} \equiv G = G_R. \tag{1.4}$$

Here G is the potential energy release rate and G_R the associated crack resistance term. This global formulation is equivalent to a local one based on the stress intensity factors of the crack tip stress field, Irwin.[10] For a mode I crack in plane strain

$$G_I = (1-\nu^2)K_I^2/E. \tag{1.5}$$

Here E and ν are the Young's modulus and Poisson's ratio respectively of the material and K_I [$MN/m^{3/2}$ or $MPa \cdot m^{1/2}$] the stress intensity factor.

The critical state at the initiation of unstable crack growth becomes

$$G_I = G_{Ic} \text{ or } K_I = K_{Ic}, \tag{1.6}$$

in this case. The critical energy release rate G_{Ic} and the <u>fracture toughness</u> K_{Ic} are formally related as G_I and K_I. K_{Ic} is defined for metallic materials by the ASTM standard E 399-81. No such standard exists for rock materials, so we rely on E 399.

<u>1.4.1 Validity of Metals Testing Criteria.</u> Some criteria of validity for K_{Ic} as per E 399 are listed

1. Specimen size requirements:
   a. thickness $\geq 2.5\ (K_{Ic}/\sigma_y)^2$
   b. crack length $\geq 2.5\ (K_{Ic}/\sigma_y)^2$
2. Specimen precracking requirements:
   a. K_{max} in fatigue $\leq 0.6 \cdot K_{Ic}$
   b. final crack length a within 45 to 55% of depth w
   c. local crack geometry
3. Test procedure and interpretation:
   a. 5% offset secant used to determine F_5 and F_Q
   b. $F_{max}/F_Q < 1.1$
   c. Loading rate

All fracture toughness tests for rock are invalid as per E 399 strictly seen. This motivates use of K_Q for a tentative value.

The specimen size criteria aim at making the plastic zone at the crack tip in metals insignificant. The main non-linear deformation mechanism in rock is micro-cracking however, Hoagland et al.[9] An estimate of the micro-crack zone can thus be based on a maximum normal stress criterion, Schmidt.[11] It looks like the plane strain plastic zone in metals but without the plastic constraint. Hence there is no transition to a different deformation pattern in plane stress and K_Q for rock should be independent of specimen thickness. Experiments confirm this, see figure 1.4, Schmidt and Lutz.[12]

K_Q for rock depends on the crack length much as for aluminium, see figure 1.5, Ingraffea and Schmidt.[13] This motivates a crack length requirement for rock like

$$a \geq 2.5(K_{Ic}/\sigma_t)^2, \tag{1.7}$$

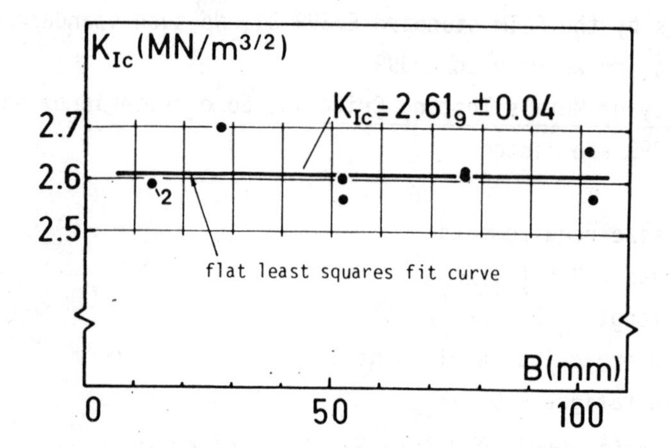

Figure 1.4: K_{Ic} or upper limit K_Q-values for CT specimens of Westerly granite with a ≈ 100 mm as function of specimen thickness.[12]

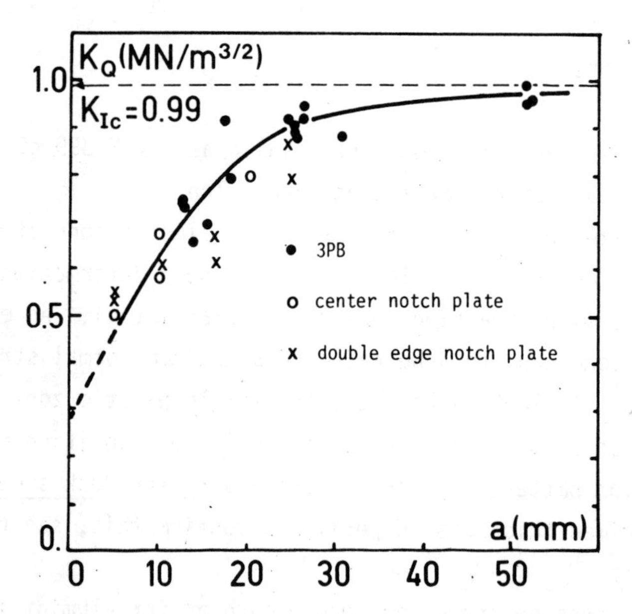

Figure 1.5: K_Q for Indiana limestone as function of crack length.[13]

based on uniaxial tensile strength σ_t instead of yield stress σ_y, Schmidt.[11] The notation K_{Ic} could perhaps be used to denote upper limit or size independent K_Q values that meet this requirement and as many of the other metals testing criteria as possible. One presumably applies to ligament size too. Another one relates to the micro-structure versus the continuum assumption, especially in the presence of high stress gradients. The average grain size must be much smaller than specimen dimensions or even the critical size of the micro-crack zone at incipient crack growth.

The specimen precracking criteria aim at producing a well defined crack outside the influence of residual stresses and premade notches. The requirement K_{max} in fatigue $\leq 0.6 \cdot K_{Ic}$ tends to avoid spuriously high K_Q-values due to crack tip blunting in metals. K_{max}/K_Q ratios that induce fatigue crack growth from a notch in rock are 0.6-0.95, Ouchterlony.[1] This ratio can be environment sensitive.

K_Q-values from first load cycles on prefatigued specimens of rock are somewhat lower than those from subsequent load cycles on the same specimens, Schmidt and Lutz.[12] The explanation may be that the side cracking is less developed for fatigue cracks than for cracks that grow under monotonic loading, Kim and Mubeen.[14] The wake of side cracked material behind the crack tip is equivalent to blunting.

Just as in metals fatigue, crack closure behind the tip occurs in rock, Schmidt and Lutz.[12] This could be due to intergranular displacements and relaxation of residual elastic strains near the crack surfaces, Friedman et al.[6]

The crack length in rock is almost without exception measured with compliance calibration techniques. Many problems are avoided in this way but there is an uncertainty involved about crack tip details, Ouchterlony.[1]

A final crack length in the specimen mid section gives material economy. It also ensures that the 5% offset secant procedure of figure 1.6, which the E 399 test interpretation dictates, corresponds to 2% apparent crack growth. In research a variable secant may help to relax the specimen size requirements, Munz.[15]

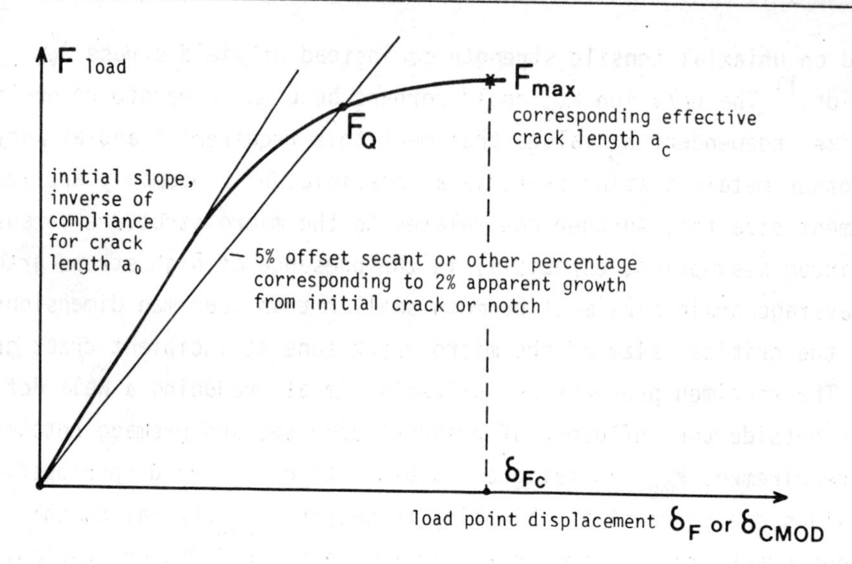

Figure 1.6: Load - displacement diagram with points that define fracture toughness evaluation. F_{max} is usually larger than F_Q for rock. E 399 requires 5% offset in F - δ_{CMOD} diagram.

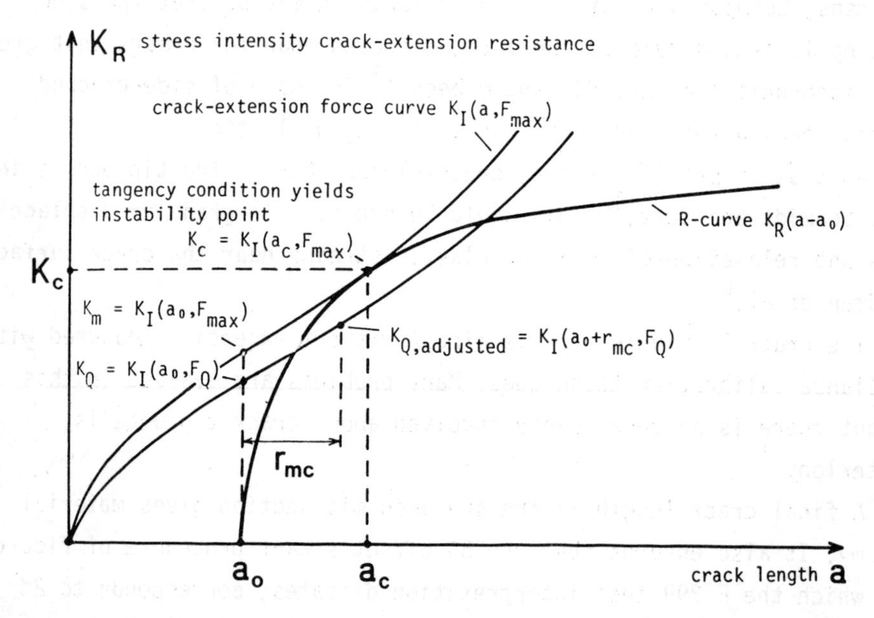

Figure 1.7: Graphic interpretation of fracture toughness parameters K_c, K_m, K_Q, and $K_{Q,adjusted}$ in terms of an R-curve.

The requirement $F_{max}/F_Q < 1.1$ recognizes that E 399 at a_o+0.02a_o serves to give a single representative point on a plane strain R-curve with an expected flat upper part. Conservatively estimated by

$$K_Q = K_I(a_o, F_Q), \tag{1.8}$$

this point should lie beyond the knee.

The loading rate is not very critical in the fracture toughness testing of rock, which is usually made 10-100 times slower than metals testing. The variance in the toughness data usually masks an increase in loading rate by a factor of ten, Ouchterlony.[1]

1.4.2 Other Aspects. The validity of K_{Ic}-tests can be approached from other angles. One is the possibility of using notched specimens without precracking them. One can either use sufficiently narrow notches in standard specimens or specimen types which produce their own precracks during the tests.

There is some basis for using the former approach. The associated approximate fracture toughness

$$K_m = K_I(a_o, F_{max}), \tag{1.9}$$

is based on initial notch length a_o and maximum load F_{max} rather than F_Q. Its main use is rock core testing where the test procedure should be kept simple. Because of core sizes, neither the crack length requirement nor $F_{max}/F_Q < 1.1$ can usually be met. There is most likely substantial micro-cracking at the crack tip which explains why K_m-values depend both on notch length and specimen type, Ouchterlony.[1]

The analogy with Irwin's plasticity adjustment factor yields one for micro-cracking in rock. Schmidt[11] gives

$$r_{mc} \approx 0.2(K_{Ic}/\sigma_t)^2, \tag{1.10}$$

from which a fracture toughness estimate for sub-size specimens

$$K_{Q,\text{adjusted}} = K_I(a_o+r_{mc},\ F_Q) \tag{1.11}$$

emerges. This is interpreted in figure 1.7 in terms of R-curves. The microcrack zone adjustment has emerged in several other forms.

The other approach was influence of specimen type. In the standard ones of E 399 the crack growth, once initiated, tends to be unstable under load control. This holds for many other specimens as well. K_Q-data for rock from these are generally specimen independent, even for sub-size ones as per E 399, Ingraffea and Schmidt [13] and Schmidt and Lutz.[12]

With sub-size standard specimens it is vital to incorporate the sub-critical crack growth to obtain representative toughness values. This is probably easier to achieve with specimens that produce their own precrack, such as the Double Torsion (DT), the Short Rod, Barker[16], and the Chevron Edge Notch Round Bar in Bending (CENRBB), Ouchterlony.[17] The DT specimen K_I is independent of crack length in a wide region and thus presumably representative toughness values can be measured without exact knowledge of the crack length.

The Short Rod is a core based specimen with a characteristic chevron notch, figures 1.8. So is the CENRBB which is introduced in section 2. When the crack grows from the apex the length of the crack front increases which tends to stabilize the crack growth at first. Thus the crack extension force curves are concave upwards with a minimum at a specific crack length, $a = a_m$ say. Beyond it the crack extension tends to become unstable. The Short Rod toughness was initially defined

$$K_{SR} = K_I(a_m,F_{max}) = K_{Imin}(F_{max}) = \beta_{SR}\cdot F_{max} \tag{1.12}$$

and likewise for the CENRBB with a different β, if so desired. See equation (2.29).

This assumes that the evaluation point a_m lies on the plateau of a mainly flat R-curve. Since significant plasticity corrections are

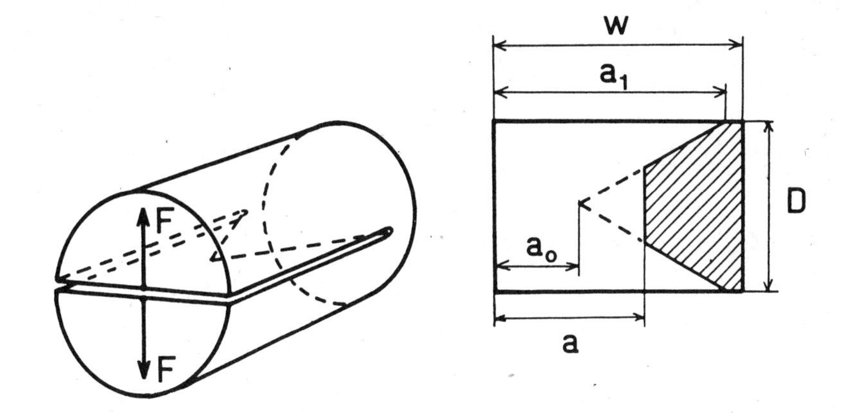

Figure 1.8a: Short Rod specimen.[16] Proportions are roughly $w = 1.5 \cdot D$, $a_o = 0.5 \cdot D$, and $a_1 = 1.4 \cdot D$.

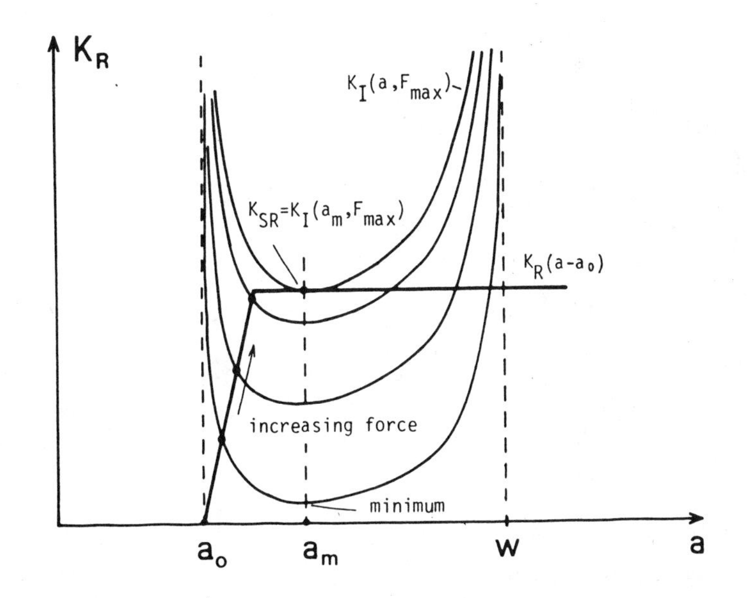

Figure 1.8b: Approximate crack extension force curves of Short Rod specimen illustrate how fracture toughness evaluation K_{SR} in equation (1.12) requires a flat R-curve.

needed to make K_{SR} correlate well with K_Q-values, Costin [18], this proposition is doubtful.

1.5 J-integral Resistance

The J-integral allows an extension of LEFM to certain non-linear material behavior, Rice.[19] Due to its path independence, J is a characteristic scalar measure of the conditions at the crack tip, much as the stress intensity factor in LEFM. It is basically limited to non-linear elasticity. From experience it also covers steady loading of stationary cracks in elasto-plastic materials, but generally not further growth since the unloading behind the tip is irreversible.

For an elastic, not necessarily linear, material

$$J = G = - \frac{\partial U}{\partial A_c} . \tag{1.13}$$

Thus it is reasonable to associate a critical value

$$J \equiv J_I = J_{Ic} \tag{1.14}$$

in mode I, with the initiation of crack extension. There is formally the theoretical equivalence

$$J_{Ic} = (1-\nu^2)K_{Ic}{}^2/E, \tag{1.15}$$

but their definitions as per the standard test procedures E 813 and E 399 will usually make them somewhat incompatible.

Under displacement controlled conditions U = W, the work performed by the external load. J_{Ic} measurements would hence require numerical differentiation of W based on several specimens. It is however possible to transform the J_I-expression to

$$J_I = \eta \cdot \frac{W}{A}, \tag{1.16}$$

such that only a single specimen is needed, in theory, Rice et al.[20] The logarithmic derivative η usually depends both on specimen geometry and material properties. Fortunately η for bend type specimens is rather insensitive to the material behavior.

1.5.1 Applicability of J_{Ic} Test Practice for Metals to Rock. The main test procedure of E 813 is depicted in figure 1.9. It involves at least four nearly identical, fatigue precracked, CT or 3PB specimens which are loaded somewhat beyond the initiation point. W and Δa are measured for each. A pseudo resistance curve is then formed from a linear fit line to valid data points ($\eta W/A$, Δa) and the blunting line. The knee defines the critical value J_{Ic} where real crack growth sets in. A test technique where a single specimen yields enough data points is recommended as an alternative.

This test requires that the governing specimen dimensions be much larger than the process zone at the crack tip, or

$$a,\ W\text{-}a,\ B \geq 25(J_{Ic}/\sigma_y) \tag{1.17}$$

for bend type specimens like the CT and the 3PB. Two factors complicate its transfer to rock, the nature of the non-linearity involved and the meaning of initiation of crack growth.

If the process zone in rock is modelled within a thin fracture zone with cohesive forces, a size requirement as above with σ_t instead will result. Thus

$$a,\ W\text{-}a,\ B \gtrsim 25(J_{Ic}/\sigma_t) = 25[(1-\nu^2)\sigma_t/E]\cdot(K_{Ic}/\sigma_t)^2 \tag{1.18}$$

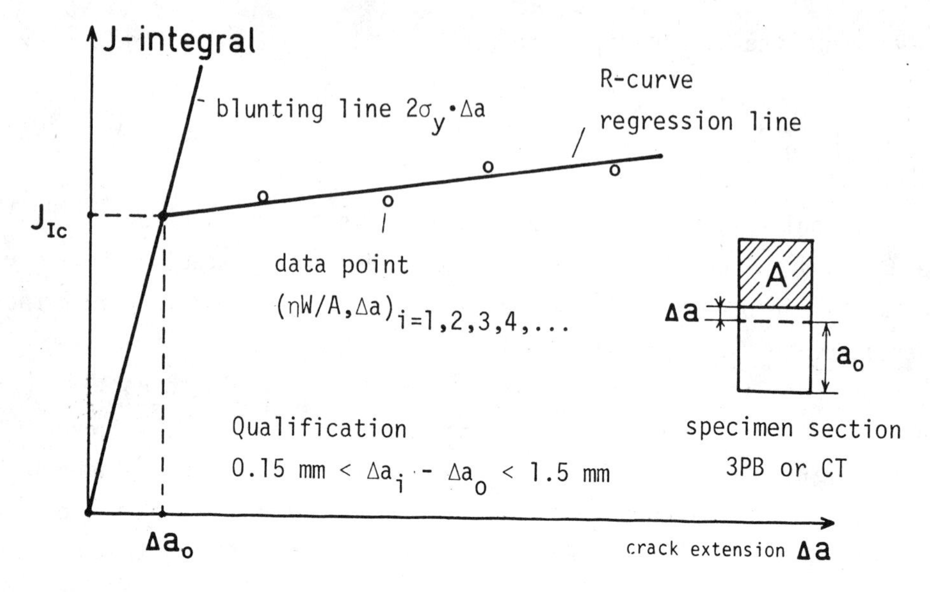

Figure 1.9: Operational definition of J_{Ic} from resistance curve as per E 813.

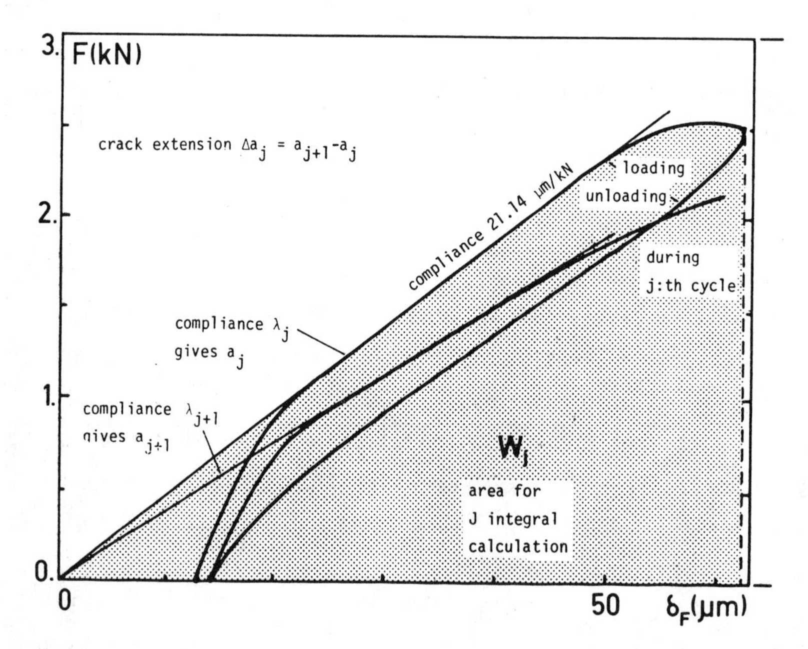

Figure 1.10: Evaluation of J-integral resistance from subcritical failure cycles on CT specimens of Westerly granite.[12]

in terms of K_{Ic}. Since $\sigma_t/E < 0.001$ typically for rock this is 100 times less restrictive than the LEFM requirement.

The surrounding micro-crack zone complicates things. Not only the cohesive force law with its locally unstable conditions in the small but also the singularities of the micro-cracks ruin the vital path independence of the J-integral. Both mechanisms operate during the steady loading of a stationary crack. The J-integral evaluation should, apparently, be made outside the micro-crack zone with a size requirement like the LEFM one as a result. Reality is not necessarily as restrictive as this, Ouchterlony.[1] The J-integral will still be limited to the prediction of onset of stable crack growth.

To determine the initiation point is thus important. It is difficult since the crack tip isn't well defined and from a practical point of view the effects of the micro-cracking are best described in average as the compliance of the specimen. Alltogether this suggests an operational definition of the initiation point and hence of J_{Ic} which incorporates a micro-crack blunting line and a work term that is not evaluated beyond maximum load, Ouchterlony.[1] The compliance offers a convenient way of measuring the crack growth in a single specimen test.

1.5.2 J_{Ic} Measurements on Rock. The basic validity check on J_{Ic}-data is that size requirements etc. are met. A second check is the compability of J_{Ic} with K_{Ic}. Most work on rock discusses both angles.

The first J-integral resistance measurements on rock are due to Finnie et al.[21] They used 3PB specimens of Solnhofen limestone, a fine grained rock with an elastic brittle fracture behavior. The influence of notch width was checked along with the compatibility.

Three crack resistance measures were recorded. One was the specific work of fracture $\overline{R}$, another K_m which was practically equal to K_Q in this case, and the final one an approximate J-integral value J_m based on maximum load.

$$J_m = J_I(a_o, W_c) = [\eta/A]_{a=a_o} \cdot W_c, \tag{1.19}$$

where W_c is the total work to F_{max}, which was assumed to coincide with the initiation of crack growth. The K_m-values for the narrower notches, in the energy rate form $G_m = (1-\nu^2)K_m{}^2/E$, are in good agreement with the J_m-values of typically 14-16 J/m². Even so J_m is not necessarily a good approximation of J_{Ic} because of the initiation assumption and because size requirements weren't discussed.

The $\overline{R}$-values are on the average more than 60% larger, which is far from the ideal elastic brittle behavior with its first sharply rising then flat R-curve. Other rocks give similar results, with $G_m < J_m$ when the load displacement record shows excessive non-linearity, Ouchterlony.[1]

Schmidt and Lutz[12] and Costin [18] present J-integral work that is closer to the intentions of E 813. The former is a comparatively complete study of the effects of specimen dimensions and of the compatibility of K_Q from CT and 3PB specimens with J_{Ic} from CT specimens, all of Westerly granite.

The J_{Ic} determination used a single CT specimen to obtain the pseudo resistance curve. A repetition of subcritical failure cycles and compliance calibration techniques were used, see figure 1.10. Each cycle defined a data point through

$$J_{Rj} = J_I(a_j,W_j) \text{ and } \Delta a_j = a_{j+1}-a_j \tag{1.20}$$

for the j:th cycle. The pseudo resistance curve $J_R(\Delta a)$ was then formed by a linear regression line to these points, without a blunting line. Thus J_{Ic} is represented by the extrapolation

$$J_{Ro} = \lim_{\Delta a \to 0} J_R(\Delta a) \tag{1.21}$$

to infinitesimal crack growth, see figure 1.11. J_{Ro} may then be converted to fracture toughness via the formula

$$K_{Jc} = [EJ_{Ro}/(1-\nu^2)]^{1/2}. \tag{1.22}$$

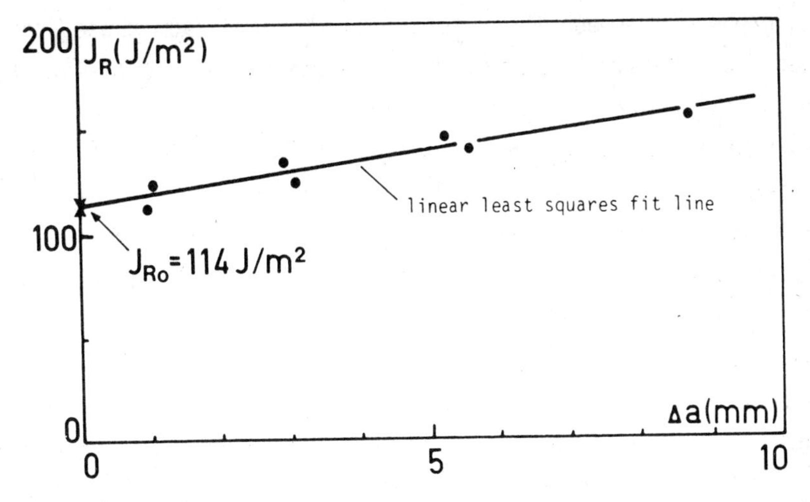

Figure 1.11: J-integral resistance curve $J_R(\Delta a)$ from single specimen of Westerly granite and intercept definition of J_{Ro}.[12]

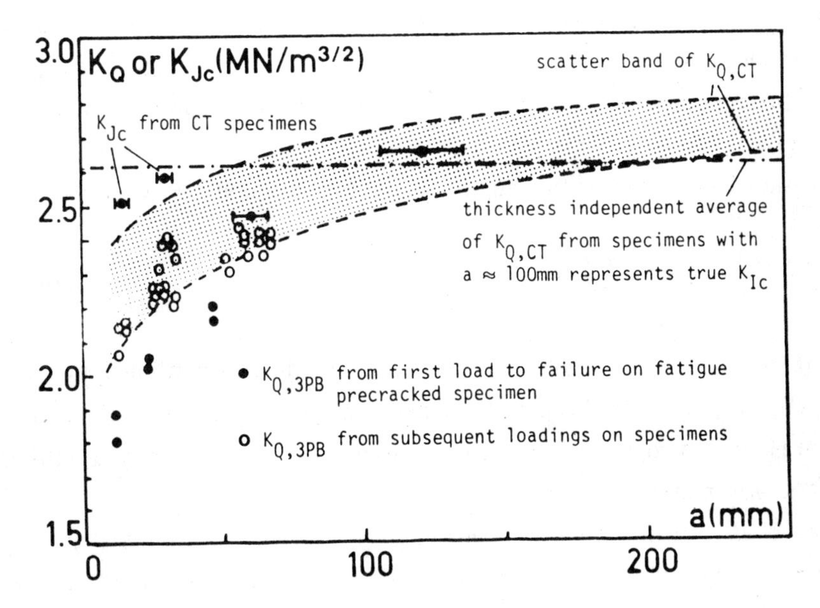

Figure 1.12: Comparison of LEFM based fracture toughness K_Q and J-integral toughness K_{Jc} for Westerly granite.[12]

The agreement between the K_{Jc} data and K_Q data from those specimens which meet the LEFM size requirement $a > 90$ mm is quite satisfactory. See figure 1.12. K_{Jc} is fairly constant when a exceeds 10 mm. The J-integral size requirement yields $a_{min} \approx 0.2$ mm though which obviously is irrelevant when the average grain size is 0.75 mm. These results show that valid fracture toughness or crack resistance data can be obtained on specimens that are substantially smaller than the LEFM size limit, providing a J-integral approach is used.

Costin[18] reports extensive crack resistance measurements on Anvil Points oil shale, a layered dolomite marlstone which is kerogen rich. Four independent test methods were used, the 3PB for K_Q and J_{Ro} measurements, the Short Rod for K_{SR} determinations, and the Charpy impact test for obtaining an equivalent dynamic K_Q.

The testing and evaluation procedure was essentially the same as that of Schmidt and Lutz.[12] One J_{Ro} and several K_Q-values per specimen were obtained. Results are shown in figure 1.13. A constant K_Q value is reached for each specimen, much like an R-curve. It depends on specimen size though, possibly because the micro-crack zone may have been comparable to specimen dimensions.

J_{Ro} was converted assuming plane strain transverse isotropy[18]

$$K_{Jc} = [EJ_{Ro}/(1-\nu_n^2 \cdot E/E_n)]^{1/2}, \tag{1.23}$$

see equation (1.24). Properties normal to the bedding planes are denoted by index n. K_{Jc} and final K_Q values are in good agreement. Thus the effect of material anisotropy on the calculation of K_Q from standard isotropy formulas seems negligable in this case.

Note that in the works discussed here the initiation of crack growth is assumed to coincide, more or less, with maximum load and thus rather closely with the onset of unstable crack growth. Thus the compatibility between J_{Ro} and K_Q is to some extent inherent and J_{Ro} would be an non-linear alternative to K_Q rather than representative of J_{Ic}. As such J_{Ro} is

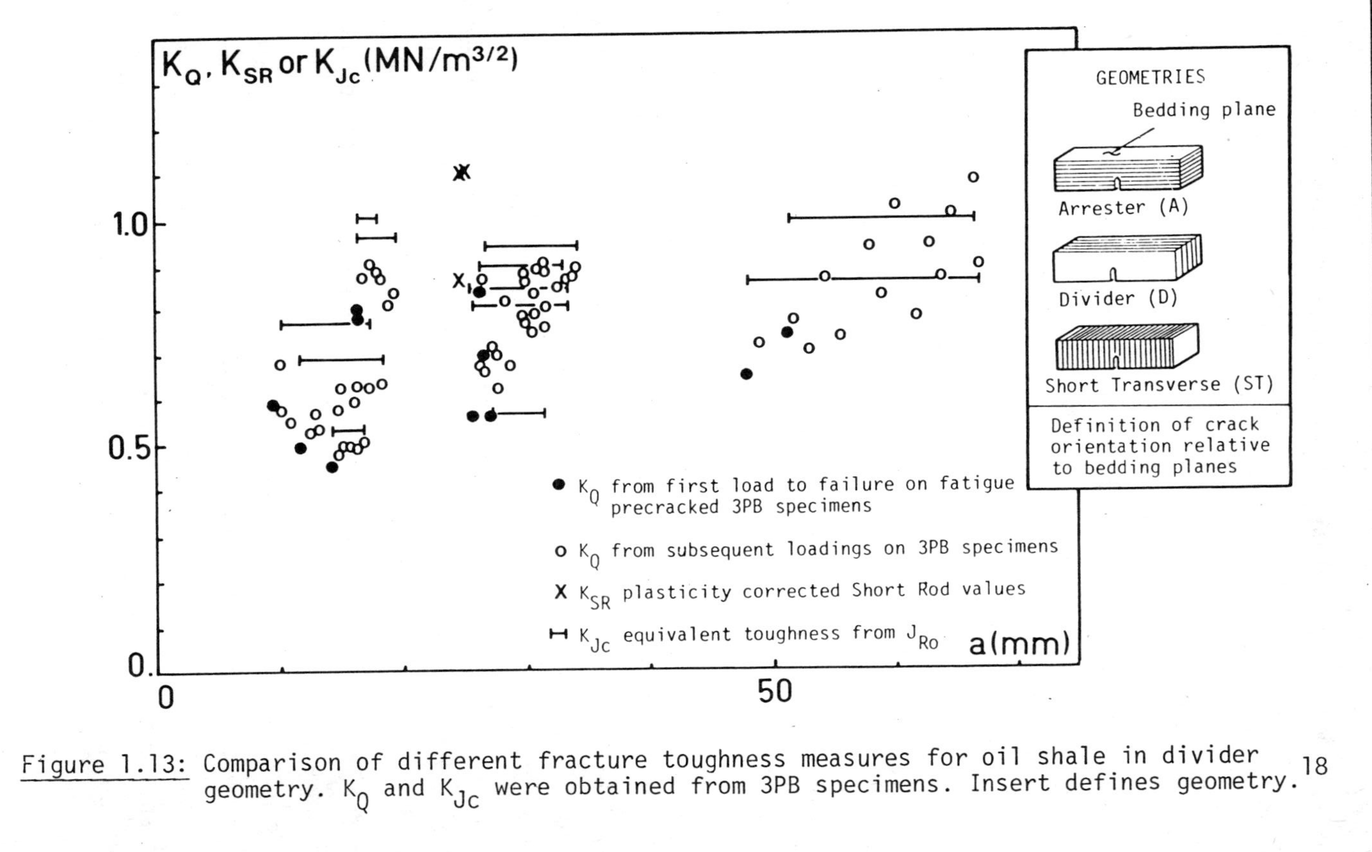

Figure 1.13: Comparison of different fracture toughness measures for oil shale in divider geometry[18]. K_Q and K_{Jc} were obtained from 3PB specimens. Insert defines geometry.

valuable since the size requirement are less stringent than for K_Q.

1.6 Anisotropy Effects

The previous discussion on oil shale brings up the matter how anisotropy influences crack growth resistance measurements. This has been treated recently by Barton [58] and Ouchterlony.[59] Since most rocks contain planar anisotropies like primary bedding, foliation, or micro crack sets, nearly all rock types investigated for anisotropy show some amount of it.

Rocks with anisotropic crack growth resistance properties include

1. Anvil Points oil shale, Berea sandstone, and Indiana limestone
2. Ekeberg marble, St Pons marble, and Corris slate
3. Barre granite, Berkeley granite, and Chelmsford granite

Appreciable non-linearity occurs in the uniaxial testing of intact rock, see Hawkes et al.[60], but in toughness testing it is more or less contained at the crack tip.

For granite a non-linear orthotropic elastic description is suitable, Douglass and Voight [61] and Peng and Johnson.[62] For oil-shale a non-linear transversly isotropic elastic description is commonly used, Cleary [63] and Chong et al.[64] Of the rocks listed above the St Pons marble and the Indiana limestone seem to be the most isotropic ones, Henry et al.[65] and Schmidt.[66]

Unfortunately matching elastic and strength properties of rocks, that are necessary to judge anisotropy effects as well as specimen size requirements etc., are seldom reported together with the crack growth resistance properties. Taken as a whole the work on oil shale is an exception. See Schmidt [67], Hangen [68], Cleary [63], Switchenko,[69] and Costin.[18,70]

Furthermore anisotropic K_I formulas are almost never used. These errors come in addition to whatever the various methods induce. The

J-integral accomodates both non-linear and anisotropic behavior to some extent however and its evaluation has the advantage of requiring smaller test specimens.

1.6.1 Elastic Description. We concentrate on oil shale as a linear transversely isotropic elastic material. As such it is completely described by 5 elastic constants: E and ν in the bedding planes and E_n, ν_n, and G_n perpendicular to them. Defining the degree of anisotropy as $n = E/E_n$, they must satisfy the inequality $1 - \nu > 2n\nu_n^2$.

To cause the mode I situation relevant to most testing, the crack must be oriented along the principal axes of the material. These crack orientations in a transversely isotropic material are shown in figure 1.13. Even so anisotropic fracture properties can make the crack branch.

For a crack in divider (D) orientation, the equations of Sih and Liebowitz [71] yield

$$G_I = \begin{cases} K_I^2/E \text{ in plane stress and} \\ (1-n\nu_n^2)K_I^2/E \text{ in plane strain,} \end{cases} \tag{1.24}$$

referring to anisotropic K_I values. The last expression parallels equation (1.23). For a crack in arrester (A) orientation

$$G_I = \begin{cases} f_{PS}\cdot K_I^2/E \text{ in plane stress and} \\ f_{PD}\cdot(1-\nu^2)K_I^2/E \text{ in plane strain,} \end{cases} \tag{1.25}$$

and for a crack in short transverse (ST) orientation

$$G_I = \begin{cases} f_{PS}\cdot K_I^2/E_n \text{ in plane stress and} \\ f_{PD}\cdot(1-n\nu_n^2)K_I^2/E_n \text{ in plane strain,} \end{cases} \tag{1.26}$$

with the same f-factors as in arrester orientation.

They are in turn given by

$$f_{PS} = [1+ \frac{\sqrt{n}}{2}(E_n/G_n-2\nu_n)]^{1/2} \text{ and}$$

$$f_{PD} = \{1+ \frac{\sqrt{n}}{2}[E_n/G_n-2\nu_n(1+\nu)]/[(1-\nu^2)(1-n\nu_n^2)]^{1/2}\}^{1/2}. \qquad (1.27)$$

Cleary [63] reports crack resistance data for two shales, a lean nahcolite bearing one with $n = 3.1$ and $\nu = \nu_n = 0.31$ and a rich one with $n = 2.1$ and $\nu = \nu_n = 0.5$. For these f_{PD} and f_{PS} are in the range 1.2 to 1.3. Obviously the plane strain factor $(1-n\nu_n^2)$ is just as important.

Concerning the use of isotropic K_I formulas, an analogue with orthotropic K_I solutions by Bowie and Freese [72] and Delale and Erdogan [73] suggest that the roughly square CT specimen is much more sensitive than the relatively slender 3PB specimen. In this analogue a CT specimen with a midsection crack requires a K_I correction factor of about 1.0 in divider, 1.10 in arrester, and 0.95 in short transverse when $n = 2$. Finite element calculations on a CT specimen by Miller and Cleary [74] indicate equally large correction factors when $n = 4$ and $\nu = \nu_n = 0.4$.

Thus it seems that treating the conversion from K_I to G_I or from fracture toughness to J-integral resistance as isotropic can incur much larger errors than using isotropic K_I formulas, especially if the Poisson's ratio is large. Both errors can be avoided by measuring the J-integral resistance directly. Then the difficulty of measuring the load line displacement correctly for a 3PB specimen and the size of the Merkle-Corten correction factor for the CT specimen must be considered.[75]

1.6.2 Crack Growth Resistance Values. This section concentrates on Anvil Points oil shale, data is given in table 1.1. A discussion of the data for the rocks listed before is given by Barton [58] and Ouchterlony.[59] The degree of anisotropy in measured fracture toughness values may amount to roughly the same as that of the matching moduli but it is often less. All rocks show some amount of it, if not in fracture toughness then in energy rate resistance. Usually the crack growth resistance is lowest for cracks propagating in weak planes like bedding planes, foliation planes, or the rift plane in granites.

For the oil shale, Schmidt [67] reports LEFM fracture toughness testing on 3PB specimens with isotropic K_I formulas. His K_Q values are claimed to approximate the true material property (K_{Ic}) adequately. He studied two grades of shale and found the highest values in divider orientation and the lowest ones in short transverse. In the arrester orientation, the crack tended to branch off into the bedding planes. He also found that the leaner shale was the toughest. In terms of energy rate resistance the reverse is probably true, but not necessarily.

Hangen [68] has made a thorough study of the deformation and fracture of oil shale as a non-linear transversely isotropic material, see also Cleary.[63] The J-integral approach was used on notched CT specimens. He employed two evaluation methods, one due to Wilkening [76], in which an initiation of crack growth value J_i is determined. See also Ouchterlony [1,2] section 6.3.1. The other is based on maximum load like J_m in equation (1.19). As discussed previously, J_m is probably more comparable to K_Q than J_i is. The latter values are considerably lower, roughly 50%.

Hangen's J_m data shows that the richer shale is much more resistant, perhaps due to the presence of weaker nahcolite crystals in leaner one. The comparison with Schmidt's [67] data is certainly not as unfavorable as suggested by Cleary.[63] Considering all sources of error, the agreement might even be considered as good as one can expect.

Switchenko's work [69] builds on Hangen's. It emphasizes the thermo-mecanical response of oil shale. The measured elastic moduli and uniaxial strengths all decrease with increasing temperature but Poisson's ratios, failure strains, and the amount of inelastic behavior all tend to increase. His grade of oil shale is quite brittle in that $J_i \approx J_m$ at room temperature, even at 93^{0}C the difference is usually less than what Hangen [68] found.

The J_m results show that the divider orientation is the most resistant one over the whole temperature range, 21^{0}C to 93^{0}C, but that the arrester values are comparable. The divider values increase with the temperature but branching into bedding planes tends to lower the

DATA / ROCK TYPE		Grain Size	Stress Intensity Resistance		Modulus E	Poisson Ratio ν	Energy Rate Resistance		Tensile Strength	Crack Length	Test Config	Ref
		[mm]	Symbol	[MN/m$^{3/2}$]	[GPa]	-	[J/m^2]	Symbol	[MPa]	[mm]	Abbrev	No
Anvil Points oil shale												
20 gpt	D		K_Q	1.076_5	20-25	-	52	K_Q^2/E	≈ 17	20-30	3PB	67
(oil content in	A		K_Q	0.977_5	20-25	-	43	K_Q^2/E	≈ 17	20-30	3PB	67
gallons per ton)	ST		K_Q	0.750_1	–	-			-	20-30	3PB	67
40 gpt	D		K_Q	0.674_5	4	-	114	K_Q^2/E	≈ 12.5	20-30	3PB	67
	A		K_Q	0.604_2	4	-	91	K_Q^2/E	≈ 12.5	20-30	3PB	67
	ST		K_Q	0.370_2	8	-	17	K_Q^2/E	≈ 3	20-30	3PB	67
34 gpt (nahcolite)	D				17	0.31	36.8_3±4.2	J_m	8.5	>50	CT	68,63
	A				17	0.31	90.5_3±34.2	J_m	8.5	>50	CT	68,63
	ST				5.5	0.31	28.9_3±1.7	J_m	2	>50	CT	68,63
65 gpt	D				8.1	0.5	122.2_1	J_m	2.5	>40	CT	68,63
	A				8.1	0.5	73.0_1	J_m	2.5	>40	CT	68,63
	ST				3.9	0.5	29.4_1	J_m	1	>40	CT	68,63
30 gpt at 21°C	D				16.5_3	0.25	55.6_2	J_m	7.6_3	≈50	CT	69
	A				16.5_3	0.21	54.1_2	J_m	7.6_3	50	CT	69
	ST				8.3_3	0.21	16.0_2	J_m	2.5_3	50	CT	69
38°C	D				9.0_3	0.33	64.2_2	J_m	6.3_3	50	CT	69
	A				9.0_3	0.28	55.4_2	J_m	6.3_3	50	CT	69
	ST				6.6_3	0.28	–	–	1.9_3	50	–	69

DATA / ROCK TYPE	Grain Size	Stress Intensity Resistance		Modulus E	Poisson Ratio ν	Energy Rate Resistance		Tensile Strength	Crack Length	Test Config	Ref
	[mm]	Symbol	[MN/m $^{3/2}$]	[GPa]	-	[J/m²]	Symbol	[MPa]	[mm]	Abbrev	No
Anvil Points oil shale											
30 gpt 93°C D				1.4_3	0.42	107.3_2	J_m	2.4_3	50	CT	69
A				1.4_3	0.35	32.5_2	J_m	2.4_3	50	CT	69
ST				1.7_2	0.35	24.5_2	J_m	0.65_2	50	CT	69
149°C D				-	0.40	150_2	J_m	-	50	CT	69
suppressed branching: 21°C A				-	-	19.0_2	J_m	-	50	CT	69
38°C A				-	-	28.9_2	J_m	-	50	CT	69
93°C A				-	-	49.2_2	J_m	-	50	CT	69
50 gpt D		K_Q	$0.5\text{-}1.0_{12}$	8-16	-	$30\text{-}120_{12}$	J_{Ro}	-	12-50	3PB	18
D		K_{SR}	$0.86\text{-}1.10_3$	-	-			-	≈35	SR	18
ST		K_{SR}	$0.42\text{-}0.52_2$	-	-			-	≈35	SR	18
D		K_Q	$0.83\text{-}1.04_{11}$	-	-			-	12	Dyn3PB	18
D		K_Q	$1.03_8 \pm 0.15$	-	-			-	4	Charpy	18
A		K_Q	$1.13_{10} \pm 0.12$	-	-			-	4	Charpy	18
ST		K_Q	$0.60_6 \pm 0.10$	-	-			-	4	Charpy	18

Table 1.1: Crack growth resistance data for Anvil Points oil shale. Values given as $mean_{no\ tests} \pm$ std deviation if possible. Some matching modulus and strength values inferred from stress-strain diagrams etc. Approximate conversion K_Q^2/E used on Schmidt's [67] data.

arrester values. With mechanically suppressed branching, the arrester values increase with the temperature instead but starting from a much lower level.

Switchenko [69] also cycled some of his notched CT specimens beyond maximum load before evaluating J_R values from subsequent subcritical failure cycles. As seen previously in section 1.5.2, a way of measuring crack length or ligament is involved. Switchenko's way of doing it isn't clear though. His results show about 40% lower resistance for reinitiation of crack growth from natural fractures, in all orientations.

A section 3 will show, my experience is the opposite. I find that J_R values for precracked specimens are higher than J_m values for notched specimens. Compare K_{Jm} with K_{Jc} in table 3.2. In this light I find that the alleged difference between Switchenko's data and Schmidt's [67] probably is given too much importance.

Costin's work [18], which was discussed in section 1.5.3, reconciles some of the apparent incompatibility between the LEFM fracture toughness approach and the J-integral resistance approach. He has tested both and made Short Rod, dynamic 3PB, and Charpy tests as well. The main point is that he found good agreement between final K_Q values and J_{Ro} equivalent K_{Jc} values from the same specimen. J_{Ro} was converted to fracture toughness using the proper conversion formula for divider orientation, equation (1.23). Thus the effect of anisotropy on the calculation of K_Q from standard isotropic formulas is negligible in this case.

Specimens with cracks in the arrester and short transverse orientations are admittedly more crucial tests. Unfortunately such experiments along the lines of Costin's work [18] seem to be lacking. To be conclusive they would have to include the matching moduli and strength data explicitely.

1.7 Conclusions

This review section shows that

1. The LEFM method of measuring K_{Ic} for metals as per ASTM E 399 seems effective for rock. The in-plane specimen size requirements are important to meet but specimen thickness and loading rate have little effects on the tentative K_Q results.
2. The approximate fracture toughness evaluation K_m, tried on notched but not precracked rock cores, usually yields results that are too low.
3. The Short Rod evaluation K_{SR}, when plasticity corrected, has the potential of yielding results that agree with K_Q.
4. A J_{Ic} evaluation modelled on ASTM E 813 for metals also seems effective for rock. It is a single specimen test method with compliance measurements that is used. It requires smaller specimens than the K_{Ic} evaluation. The generally acceptable compatibility with K_Q values found seems to be largely inherent in the method.
5. None of the other crack resistance parameters appear to be compatible with the specific work of fracture $\overline{R}$.
6. Anisotropy effects, as well as non-linear effects, are best avoided by measuring J-integral resistance. The apparent isotropic LEFM fracture toughness may yield reliable results too. The K to J conversion formula can produce considerable errors however.
7. Matching elastic moduli and strength properties should always accompany crack resistance data so that specimen size criteria and anisotropy effects can be evaluated.

Many crack resistance values for rock are listed by Ouchterlony.[1] They are generally not comparable even in relative terms because of the different test methods. The standard, which is underway through subcommittee E 24.07 of the ASTM, is certainly welcome. It must prescribe core specimens to become widely accepted. Next, in section 2, I will describe the development of two core bend specimens, Ouchterlony.[17,22,23,24]

Core specimens will nearly always be sub-size with respect to LEFM size criteria but not necessarily with respect to J-integral method requirements. R-curve methods could also be used on them to follow the crack growth as the full crack resistance develops. These laboratory oriented methods will be treated in section 3.

If the R-curves have a relatively flat post-critical part this level could perhaps be used as a representative resistance value. An a priori assumption of an R-curve with this simple form has also been applied to the fracture toughness testing of sub-size core specimens, Ouchterlony.[25, 26] This approach, with which I hoped to trade the complications of compliance measurements for a multi specimen test method, will also be described in section 3.

An ultimate field test method would use only a single specimen with known dimensions and the maximum load during the test to produce a representative toughness value directly. Two candidates are the Short Rod and the CENRBB specimens since they permit a certain amount of crack growth from the chevron notch before the evaluation is made. If the question of plasticity corrections is settled, portable instruments for the testing are already available, Swan and Olofsson.[27] One application could be prediction of tunnel boring machine performance, Ingraffea et al.[28]

2. DEVELOPMENT OF CORE BEND SPECIMENS

This section describes the development of two core specimen types with transverse notches for three point bending. Such specimens together with the Short Rod and a Round CT-type specimen like in E 399 would make it possible to measure the fracture toughness of rock in three mutually orthogonal directions from one core.

2.1 Single Edge Crack Round Bar in Bending (SECRBB)

This specimen was introduced in several versions of aluminum bars, Bush.[29] The most compact one is a solid bar with S/D = 3.33, see figure 2.1.

SECRBB configuration

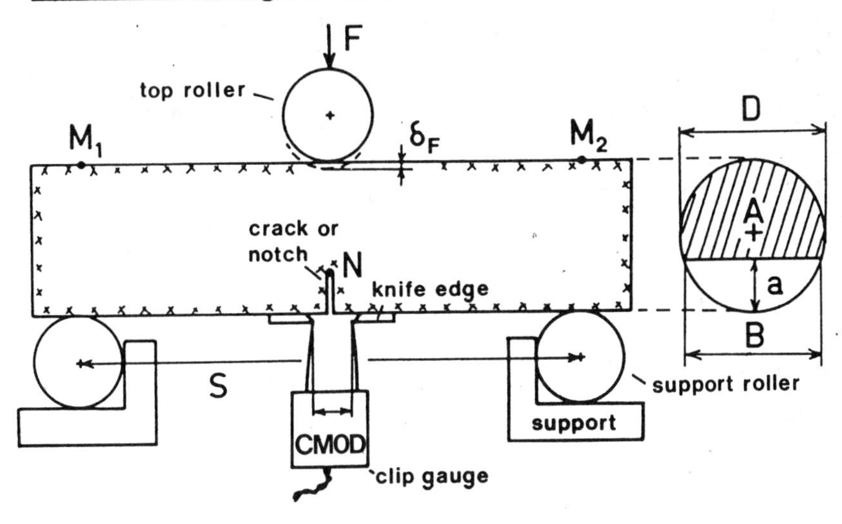

Basic notation:

D = diameter of round bar specimen

S = distance between support points, or support span

a = (maximum) crack length

a_o = initial crack length or notch depth

A = reduced cross sectional area

B = crack front length

F = load on specimen

δ_F = (ideal) load point displacement (LPD)

δ_{CMOD} = crack mouth opening displacement (CMOD)

N = notch tip or front

M_1, M_2 = saddle points of measuring base

Figure 2.1: The single edge crack round bar in bending (SECRBB) specimen with basic notation.

Here I discuss briefly its adaption to rock testing, Ouchterlony.[17,22,24] Cores of Ekeberg marble were used for a compliance calibration on which a thorough fracture mechanics analysis is based. The latter includes matters such as crack growth stability, a secant offset procedure, and the relation between J-integral and work done.

The load point compliance of the specimen is

$$\lambda_F = \delta_F/F. \tag{2.1}$$

where δ_F is the load point displacement (LPD). Within the limits of slender beam theory, a dimensionless compliance function

$$g(a/D) = \lambda_F \cdot ED \tag{2.2}$$

is valid for all scaled SECRBB configurations of other linear elastic materials. Since S/D = 3.33 is not exactly slender, a displacement component for unnotched beams that is due to shear will be included. The λ_F formula for aluminum, Bush[29], may be recast to include both an E and a ν dependence. Thus with $\alpha = a/D$,

$$g(\alpha,\nu) = 15.6719[1+0.1372(1+\nu)+11.5073(1-\nu^2)\alpha^{2.5}], \tag{2.3}$$

valid when $0 \le \alpha < 0.35$. These expressions have been used to determine the Young's modulus E of several rock types.

Similarly the crack mouth opening displacement δ_{CMOD} or just CMOD defines a compliance term through

$$\lambda_{CMOD} = \delta_{CMOD}/F. \tag{2.4}$$

Assuming linear elasticity a non-dimensional form

$$h(\alpha) = \lambda_{CMOD}/\lambda_F = \delta_{CMOD}/\delta_F \tag{2.5}$$

results when the ν-dependence of g is treated as insignificant.

The stress intensity factor is directly obtainable from Kies-Irwin's potential energy release rate

$$G = -\frac{\partial U}{\partial A_C} = \frac{F^2}{2B} \cdot \frac{d\lambda_F}{da}, \tag{2.6}$$

where $dA_C = B \cdot da$ was used and the notch front length $B = 2D[\alpha-\alpha^2]^{0.5}$. Assuming plane strain, mode I, and constant K_I conditions along the crack front the conversion equation (1.5) yields K_I. Writing like Bush [29]

$$K_I = 0.25 \left(\frac{S}{D}\right) \cdot Y' \cdot F/D^{1.5}, \tag{2.7}$$

the dimensionless stress intensity factor is given by

$$Y'(\alpha) = 2\left(\frac{D}{S}\right)\left[\frac{dg}{d\alpha}/(1-\nu^2)\right]^{0.5}/(\alpha-\alpha^2)^{0.25}. \tag{2.8}$$

This form is independent of ν and varies only slightly with S/D.

Inserting $g(\alpha,\nu)$ from above yields

$$Y'(\alpha) = 12.7527\ \alpha^{0.5}/(1-\alpha)^{0.25}, \tag{2.9}$$

valid when $0 \leq \alpha < 0.35$. This range is too small for rock testing with R-curve methods, hence an extension became necessary.

2.1.1 Experimental Procedure. Details are given by Ouchterlony.[22] The experiments were run in a servohydraulic testing machine in strain control. A clip gage measuring CMOD gave the strain signal. The knifes edges were affixed with cyanoacrylate adhesive. Curves $F(\delta_F)$ and $F(\delta_{CMOD})$

were registered on a two-channel X-Y recorder. Typical test records for notched SECRBB specimens are shown in figure 2.2.

Based on earlier experience, Finnie et al.[21] and Bush [29], some parasitic factors which may affect LPD measurements are:

1. Friction effects from the support rollers
2. Twisting caused by non-ideal test geometries
3. Non-linear deformation of roller to specimen contacts
4. Local crushing at roller to specimen contacts

Friction is avoided by mobile rollers as per E 399. A cylindrical specimen minimizes twisting but aggravates the contact factors. These may however be virtually eliminated.

My arrangement to avoid the contact problems is shown in figures 2.1, 2.3, and 2.4. It is the displacement of the notch tip N relative to the saddle points M_1 and M_2 which is measured. The measuring base is aligned to the testing machine and it rides over the support points. A yoke is pulled into the notch tip from the base by rubber bands. Their relative displacements are measured by two gaging LVDT:s whose signals are summed to eliminate potential twisting effects. None of the $F(\delta_F)$ curves in figure 2.2 show any sign of non-linear contact deformation or crushing, nor have any other registered curves.

When the initial part of the $F(\delta_F)$ curve was too steep to permit accurate reading a low load cycle with higher LPD magnification was run before the complete failure curve. The initial compliance $\lambda_{Fo} = \lambda_F(a_o)$ equals the inverted slope value. Twenty D ≈ 41 mm cores of 0.5 m length were made from one block of marble, each providing 5 specimens with a notch width of 1.6 mm. 56 out of 65 prepared specimens at this stage were judged to give valid LPD and CMOD compliance data in the interval $0 \leq a_o/D < 0.6$. The fracture mechanics data obtained from these test records will be discussed later in section 3.

Ekeberg marble is fine to medium grained with a well recrystallized texture of mainly spherical carbonate grains. Its density is ≈ $2.9 \cdot 10^3$ kg/m³

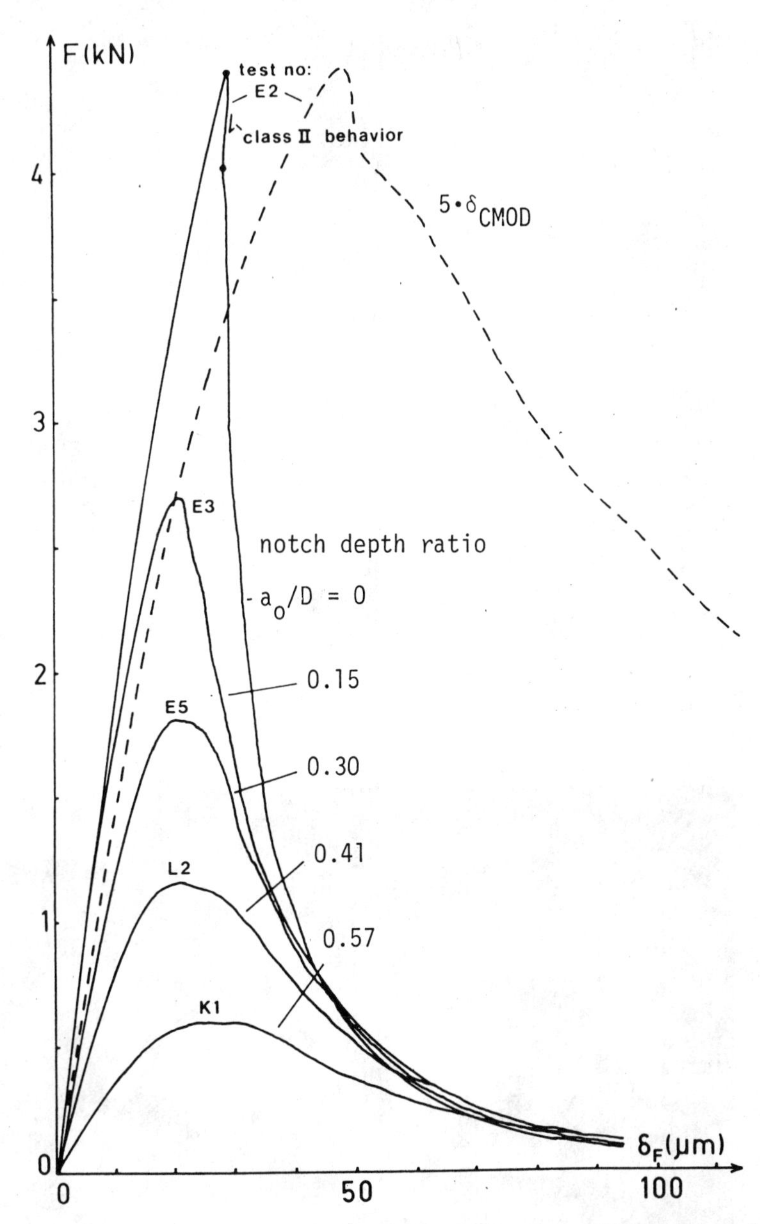

Figure 2.2: Set of test records $F(\delta_F)$ for $D \approx$ 41 mm SECRBB specimens of Ekeberg marble with different notch depths. Broken curve shows $F(5 \cdot \delta_{CMOD})$ for unnotched specimen E2.

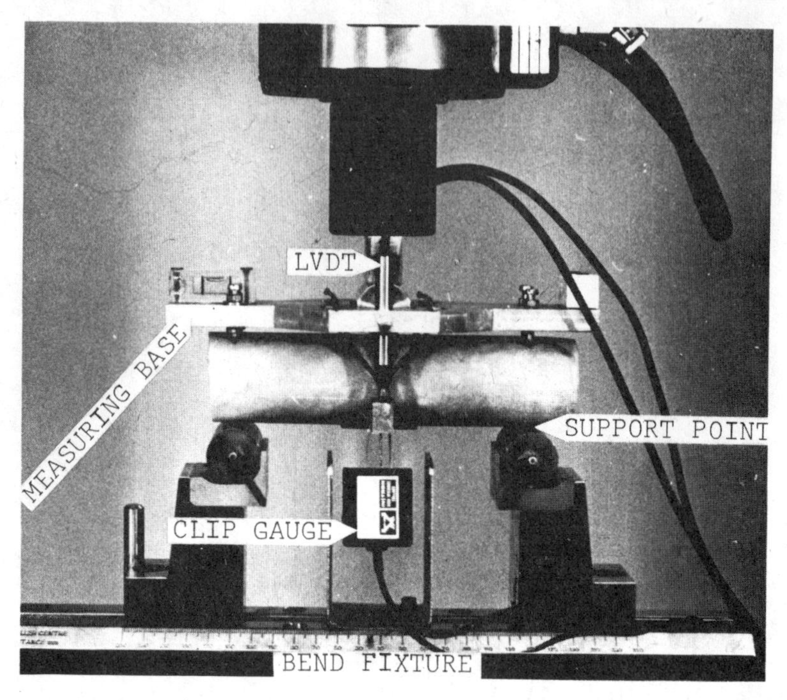

Figure 2.3: Details of experimental set up for accurate measurement of load point displacement on SECRBB specimen of Ekeberg marble.

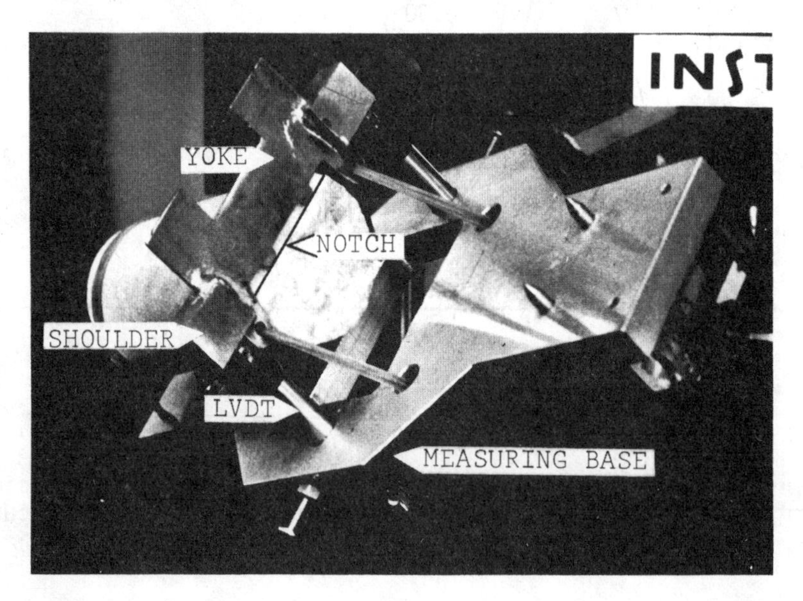

Figure 2.4: Measuring base with yoke and LVDT:s.

and its porosity ≈ 0.25%. The mechanical properties are relatively isotropic. A compilation of mechanical properties is given by Ouchterlony.[22]

2.1.2 Results. Poisson's ratio for Ekeberg marble is $\nu_{Em} \approx 0.22$. Given this and the initial notch depth ratio $\alpha_o = a_o/D$, a value $g_{Em} = g(\alpha_o,\nu_{Em})$ is given by equation (2.3) for each specimen. With $\lambda_F(a_o)$ this gives the Young's modulus from equation (2.2). Data from the notch interval $0 \leq \alpha_o < 0.35$ is shown in figure 2.5. It shows no significant difference between the E values from the different cores nor any systematic variation with notch depth.

The plane stress assumption of simple beam theory gives an estimate of the transverse contraction in an unnotched bar that is included because the LPD isn't measured at the neutral axis. The correction is

$$\Delta g = -\frac{2}{\pi}\left(\frac{S}{D}\right)\eta^2\nu = -2.12\eta^2\nu \approx -1.717\ \nu \qquad (2.10)$$

in our case because the measuring points are $\eta R \approx 0.9R$ from the neutral axis. The minus sign indicates that the contraction makes the beam seem stiffer.

A final estimate of the Young's modulus including errors is

$$E_{Em} = 85.7 \pm 3.3 \text{ GPa}. \qquad (2.11)$$

This is insignificantly lower than strain gage data from uniaxial tests. The results for Bohus granite, Ouchterlony [25], and Stripa granite, Ouchterlony et al.[30] are just as encouraging. Since the measured modulus values are both precise and reliable, the displacement measuring method is clearly valid.

The LPD compliance data over the extended interval $0 \leq \alpha_o < 0.6$ was then used for curve fitting with a weighted least squares procedure with a smoothness constraint, Ouchterlony.[22] The extension of g is

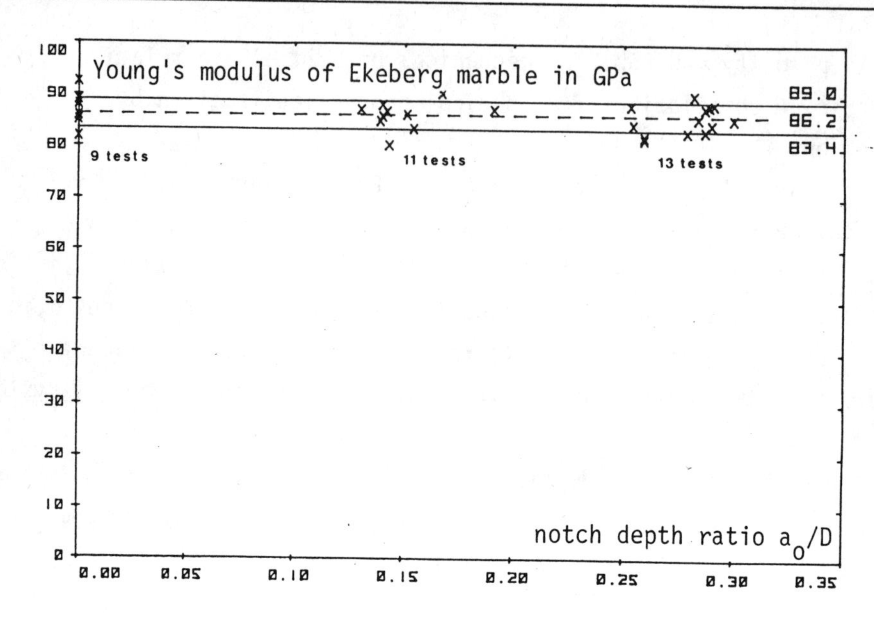

Figure 2.5: Young's modulus of Ekeberg marble from compliance measurements on D ≈ 41 mm SECRBB specimens. Lines show mean ± standard deviation.

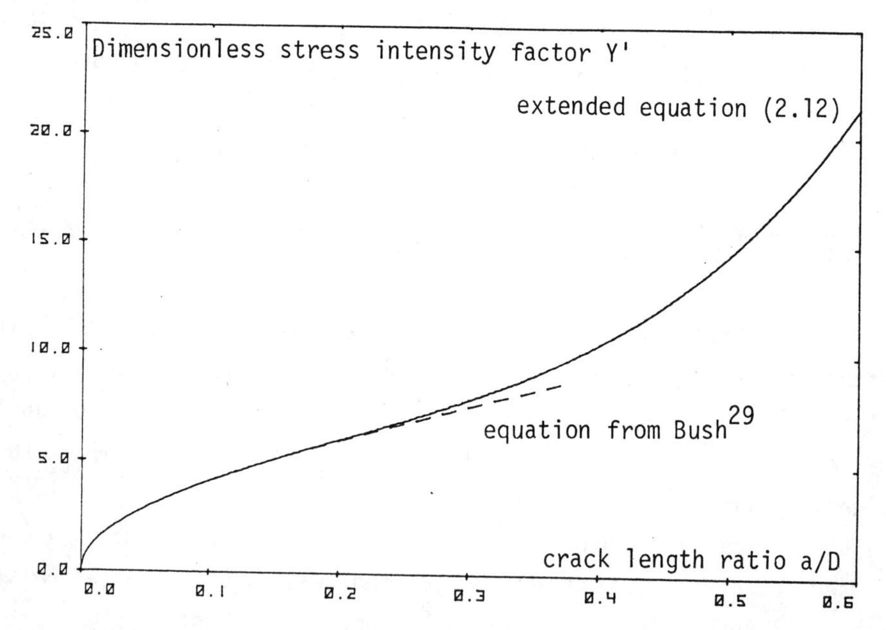

Figure 2.6: Dimensionless stress intensity factor Y'(a/D) of SECRBB specimen. Broken line shows old equation.[29]

$$g(\alpha,\nu) = 15.6719\{1+0.1372(1+\nu)+11.5073(1-\nu^2)\alpha^{2.5}\cdot[1+7.0165\alpha^{4.5}]\} \qquad (2.12)$$

and the corresponding stress intensity factor

$$Y'(\alpha) = 12.7527\alpha^{0.5}(1+19.646\alpha^{4.5})^{0.5}/(1-\alpha)^{0.25}, \qquad (2.13)$$

valid when $0 \leq \alpha < 0.6$. $Y'(\alpha)$ is shown in figure 2.6.

The CMOD compliance data over the interval $0 \leq \alpha_0 < 0.6$ was later used for curve fitting in the non-dimensional form $h = \lambda_{CMOD}/\lambda_F$ of equation (2.5). Including a monotonicity requirement the result was

$$h(\alpha) = 0.2475+0.7351\alpha+1.4257\alpha^2-1.1834\alpha^3, \qquad (2.14)$$

valid when $0 \leq \alpha < 0.6$, Ouchterlony.[24]

2.1.3 Further Fracture Mechanics Formulas. The dimensionless compliance functions contain more information relevant to fracture mechanics testing. First consider crack growth stability. It may judged by G·B, which is also called the crack driving force. If G·B increases with crack length, excess energy is available and the crack tends become unstable. A decreasing driving force has the opposite effect.

Thus testing in an infinitely soft machine where load is controlled, a monotonically increasing $g'(\alpha)$ shows that the SECRBB specimen is potentially unstable once crack growth has initiated.

For testing in a stiff machine, where LPD is controlled, we may use the non-dimensional crack driving force as a stability function. Equations (2.2) and (2.6) yield

$$S_{LPD} = GB/E\delta_F^2 = \frac{1}{2g} \cdot \frac{d\ell ng}{d\alpha} \qquad (2.15)$$

$S_{LPD}(\alpha)$ is plotted in figure 2.7. It has a maximum at $\alpha = 0.325$ beyond which the crack growth tends to be stable.

Our testing was done in a servohydraulic machine under CMOD control. Under such conditions

$$S_{CMOD} = GB/E\delta^2_{CMOD} = S_{LPD}/h^2. \qquad (2.16)$$

It is also plotted in figure 2.7. The maximum occurs at $\alpha = 0.179$. As expected this implies that CMOD controlled testing is more stable than LPD controlled.

The figure shows that an ideal elastic brittle (elastic-brittle) material, one with a flat R-curve and without sub-critical crack growth would make even a CMOD controlled test unstable if $\alpha < 0.179$. All such tests are stable in practice however, even the unnotched ones. It would appear that a certain amount of crack growth from the initial notch is necessary before the crack extension resistance is fully developed, Ouchterlony.[25]

Next consider a secant offset procedure analogous to E 399. Logarithmic differentiation of equations (2.1) and (2.2) yield

$$\Delta\lambda_F/\lambda_F = \Delta\delta_F/\delta_F \approx H(\alpha)\cdot\Delta a/a \text{ with } H = \alpha\cdot \frac{d\ell ng}{d\alpha}. \qquad (2.17)$$

Curves are given by Ouchterlony.[22,24] With the crack in midsection as per E 399, $H(0.5) \approx 2.5$. Thus 2% apparent crack growth, $\Delta a/a = 0.02$, yields the same 5% offset secant procedure. If $\alpha = 0.3$ however one would use a 2% offset secant procedure. This equation may well be used in a variable secant offset procedure, Munz.[15] One relating to CMOD is also easy to derive.

Third and last consider the relation between J-integral and work done as expressed by equation (1.16), $J_I = \eta\cdot W/A$. For a linear elastic material equation (1.13) with $U = W = \delta_F{}^2/2\lambda_F$ yields

$$\eta_{el} = \frac{A}{BD}\cdot\frac{d\ell ng}{d\alpha}. \qquad (2.18)$$

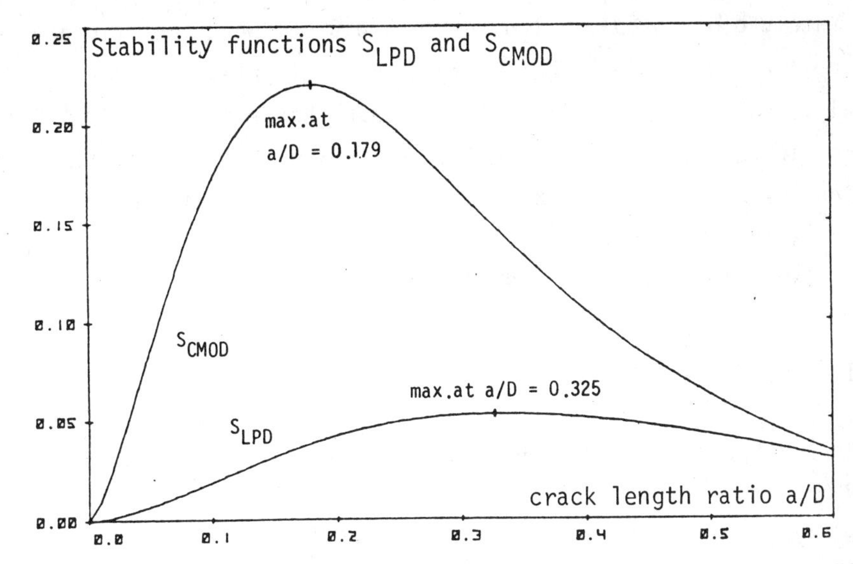

Figure 2.7: Dimensionless stability functions S_{LPD} and S_{CMOD} of SECRBB specimen.

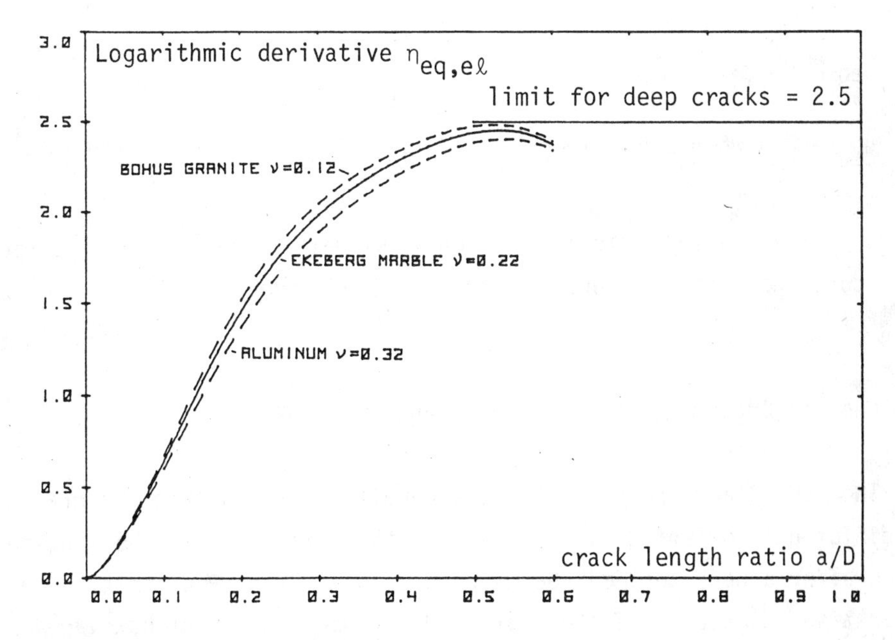

Figure 2.8: Equivalent logarithmic derivative $\eta_{eq,e\ell}$ of SECRBB specimen. Elastic and plastic limits for deep cracks coincide at 2.5.

Unfortunately $g(\alpha)$ is only valid when $\alpha < 0.6$ but an asymptotic analysis yields $\eta_{el} \to 5/3$ when $\alpha \to 1$. For a rigid perfectly plastic material the J-integral is related to the limit load rather than work done. Again limit analysis yields $\eta_{p\ell} \to 5/3$ when $\alpha \to 1$, Ouchterlony.[22, 23] This limit value would thus perhaps also be representative of real material behavior, Srawley [31], in the interval $\alpha > 0.5$ say.

To give η a suitable form we define

$$J_I = \eta_{eq} \cdot W/A_{eq} \text{ with } A_{eq} = B \cdot (D-a), \tag{2.19}$$

being an equivalent uncracked area. Hereby

$$\eta_{eq,e\ell} = (1-\alpha) \cdot \frac{d\ell ng}{d\alpha} \text{ with } \eta_{eq,e\ell} \to 2.5 \text{ when } \alpha \to 1. \tag{2.20}$$

Curves are given in figure 2.8. Usually $\eta_{p\ell} > \eta_{el}$, Turner [32], and I suggest that equation (2.20) be used as a tentative lower limit for real material behavior when $\alpha < 0.5$ such that

$$\left.\begin{aligned} \eta_{eq} &\approx \eta_{eq,e\ell} \text{ when } 0 < \alpha < 0.5 \text{ and} \\ \eta_{eq} &= 2.5 \text{ when } 0.5 < \alpha < 1. \end{aligned}\right\} \tag{2.21}$$

Their application in a highly non-linear situation needs further support. This concludes the post analysis of the SECRBB specimen.

2.2 Chevron Edge Notch Round Bar in Bending (CENRBB)

The SECRBB and the Short Rod are complementary in that the cracks are differently oriented with respect to the core axis and the specimen halves after a bend test may be used in ensuing Short Rod tests. The comparative advantages of the Short Rod are more important however.

The CENRBB was designed to incorporate these within the same outer proportions as the SECRBB, Ouchterlony.[17] It is shown in figures 2.9.

The notched section of the CENRBB specimen is a chevron V with a subtended angle 2θ. The apex position is given by a_0 and the chevron ends at a_1 where the crack front length

$$B_1 = B(\alpha_1 = a_1/D) = 2D(\alpha_1 - \alpha_1^2)^{0.5}. \tag{2.22}$$

Only two of the parameters θ, α_0, and α_1 are independent since

$$\theta = \text{atan}[B_1/2(a_1 - a_0)] = \text{atan}[(\alpha_1 - \alpha_1^2)^{0.5}/(\alpha_1 - \alpha_0)] \tag{2.23}$$

The position of the assumedly straight crack is given by a. In the chevron area $\alpha_0 < \alpha \leq \alpha_1$ the crack front length becomes

$$B = \frac{\alpha - \alpha_0}{\alpha_1 - \alpha_0} \cdot B_1. \tag{2.24}$$

When $\alpha > \alpha_1$, the CENRBB becomes identical with the SECRBB specimen.

Using subscript C for CENRBB, the previous SECRBB equations yield

$$Y'_C = Y'_{Co} \cdot \left[\frac{(\alpha - \alpha^2)^{0.5}}{\alpha - \alpha_0} / \tan\theta\right]^{0.5}, \tag{2.25}$$

where in analogy with equation (2.8)

$$Y'_{Co} = 0.6006[(dg_C/d\alpha)/(1-\nu^2)]^{0.5}/(\alpha - \alpha^2)^{0.25}, \tag{2.26}$$

since $S/D = 3.33$. Y'_{Co} is a function of α, α_0, and θ.

An exact analysis would perhaps best rely on an experimental determination of g_C, like the compliance calibration of the SECRBB. Satisfactory approximate results may however be obtained with the

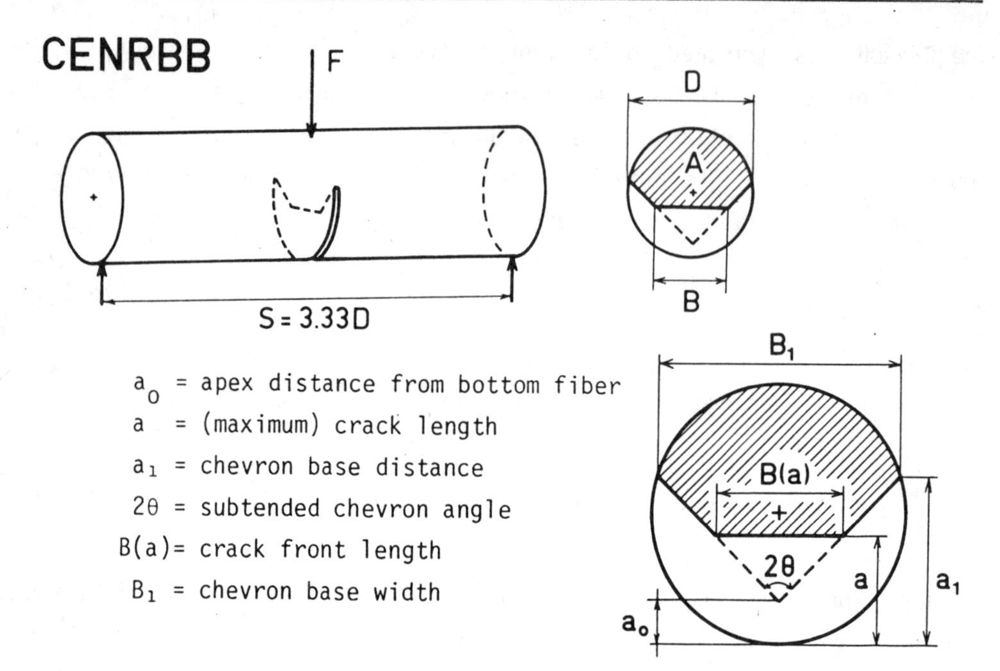

Figure 2.9: The chevron edge notch round bar in bending (CENRBB) specimen with the extra notation required.

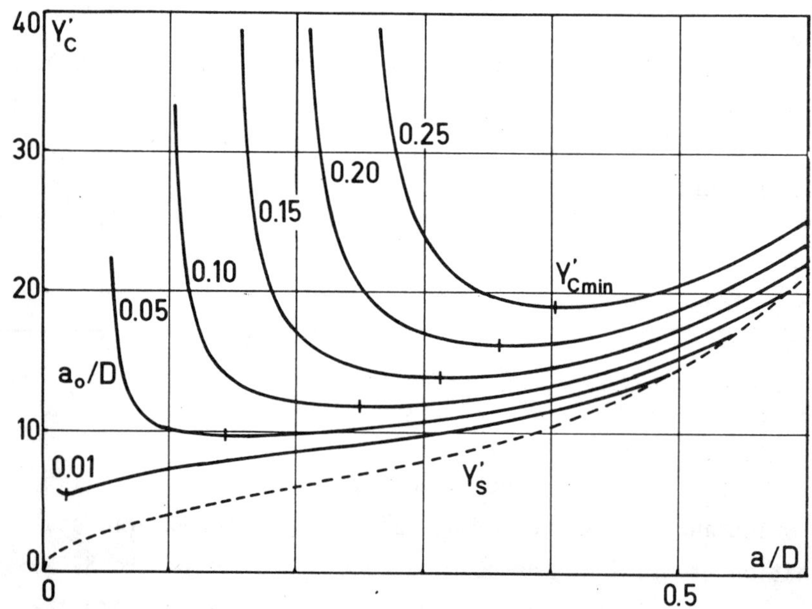

Figure 2.10: Dimensionless stress intensity factor $Y'_C(a/D)$ of CENRBB specimen with right angle chevron $2\theta = 90^\circ$.

Straight-Through Crack Assumption (STCA), Munz et al.[33, 34] The STCA states that $dg_C/d\alpha$ for a crack in a chevron notch may be approximated by $dg/d\alpha$ for a crack straight-through the net section. Thus

$$\text{STCA: } Y'_{Co}(\alpha) \simeq Y'(\alpha) \text{ when } \alpha_o < \alpha \leq \alpha_1, \tag{2.27}$$

where Y' is the SECRBB formula of equation (2.13).

The desired property of the CENRBB is that K_I and hence $Y'_C(\alpha)$ have a minimum inside the chevron area. This will occur if $dY'_C/d\alpha$ changes sign in the interval $\alpha_o < \alpha < \alpha_1$. Since

$$\frac{dY'_C}{d\alpha}/Y'_C = \frac{1}{4}\left[\frac{3}{\alpha} + \frac{9}{\alpha}[1-1/(1+19.646\alpha^{4.5})] - \frac{2}{\alpha-\alpha_o}\right] \tag{2.28}$$

this requires that $\alpha_o > 0$. If so, the change of sign at $\alpha = \alpha_m$ yields a minimum $Y'_C(\alpha_m) = Y'_{Cmin}$. This coincides with the instability point of an ideal elastic brittle material with a flat R-curve, such as was assumed in the Short Rod equation (1.12).

The analogous CENRBB evaluation follows from equation (2.7) as

$$K_{CN} = K_{Imin}(F_{max}) = 0.8325\ Y'_{Cmin}\ F_{max}/D^{1.5}. \tag{2.29}$$

Right angled chevron notches, $2\theta=\pi/2$, should be of greatest practical interest. In this case the chevron base position becomes $\alpha_1 = 0.25[1+2\alpha_o+(1+4\alpha_o-4\alpha_o^2)^{0.5}]$ from equation (2.23). Corresponding curves of $Y'_C(\alpha)$ for different apex positions are plotted in figure 2.10. All curves display minima inside the chevron. We may expect the STCA based Y'_{Cmin} values to be within perhaps 5% of the true ones, Munz et al.[34]

The amount of crack extension at the minimum, $(\alpha_m-\alpha_o)\cdot D$, should be as large as possible. According to equation (2.28) it depends only on α_o and it has a relatively broad maximum $\alpha_m-\alpha_o = 0.1625$ at $\alpha_o = 0.1625$.

This is only half of the crack extension obtained in a standard Short Rod specimen of the same diameter, Bubsey et al.[35]

An analytical expression for $Y'_{Cmin}(\alpha_o)$ is valuable in evaluating K_{CN} since α_o for a specimen won't always coincide with the selected ones. A polynominal regression yields

$$Y'_{Cmin} = 7.2984+54.026\alpha_o-122.34\alpha_o^2+374.67\alpha_o^3, \qquad (2.30)$$

in the interval $0.05 \leq \alpha_o \leq 0.25$.

The assumed flatness of the R-curve should not be taken for granted though. If a compliance calibration were made the effective crack length during a test could be measured, an R-curve registered, and its flatness checked. This would also make it possible to derive J-integral expressions, secant offset formulas etc.

3. CRACK RESISTANCE MEASUREMENTS ON CORE SPECIMENS

As shown in the review part, section 1, K_{Ic}-testing of rock in the metals sense requires above all certain minimum specimen dimensions. Core specimens will nearly always be sub-size in comparison and yield erroneously low toughness results if the sub-critical cracking is disregarded.

This section is mainly devoted to investigating whether representative toughness values can be obtained from sub-size core specimens. J-integral methods and R-curve measurements on SECRBB specimens are reported in detail. They rely on compliance measurements to follow the crack growth and are thus relatively complicated or impractical. Crack-extension resistance or toughness data from them are compared with some simplified measures and CENRBB data.

The Short Rod and CENRBB toughness evaluations, K_{SR} of equation (1.12) and K_{CN} of equation (2.29), avoid the compliance measurements but require instead that the test material have a mainly flat R-curve. This proposition is doubtful both because of required plasticity corrections, Costin [18], and because of previous R-curve measurements, Kobayashi et al.[36] Here its predictions as to the approximate crack resistance parameters K_m, J_m, and $\overline{R}$ are tested against data from SECRBB specimens. These preliminary tests, Ouchterlony [25, 26], will be presented first.

3.1 Simple R-curve Approach to SECRBB Testing

The basic proposition (in metals) is that R-curves are regarded as material properties, independent of starting crack length a_0 and the specimen in which they are developed. Thus the crack-extension resistance in stress intensity terms becomes $K_R(a-a_0)$. It may depend on strain rate and temperature though. To predict crack instability K_R is compared with the crack-extension force or $K_I(a,F)$ curves of the specimen. The unique K_I-curve that develops tangency with K_R defines the critical load, which is F_{max} in a load controlled test. See ASTM Standard Practice for R-Curve Determination E 561 for further details and also figure 1.7 in section 1.

3.1.1 Prediction Formulas. Here two linear elements are joined to a simple R-curve with a well defined knee and a flat post-critical part,

$$K_R = \begin{cases} (a-a_0)/\Delta a_c \cdot K_{Rc} & \text{when } a_0 \leq a \leq a_0+\Delta a_c \text{ and} \\ K_{Rc} & \text{when } a > a_0+\Delta a_c. \end{cases} \quad (3.1)$$

The critical crack resistance K_{Rc} and the amount of sub-critical crack growth Δa_c are assumed properties of an otherwise elastic-brittle material. Thus K_{Rc} ideally represents a proper K_{Ic} value. Swan [37,38] supports the linear sub-critical part of the K_R-curve and the knee but not necessarily the flat part which was a prerequisite for K_{SR}.

This curve and the critical crack-extension force curve of the SECRBB specimen are plotted in figure 3.1. The approximate fracture

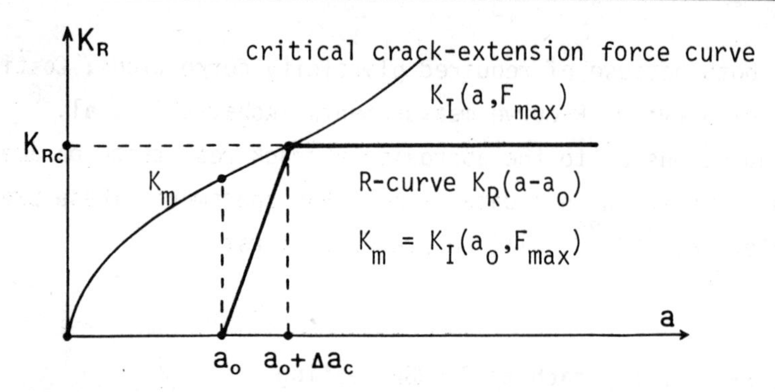

Figure 3.1: Simple bilinear R-curve and critical K_I-curve illustrate evaluation of K_m.

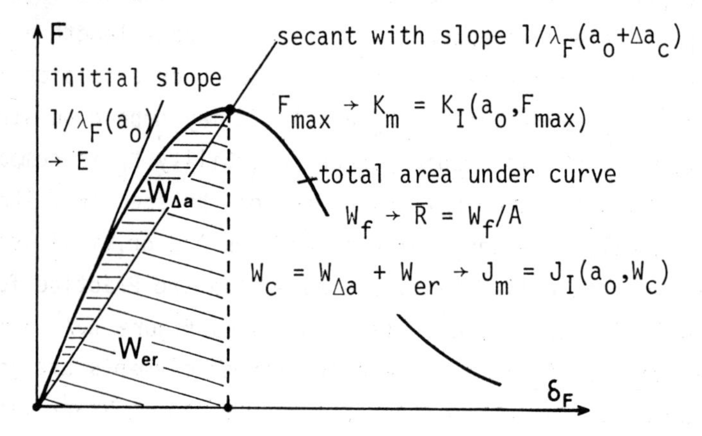

Figure 3.2: Complete failure curve of SECRBB specimen of elastic-brittle material with sub-critical crack growth.

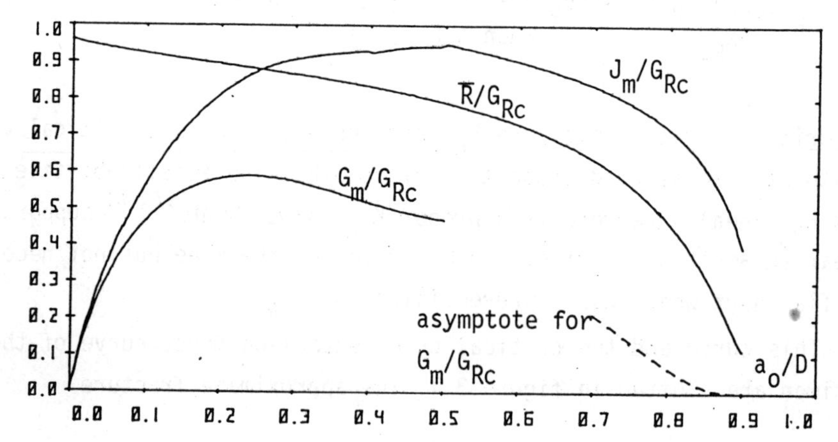

Figure 3.3: Comparison of energy rate crack resistance measures G_m, J_m, and $\overline{R}$ for $\Delta a_c/D = 0.1$.

toughness K_m of equation (1.9) then follows as

$$K_m/K_{Rc} = Y'(\alpha_o)/Y(\alpha_o+\Delta\alpha), \tag{3.2}$$

where $\alpha_o = a_o/D$ and $\Delta\alpha = \Delta a_c/D$ have been used.

The approximate J-integral J_m from equation (1.19) requires more analysis, see figure 3.2. For convenience we write

$$J_m = [\eta_{eq}/A_{eq}]_{a=a_o} \cdot W_c = \gamma(\alpha_o) \cdot W_c/D^2, \tag{3.3}$$

where the energy W_c consist of two parts, the recoverable energy W_{er} and the consumed fracture energy $W_{\Delta a}$. There follows that

$$\left.\begin{aligned} W_{er} &= \frac{1}{2}\lambda_F(a_o+\Delta a_c)F_{max}^2 = G_{Rc} \cdot D^2/\gamma(\alpha_o+\Delta\alpha) \text{ and} \\ W_{\Delta a} &= \frac{1-\nu^2}{E}\int_{a_o}^{a_o+\Delta a_c} K_R^2(a-a_o) \cdot B da = G_{Rc} \cdot 2D^2 \cdot \int_{\alpha_o}^{\alpha_o+\Delta\alpha} (\alpha-\alpha_o)^2\sqrt{\alpha-\alpha^2} d\alpha/\Delta\alpha^2 . \end{aligned}\right\} \tag{3.4}$$

Let $I_2(\alpha_o,\Delta\alpha)$ denote the integral factor in the last member, index 2 referring to the exponents of $(\alpha-\alpha_o)$ and $\Delta\alpha$. Then with $W_c = W_{\Delta a} + W_{er}$

$$J_m/G_{Rc} = \gamma(\alpha_o)/\gamma(\alpha_o+\Delta\alpha) \cdot [1+2I_2 \cdot \gamma(\alpha_o+\Delta\alpha)], \tag{3.5}$$

where equation (1.5) was used to convert K_{Rc} to G_{Rc}. Being a semi-integrated measure J_m should give a better prediction of K_{Ic} than K_m does.

Converting K_m to energy rate terms through $G_m = (1-\nu^2)K_m^2/E$, equations (3.2), (3.3), and (1.5) yield

$$G_m/G_{Rc} = \gamma(\alpha_o)/\gamma(\alpha_o+\Delta\alpha) \cdot g(\alpha_o)/g(\alpha_o+\Delta\alpha). \tag{3.6}$$

Since the g-ratio is always less than one it follows that $G_m < J_m$.

The specific work of fracture $\overline{R}$ finally is given by equations (1.3) and (3.1) as

$$A \cdot \overline{R} = \frac{1-\nu^2}{E} \int_{a_o}^{D} K_R^2(a-a_o) \cdot B da. \qquad (3.7)$$

With the subcritically swept section area being $2D^2 \cdot I_o(\alpha_o, \Delta\alpha)$

$$\overline{R}/G_{Rc} = 1 - 2(I_0 - I_2)D^2/A. \qquad (3.8)$$

In contrast to G_m and J_m, $\overline{R}$ doesn't depend on the elastic properties of the specimen.

A comparison of G_m, J_m, and $\overline{R}$ for $\Delta a_c/D = 0.10$ is given in figure 3.3. Both $\overline{R}$ and J_m are superior to G_m as G_{Ic} or K_{Ic} predictors. G_m may be as much as 50% lower than G_{Rc}.

Since a well established crack is desirable, J_m is the best predictor. For $0.3 \leq \alpha_o \leq 0.6$ it falls within 10% of G_{Rc}. For smaller $\Delta a_c/D$-values all curves fall closer to G_{Rc}.

3.1.2 Energy Rate Crack Resistance Data. The same $D \approx 41$ mm SECRBB specimens of Ekeberg marble that were used for the compliance calibration in section 2.1 gave the data for this part. An additional 22 SECRBB specimens of Bohus granite were tested. Each with its own modulus value from $\lambda_F(a_o)$. As noted these moduli are independent of notch depth and agree closely with strain gage data. The data was obtained from complete failure curves $F(\delta_F)$ such as those in figure 2.2. Data for Stripa granite, Swan[37, 38], has also been reinterpreted.

The elastic properties for these rocks, ± standard deviation essentially, are

$E = 85.7 \pm 3.3$ GPa and $\nu \approx 0.22$ for Ekeberg marble,

$E = 41.4 \pm 2.7$ GPa and $\nu \approx 0.10$ for Bohus granite, and

$E = 57 \pm 4$ GPa and $\nu \approx 0.12$ for Stripa granite.

Further mechanical properties are given by Ouchterlony [22], Wijk [39], and Swan [40] for example.

The energy rate crack resistance data is plotted in figures 3.4 to 3.6. All results are similar. $\overline{R}$ is always the largest and on average $\overline{R}:J_m:G_m$ is roughly 4:2:1. $\overline{R}$ depends considerably on a_o/D, G_m depends on a_o/D, and so does J_m for Bohus granite. This dependence and their numerical incompatibility make all three quantities unfit as single parameter material properties. Use of the present two parameter R-curve model to interpret the data instead yields that:

The best least squares fit G_m-curves, Ouchterlony [25] Appendix B, were obtained when

G_{Rc} = 39.2 J/m² and $\Delta a_c/D$ = 0.10 for Ekeberg marble,

G_{Rc} = 66.8 J/m² and $\Delta a_c/D$ = 0.11 for Bohus granite, and

G_{Rc} = 93.2 J/m² and $\Delta a_c/D$ = 0.10 for Stripa granite.

These G_m-curves are shown in figures 3.4 to 3.6.

The G_{Rc}-results are clearly reasonable, if somewhat low compared to the J_m levels. To achieve a good fit for the G_m-curve is the least demanding test of the simple R-curve model since only the sharp knee is involved. An assessment of the minimum required crack length as per equation (1.7) yields $a_{min} \approx 130$ mm for all three rocks, emphasizing the need to predict representative toughness values from sub-size specimens.

The $\Delta a_c/D$ values are practically equal. Additional K_m or G_m data from D ≈ 41 mm SECRBB specimens of several other rocks, Sun Zongqi [41], can be interpreted such that $\Delta a_c \approx 0.1$ always holds, independent of rock type. This points to a geometrical explanation of the sub-critical crack growth, as opposed to the traditional time independent R-curve approach which would require Δa_c constant and material dependent instead.

A plausible mechanism is crack tunneling, that is growth from the notch (crack) starts locally at the middle where the stress intensity (factor) is largest, Blackburn.[42] Gradually the whole crack starts extending but its front is now curved not straight. The equilibrium shape of the crack front changes during growth and may not necessarily be one of constant K_I conditions, Wanhill et al.[43] Time dependent sub-critical crack growth could well be involved, Rummel and Winter [44] and Atkinson.[45]

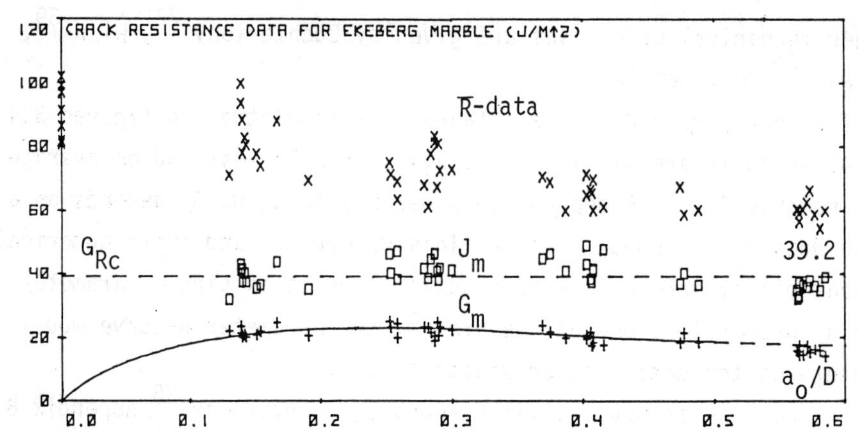

Figure 3.4: Energy rate crack resistance data for Ekeberg marble.

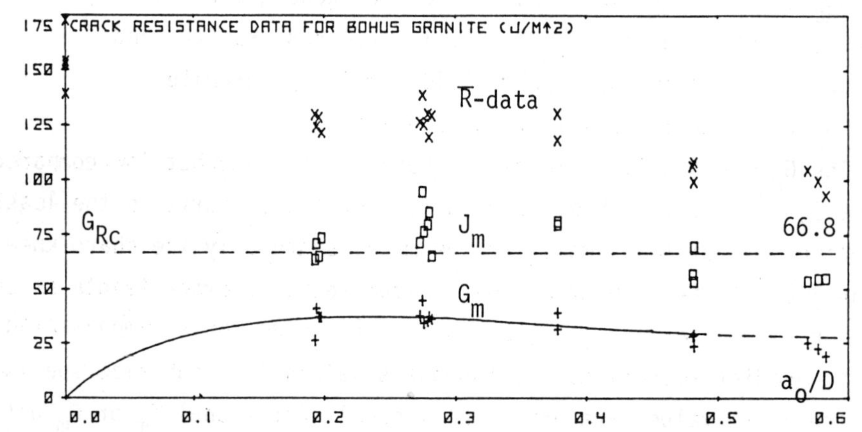

Figure 3.5: Energy rate crack resistance data for Bohus granite.

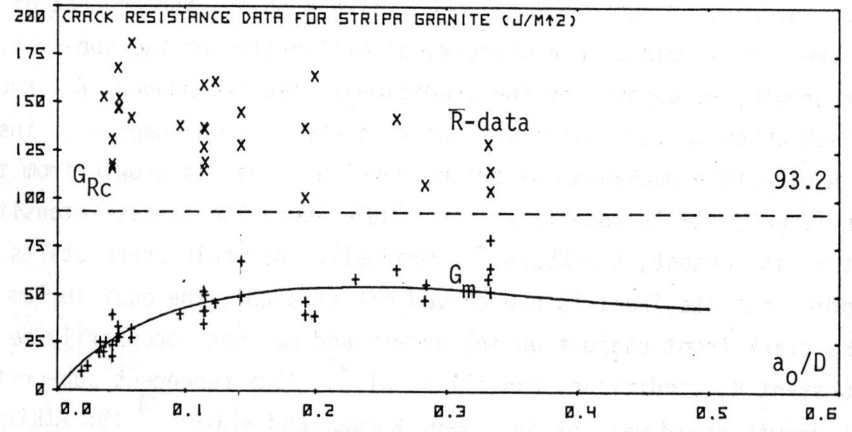

Figure 3.6: Energy rate crack resistance data for Stripa granite.

The present material is not sufficient to draw definite conclusions as to the nature of the sub-critical crack growth but the matter is being investigated.

A J_m-curve depends both on the sub-critical part of the R-curve and on the knee. The curve fit is actually worst around $\Delta a_c/D \approx 0.10$. The predicted G_{Rc}-values are not bad but the simple R-curve model essentially fails to account for the J_m-data, Ouchterlony.[25]

The trend of the $\overline{R}$-data with a_0/D is correctly reproduced by the $\overline{R}$-curve in figure 3.3 whose level is roughly 50% too low. The $\overline{R}$-curve integrates the whole K_R-curve. A bilinear model with sharply increasing post-critical part would raise the predicted $\overline{R}$-curve, possibly to the level of the data, just as for alumina bend bars, Kleinlein and Hübner.[46] Such a K_R-curve could maybe also explain why uncorrected Short Rod values are too low, Ouchterlony.[25] It is a better approximation to measured ones, Kobayashi et al.[36] and Swan and Alm [47], but the problem with J_m would remain.

3.1.3 Conclusions. The simple R-curve model with a flat post-critical part essentially fails to account for the experimental energy rate resistance data. The knee of the K_R-curve is supported by the G_m-data but the curve doesn't account for the J_m or the $\overline{R}$-data.

Crack tunneling as an explanation of the sub-critical crack growth would make the SECRBB specimen more like the ones with chevron notches in that some crack growth would occur before instability even during constant crack resistance conditions. Since the factual R-curves of rock aren't so clear cut however, a direct measurement of them seems better than indirect conclusions based on their assumed form.

3.2 Direct R-curve Measurements on SECRBB Specimens

Two different ways of testing have been used to obtain the R-curves. In the first one semi-continous failure curves were recorded and interpreted in the vein of ASTM E 561, Standard Practice for R-curve Determination. In the second one a succession of sub-critical failure cycles were used, much as described in section 1.5.2 for J-integral testing. The displacement measurement method developed in section 2.1 was used to measure compliance and crack length. A preliminary presentation of this work was made by Ouchterlony.[48]

Though relatively complicated, these test methods yield a whole R-curve from a single specimen where the previous approach may need several specimens to yield a toughness estimate. Hereby one also avoids the variations in material properties between specimens which are inherently larger in rock than in metals, especially if investigation site cores were to be used.

3.2.1 R-curves from Complete Failure Curves. This approach relies heavily on E 561. Thus the appropriate crack length to use in K_I and J-integral formulas is an effective one,

$$a \equiv a_{eff} = a_{phys}+r_{mc} = a_o+\Delta a+r_{mc}. \tag{3.9}$$

The physically observable crack length a_{phys} consist of the starting crack (notch) length a_o and the physical crack growth Δa at the tip. It is augmented by a micro-crack zone adjustment r_{mc} like in equation (1.10), see figure 3.7. If this zone were ideally plastic it would display zero incremental stiffness during loading and then act as a direct addition to the crack length.

LEFM size requirements aim at making r_{mc} so small that $a_{eff} \simeq a_{phys}$, a nearly ideal elastic-brittle situation. For a specimen of such material the unloading from a failure curve coincides with the secant through its unstressed state, see figure 3.7. In E 561 this constructed secant compliance is a preferred method of measuring a_{eff} along a failure curve even when $a_{phys} \neq a_{eff}$. Hereby the crack length would become micro-crack zone corrected automatically even during growth, $\dot{a} > 0$.

In sub-size specimens the micro-crack zone won't be negligible during loading. In the plasticity analogy it would behave elastically

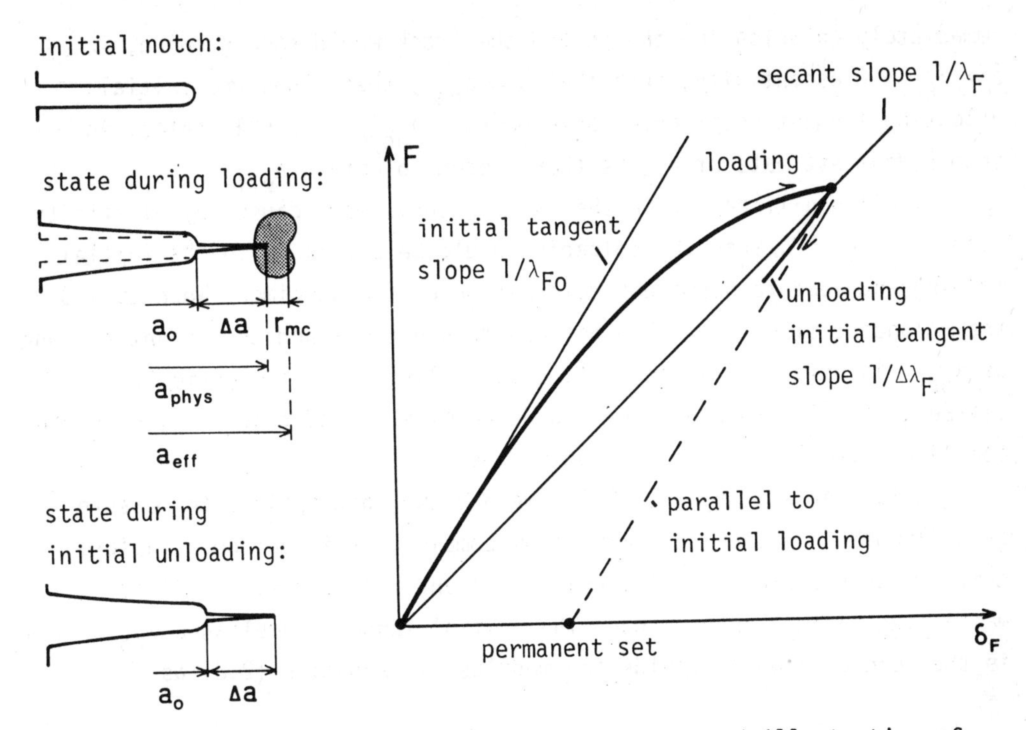

Figure 3.7: Definition of crack length parameters and illustration of crack length measures in single compliance method.

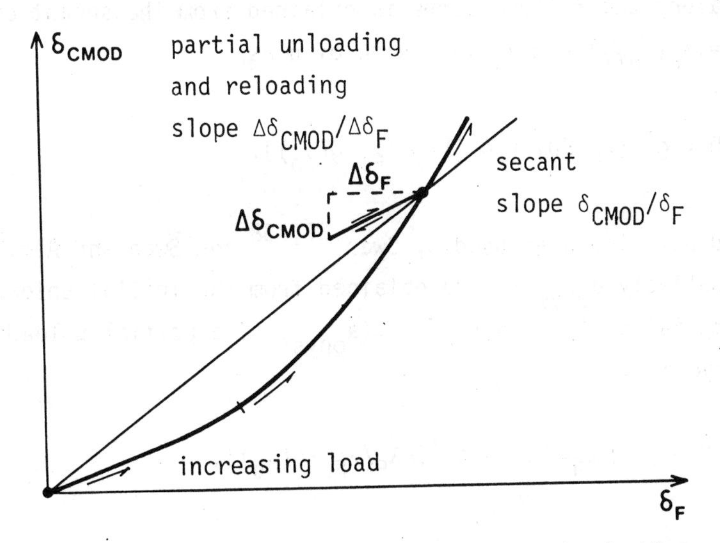

Figure 3.8: Illustration of crack length measures in double compliance method.

immediately on unloading though and the crack would stop growing, $\dot{a} = r_{mc} = 0$ essentially, such that $a \approx a_{phys}$ then. Thus the initial unloading tangent compliance would measure a_{phys} as E 561 states. An independent estimate of r_{mc} is then needed to obtain a_{eff}.

If all non-linearity in the failure curve were caused by plasticity such that $\Delta a \approx 0$, then the unloading would be parallell to the initial loading and lead to permanent deformation of the specimen. When $\Delta a > 0$ the unloading slope would lie between this extreme and the secant as long as a_{phys} is effectively constant. Effects like reverse plasticity, Clarke et al.[49] and Budiansky and Hutchinson [50], and crack surface closure complicate the later stages of unloading.

In further analogy with E 561 we had two main options to measure crack length in SECRBB specimens from complete semi-continous failure curves with recurrent partial unloadings and reloadings. The first option was a <u>single compliance method</u>. The initial tangent compliance $\lambda_{Fo} = \lambda_F(a_o)$ is the reference which yields the modulus via equation (2.2) as

$$E = g(\alpha_o)/[D \cdot \lambda_{Fo}] \tag{3.10}$$

Then a_{eff} along the failure curve is obtained from the secant compliance $\delta_F/F \equiv \lambda_F = \lambda_F(a_{eff})$ and an inversion of g as

$$a_{eff}/D = g^{-1}[\lambda_F \cdot ED] = g^{-1}[\lambda_F/\lambda_{Fo} \cdot g(\alpha_o)]. \tag{3.11}$$

This method has also been used by Swan [37, 38] and Swan and Alm.[47]

Alternatively a_{phys} may be obtained from the initial unloading tangent compliance $\Delta\delta_F/\Delta F \equiv \Delta\lambda_F = \lambda_F(a_{phys})$ of a partial unloading reloading cycle as

$$a_{phys}/D = g^{-1}[\Delta\lambda_F \cdot ED] = g^{-1}[\Delta\lambda_F/\lambda_{Fo} \cdot g(\alpha_o)], \tag{3.12}$$

providing $\dot{a} = 0$. For steel $\Delta F \leq 0.1 \cdot F_{max}$ keeps $a \approx a_{phys}$ with negligible reverse plastic zone, Clarke et al.[49] An r_{mc} adjustment is then needed to obtain a_{eff}. Conversely $a_{eff} - a_{phys}$ from the above equations may be used to check the r_{mc} estimate.

The second option was a double compliance method such as used by Kobayashi et al.[36] In an SECRBB adaption LPD and CMOD are plotted together, best as δ_{CMOD} versus δ_F. See figure 3.8. The secant along this curve δ_{CMOD}/δ_F then yields the crack length

$$a_{eff}/D = h^{-1}[\delta_{CMOD}/\delta_F]. \qquad (3.13)$$

Alternatively the unloading tangent ratio $\Delta\delta_{CMOD}/\Delta\delta_F$ yields

$$a_{phys}/D = h^{-1}[\Delta\delta_{CMOD}/\Delta\delta_F], \qquad (3.14)$$

providing $\dot{a} = 0$. Then a_{phys} plus the r_{mc} correction gives a_{eff}.

By plotting $F(\delta_F)$ and simultaneously either $\delta_{CMOD}(\delta_F)$ or $F(\delta_{CMOD})$ on a two channel X-Y recorder both compliance methods could be used together. 11 SECRBB specimens of Ekeberg marble with $D \approx 41$ mm from the same lot as before were tested this way. 8 tests were successful. The test record of specimen R1 is shown in figure 3.9. It has been evaluated according to equations (3.9) to (3.14) at every unloading tic which corresponds to crack lengths $a_{eff}/D < 0.6$. So have five other test records, two didn't yield CMOD data.

All specimens had the same notch width of 1.6 mm, just as before. The average Young's modulus was $87.8_{10} \pm 3.4$ GPa which is insignificantly higher than in equation (2.11). The calculated crack lengths were then used to generate stress intensity resistance K_R via equation (2.7). The results of the double compliance method given in equations (3.13) and (3.14) predicted a decrease in crack length from the tip of the starting notch. This is impossible and the results were not analyzed further. The method in its original version yields better results, Kobayashi et al.[36]

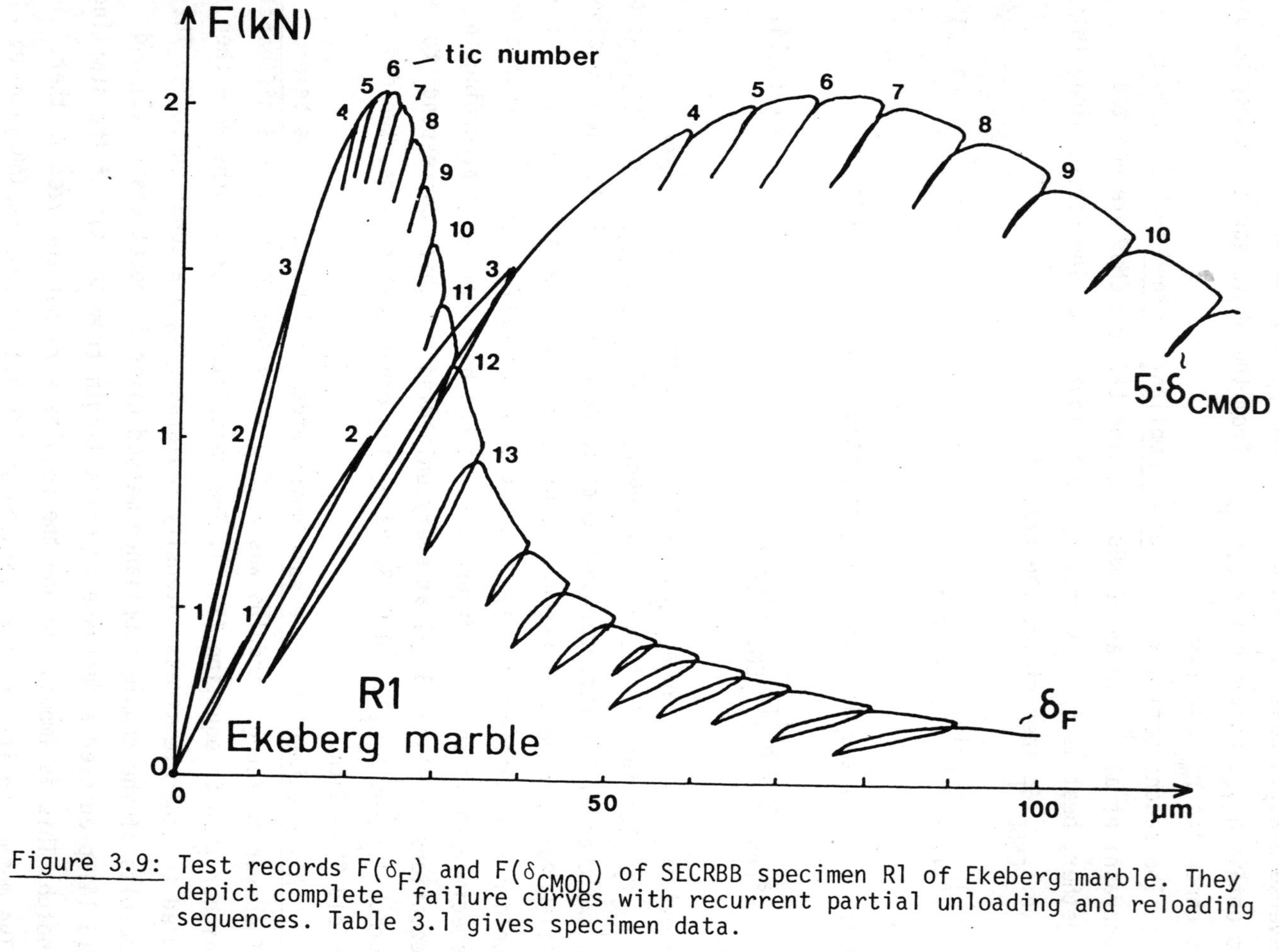

Figure 3.9: Test records $F(\delta_F)$ and $F(\delta_{CMOD})$ of SECRBB specimen R1 of Ekeberg marble. They depict complete failure curves with recurrent partial unloading and reloading sequences. Table 3.1 gives specimen data.

The results of the single compliance method were more promising. The $K_R(a_{eff})$ curves in figure 3.10 and 3.12 look much the same. Especially the initial sub-critical parts seem independent of the starting notch length a_o such that $K_R = K_R(a_{eff}-a_o)$. The K_R-curves from the secant compliance seem to level out in the post-critical regime as opposed to those of Swan [37,38] but those that are based on the initial unloading tangent compliance tend to become unstable. The micro-crack adjustment was made iteratively,

$$r^*_{mc} = 0.05(K_I/\sigma_t)^2 \tag{3.15}$$

with σ_t = 10 MPa and a_{phys} as the starting crack length for K_I, then $a_{phys} + r^*_{mc}$ until a_{eff} increased less than 0.0001·D from one step to the next.

The factor 0.05 above is considerably smaller than the 0.2 of equation (1.10) but larger values destabilize all $K_R(a_{phys}+r^*_{mc})$-curves, not only that of specimen S3. If we instead compute

$$\rho = (a_{eff}-a_{phys})/[K_R(a_{eff})/\sigma_t]^2, \tag{3.16}$$

we find that ρ has a minimum of roughly 0.05 around F_{max}, the point in figure 3.12 where K_R tends to become unstable. The corresponding $\Delta a+r_{mc}$ estimate $a_{eff}-a_o \approx 0.1$ to 0.2·D, see table 3.1 which gives values for specimen R1.

The inherent instability of the $K_R(a_{eff})$-curves in figure 3.12 reflects the shape of $K_R(a_{phys})$ in figure 3.11, which tends to have a maximum K_{Rmax} beyond F_{max} and then a descending part. K_{Rmax} tends to decrease with increasing a_o/D, as in aluminum where K_{Rmax} coincides with net section yielding, Schwalbe and Setz.[51] Their conclusion that: "Beyond these points K is assumed to loose its correlation with stable crack growth, below all data falls on a single curve $K_R(a_{phys}-a_o)$." could be relevant here too.

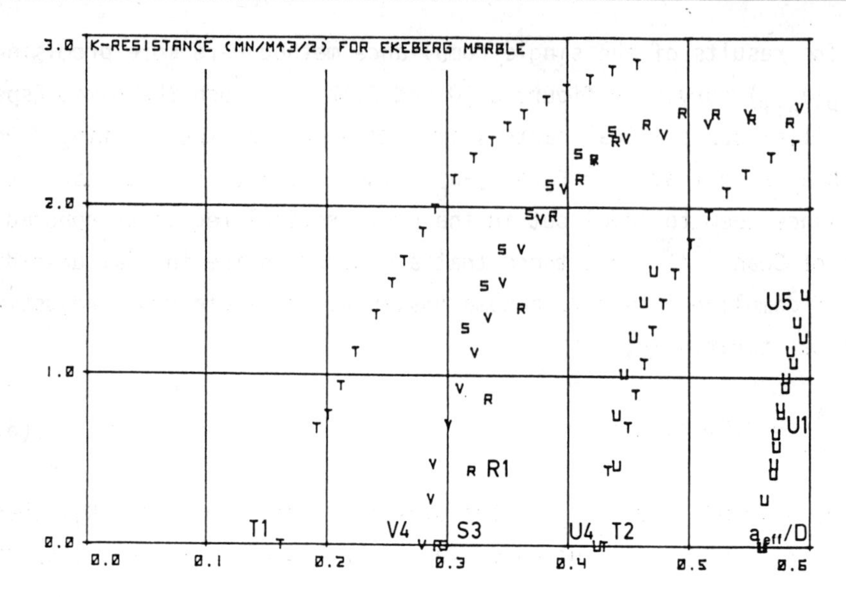

Figure 3.10: K_R-curves from SECRBB specimens of Ekeberg marble based on single compliance method. The crack length a_{eff} was computed using the secant compliance equation (3.11).

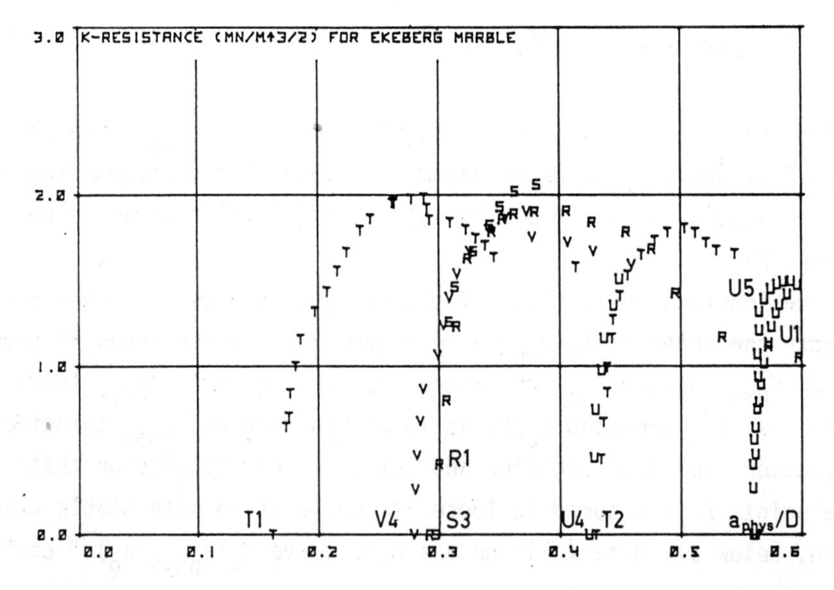

Figure 3.11: K_R-curves from SECRBB specimens of Ekeberg marble based on single compliance method. The crack length a_{phys} was computed using the initial unloading tangent compliance equation (3.12).

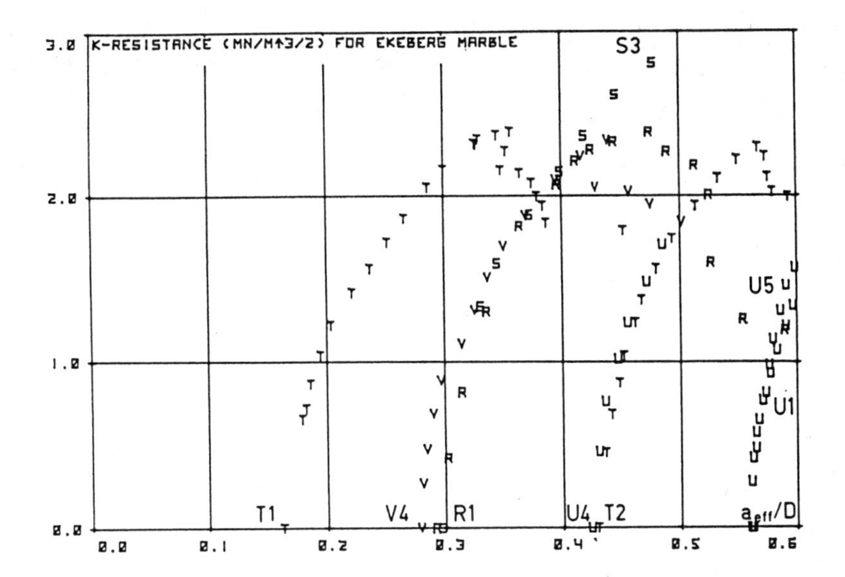

Figure 3.12: K_R-curves from SECRBB specimens of Ekeberg marble based on single compliance method. The crack length $a_{eff} = a_{phys} + r^*_{mc}$ was computed iteratively using the initial unloading tangent compliance and the micro-crack adjustment of equation (3.15).

tic no.	a_{eff}/D	a_{phys}/D	ρ	$K_R(a_{eff})$ [$MN/m^{3/2}$]	$G_R(a_{eff})$ [J/m^2]	$J_R(a_{eff})$ [J/m^2]	
0	0.292	0.292	0	0	0	0	
1	0.320	0.300	0.424	0.44	2.1	2.1	
2	0.334	0.306	0.154	0.86	8.2	8.3	
3	0.361	0.315	0.098	1.40	21.5	22.9	
4	0.388	0.324	0.070	1.95	41.7	46.1	
5	0.410	0.344	0.058	2.16	51.3	59.9	
6	0.422	0.353	0.055	2.28	57.3	68.9	max.load
7	0.440	0.363	0.056	2.39	62.8	-	
8	0.465	0.380	0.057	2.49	68.2	-	
9	0.495	0.406	0.056	2.56	72.3	-	
10	0.523	0.427	0.060	2.56	71.9	-	
11	0.552	0.455	0.063	2.53	70.2	-	
12	0.584	0.477	0.071	2.51	69.1	-	
13	-	0.496	-	-	-	-	
14	-	0.534	-	-	-	-	
15	-	0.572	-	-	-	-	
16	-	0.598	-	-	-	-	
m	0.292	0.292	-	1.58	27.5	51.3	K_m etc.

Table 3.1: Crack length and crack resistance data for SECRBB specimen R1 of Ekeberg marble, obtained with single compliance method of equations (3.10) to (3.12). The ρ-factor of equation (3.16) is also given as are the approximate crack resistance parameters K_m, G_m, and J_m. Specimen reference data: D=41.4 mm, λ_{Fo} = 7.38μm/kN, $\nu \approx 0.22$, and E = 86.5 GPa.

The micro-crack parallel to net section yielding may be seen as an influence of the ligament which prevents the micro-crack zone and hence K_R from developing fully before crack growth sets in. Yet $K_{Rmax} \approx$ 2.3 to 2.4 $MN/m^{3/2}$ in figure 3.12 agrees fairly well with the post-critical level in figure 3.10. This would then be a low estimate of the fully developed crack resistance.

Table 3.1 also contains $J_R(a_{eff})$ up to F_{max} from secant compliance data, as defined by equation (2.19). The energy rate equivalent of K_R,

$$G_R = (1-\nu^2)K_R^2(a_{eff})/E, \tag{3.17}$$

which depends on E is also given. J_R is also modulus dependent because the initial compliance $\lambda_{Fo} \propto 1/E$ can be extracted as a scale factor from the work factor in J_R. The J_R-data for Ekeberg marble is significantly higher than the G_R-data. This is disconcerting since the compatibility $J_R \approx G_R$ can be achieved from secant compliance data for steel and aluminum even during large scale plasticity, up to the limit of net section yielding, McCabe and Landes.[52]

The incompatibility between G_R and J_R in this case outweighs both their higher maximum levels, as compared with the G_m and J_m-data and the G_{Rc} estimate of figure 3.4, as well as the fact that $K_R = K_R(a_{eff}-a_o)$ in the sub-critical regime. The latter argument loses weight if crack tunneling occurs.

3.2.2 R-curves from Sub-critical Failure Cycles. One version of this technique, Schmidt and Lutz [12] and Costin [18], was presented in section 1.5.2, see equations (1.20) to (1.22) and figure 1.10. The one used here is modified to suit the previous SECRBB testing and the conclusions of the review section.

The notched SECRBB specimens were cycled manually in the testing machine with maximum LPD increasing from cycle to cycle. The CMOD rate was kept constant during each cycle but varied within 0.1 - 1.0 µm/s, roughly in proportion to increasing compliance. A small bias load was maintained during the whole process and the LPD was offset manually before each cycle in the pre-failure region to avoid double traces on the X-Y recorder. See figure 3.13 which shows some failure cycles of specimen S4 of Ekeberg marble.

The first cycles are nearly ideal, linear elastic. The compliance of the first two is the same, then it increases somewhat with each

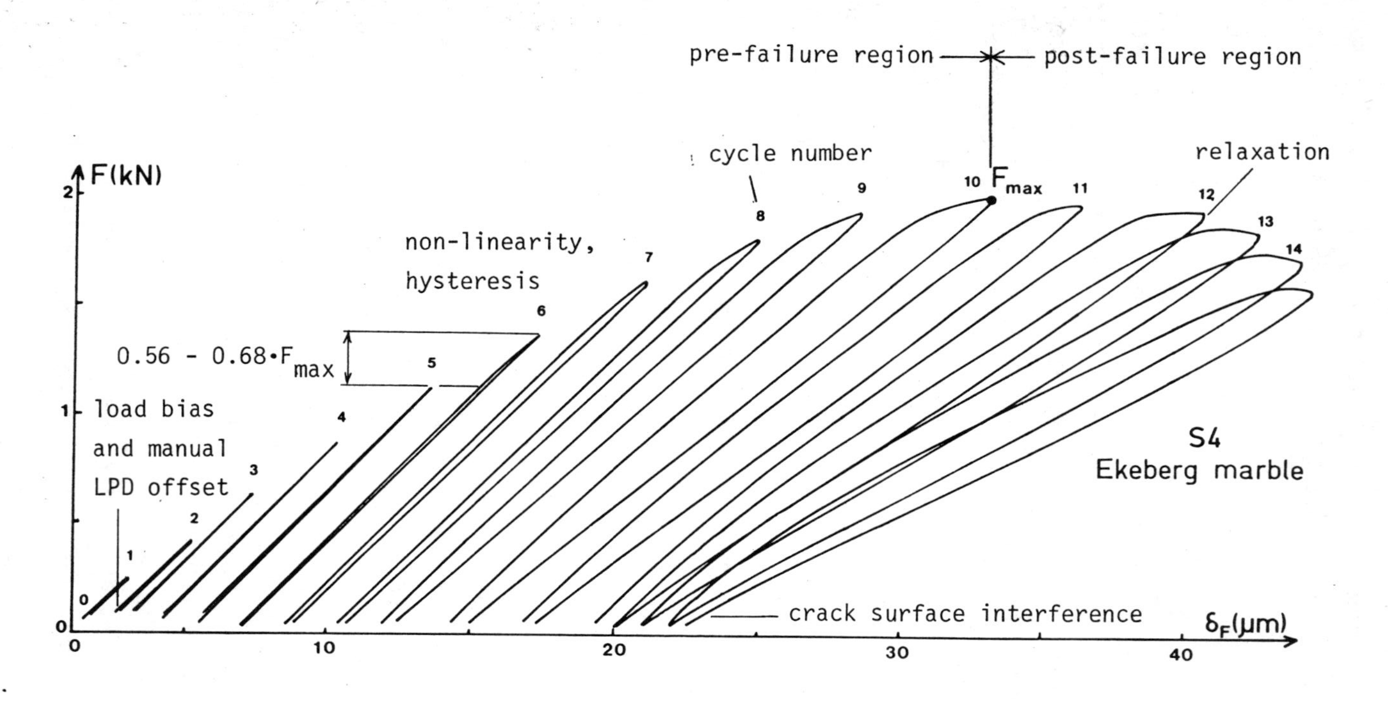

Figure 3.13: Complete sub-critical failure cycles of specimen S4 of Ekeberg marble. Specimen reference data are D = 40.9 mm, a_o = 11.5 mm, λ_{Fo} = 7.5 μm/kN, $\nu \approx 0.22$, and E = 83.6 GPa.

successive cycle. Non-linearity during loading and hysteresis becomes apparent during cycle 6. The new load range it covers is $0.56 < F/F_{max} < 0.68$. This correlates well with the onset of acoustic emission at about $0.5 \cdot F_{max}$ in SECRBB tests on Ekeberg marble, Stillborg [53], and with the onset of micro-cracking in an SEM when $0.5 < F/F_{max} < 0.7$, Olofsson and Lindqvist.[54] During the next two cycles the hysteresis increases but both loading and unloading are essentially linear. An increasing amount of permanent deformation is present.

From cycle 9 and on the initial loading is considerably stiffer than the main part and from cycle 10, which yields F_{max}, the final unloading also gets stiffer. In all tested specimens this behavior was well established in the post-failure range. It was perhaps first observed in Indiana limestone, Schmidt [55], and several others report it too. It occurs in Westerly granite, see figure 1.10, and it has been interpreted as crack surface interference or crack closure. Intergranular displacements and relaxation of residual elastic strains adjacent to the formed crack surfaces, Friedman et al.[6], which create a surface mismatch could be the reason for it. A relaxation component is probably present in the initial unloading from cycle 12 and on. Both independent observations, Rummel and Winter [44] and a test that was made on a foil gage mounted specimen points at crack growth $\dot{a} > 0$ as a cause. If so it could explain why the initial unloading tangent compliance, that was used to predict a_{phys} from equation (3.12) under the condition $\dot{a} = 0$, failed to provide reliable K_R-data beyond K_{Rmax}.

Without going into a detailed analysis, along the lines of Budiansky and Hutchinson [50] for example, it is obvious that the failure cycles in the post-failure region contain several features which normally would put their evaluation outside the range of both LEFM and J-integral techniques. These are also apparent in the curves of Schmidt and Lutz [12], see figure 1.10. They resolve one complication by noting that the linear portion of the loading extrapolates back to the origin such that no permanent deformation would remain in the unloaded specimen, but for the crack surface interference. They also found that the slope at the beginning of each cycle was identical to that of the first one, indicating a completely closed crack then. My results essentially support this.

Thus the best linear loading compliance would represent the maximum crack length of each cycle, at least to begin with. But as the crack growth increment during a cycle increases some other compliance should be a better measure of this crack length. Candidates are the secant compliance, the best linear unloading compliance, and even the best linear reloading compliance, see figure 3.14. The last two are nearly equal and they fulfil $\dot{a} = 0$ but include any crack growth during relaxation. This crack growth will probably be closely related to the size of the micro-crack zone. Thus these compliances tend to overestimate a_{phys} and be better estimates of a_{eff}.

Because of the inflection point of the unloading in the post-failure region, the best linear unloading compliance $\lambda_F^{u\ell}$ is well defined. It was tried first on 6 remaining SECRBB specimens of Ekeberg marble from the same lot as before. The stress intensity resistance for cycle j was evaluated as

$$K_{Rj} = K_I(a_j^{u\ell}, F_j) \text{ where } a_j^{u\ell} = g^{-1}[\lambda_{Fj}^{u\ell}/\lambda_{Fo} \cdot g(\alpha_o)], \tag{3.18}$$

and $a_{eff} \approx a > a_{phys}$ is implied. The resulting $K_R(a_{eff})$ curves are shown in figures 3.15 and 3.16. The first shows that $K_R = K_R(a_{eff}-a_o)$ in the sub-critical regime and that K_R tends to level out in the post-critical regime, roughly at the same level as in figures 3.10 and 3.12. There is some scatter in this level but no apparent dependence on the initial notch depth a_o.

A comparison of figures 3.15 and 3.16 shows that $K_R(a_{eff})$ of a notched virgin specimen on the average looks quite different from $K_R(a_{eff})$ of a precracked one. The crack surface interference stretches the latter towards a_o. Extracting it we see that a precracked specimen displays only half the amount of sub-critical crack growth that the notched virgin one does, $\Delta a_c \approx 0.05 \cdot D$ versus $0.10 \cdot D$ roughly. Thus at least

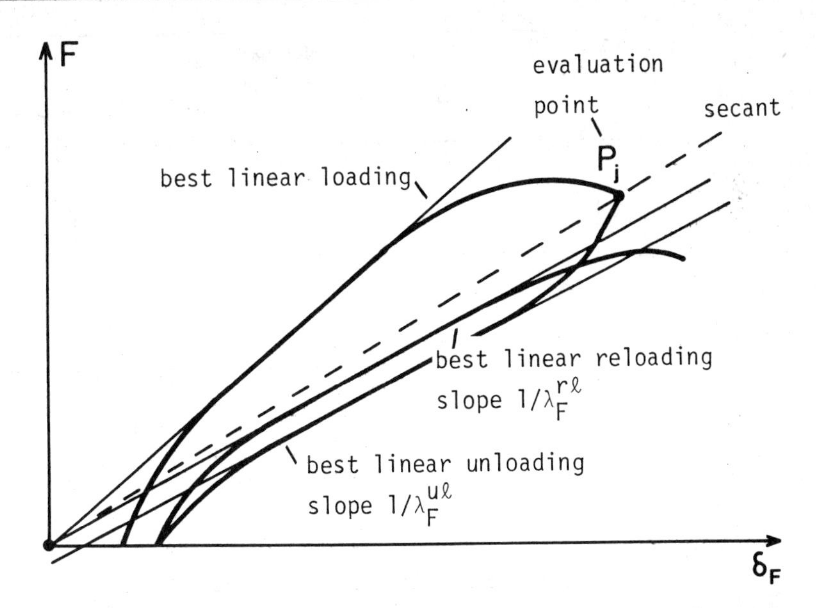

Figure 3.14: Illustration of crack length measures and evaluation point of a complete sub-critical failure cycle in the post-failure range.

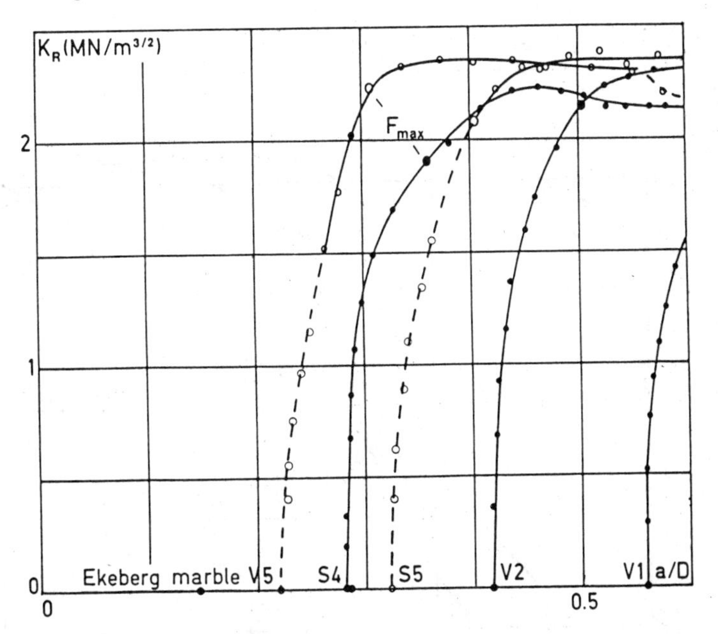

Figure 3.15: K_R-curves from SECRBB specimens of Ekeberg marble, based on best linear unloading compliance to compute crack length. Dashed portions have been adjusted.

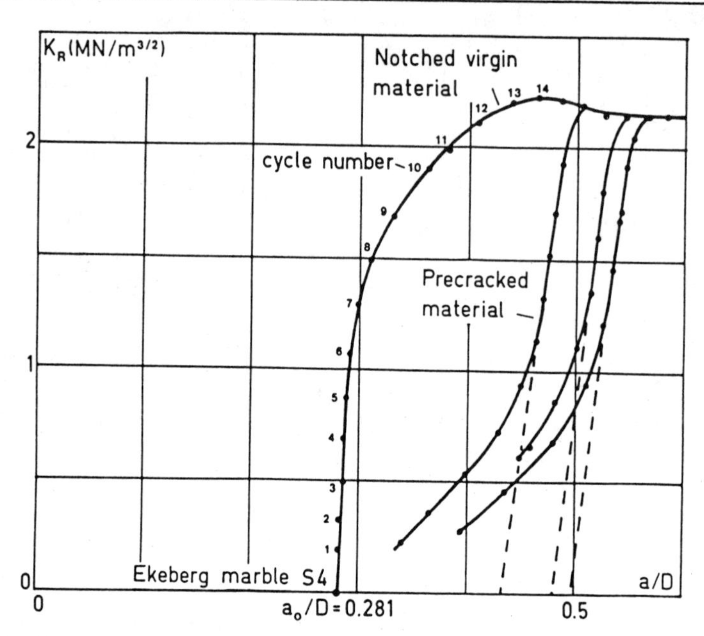

Figure 3.16: K_R-curves from specimen S4. Curve with numbers represents notched virgin material, other curves represent subsequent sets of sub-critical failure cycles on same specimen. Numbers refer to cycles in figure 3.13.

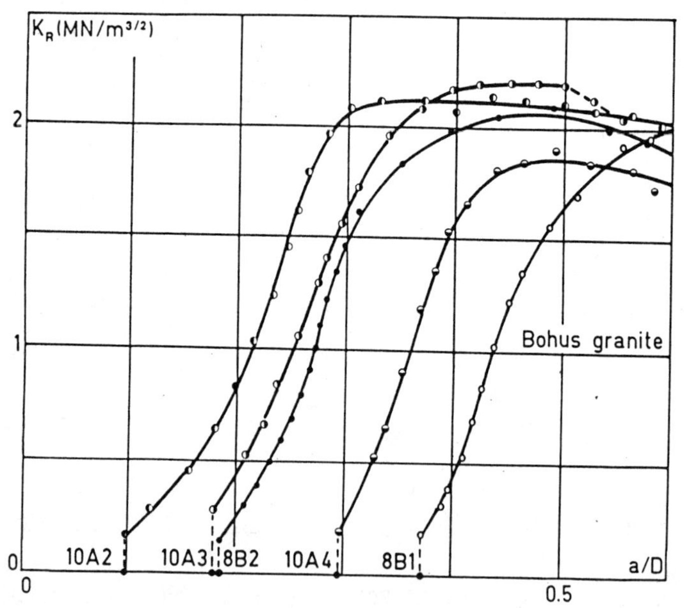

Figure 3.17: K_R-curves from SECRBB specimens of Bohus granite, based on best linear unloading compliance to compute crack length.

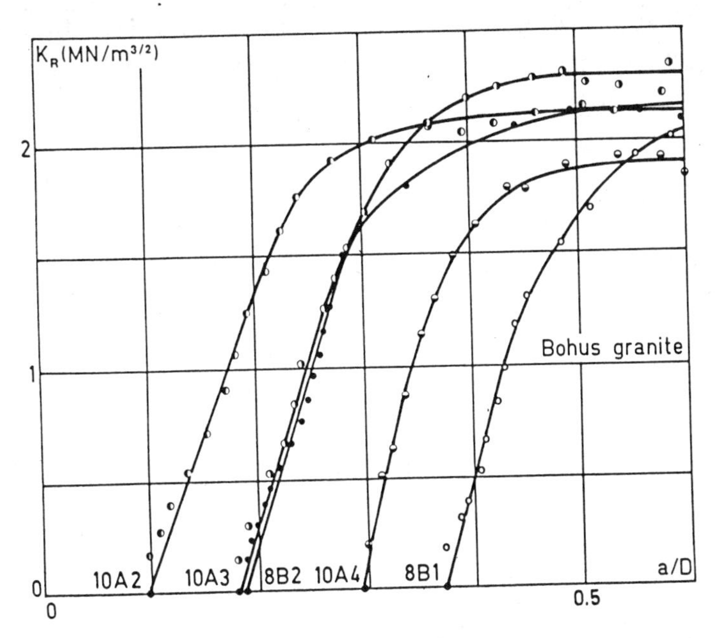

Figure 3.18: K_R-curves from SECRBB specimens of Bohus granite, based on best linear reloading compliance to compute crack length.

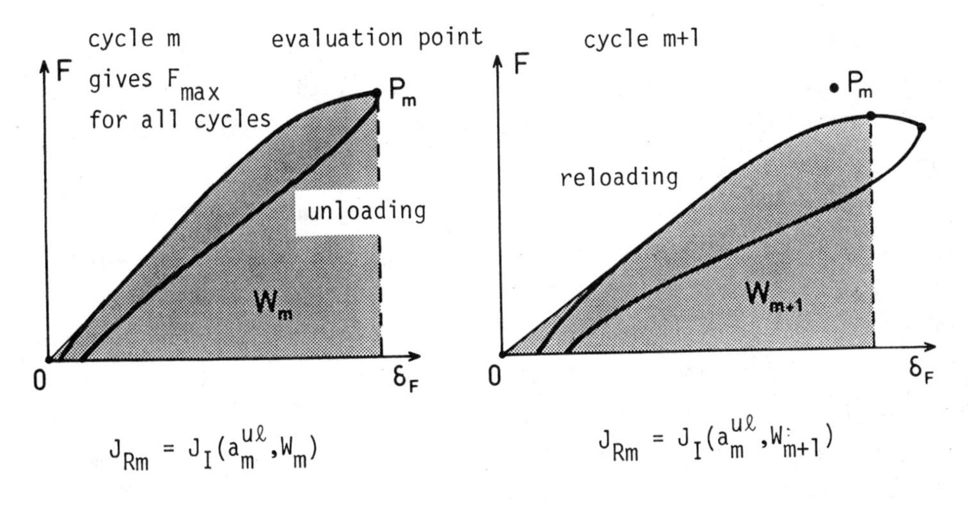

Figure 3.19: Illustration of J_R-evaluation from sub-critical failure cycles shows difference between pre- and post-failure regions.

a part of Δa_c could have a geometrical explanation, like thumbnailing, in the SECRBB specimen. The exact shape of the sub-critical part of K_R is relatively unimportant though when it comes to finding a representative toughness value.

Then $D \approx 41$ mm SECRBB specimens of Bohus granite from the same blocks as before were tested. Figures 3.17 and 3.18 show $K_R(a_{eff})$ curves from 4 of these, obtained using both the best linear unloading and reloading compliances $\lambda_F^{u\ell}$ and $\lambda_F^{r\ell}$ as crack length measures in equations (3.18). The K_R-curves based on the best linear reloading compliance look more appealing in that the sub-critical regime is nearly linear and in that the post-critical one doesn't tend to drop off.

The main crux with the previous direct method of obtaining R-curves was however the discrepancy between $G_R = (1-\nu^2)K_R^2/E$ and J_R. The evaluation of J_R used here is shown in figure 3.19. Based on $\lambda_F^{u\ell}$ the points on the J_R-curve are defined by

$$J_{Rj} = J_I(a_j^{u\ell}, W_j) \text{ and } \Delta a_j = a_j^{u\ell} - a_o, \qquad (3.19)$$

for the j:th cycle in the pre-failure region. Again $a \approx a_{eff}$ is implied. In the post-failure region W_j is replaced by W_{j+1} to compensate for the crack growth between the local F_{max} and the evaluation point P_j. $\lambda_F^{r\ell}$ may be used similarly.

Here Δa_j defines the total effective crack growth up to the j:th cycle. This is different from equation (1.20) but it follows the build up of crack resistance in the pre-failure region better. Both methods give pseudo resistance curves since they don't describe continous growth but a sequence of separate growth increments. This is best brought out by equation (1.20), but naturally the single specimen technique leads to an interdependence of these increments so neither definition would seem to be superior in the post-failure region.

A comparison of G_R and J_R values from best linear reloading data on Bohus granite is shown in figure 3.20. It is made in stress intensity terms, that is K_R versus K_J,

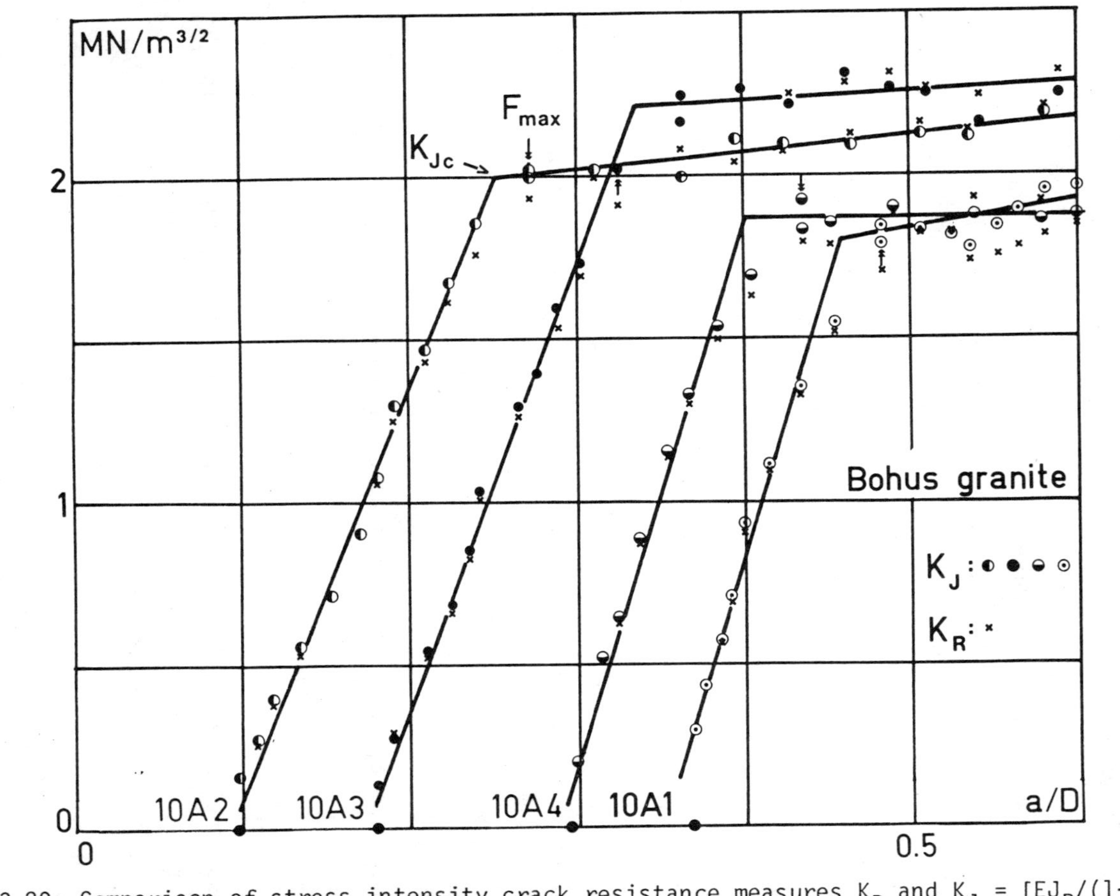

Figure 3.20: Comparison of stress intensity crack resistance measures K_R and $K_J = [EJ_R/(1-\nu^2)]^{1/2}$ as evaluated on the same SECRBB specimens of Bohus granite. Points corresponding to F_{max} in failure curves are shown. Knee of fitted bilinear curve defines critical stress intensity resistance value K_{Jc}.

where

$$K_J = [EJ_R/(1-\nu^2)]^{1/2}, \tag{3.20}$$

as in equation (1.22). The agreement is excellent except that $K_J(a_{eff})$ has a sharper knee than $K_R(a_{eff})$, compare figures 3.18 and 3.20. The agreement is just as good for best linear unloading data on Bohus granite and for Ekeberg marble.

Let the knee of a bilinear curve fit to the K_J data define a critical value K_{Jc} in analogy with the metals test practise in figure 1.9 and let the inverse of equation (3.20),

$$J_{Rc} = (1-\nu^2)K_{Jc}^2/E, \tag{3.21}$$

define the critical J-integral value J_{Rc}.

Then the results from 10 specimens of Bohus granite yield

$$K_{Jc} = 2.03_{10}\pm 0.16 \text{ MN/m}^{3/2} \text{ or } J_{Rc} = 94.6\pm 12.2 \text{ J/m}^2, \tag{3.22}$$

and an average amount of sub-critical crack growth

$$\Delta a_c/D = 0.13_{10}\pm 0.03. \tag{3.23}$$

The less conclusive results from 5 specimens of Ekeberg marble yield

$$K_{Jc} \approx 2.1\text{-}2.3 \text{ MN/m}^{3/2} \text{ or } J_{Rc} \approx 50\text{-}60 \text{ J/m}^2, \tag{3.24}$$

and $\Delta a_c/D \approx 0.10$.

These J_{Rc} values are substantially higher than the G_{Ic} predictions of the simple R-curve approach in section 3.1 but fit well with the energy rate results of figures 3.4 and 3.5. Thus J_{Rc} above is an approximate lower limit to the $\overline{R}$ data and considerably higher than the J_m and G_m data as expected. K_{Jc} for Ekeberg marble is somewhat lower than the K_R data from the complete failure curves in section 3.2.1.

3.2.3 Conclusions. Results of two direct methods of obtaining complete R-curves from a single prenotched SECRBB specimen have been presented here, the first one based on complete failure curves and the second one on sub-critical failure cycles. These methods yield K_R-values that are relatively independent of the initial notch depth such that $K_R(a_{eff}-a_o)$ is a fair description especially in the sub-critical range but also in the post-critical one. In the latter range, the first method must employ the secant compliance however and it yields K_R-values that are somewhat higher.

These K_R-curves represent the crack extension resistance of notched virgin material and there appears to be a certain amount of geometry induced sub-critical crack growth present. The K_R-curve of a precracked specimen looks different but this doesn't influence the determination of a representative toughness value.

The complete failure curves do not yield an acceptable agreement between G_R, the energy rate version of K_R, and J_R but the sub-critical failure cycles do. Thus the latter method is clearly preferred.

3.3 Conclusions from Core Toughness Data

The crack-extension resistance data presented in sections 3.1 and 3.2 from SECRBB specimens of Ekeberg marble, Bohus granite, and Stripa granite are summarized in table 3.2 under the old lot headings. The table also contains some new data from CENRBB and SECRBB specimens of all three rocks. The aim is to point out promising quantities that could yield representative fracture toughness values from sub-size core specimens.

Stress intensity resistance	Ekberg marble		Bohus granite		Stripa granite	
	old lot	new lot	old lot	same lot	old lot	new lot
K_m eq.(1.9)	1.42_{13} ±0.06	1.62_{4} ±0.04	1.26_{6} ±0.06	1.29_{2}	1.83_{9} ±0.18	1.87_{4} ±0.09
K_{CN} eq.(2.29)		2.09_{16} ±0.14		1.62_{5} ±0.08		2.40_{3} ±0.08
K_m^{sec} eq.(3.25)	2.05_{13} ±0.10	2.22_{4} ±0.11	1.88_{6} ±0.16	1.95_{2}		2.32_{4} ±0.04
K_{Jm} eq.(1.19)	1.90_{47} ±0.10		1.71_{18} ±0.15			
K_{Rc} eq.(3.1)	1.88_{37} ±0.08		1.67_{15} ±0.11		2.32_{37} ±0.23	
K_{Jc} eq.(3.20)	2.1-2.3		2.03_{10} ±0.16			
$K_{\overline{R}}$ eq.(1.1)	2.57_{55} ±0.22		2.29_{22} ±0.18		2.81_{28} ±0.21	

Table 3.2: Comparison of stress intensity crack-extension resistance measures from core bend specimens. Defining equations of basic quantities given, K-conversion through an equation like (3.18). Values in $MN/m^{3/2}$ given as $\text{mean}_{\text{no tests}}$ ± standard deviation. K_m and K_m^{sec} values given for specimens in which $0.20 < a_o/D < 0.35$.

One new toughness measure has been added, namely a secant corrected K_m value from the single compliance method in section 3.2.1,

$$K_m^{sec} = K_I(a_c^{sec}, F_{max}), \text{ where } a_c^{sec} = g^{-1}[\lambda_{Fmax}/\lambda_{Fo} \cdot g(\alpha_o)] \qquad (3.25)$$

in analogy with equation (3.11) estimates the crack length at F_{max} via the secant slope $1/\lambda_{Fmax}$ through that point. K_m^{sec} is thus relatively easy to compute but requires load point displacement (LPD) measurements. Short Rod data K_{SR} for these rocks are presently lacking but will be available soon from the University of Luleå and from the Swedish Detonic Research Foundation (SveDeFo). The comparison of K_{CN} with K_{SR} may become complicated both by anisotropy and by the different amounts of assumed sub-critical crack growth however.

The final validity check on any measure that is based on sub-size specimens is a comparison with K_{Ic} or upper limit K_Q data from specimens that fulfil the LEFM size requirements. Such tests on these rocks are also underway or planned at the University of Luleå and at SveDeFo. With these reservations in mind it appears that K_{Jc} or lower limit $K_{\overline{R}}$ values would be the most representative ones. They require both force and LPD measurements however.

Of the quantities that require only force registration, K_m both yields erroneously low values and depends strongly on a_o/D. K_{CN} gives significantly higher results, but it is conceivable that something like a plasticity correction is needed in general to reach the K_{Jc} level. Note that the new lot of Ekeberg marble is tougher than the old one so that K_{CN} is comparable to K_{Rc} and K_{Jm} rather than to K_{Jc}, seen relative to K_m. The difference between the K_m values of the lots is within the limits of anisotropy previously found, Ouchterlony.[55]

K_{Rc} from the simple R-curve approach doesn't require LPD measurements either but relies on many specimens to get one value. Together with K_{CN} and K_{SR} it shares the basic assumption of a mainly flat K_R-curve, which the evaluation doesn't verify. K_{Rc} is practically indistinguishable from K_{CN} for all three rocks. Irregardless of whether plasticity corrections or a bilinear K_R-curve with increasing post-critical part are necessary to obtain values at the K_{Jc} level, LPD measurement become necessary.

The two remaining quantities in table 3.2, K_m^{sec} and K_{Jm}, both require LPD measurements. K_{Jm} is indistinguishable from K_{Rc} or K_{CN} and thus holds no advantage. K_m^{sec} gives somewhat higher results but it depends strongly on a_o/D and it is therefore ruled out.

As for K_{SR} data on other rocks the results are mixed. According to Costin [18] significant plasticity corrections are needed to obtain agreement with K_Q data for oil shale which has a relatively low modulus. For the somewhat stiffer Indiana limestone Ingraffea et al.[28] report a general agreement between uncorrected K_{SR} values and previous K_Q and K_{Ic} values. Barker's [56] K_{SR} data for Indiana limestone did however require such a correction. Beech and Ingraffea [57] convey data for Westerly granite, $K_{SR} = 2.41_3$ MN/m$^{3/2}$ with a very tight spread. It should be compared with $K_{Ic} = 2.61$ MN/M$^{3/2}$ from figure 1.10 in section taken from Schmidt and Lutz [12] who also give a practical upper limit of $K_Q = 2.7$ MN/m$^{3/2}$. The Westerly granite data closely resembles that of Stripa granite.

A preliminary conclusion then is that it seems very difficult and perhaps impossible to obtain representative fracture toughness values from sub-size core specimens without measuring LPD or some other displacement. This conclusion is not likely to change much from the coming Short Rod data but a final one can't be made until table 3.2 is supplemented by K_{Ic} data and more rocks have been investigated.

The CENRBB and the Short Rod specimens would still be useful, especially if K_{CN} and K_{SR} yield identical results such that core orientation is relatively unimportant or such that anisotropy can be measured from a single core. The different amounts of correction that seem necessary for various rocks will however tend to exaggerate the span of measured fracture toughness values. Despite the need for one, it would be too early to make any specimen or evaluation procedure a standard until such problems have been resolved.

4. REFERENCES

1. Ouchterlony, F., Review of fracture toughness testing of rock, Swedish Detonic Research Foundation (SveDeFo) report DS 1980:15, Stockholm, Sweden, 1980.

2. Ouchterlony, F., Review of fracture toughness testing of rock, *Solid Mechanics Archive,* 7, 131, 1982.

3. Tseng, A.A., A three-dimensional finite element analysis of the three-point bend specimen, *Engng Fract Mechs,* 13, 939, 1980.

4. Awaji, H. and Sato, S., Combined mode fracture toughness measurement by the disc test, *J Engng Mtrls Techn,* 100, 175, 1978.

5. Ingraffea, A.R., Mixed mode fracture initiation in Indiana limestone and Westerly granite, in *Proc 22nd US Symp Rock Mechs*, MIT, Cambridge MA, 1981, 186.

6. Friedman, M., Handin, J., and Alani, G., Fracture surface energy of rocks, *Int J Rock Mech Min Sci*, 9, 757, 1972.

7. Forootan-Rad, P. and Moavenzadeh, F., Crack initiation and propagation in rock, Research report R68-29, Dept Civ Engng, MIT, Cambridge MA, 1968.

8. Griffith, A.A., The phenomena of rupture and flow in solids, *Phil Trans Royal Soc London,* A221, 163, 1921.

9. Hoagland, R.G., Hahn, G.T., and Rosenfield, A.R., Influence of microstructure on fracture propagation in rock, *Rock Mechanics,* 5, 77, 1973.

10. Irwin, G.R., Analysis of stresses and strains near the end of a crack traversing a plate, *J Appl Mechs,* 24, 361, 1957.

11. Schmidt, R.A., A microcrack model and its significance to hydraulic fracturing and fracture toughness testing, in *Proc 21st US Symp Rock Mechs*, Univ Missouri, Rolla MO, 1980.

12. Schmidt, R.A. and Lutz, T.J., K_{Ic} and J_{Ic} of Westerly granite - effects of thickness and in-plane dimensions, in *ASTM STP 678,* Philadelphia PA, 1979, 166.

13. Ingraffea, A.R. and Schmidt, R.A., Experimental verification of a fracture mechanics model for tensile strength prediction of Indiana limestone, in *Proc 19th US Symp Rock Mechs,* Univ Nevada, Reno NV, 1978, 247.

14. Kim, K. and Mubeen, A., Relationship between differential stress intensity factor and crack growth in cyclic tension in Westerly granite, in *ASTM STP 745*, Philadelphia PA, 1981, 157.

15. Munz, D., Minimum specimen size for the application of linear-elastic fracture mechanics, in *ASTM STP 668*, Philadelphia PA, 1979, 406.

16. Barker, L.M., A simplified method for measuring plane strain fracture toughness, *Engng Fract Mechs*, 9, 361, 1977.

17. Ouchterlony, F., A new core specimen for the fracture toughness testing of rock, SveDeFo report DS 1980:17, Stockholm, Sweden, 1980.

18. Costin, L.S., Static and dynamic fracture behavior of oil shale, in *ASTM STP 745*, Philadelphia PA, 1981, 169.

19. Rice, J.R., Mathematic analysis in the mechanics of fracture, in *Fracture, an Advanced Treatise*, Liebowitz, H., Ed., Academic Press, New York, 1968, 191.

20. Rice, J.R., Paris, P.C., and Merkle, J.G., Some further results of J-integral analysis and estimates, in *ASTM STP 536*, Philadelphia PA, 1973, 231.

21. Finnie, I., Cooper, G.A., and Berlie, J., A comparison of fracture toughness measurements using Solnhofen limestone, Internal report 403:3, Centre Européen des Recherches Atlas Copco, Ecublens, Switzerland, 1976.

22. Ouchterlony, F., Compliance measurements on notched rock cores in bending, SveDeFo report DS 1980:2, Stockholm, Sweden, 1980.

23. Ouchterlony, F., Extension of the compliance and stress intensity formulas for the single edge crack round bar in bending, in *ASTM STP 745*, Philadelphia PA, 1981, 237.

24. Ouchterlony, F., The crack mouth opening displacement of the SECRBB specimen, SveDeFo report DS 1981:21, Stockholm, Sweden, 1981.

25. Ouchterlony, F., A simple R-curve approach to fracture toughness testing of rock with sub-size SECRBB specimens, SveDeFo report DS 1981:18, Stockholm, Sweden, 1981.

26. Ouchterlony, F., A simple R-curve approach to fracture toughness testing of rock core specimens, in *Proc 23rd US Symp Rock Mechs*, Univ California, Berkeley CA, 1982, 515.

27. Swan, G. and Olofsson, T., Towards a multi-purpose rock core testing facility, manuscript submitted to *Geotech Testing Journal*.

28. Ingraffea, A.R., Gunsallus, K.L., Beech, J.F., and Nelson, P., A fracture toughness testing system for prediction of tunnel boring machine performance, in *Proc 23rd US Symp Rock Mechs*, Univ California Berkeley CA, 1982, 463.

29. Bush, A.J., Experimentally determined stress-intensity factors for single-edge crack round bars loaded in bending, *Exp Mechs*, 16, 249, 1976.

30. Ouchterlony, F., Swan, G., and Sun, Z., A comparison of displacement measurement methods in the bending of notched rock cores. SveDeFo report manuscript, Stockholm, Sweden, 1982.

31. Srawley, J.E., On the relation of J_I to work done per unit area: "Total", or component "Due to crack", *Int J Fracture*, 12, 1976, 470.

32. Turner, C.E., Methods for post-yield fracture safety assessment, in *Post-Yield Fracture Mechanics*, Latzko, D.G.H., Ed., Applied Science Publishers, London, 1979, 23.

33. Munz, D., Bubsey, R.T., and Shannon Jr, J.L., Fracture toughness determination of Al_2O_3 using four-point-bend specimens with straight-through and chevron notches, *J Am Ceramic Soc*, 63, 30, 1980.

34. Munz, D., Bubsey, R.T., and Srawley, J.E., Compliance and stress intensity coefficients for short bar specimens with chevron notches, *Int J Fracture*, 16, 359, 1980.

35. Bubsey, R.T., Munz, D., Pierce, W.S., and Shannon Jr, J.L., Compliance calibration of the short rod chevron-notch specimen for fracture toughness testing of brittle materials, *Int J Fracture*, 18, 125, 1982.

36. Kobayashi, T., Fourney, W.L., and Holloway, D.C., Further examination of microprocess zone in pink Westerly granite, presented at Am Soc Ceramics meeting, Cincinnati OH, May 1979.

37. Swan, G., Some observations concerning the strength-size dependency of rock, Univ Luleå research report TULEA 1980:01, Luleå, Sweden, 1980.

38. Swan, G., Fracture stress scale effects, *Int J Rock Mechs Min Sci*, 17, 317, 1980.

39. Wijk, G., The uniaxial strenght of rock material, *Geotech Testing Journal*, 3, 115, 1980.

40. Swan, G., The mechanical properties of Stripa granite, Technical report LBL-7074 SAC-03, Lawrence Berkeley Laboratory, Berkeley CA, 1978.

41. Sun Zong-qi., Univ Luleå, Luleå, Sweden, 1981. Private communication.

42. Blackburn, W.S., Calculation of stress intensity factors for straight cracks in grooved and ungrooved shafts, *Engng Fract Mechs*, 8, 731, 1976.

43. Wanhill, R.J.H., Bartelds, G., and de Koning, A.U., Fatigue crack front shape, *Int J Fatigue*, 4, 52, 1982.

44. Rummel, F., and Winter, R.B., Basisdaten zur Rissausbretung bei Stimulationsversuchen in gering permeablen Sedimenten vornehmlich der Karbon-und Rotliegend Formationen, Teilprojekt im Rahmen des BMFT-DGMK-Forschungsvorhabens ET 3068 A, Dept Geophysics, Ruhr Univ, Bochum, Fed Rep Germany, 1981. In German.

45. Atkinson, B.K., Subcritical crack propagation in rocks; theory, experimental results and applications, *J Structural Geology*, 4, 41, 1982.

46. Kleinlein, W.F. and Hübner, H., The evaluation of crack resistance and crack velocity from controlled fracture experiments on ceramic bend specimens, in *Proc 4th Intnl Conf on Fracture*, ICF4, vol 3, 883 Taplin, D.M.R., Ed., Univ Waterloo Press, Ontario, Canada, 1977.

47. Swan, G. and Alm, O., Sub-critical crack growth in Stripa granite: direct observations, in *Proc 23rd US Symp Rock Mechs*, Univ California, Berkeley CA, 1982, 542.

48. Ouchterlony, F., The crack extension resistance of rock as measured on core specimens, presented at Swedish National Committee for Mechanics meeting, Luleå, Sweden, Oct 1980.

49. Clarke, G.A., Andrews, W.R., Paris, P.C., and Schmidt, D.W., Single specimen tests for J_{Ic} determination, in *ASTM STP 590*, Philadelphia PA, 1976, 27.

50. Budiansky, B. and Hutchinson, J.W., Analysis of closure in fatigue crack growth, *J Appl Mechs*, 45, 267, 1978.

51. Schwalbe, K.-H. and Setz, W., R-curve and fracture toughness of thin sheet materials, *J Testing Eval*, 9, 182, 1981.

52. McCabe, D.E. and Landes, J.D., An evaluation of elastic-plastic methods applied to crack growth resistance measurements, in *ASTM STP 668*, Philadelphia PA, 1979, 288.

53. Stillborg, B., Acoustic emission and fracture of rock cores as a function of load increase up to failure, in *Failure of Rock* - some

results from a post graduate workshop, Swan, G., Ed., Univ Luleå technical report 1980:65T, Luleå, Sweden, 1980, 15. In Swedish.

54. Olofsson, T. and Lindqvist, P.-A., SEM investigation of mode I crack propagation in Ekeberg marble, Ibid, 2. In Swedish.

55. Ouchterlony, F., Fracture toughness testing of rock cores 1. Three point bend tests on Ekeberg marble, SveDeFo report DS 1978:8, Stockholm, Sweden, 1978. In Swedish.

56. Barker, L.M., Process zone effects in fracture toughness tests of brittle non-metallic materials, presented at the ASTM Symposium on fracture Mechanics Methods for Ceramics, Rock; and Concrete, Chicago IL, June 23-24, 1980.

57. Beech, J.F. and Ingraffea, A.R., Three-dimensional finite element calibration of the short-rod specimen, *Int J Fracture*, 18, 217, 1982.

58. Barton, C.R., Variables in fracture energy and toughness testing of rock, in *Proc 23rd US Symp Rock Mechs*, Univ California, Berkeley CA, 1982, 449.

59. Ouchterlony, F., Anisotropy effects in the fracture toughness testing of rock, SveDeFo report manuscript, Stockholm, Sweden, 1982.

60. Hawkes, I., Mellor, M., and Gariepy, S., Deformation of rocks under uniaxial tension, *Int J Rock Mechs Min Sci*, 10, 493, 1973.

61. Douglass, P.M. and Voight, B., Anisotropy of granites: a reflection of microscopic fabric, *Géotechnique*, 19, 376, 1969.

62. Peng, S. and Johnson, A.M., Crack growth and faulting in cylindrical specimens of Chelmsford granite, *Int J Rock Mechs Min Sci*, 9, 37, 1972.

63. Cleary, M.P., Some deformation and fracture characteristics of oil shale, in *Proc 19th US Symp Rock Mechs* vol 2, Univ Nevada, Reno NV, 1978, 72.

64. Chong, K.P., Uenishi, K., Smith, J.W., and Munari, A.C., Non-linear three dimensional mechanical characterization of Colorado oil shale, *Int J Rock Mechs Min Sci*, 17, 339, 1980.

65. Henry, J.P., Pacquet, J., and Tancrez, J.P., Experimental characterization of crack propagation in calcite rocks, *Int J Rock Mechs Min Sci*, 14, 85, 1977.

66. Schmidt, R.A., Fracture toughness testing of limestone, *Exp Mechs*, 16, 161, 1976.

67. Schmidt, R.A., Fracture mechanics of oil shale - unconfined fracture toughness, stress corrosion cracking, and tension test results, in *Proc 18th US Symp Rock Mechs*, paper 2A2, Colorado School of Mines, Golden CO, 1977.

68. Hangen, J.A., A study of determination and fracture of oil shale, M Sc thesis, Dept Mech Engng, MIT, Cambridge MA, 1977.

69. Switchenko, P.M., The thermomechanical response of oil shale, M Sc thesis, Dept Mech Engng, MIT, Cambridge MA, 1979.

70. Costin, L.S., A failure surface model for oil shale, report SAND81-1128, Sandia National Laboratories, Albuquerque NM, 1981.

71. Sih, G.C. and Liebowitz, H., Mathematical theories of brittle fracture, in *Fracture, an Advanced Treatise*, Liebowitz, H., Ed., Academic Press, New York, 1968, 67.

72. Bowie, O.L. and Freese, C.E., central crack in plane orthotropic rectangular sheet, *Int J Fract Mechs*, 8, 49, 1972.

73. Delale, F. and Erdogan, F., The problem of internal and edge cracks in an orthotropic strip, *J Appl Mechs*, 44, 237, 1977.

74. Miller, B.L. and Cleary, M.P., Finite element analysis of fracture geometries for toughness testing in rock, Report RELR-80-1, Dept Mech Engng, MIT, Cambridge MA, 1980.

75. Merkle, J.G. and Corten, H.C., A J-integral analysis for the compact specimen considering axial force as well as bending effects, *J Press Vess Techn*, 4, 286, 1974.

76. Wilkening, W.W., J-integral measurements in geological materials, in *Proc 19th US Symp Rock Mechs*, vol 1, Univ Nevada, Reno NV, 1978, 254.

NUMERICAL MODELLING OF FRACTURE PROPAGATION

Anthony R. Ingraffea
Associate Professor and Manager of Experimental Research
Hollister Hall
Cornell University
Ithaca, New York 14853

1. *Introduction*

Why should one study fracture *propagation?* Is not prediction of fracture initiation the object of fracture mechanics? In his papers on rupture under tensile and compressive loading, Griffith[1,2] proposed conditions for fracture initiation which he presumed to be coincident with structural instability. The vast majority of the fracture mechanics research since Griffith has addressed the problem of predicting structural failure as the immediate consequence of fracture initiation. Yes, considerable attention has been focused on sub-critical crack growth as in fatigue and ductile fracture. But there the amount of propagation before fracture initiation is typically small compared to that which potentially occurs after. Why, then, should one be interested in modelling propagation: where a crack goes, what it does along the way, and how much energy it takes to get there?

From the rock mechanics point-of-view, there are many answers to this question. By looking at some of these, we will begin to acquire a perspective on fracture quite different from that held by, for example, an aerospace engineer. First, he designs to minimize the chance of fracture initiation. In many cases, in rock mechanics, the objective is to generate propagation, and much of it. The sketches in Fig.1 illustrate two important rock fracture problems of extremely different scale. In both, however, the object is to propagate, not inhibit, fractures. The hot, dry rock geothermal energy extraction scheme depicted in Fig.1a demands the formation of surface area by hydrofracturing. In rock comminution problems, such as the tunnel boring machine roller cutter mechanism seen in Fig.1b, one must not only initiate fracture but also drive these to free surfaces, hopefully with a minimal amount of energy expended. As will be shown later, there is nothing in the rules of rock mechanics (or fracture mechanics, either) which says that a fracture, once initiated, is always unstable. It may stop. Where? Why? What was its trajectory? What must be done to get it going again? And, notice, we said fractures: the aerospace engineer generally only worries about one crack at a time, while in rock mechanics study of the propagation of

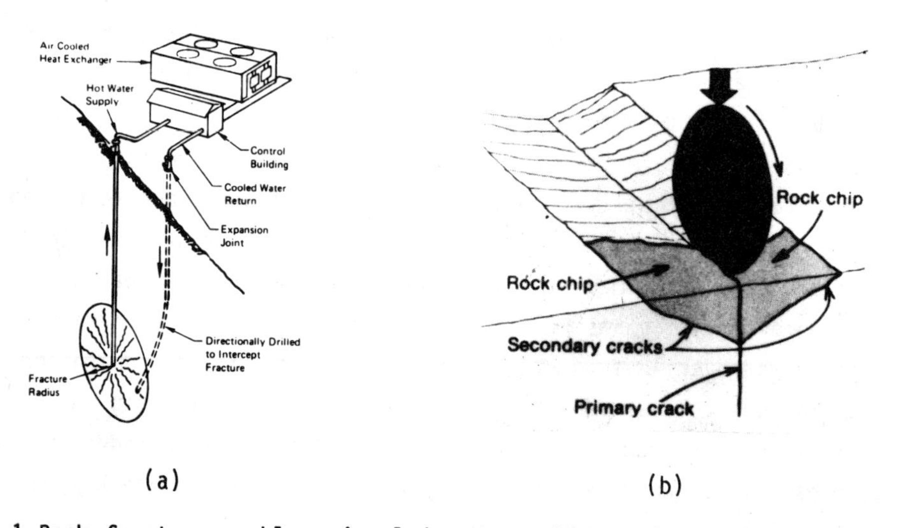

Fig.1 Rock fracture problems involving controlled propagation.
a) schematic of a hot-dry rock geothermal energy recovery method;
b) idealization of comminution beneath tunnel boring machine roller cutter

a single crack is less often the case. So, already we have:

Observation #1: Fracture propagation can be stable in the load control sense. Assuming linear elastic fracture mechanics (LEFM) conditions (as we will in this chapter, unless otherwise noted), this implies that stress-intensity is decreasing with increasing crack length. As we will see, this situation is generally due to the preponderance of compressive in situ stresses and loadings in geotechnical engineering. Obviously, it will not be sufficient just to compute stress-intensity factors for an initial crack configuration. A good numerical model should be able to update stress-intensity factors as crack length changes.

Observation #2: Propagation of multiple cracks is common in realistic problems of rock mechanics. A numerical model should be versatile enough to accommodate propagation of many cracks.

This latter observation leads to another reason for modelling fracture propagation in rock. With each growth increment of a given crack, a new boundary value problem is generated. Displacement and traction boundary conditions may change, stress trajectories are altered, even loading may change in direction and intensity. As a consequence, propagation of one crack may cause initiation of another, and cracks may influence each other's stability and trajectory. This is clearly the case in dynamic fracture because of stress wave reflections. (This problem, along with other propagation phenomena particular to dynamic fracture, is fully treated in another chapter. Only quasi-static problems are studied here.) But without a periodic "look" at the full stress field during quasi-static propagation, one might overlook:

Observation #3: Each increment of fracture changes the structure. One should be able to predict the effects of this change on the stress field and on other cracks.

Now that we have admitted the reality of multiple cracks propagating quasi-statically we must dispense with another simplification upon which the aerospace engineer usually relies. Curvilinear (mixed-mode, to some people) crack propagation is common in rock mechanics. Therefore, we must admit:

Observation #4: Cracks curve during propagation in response to a changing stress field. A numerical model should be able to predict a changing crack trajectory.

Moreover, if we accept that mixed-mode stress intensity factors, for example K_I and K_{II}, can be present along a crack front, then we have:

Observation #5: Theoretically, mixed-mode fracture initiation can occur when $K_I \leq K_{Ic}$. Consequently, a numerical model must incorporate an interaction theory which accurately predicts the critical mixture of stress intensity factors which will cause the next increment of propagation.

Of course, fracture can be foe as well as friend to the geotechnical engineer. Mine, tunnel, dam abutment, and rock foundation instability problems are often the result of unpredicted fracture propagation. A numerical model which can predict the likelihood of a roof fall or rock burst, and can suggest a method of fracture stabilization or an alternate form of energy release can be an invaluable design tool.

Clearly, then, the problems of crack propagation modelling, even with the simplifying assumptions of LEFM, are manifold. The problem *begins* with fracture initiation, so we have to go a bit further than classical LEFM takes us. This is the purpose of the present chapter. Within the context of modern techniques of stress analysis, the finite and boundary element methods, we will study:

1. Methods for efficient, accurate calculation of stress intensity factors for substitution into,
2. Mixed-mode fracture initiation theories for critical mixture and angle change predictions.
3. Methods for crack increment length prediction for a given load change, or, conversely, the prediction of the load required to drive a crack a given distance.
4. Algorithms for incorporating the above into efficient computer programs.
5. Techniques for efficient remeshing to accommodate discrete crack growth.

6. The use of interactive computer graphics in the highly adaptive, nonlinear field of fracture propagation modelling.

This will be accompanied with example problems whenever possible. The first of these will be presented in the next section to provide a physical basis for the observations particular to rock fracture just presented.

2. *The Nature of Fracture Propagation in Rock*

Example 1: Observations on Fracture Propagation Under Compression

The following example problem will serve a number of purposes. As mentioned above it will lend physical insight into characteristics particular to fracture propagation in rock. It will clarify some misconceptions and their implications regarding the theoretical fracture resistance of rock structures. Finally, it will serve as a basis for development and comparison on the techniques required for modelling fracture propagation.

The problem is shown in Fig.2 and is recognized to be that addressed by Griffith in his second paper.[2] Rock plates like those shown in Fig.2 were tested by the author[3,4,5] with the following results:

1. As predicted by Griffith,[2] first crack growth occurs from points initially under tensile stress concentration on the notch (see Fig.3a). This set of two, symmetrically placed cracks was labelled *primary*.
2. Primary crack trajectory was curvilinear.
3. In contrast to what Griffith[2] expected, propagation of primary cracks was observed to be stable: an ever increasing load was required to increase crack length.
4. After considerable primary crack propagation, a second set of two, symmetrically placed, cracks appeared. These were labelled *secondary*, and originated in the interior of the plate in a newly formed tensile stress zone (see Fig.3b).
5. Failure of the plate, defined as a through-going rupture, was a

Fig.2 Angle-notched plate loaded in uniaxial comparison

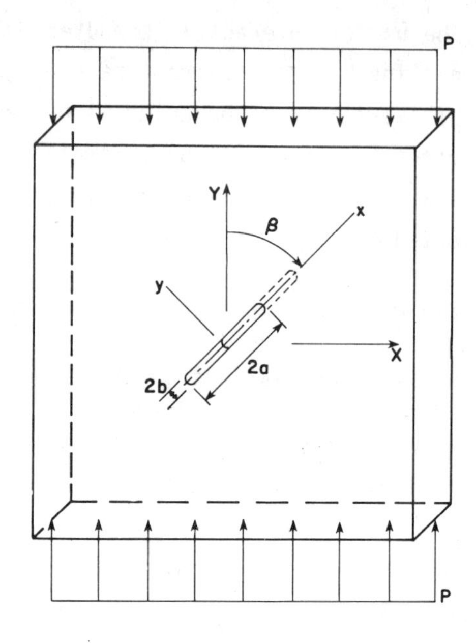

Fig.3a Results of test on Indiana limestone plate. Primary cracks to points A. Secondary cracks nucleated near points B. From Reference 3.

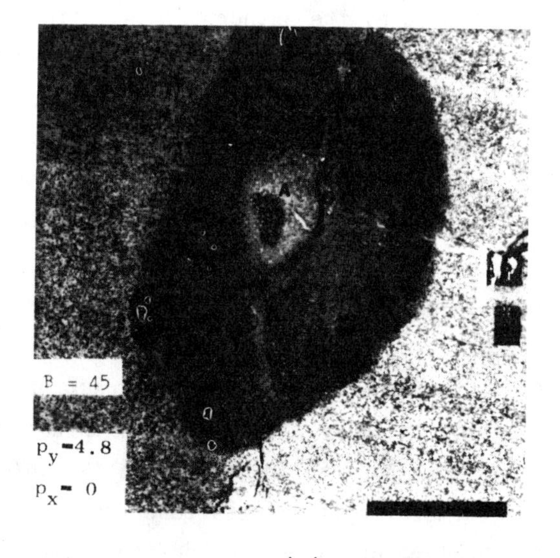

(a)

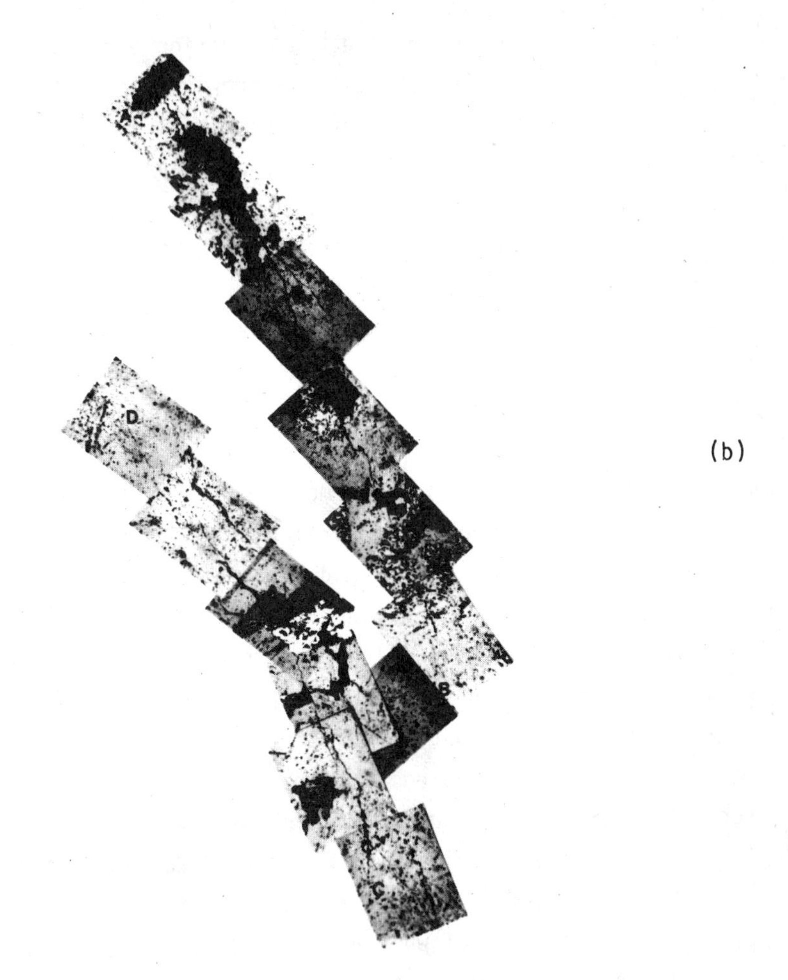

Fig.3b Composite micrograph of granodiorite fractures. Notch tip, A; primary crack tip, B; nucleating secondary crack, C to D. From Reference 3.

result of unstable secondary crack propagation, at a load level in the range of 3 to 5 times the primary crack initiation load. These observations were typical for plates of Indiana limestone and St. Cloud charcoal granodiorite, with $30° \leq \beta \leq 90°$.

Observations 2 and 3 differentiate the observed fracture response of this configuration from that usually observed in tension-loaded structures. Stable primary crack propagation indicates that, within the assumption of LEFM, the associated energy release rate, G, decreases with increasing crack length for a constant load. The curvilinear nature of the primary crack path is a result of a variable, mixed-mode stress intensity being applied to the incrementally advancing crack tip.

Items 3 through 5, above, deserve special attention. The often quoted, and quite incorrect, theoretical ratio of compressive to tensile strength of rock is based on the supposition that the *initiation* of what are here called primary cracks is synonymous with rupture. That such is not the case has been observed by many experimentalists (e.g. 6, 7, 8) and digested by few of anybody else: sub-critical crack growth can occur under ideal LEFM conditions and monotonically increasing load.

Observations 4 and 5 are particular to rock. In tests on glass, polymethylmethacrilate (PMMA) and CR39 in the same configuration,[6,7,8] only primary cracking was evident and rupture did not occur. This phenomenon is shown in Fig.4 which depicts the primary crack behavior of a PMMA plate loaded to near its compressive yield stress. Results 4 and 5 therefore indicate a fundamental difference in the fracture response of rock structures as compared to glass, plastic, and metals. As we shall see later, the high (though not theoretically predictable!) compressive to tensile strength ratio of rock compared to those materials leads to what the author has called the strength ratio effect.[9] This effect explains the initiation of what are called secondary cracks in the present problem, and is the proximate cause of rupture in this as well as many other problems in rock fracture. The strength ratio effect is actually a corollary to observation #3 mentioned in the Introduction, but bears individual emphasis:

> Observation #6: Due to fracture propagation, new regions of tensile stress can be generated. Although the magnitude of those tensile stresses may be low compared to an applied compressive stress, the relatively low tensile strength of rock makes such regions potential sites for nucleation of additional cracks. A model for fracture propagation should be capable of predicting formation of such sites.

Fig.4 Results of test on PMMA plate. Primary cracking only; no rupture occurs. From Reference 3.

With some experimental observations behind us, and with a list of requirements for modelling of fracture propagation in hand we are now ready to proceed toward model formulation.

3. *Stress Intensity Factor Computation*

The prediction of load level, angle change, and length corresponding to each increment of fracture propagation requires accurate computation of mixed-mode stress intensity factors. Their efficient computation is also desirable since, as noted above, many analyses may need to be performed in a single problem. This section presents background and formulation for a technique which is accurate and efficient. Moreover, it will be shown to be comparatively trivial algorithmically, and amenable to both finite and boundary element analyses.

3.1. *An Historical Overview*

Three methods are generally available for computing stress intensity factors using the Finite Element Method. These techniques can be labeled:

1. Global energy release method
2. Hybrid-direct method
3. Displacement or stress correlation method.

In the global energy release method, the total potential energy of the problem is computed before and after a small increment of length is given to a crack. Since it has been shown that the rate of change of potential energy can be related to the stress intensity factors, these can theoretically be computed at the tip of the original crack by using very small increments. There are, however, numerous drawbacks to this technique. First, at least two computer runs are required to compute the stress intensity factor for pure Mode I loading. Second, for mixed-mode loading the total energy release is proportional to a function of all stress intensity factors. Although improved techniques are available,[10] and decomposition of the total energy change into component parts for separating, for example, K_I from K_{II}, is possible, such decomposition involves at least four computer runs for a single computation.

The hybrid-direct method takes its name from the fact that it computes stress intensity factors directly by making them nodal variables, along with displacements, in a hybrid finite element method technique.[11] Although this is perhaps the most accurate and efficient of the three methods, its main drawback is that it involves a special element stiffness formulation and is not available in most general purpose finite element method programs.

The correlation technique is the most versatile and popular and will be reviewed in more detail. In the displacement version of this method, displacements computed at nodes near the crack tip are correlated with the theoretical values. For example, under pure Mode I, near tip displacements relative to the crack tip are,

$$u = \frac{K_I}{4G}\sqrt{\frac{r}{2\pi}}\,[(2\kappa-1)\cos\frac{\theta}{2} - \cos\frac{3\theta}{2}] + \ldots$$

$$v = \frac{K_I}{4G}\sqrt{\frac{r}{2\pi}}\,[(2\kappa+1)\sin\frac{\theta}{2} - \sin\frac{3\theta}{2}] + \ldots \qquad (1)$$

where,

u = displacement parallel to crack axis

v = displacement normal to crack axis

ν = Poisson's ratio

G = shear modulus

κ = $(3-4\nu)$ for plane strain

κ = $(3-\nu)/(1+\nu)$ for plane stress

Taking θ = 180°, and $r = r_{AB}$, in Fig.5, the K_I computed by the finite element method is expressed by,

$$K_I = \sqrt{\frac{2\pi}{r_{AB}}}\,\frac{2G}{(\kappa+1)}\,(v_B - v_A) \qquad (2)$$

For pure Mode II and plane stress,

$$u = \frac{K_{II}}{4G}\sqrt{\frac{r}{2\pi}}\,[(2\kappa+3)\sin\frac{\theta}{2} + \sin\frac{3\theta}{2}] + \ldots$$

$$v = \frac{-K_{II}}{4G}\sqrt{\frac{r}{2\pi}}\,[(2\kappa-3)\cos\frac{\theta}{2} + \cos\frac{3\theta}{2}] + \ldots \qquad (3)$$

And an expression for K_{II} is, for θ = 180°, $r = r_{AB}$,

$$K_{II} = \sqrt{\frac{2\pi}{r_{AB}}}\,\frac{2G}{(\kappa+1)}\,(u_B - u_A) \qquad (4)$$

For mixed-mode loading, the near-tip displacements are given by linear combinations of equ.(1) and equ.(3). However, it can be seen that the combined expressions for u and v uncouple at θ = ± 180°. Consequently, for cracks in symmetrical structures, eqs.(2) and (4) are applicable for mixed-mode loading. In the general, asymmetrical case, the stress intensity factors must be computed from crack-opening and crack-sliding

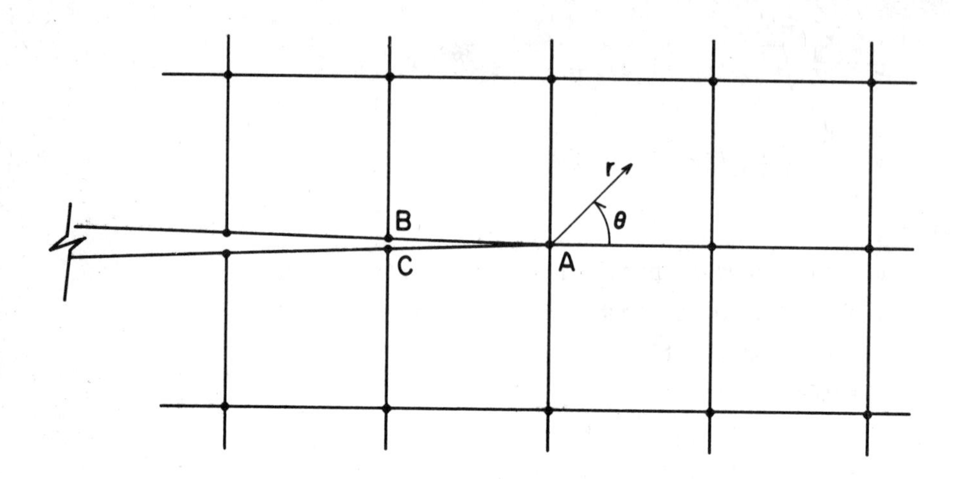

Fig.5 Nodal lettering for stress intensity computation with eqs. 2 and 4

displacements, COD and CSD, respectively, where, for $r = r_{AB} = r_{AC}$,

$$\begin{aligned} COD &= v_B - v_C \\ CSD &= u_B - u_C \end{aligned} \tag{5}$$

Substituting eqs.(1) and (3) into equ.(5), and solving for K_I and K_{II} yields,

$$\begin{aligned} K_I &= \sqrt{\frac{2\pi}{r_{AB}}}\,\frac{G}{(\kappa+1)}\,(v_B - v_C) \\ K_{II} &= \sqrt{\frac{2\pi}{r_{AB}}}\,\frac{G}{(\kappa+1)}\,(u_B - u_C) \end{aligned} \tag{6}$$

Eqs.(6) are the most general form for computing stress intensity factors using the displacement correlation technique.

Obviously, the success of this technique depends on an accurate modeling of the $r^{1/2}$ displacement variation near the crack tip. Various workers have applied a number of finite element method techniques to this end. Earliest efforts employed brute force. Constant strain triangles

(CST) were used with extremely fine meshes near the crack tip. The obvious major drawback to use of the CST in fracture problems is the very large number of degrees of freedom required for engineering accuracy even for simple geometry, loading, and a single crack. Reference 12 shows typical, early use of CST's and the displacement correlation technique.

The next step in finite element method development for the displacement correlation technique was to take a hint from the elasticity solution for the crack tip displacements and develop an element containing an $r^{1/2}$ interpolation function. In practice, this has been done by a number of workers using elements of various shapes. Wilson[13] appears to have produced the first so-called singularity element in 1969. He used a circular element centered at the crack tip and bordered by CST's.

Tracy[14] embedded the elastic singularity in the shape functions of isoparametric four-node quadrilaterals which were degenerated into triangles focused on the crack tip. In contrast to Wilson's element,[13] Tracey's singularity elements were fully displacement compatible. This element produced engineering accuracy with an order of magnitude fewer degrees-of-freedom than were previously required.

To this point in their development, singularity elements were distinct from the remainder of the elements in a mesh, necessitating separate element stiffness formulation routines. They also suffered from one or more of the following drawbacks:

1. Some edge displacement incompatibility existed between singular and non-singular elements in a mesh.

2. Special intermediate elements were necessary to provide complete compatibility.

3. The constant strain mode, necessary for problems involving thermal loading, was absent.

A now popular singularity element, discovered accidentally, overcame all these problems. In studying the effect of geometrical distortion on the isoparametric eight-node quadrilateral, Jordan[15] discovered that placing the mid-side nodes on or outside the one-fourth point of the sides caused the Jacobian of transformation to become non-positive

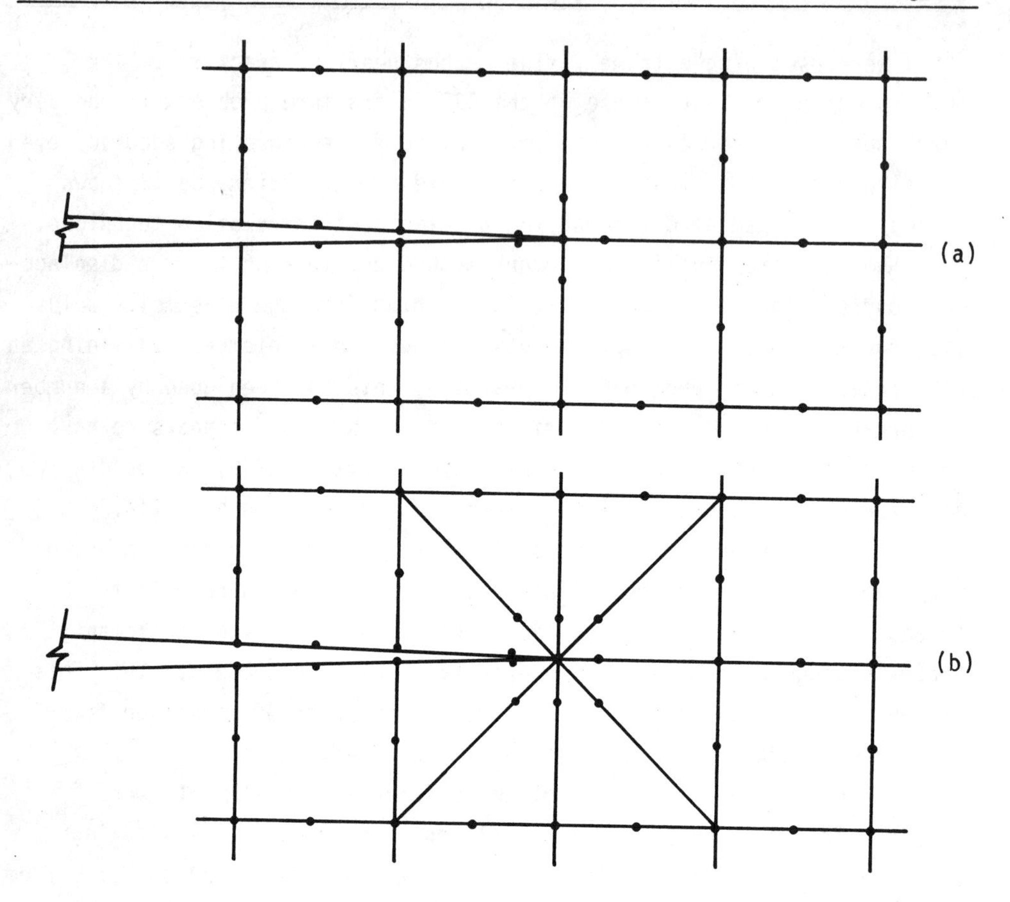

Fig.6 a) Quarter-point singular quadrilateral elements, b) quarter-point singular triangular elements.

definite. In fact, if the mid-side nodes are arranged as shown in Fig.6a, the Jacobian becomes singular at the crack tip. Barsoum[16] claims to be the first to have examined the applicability of this situation to fracture mechanics, and to show that this singularity is of the desired $r^{-1/2}$ form in stress.

Since the quadratic displacement interpolation functions are not in any way altered, use of this offset-node geometry does not change the element's convergence characteristics. As shown by Barsoum,[16] the element still satisfies all essential convergence criteria, namely: inter-element displacement compatibility and intra-element displacement continuity, rigid body modes, and constant strain modes. Moreover, the element stiffness matrix is formulated by the same subroutine as usual; only the nodal coordinate input data are altered.

Barsoum[17] has also shown that if the 8-node isoparametric quadralateral is first degenerated to an isoparametric triangle with 3 nodes at the crack tip and then distorted by moving the side nodes, Fig.6b, improved results are obtained. In fact, one can start with the isoparametric linear strain triangle (LST), move its midside nodes to the quarter-point position, and achieve the same effect, with some additional benefits.[18]

3.2. *Computation by Finite Element Method*

It is the author's opinion that stress intensity factor computation for fracture propagation problems, repetitive and mixed-mode, currently is best done using quarter-point singular elements. The advantages of the element itself are mentioned above. Combination of this singular element with the extraction technique outlined below makes programming for stress intensity factor computation almost trivial compared to the efforts of less than a decade ago.

As mentioned above, a number of methods for stress intensity factor extraction are available to the finite element analyst. The method detailed here, and used by the author in all his programs, is a displacement correlation technique, and its particular application to quarter-point singular elements was first proposed by Shih et al[19] for Mode I problems. The generalization to mixed-mode is straightforward[3] and goes as follows. First, let a crack tip be surrounded by quarter-point singular elements and the crack face nodes be lettered as shown in Fig.7. The only meshing constraint at this point is that the lengths of the two elements containing the lettered nodes be the same, i.e. $\overline{AC} = \overline{AE}$. Next, expand the lettered nodal displacements in terms of the element shape

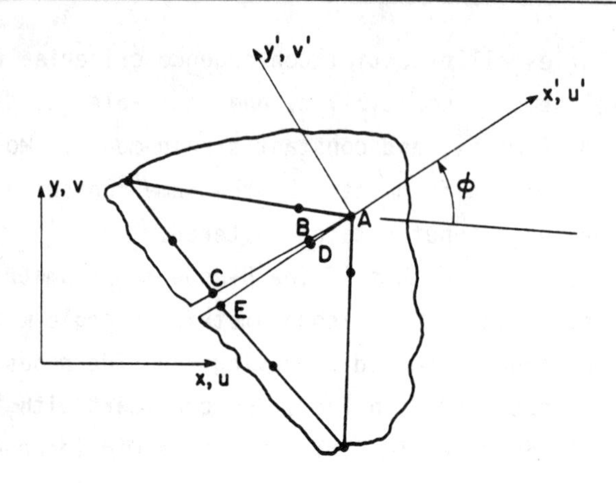

Fig.7 Nodal lettering for stress intensity factor computation using equ. 8.

functions and with respect to the crack tip coordinate system shown in Fig.7. This leads to expressions of the following form:

$$\begin{aligned} v' &= v'_A + (-3v'_A + 4v'_B - v'_C)\sqrt{r/L} + (2v'_A - 4v'_B + 2v'_C)r/L \\ u' &= u'_A + (-3u'_A + 4u'_B - u'_C)\sqrt{r/L} + (2u'_A - 4u'_B + 2u'_C)r/L \end{aligned} \tag{7}$$

Similar expressions are obtained along the $\overline{ADE}$ ray. Taking the difference between the expressions along the two rays yields a computed crack opening or crack sliding profile.

Now the displacement expansions from the asymptotic analytical solution, equ.(1), are evaluated at $\theta = \pm 180°$ to compute a theoretical crack opening or crack sliding profile. Equating like powers of r in the computed and theoretical profiles leads directly to the simple expressions,

$$\begin{aligned} K_I &= \sqrt{\frac{2\pi}{L}}\frac{G}{\kappa+1}\left[4(v'_B - v'_D) + v'_E - v'_C\right] \\ K_{II} &= \sqrt{\frac{2\pi}{L}}\frac{G}{\kappa+1}\left[4(u'_B - u'_D) + u'_E - u'_C\right] \end{aligned} \tag{8}$$

in which,

L = length of singularity element side along the ray,

v' = crack-opening nodal displacements,

u' = crack-sliding nodal displacements.

The primes indicate that the global coordinate nodal displacements have been transformed to the crack-tip coordinate system defined in Fig.7. The above procedure was generalized to the three-dimensional case by Ingraffea and Manu.[20]

Algorithmically, the displacements and coordinates of the crack face nodes belonging to the quarter-point elements need to be flagged and retrieved for each crack increment solution. These are then transferred to a simple subroutine which codes eqs.8. With the efficiency of the method obvious, its accuracy will be discussed with reference to the meshes shown in Fig.8. These are for the ASTM standard three-point bend specimen.

The accuracy of eqs.8 depends on a number of mesh characteristics. For example, Fig.9 shows two- and three-dimensional finite element results compared to the ASTM solution, known to be accurate within 1 percent, with L/a as a parameter. The differences in the two-dimensional results vary from -8 percent at L/a = 0.20 to -1 percent at L/a = 0.03. The three-dimensional results bracket the ASTM value, and centerline values are within ± 5 percent difference over the entire L/a range tested. The conclusion here is that there is an optimum L/a ratio; however, it is problem and mesh dependent. Moreover, in the three-dimensional case, optimum L/a ratio also depends on Poisson's ratio, as shown in Fig.10. From the author's experience on a large number of two- and three-dimensional analyses, it is safe to say that, other conditions being met as discussed below, engineering accuracy is assured with L/a ratios in the 0.05 to 0.15 range.

A second mesh characteristic which can strongly influence accuracy is the number of singular elements positioned around the crack tip. More specifically, the maximum circumferential angle subtended by any of these elements must be limited to less than about 60 degrees, with 45 degrees

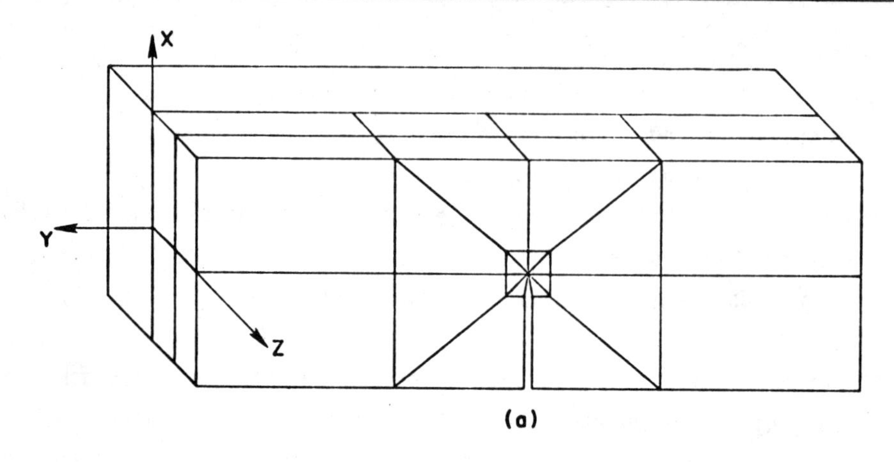

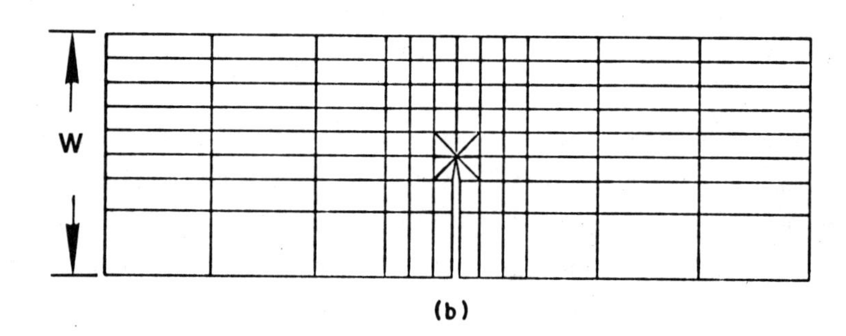

Fig.8 Meshes for analysis of three-point bend specimen. a) Three-dimensional case, 20 elements, 139 nodes. b) Two-dimensional case, 50 elements, 175 nodes.

seeming to be an optimal choice. The reason for this is that the theoretical circumferential variation in displacement is trigonometric and has a number of inflection points; the quarter-point elements are piecewise quadratic in their approximation to this variation. Take the three-point bend problem of Fig.6b and the results shown in Fig.9, for example. If the number of singular elements around the crack tip were reduced from 8 to 6 (4 to 3 using the symmetry of the problem), the error for any given value of L/a would double, approximately. Although the result might still be acceptable for small L/a, one would, of course,

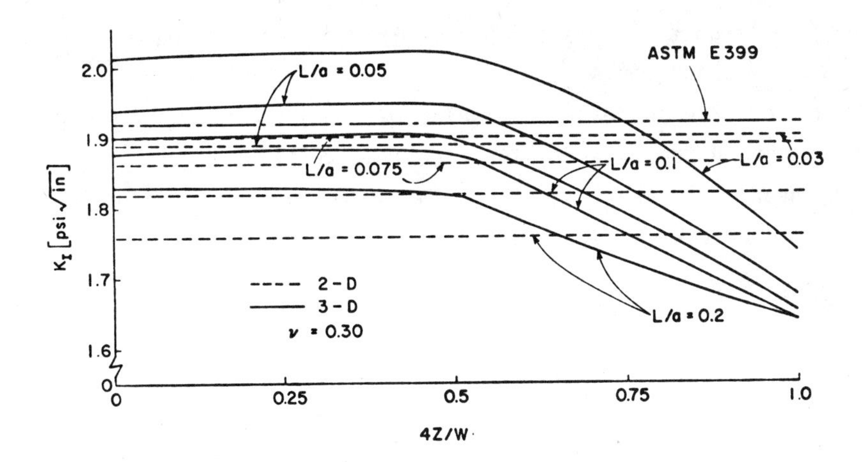

Fig.9 Stress intensity factor results from meshes of Fig.8. (1 psi $\sqrt{in}$ = 1.1 kPa $\sqrt{m}$) From Reference 20.

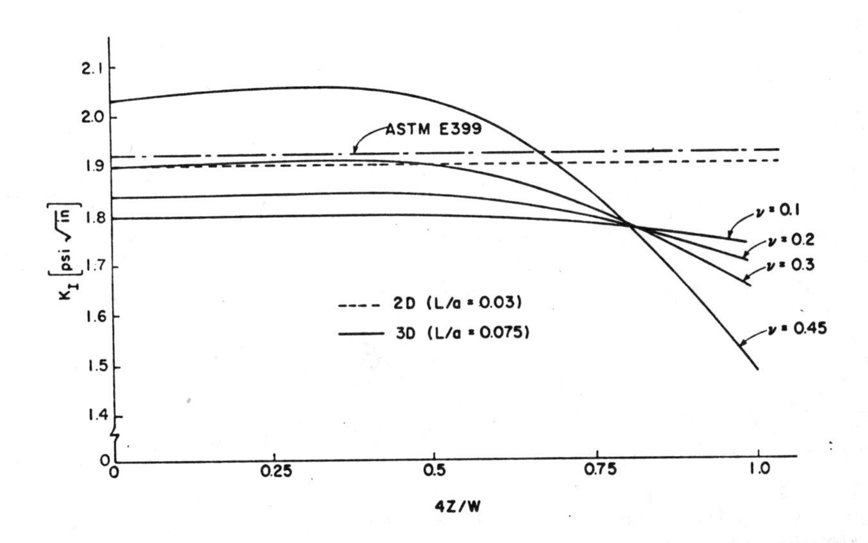

Fig.10 Effect of Poisson's ratio in three-dimensional computations with fixed L/a. (1 psi $\sqrt{in}$ = 1.1 kPa $\sqrt{m}$) From Reference 20.

rather use the largest L/a practicable so that the mesh is minimally disturbed by the crack tip.

With proper care in near-tip meshing with quarter-point modified elements, eqs.8 and their three-dimensional generalizations can easily produce engineering accuracy in stress intensity factor calculations. No special program or element is required. How convenient!

3.3. *Computation by Boundary Element Method*

The boundary element method is rapidly becoming a viable alternative to the finite element method in some classes of continuum mechanics problems. Because of its inherent reduction in problem dimensionality, it has particular advantages in crack propagation analyses, especially in three-dimensional models. It will be shown here that, with one additional modification, the method just outlined for stress intensity factor calculation using finite elements is also applicable to boundary elements analyses.

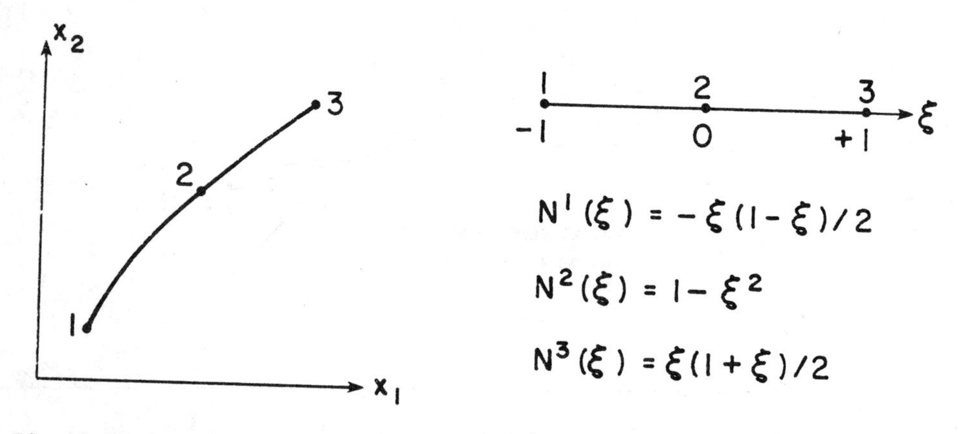

Fig.11 Quadratic, isoparametric boundary element and shape functions. From Reference 22.

Consider the quadratic, isoparametric boundary element shown in Fig.11. Its shape functions are written in a normalized natural coordinate, like the isoparametric Q8 and LST finite elements, and they are of the same algebraic form. A key difference between the boundary and finite element formulations, however, is the following. For finite elements, only the displacement field is directly approximated by the chosen shape functions. The stress field and unknown boundary tractions are obtained from a subsidiary calculation based on the relevant strain-displacement and constitutive relations. In boundary element theory, however, both the unknown boundary displacements and the tractions are directly, and independently, approximated by the chosen shape functions. This leads to the following interim result.

It has been shown above that, for finite elements the $\sqrt{r}$ displacement variation at the crack tip can be represented using isoparametric quadratic elements by placing the midpoint node at the quarter-point. Using the shape function definitions for the boundary element shown in Fig.11, the variation of the displacements and tractions would become:

$$\left.\begin{matrix} u_i \\ t_i \end{matrix}\right\} = A_i^1 + A_i^2 \sqrt{r} + A_i^3 r \qquad (9)$$

where

u_i = displacements normal and tangential to the boundary
t_i = tractions normal and tangential to the boundary
A_i = coefficients which are functions of nodal values of u_i or t_i

In finite element applications, the tractions are obtained by differentiating the displacement interpolation functions, which results in the correct $1/\sqrt{r}$ singularity in the traction variations being obtained. However, in the boundary element method, this is not the case since the displacement and traction variations are independently represented in equ.(9).

The displacement variation given by equ.(9) is analytically correct for the first terms of the theoretical series expansion about the crack tip. However, the traction variation does not possess the correct $1/\sqrt{r}$ singularity. The inclusion of this singularity is accomplished by multiplying the right hand side of equ.(9), traction, by $\sqrt{\ell/r}$ so that:[21]

$$t_i = (A_i^1 + A_i^2 \sqrt{r} + A_i^3 r)\sqrt{\ell/r}$$
$$= B_i^1/\sqrt{r} + B_i^2 + B_i^3 \sqrt{r} \qquad (10)$$

This corresponds to the correct traction variations in the vicinity of the crack tip. The expression $\sqrt{\ell/r}$ in equ.(10) is written in the non-dimensionalized coordinate system as shown in Fig.12.

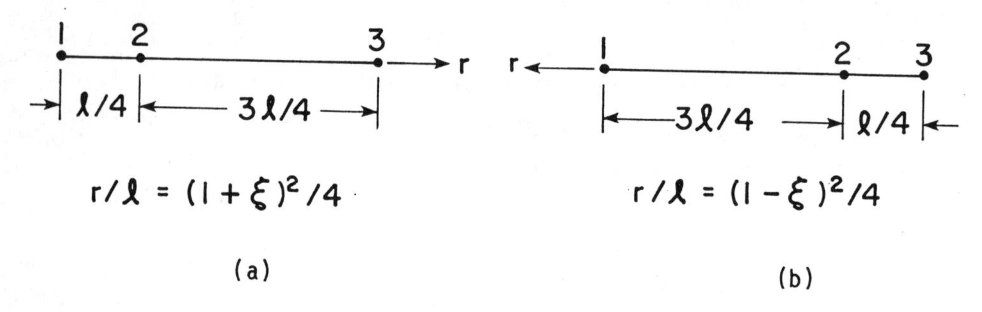

Fig.12 Traction singular quarter-point boundary elements. a) Crack tip at node 1; b) crack tip at node 3. Reference 22.

The displacement and traction variations for the elements adjoining the crack tip are given by eqs.(9) and (10). The resulting boundary elements are "traction singular quarter-point" boundary elements.[22]

Obviously, the displacement correlation technique described above is equally applicable for the case of traction singular quarter-point boundary elements.[22] The problem of Fig.13 is offered as an example of accuracy. Shown is a boundary element mesh of one-half of a beam in symmetric four-point bending.

The problem was analyzed using a fixed number of degrees of freedom and various ℓ/a ratios. The two end elements along the crack were fixed in length to be 0.50a. The results of the analysis are given in Table 1. These results are compared with the results of Brown and Srawley[23] which are accurate to within ± 1% for the range $0 < a/W \leq 0.60$. The results given in Table 1 reveal that the boundary element method yields accurate stress intensity factors for the range $0.10 \leq \ell/a \leq 0.20$ as contrasted

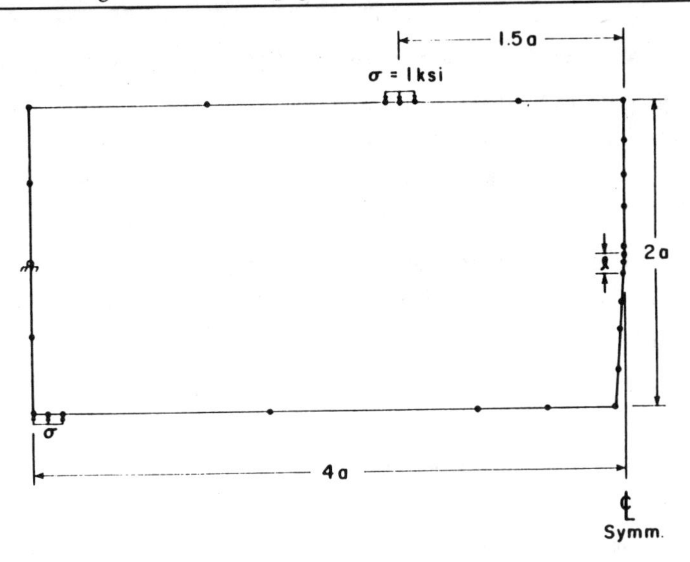

Fig.13 Boundary element mesh for four-point bend problem (a = 1 inch). (1 inch = 25.4 mm, 1 ksi = 6.9 MPa) From Reference 22.

Table 1

Boundary Element Solutions for the Four-Point Bend Crack Problem, Figure 13 (1 ksi $\sqrt{in}$ = 1.1 MPa $\sqrt{m}$)

ℓ/a	Result (ksi $\sqrt{in}$)	% Diff.
0.05	3.883	1.70
0.10	3.870	1.36
0.15	3.816	-0.05
0.20	3.759	-1.55
0.25	3.705	-2.96
Brown and Srawley,[23] 1966		3.818

with finite element results for bending problems in which smaller ℓ/a ratios were required for similar accuracy.

It is interesting to note that the reduced dimensionality afforded by the boundary element method precludes the problem, mentioned above with reference to the finite element method, of placing a sufficient number of singular elements around the crack tip. In fact, an accuracy advantage of the boundary element method derives from its ability to represent the interior solution, here the circumferential stress and displacement variations around the crack tip, exactly, if the boundary solution is exactly represented. Since traction singular quarter-point elements are being used, this condition is fulfilled close to the crack tip.

Many more example stress intensity factor problems, both Mode I and mixed mode, are solved using the boundary element methods just outlined in Reference 22.

We have now seen that an accurate, efficient technique for computation of stress-intensity factors is available for both finite and boundary element analysis. In a crack propagation analysis, this computation would be done for each crack tip after each load or growth increment. The computed stress intensity factors would determine the stability and angle change of each crack tip according to one of the theories outlined in the next section.

4. *Theories of Mixed-Mode Fracture*

The determination of fracture initiation from an existing flaw in pure Mode I or Mode II requires knowledge of two parameters: The stress intensity factor, determined analytically and a function of geometry and load, and the appropriate fracture toughness, a material state property, determined experimentally. The relationship between stress intensity factor, K_I and K_{II}, and the critical stress intensity factor, K_{Ic} and K_{IIc}, is analogous to the relationship between stress, σ, and a strength measure, say, the yield stress, σ_{yld}, in an uncracked, ductile specimen.

For pure Mode I or Mode II loading, as long as,

$$K_I < K_{Ic} \tag{11}$$

$$K_{II} < K_{IIc}$$

there will be no fracture propagation from the flaw. The analogy here is to the no-yield, uniaxial stress condition,

$$\sigma < \sigma_{yld} \tag{12}$$

Continuing the analogy, two-dimensional, mixed-mode loading on a crack is likened to biaxial loading on ductile, uncracked specimens. In the latter case, a yield surface of the type,

$$F(\sigma_1, \sigma_2, \sigma_{yld}) = 0 \tag{13}$$

where σ_1, σ_2 are principal stresses

can be postulated.

It follows then that in the mixed-mode, fracture initiation case, a fracture surface of the form,

$$F(K_I, K_{II}, K_{Ic}, K_{IIc}) = 0 \tag{14}$$

is of interest. Of course, in the plasticity relationship, equ.(13), the form of yield function is material dependent. Similarly, theoretical relationships of the form of equ.(14) must also be verified experimentally.

In this section, two theories for mixed-mode fracture initiation will be examined. It will be shown that, although the theories spring from a distinctly different fracture-governing parameter, in the two-dimensional case each can be expressed in the form of a fracture surface in the K_I-K_{II} plane.

4.1. *The* $\sigma_{\Theta_{max}}$ *Theory*

The first of the mixed-mode fracture initiation theories to be discussed was first formulated by Erdogan and Sih.[24] The parameter governing fracture initiation in their theory is the maximum circumferential tensile stress, $\sigma_{\Theta_{max}}$, near the crack tip.

Given a crack under mixed-mode conditions, the stress state near its tip can be expressed in polar coordinates as,

$$\sigma_r = \frac{1}{\sqrt{2\pi r}} \cos\frac{\Theta}{2}\left[K_I\left(1+\sin^2\frac{\Theta}{2}\right) + \frac{3}{2}K_{II}\sin\Theta - 2K_{II}\tan\frac{\Theta}{2}\right] + \ldots$$

$$\sigma_\Theta = \frac{1}{\sqrt{2\pi r}} \cos\frac{\Theta}{2}\left[K_I\cos^2\frac{\Theta}{2} - \frac{3}{2}K_{II}\sin\Theta\right] + \ldots \qquad (15)$$

$$\tau_{r\Theta} = \frac{1}{\sqrt{2\pi r}} \cos\frac{\Theta}{2}\left[K_I\sin\Theta + K_{II}(3\cos\Theta - 1)\right] + \ldots$$

These stress components are shown in Fig.14. The $\sigma(\Theta)_{max}$ theory states that:

1. Crack extension starts at the crack tip and in a radial direction.

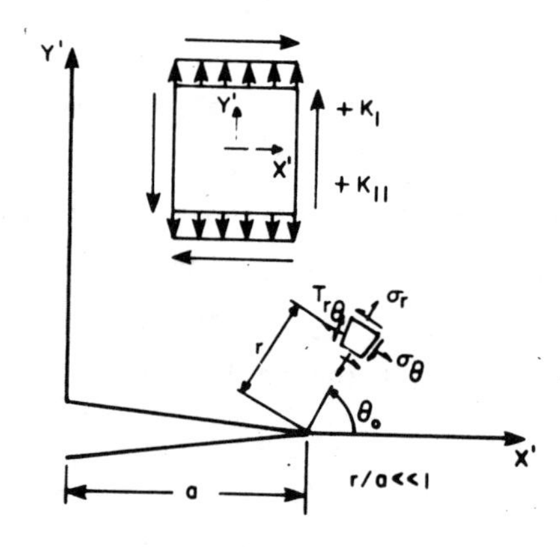

Fig.14 Crack tip polar coordinate system and sign convention

2. Crack extension starts in a plane normal to the direction of greatest tension, i.e., at Θ_0 such that $\tau_{r\Theta} = 0$.

3. Crack extension begins when $\sigma_{\Theta_{max}}$ reaches a critical, material constant value.

The theory is stated mathematically using eqs.(15),

$$\sigma_\Theta \sqrt{2\pi r} = \text{constant} = \cos\frac{\Theta_0}{2}\left[K_I \cos^2\frac{\Theta_0}{2} - \frac{3}{2}K_{II}\sin\Theta_0\right] = K_{Ic} \tag{16}$$

or

$$1 = \cos\frac{\Theta_0}{2}\left[\frac{K_I}{K_{Ic}}\cos^2\frac{\Theta_0}{2} - \frac{3}{2}\frac{K_{II}}{K_{Ic}}\sin\Theta_0\right] \tag{17}$$

and,

$$\tau_{r\Theta} = 0 \Rightarrow \cos\frac{\Theta_0}{2}\left[K_I \sin\Theta_0 + K_{II}(3\cos\Theta_0 - 1)\right] = 0. \tag{18}$$

Eqs.(17) and (18) are the parametric equations of a general fracture initiation locus in the K_I-K_{II} plane, shown in Fig.15. Also,

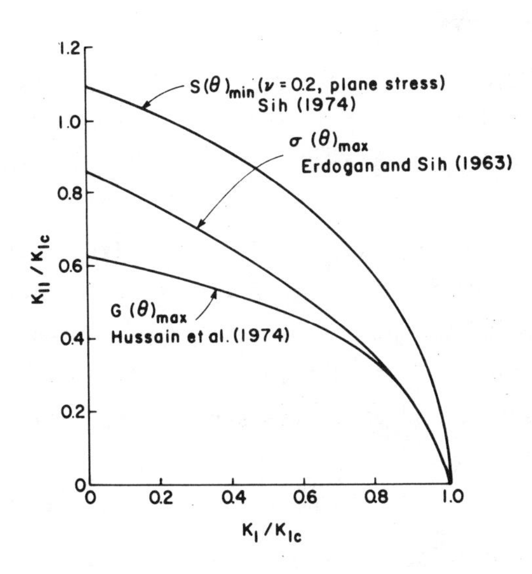

Fig.15 Fracture initiation loci for mixed-mode theories

the direction of the initial fracture increment, θ_0, can be found from equ.(18) which gives,

$$\Theta_0 = \pm\,\pi \text{ (trivial)}$$

$$K_I \sin\Theta_0 + K_{II}(3\cos\Theta_0 - 1) = 0. \qquad (19)$$

For example, for pure Mode I,

$$K_{II} = 0$$

$$K_I \sin\Theta_0 = 0 \Rightarrow \Theta_0 = 0° \qquad (20)$$

and the crack would propagate in its own plane. However, for pure Mode II,

$$K_I = 0$$

$$K_{II}(3\cos\Theta_0 - 1) = 0 \Rightarrow \Theta_0 = \pm\cos^{-1}\frac{1}{3} \qquad (21)$$

$$\Theta_0 = -70.5°.$$

That is, under pure Mode II, or any combination of Modes I and II, the crack would not propagate in its own plane.

In summary, the governing equations of the $\sigma_{\Theta_{max}}$ theory are (17) and (18). Algorithmically, the stress intensity factors for a given crack tip location and loading are first substituted into equ.(18) to obtain the new angle of propagation, Θ_0. The stress intensity factors and the angle Θ_0 are then substituted into Eq. (17). If it is not satisfied, the stress intensity factor pair plots either within or outside the fracture locus shown in Fig.15. If within, then that crack cannot propagate without a sufficient increase in stress intensity factors. If outside, then the crack is unstable and can continue to propagate until it reaches a free surface or until the stress intensity factor pair returns to within the locus.

Much more can be said about the $\sigma_{\Theta_{max}}$ theory. The only additional comment warranted here arises from equ.(16). If the right-hand-side of this equation, K_{Ic}, is a material constant, then so is the left-hand-side. That is, it is at some material characteristic value of r, say r_0, that $\sigma_{\Theta_{max}}$ reaches a material constant value. This observation, the reaching of a critical value of a field variable at a characteristic distance, is common to the $\sigma_{\Theta_{max}}$ and the second theory discussed here.

4.2. *The $S(\Theta)_{min}$ Theory*

The second of the mixed-mode fracture initiation theories to be discussed was formulated by Sih.[25,26,27,28] The parameter governing fracture initiation in his theory is the strain energy density near the point of initiation.

Sih has shown[26] that the strain energy density variation at a distance r from a crack tip is,

$$\frac{\partial U}{\partial V} = \frac{1}{r}\left(\frac{a_{11}K_I^2 + 2a_{12}K_IK_{II} + a_{22}K_{II}^2}{\pi}\right) \tag{22}$$

where,

$$\begin{aligned} a_{11} &= \frac{1}{16G}\left[(1 + \cos\Theta)(\kappa - \cos\Theta)\right] \\ a_{12} &= \frac{\sin\Theta}{16G}\left[2\cos\Theta - (\kappa - 1)\right] \\ a_{22} &= \frac{1}{16G}\left[(\kappa + 1)(1 - \cos\Theta) + (1 + \cos\Theta)(3\cos\Theta - 1)\right]. \end{aligned} \tag{23}$$

If the quantity in parentheses in equ.(22) is called S, i.e.,

$$\frac{\partial U}{\partial V} = \frac{S}{r_0}\,, \tag{24}$$

it can be seen that, at a constant value of r, S is the varying intensity of the strain energy density around the crack tip.

The $S(\Theta)_{min}$ theory proposes that:

1. Crack extension occurs in the direction along which $\partial U/\partial V$ possesses a minimum value, i.e., Θ_0 such that,

$$\frac{\partial S}{\partial \theta} = 0 \qquad \frac{\partial^2 S}{\partial \theta^2} \geqq 0. \tag{25}$$

2. Crack extension occurs when $S(\Theta_0)$ reaches a critical, material value, S_c.

3. $S(\Theta)$ is evaluated along a contour $r = r_0$, where r_0 is a material constant.

Combining 2) and 3) shows that,

$$\left(\frac{\partial U}{\partial V}\right)_c = \frac{S_c}{r_0} . \tag{26}$$

That is, specifying S_c and r_0 to be material constants is equivalent to specifying a materal critical strain energy density.

A relationship between S_c and fracture toughness can be obtained in the following manner. For $K_{II} = 0$, equ.(25) predicts that $\Theta_0 = 0°$. Then, for $K_I = K_{Ic}$, equating equ.(26) to equ.(22) yields,

$$S_c = \frac{(\kappa-1)K_{Ic}^2}{8\pi G} \tag{27}$$

A fracture initiation locus in the K_I-K_{II} plane is then obtained from,

$$\frac{S_c}{r_0} = \frac{1}{\pi r_0}\left(a_{11}K_I^2 + 2a_{12}K_I K_{II} + a_{22}K_{II}^2\right)$$

or

$$1 = \frac{8G}{(\kappa-1)}\left[a_{11}\left(\frac{K_I}{K_{Ic}}\right)^2 + 2a_{12}\frac{K_I K_{II}}{K_{Ic}} + a_{22}\left(\frac{K_{II}}{K_{Ic}}\right)^2\right] \tag{28}$$

where Θ_0 is obtained from the conditions equ.(25).

Although the direction $\Theta_0 = 0°$ for fracture initiation in pure Mode I is not elasticity constant dependent, Θ_0 for all other cases is a function of Poisson's ratio in the $S(\Theta)_{min}$ theory.

An ambiguity of fracture direction sometimes arises in the application of the $S(\Theta)_{min}$ theory. Under certain combinations of K_I-K_{II}, the direction of *absolute* minimum S is associated with negative σ_Θ. That is, were the crack to propagate in the direction $S(\Theta)_{min}$, it would be extending in a direction of negative K_I, a physically untenable situation. However, under these combinations there are angles corresponding to *local* minima in $S(\Theta)$ along which a crack could propagate in positive K_I. Consequently, the first of Sih's hypotheses, equ.(25), should be amended to read,

1. Crack extension occurs in the direction along which $\partial U/\partial V$ is minimal over those values of Θ for which $K_I(\Theta)$ is positive.

Again, to summarize, eqs.(25) and (28) govern the $S(\Theta)_{min}$ theory. The algorithm begins with satisfaction of equ.(25) to obtain Θ_0. It and the stress intensity factors are then substituted into equ.(28). The crack stability implications are then the same as outlined in the section describing the $\sigma_{\Theta_{max}}$ theory.

Of course, in a quasi-static fracture propagation analysis the governing equations for one of the theories would be applied at the end of each growth step or load step. Recall Observation #3: It may not be necessary to increase loads to bring the stress intensity factors of a previously stable crack tip onto the fracture locus. The propagation of another crack may cause the same effect. Algorithmically, this implies that the interaction factor for each crack tip, the right-hand-side of equ.(17) or (28), be updated in memory after each crack or load increment. As we will see later, depending on the mode of interaction between the program and the user, the former or the latter will use the interaction factors to decide which one or more of the crack tips should be propagated in a given fracture step.

4.3. *Comparison of Mixed-Mode Fracture Theories*

Fig.15 compares the interaction effects predicted by the $\sigma_{\Theta_{max}}$ and $S(\Theta)_{min}$ theories. Also shown is a prediction from one of the many other existing theories: The maximum energy release rate formulation of Hussain, Pu, and Underwood.[30] It can be seen that the $S(\Theta)_{min}$ theory is the least conservative of the three shown. It also predicts that the Mode II fracture toughness of most rocks ($\nu < 0.3$) is larger than K_{Ic}, while the other theories predict a smaller value.

How does theory compare to experiment? References 3 and 31 contain much data and lengthy discussion relevant to this question, but the comparisons are all based on materials other than rock. Recently, the author has performed wide spectrum, mixed-mode fracture initiation tests on Indiana limestone and Westerly granite.[32] The results are shown in Fig.16a and 16b for the limestone and granite, respectively. It appears, based on this somewhat limited data, that the $S(\Theta)_{min}$ theory is the most accurate of the three theories used for comparison.

It should be emphasized, however, that a crack finding itself under substantial Mode II loading does not long remain in the high K_{II}/K_I domain of interaction. Such a crack quickly changes trajectory to minimize or eliminate the K_{II} component. Consequently, the life of a crack propagating quasi-statically is spent in the high K_I/K_{II} region of the

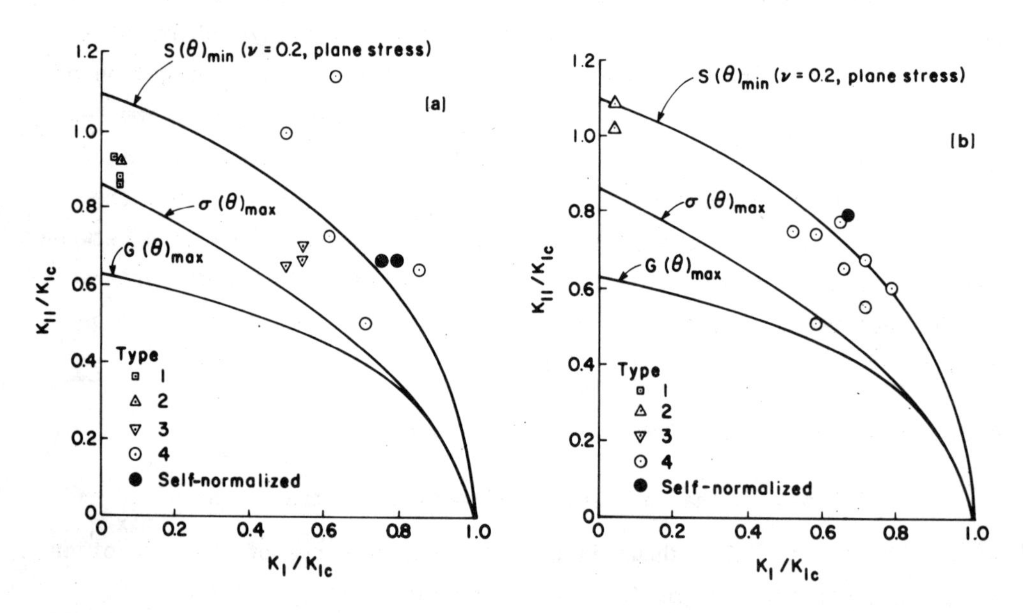

Fig.16 Results of mixed-mode fracture initiation tests. a) Indiana limestone; b) Westerly granite. From Reference 32.

interaction plane where the differences among the theories are minimal. It is the author's opinion that, except for fracture increments under high K_{II}, use of any of the referenced theories would result in substanially the same trajectory and load history. This will be seen in example problems to follow.

4.4. *Predicting Crack Increment Length*

We have seen how to compute stress intensities, and how to use them to predict local stability and angle change. However, Observation #1 reminds us that to complete our fracture propagation model we must also be able to predict either, a) the length of a fracture increment for a given load change, or, b) the load change required to drive a crack a specified length. These predictions are relatively simple and straightforward.

The fundamental principle here is that a fracture, once initiated, will continue to propagate as long as there is sufficient energy or, equivalently, effective stress intensity, available. Effective stress intensity, K^*, here refers to a mixed-mode case and is the combination of Mode I, II, and III stress intensity factors required by the particular mixed-mode theory in use. The right-hand sides of eqs.(17) and (28) can, therefore, be viewed as normalized effective stress intensity factors.

We must consider a number of possible stability cases in creating an algorithm for predicting fracture increment length. Let's look at some examples. We will reference Figs.17-18 in these examples and assume, for simplicity, that LEFM applies for all crack lengths. Let's also assume that we are investigating propagation along some predicted direction θ_0.

Case 1: Effective stress intensity increases monotonically with crack length, curve OA in Fig.17. If the initial flaw size is less than a_i, no propagation occurs. For $a = a_i$, propagation can occur and it will continue at $P = P_i$; that is, a condition for local instability has been met. Of course, an algorithm could be written which would

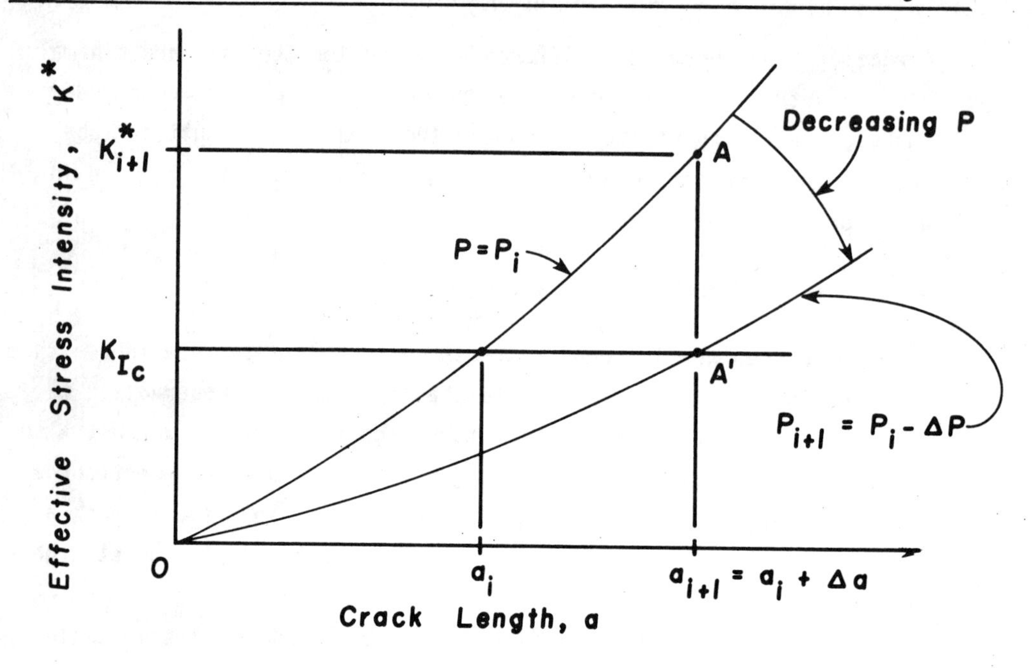

Fig.17 Stress intensity factor variation for Case 1.

place such a scenario in displacement or crack-length control: A crack increment, Δa, could be specified and the load decrement required to just bring the crack tip to $a_i + \Delta a$ could be computed. This situation is depicted by curve OA' in Fig.17. To compute P_{i+1} recall that LEFM specifies that at instability,

$$K_{Ic} = \alpha_i P_i \sqrt{a_i} = \alpha_{i+1} P_{i+1} \sqrt{a_{i+1}} \tag{29}$$

where,

α = factor depending on geometry and interaction theory

Therefore,

$$P_{i+1} = \frac{\alpha_i}{\alpha_{i+1}} P_i \sqrt{\frac{a_i}{a_{i+1}}} \tag{30}$$

Equ.(30) is only directly useful, however, if the α_{i+1} coefficient is known at Step i. For arbitrary problems, this is certainly not the case. An alternative is to propagate the fracture an amount Δa (remember, in the direction Θ_0) and compute K^*_{i+1} at load level P_i. The new load level is then,

$$P_{i+1} = \left(\frac{K_{Ic}}{K^*_{i+1}}\right) P_i \tag{31}$$

as can be seen in Fig.17. Behavior described in this case is typical of many of the Mode I fracture specimens used to measure K_{Ic} of rock. As we shall see in the example problems later in this chapter, it can also occur in a variety of circumstances in practical rock fracture problems.

Case 2: Effective stress intensity increases, reaches a maximum value, and then decreases with increasing crack length, curve OA in Fig.18a. For the value of K_{Ic} shown and at load level P_1, no crack propagation is possible. At load level P_2, propagation is possible only at crack length $a = a_2$, but the corresponding, theoretical fracture increment length is $\Delta a = 0$. At load level P_3, propagation can occur for a crack of length a_1, and it would be unstable in load control. Again, as in case 1, above, using a crack length or displacement control algorithm the crack of initial length a_1 could be propagated stably to length a_2 by decreasing the load incrementally from level P_3 to level P_2, as shown in Fig.18b.

For crack lengths longer than a_2, fracture propagation is stable in the load control sense. An effective stress intensity monotonically decreasing with increasing crack length implies that a monotonically increasing load is required for continued propagation. In Fig.18, it can be seen that if the load is again increased to P_3 propagation to crack length a_3 is possible.

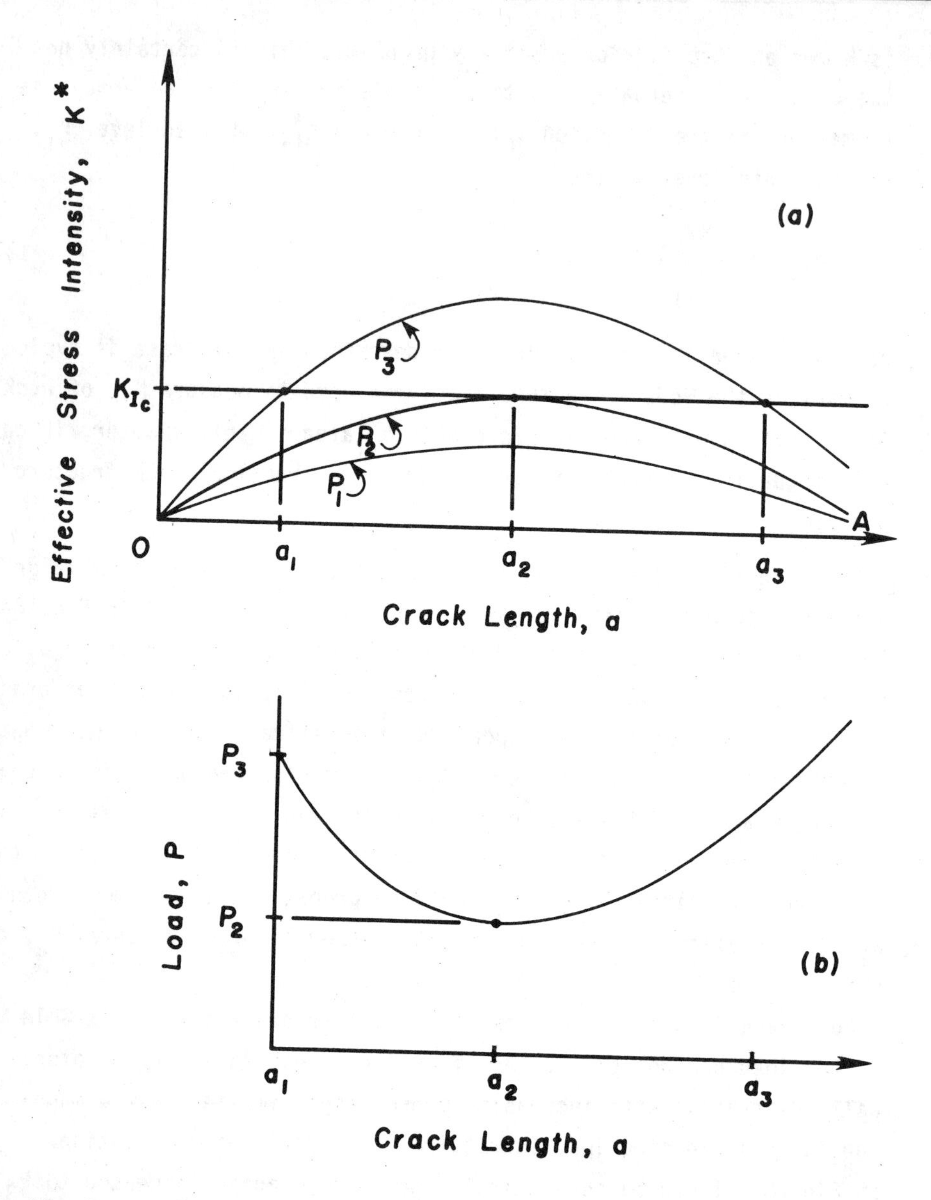

Fig.18a Stress intensity factor variation for Case 2; b) load variation for Case 2.

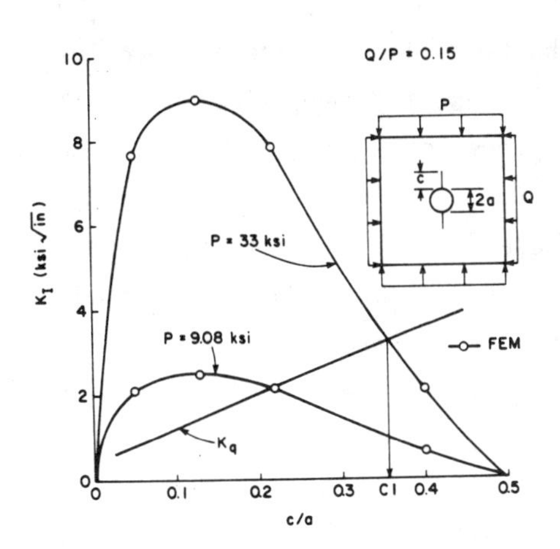

Fig.19 Stress intensity factor variation for crack propagating from circular hole. (1 ksi = 6.9 MPa, 1 ksi $\sqrt{in}$ = 1.1 MPa $\sqrt{m}$) From Reference 9.

An example of this behavior for a pure Mode I case is shown in Fig.19 which is taken from a study on fracture propagation around underground openings.[9] Cracks are induced at crown and invert of a circular opening in a plate under the indicated biaxial compression. The computed K^* (here $K^* = K_I$) versus crack length relationship at two load levels is shown. Although it was assumed that the effective toughness, K_Q, increased with crack length, the propagation scenario is the same as described above. Propagation is unstable at first, but ceases each time the K_I and K_Q curves intersect. This type of propagation behavior is very common in rock mechanics. It occurs, for example, in both of the examples shown in Fig.1.

Obviously, if we are starting with crack length a_1, and load level P_3 (Fig.18), the prediction technique is the same as described under Case 1: Propagate the fracture an amount Δa in the direction Θ_0 at load P_3, compute the effective stress intensity for the new crack length, and apply equ.(31).

Let's suppose, however, that we are at load level P_2 and crack length a_2. We can still use the same algorithm: the only difference is that the quantity in parentheses in equ.(31) will now always be less than one.

The reverse of Case 2 is also possible: Effective stress intensity can at first decrease and then begin to increase with increasing crack length. (See, for example, References 33 and 34.) This implies nothing new algorithmically, however, since the implications of this situation are handled by the techniques described in Cases 1 and 2.

A number of alternative numerical techniques for fracture increment length prediction are available.[3,35,36] Some are based on energy balance, some are more approximate than others. The simple technique described here is theoretically exact for pure Mode I, colinear propagation. Recalling Observation #4, however, we can see that any technique which employs finite, straight fracture increments will be approximate. One is updating effective stress intensity incrementally, rather than continuously. A curvilinear trajectory is being modeled piecewise by straight segments. Stress intensity factors and angle changes will be somewhat in error. The error depends on the specified length of the fracture increment.

The analogy here is with dynamic analysis where the time step controls accuracy and stability of the solution. It is the author's experience with his codes that predicted trajectories sometimes oscillate about an average path. This is a manifestation of error in K_{II} simulation which is a result of "kinking" the crack path rather than allowing it to continuously curve. Spuriously high K_{II} values are computed which, alternating in sign with each increment, zig-zag the crack. However, it is quite possible that if too large an increment is used divergence of predicted trajectory could occur.

All the theoretical ingredients for fracture propagation modelling under mixed-mode, LEFM assumptions have now been presented. These ingredients have been combined in computer programs developed by the author and his students at Cornell University. In the next section, some general observations concerning these programs will be presented.

5. *Fracture Propagation Programs*

Research and application thrusts into fracture propagation modelling at Cornell University can be divided into areas of numerical method and user-computer interface.

5.1. *Numerical Methods*

Incremental fracture propagation codes have been developed using both the finite and boundary element methods. As will be shown, each of these methods has characteristics which make it the appropriate choice for given structure, dimensionality of model, or interface hardware.

In general, the boundary element method is suited to elastic, homogeneous structures containing few propagating cracks. The boundary element method is superior in efficiency and accuracy to the finite element method for modelling of three-dimensional crack propagation problems. Since only the boundaries of the structure, including the crack faces, need to be discretized, the data base for a boundary element method analysis is much smaller than that of a finite element method analysis. Also, since perspective views of three-dimensional meshes are not encumbered with all the interior nodes and elements of a finite element mesh, effective user-computer interface can be obtained with low-level computer graphics equipment.

A two-dimensional code, Boundary Element Fracture Analysis Program, BEFAP,[35,37] is operational, and an example of its use is described later. The three-dimensional version is under development in connection with the Cornell Program for Computer Graphics for use with high-level computer graphics hardware (see next section for description of high- and low-level computer graphics hardware).

For problems involving inhomogeneities, interfaces, or many cracks, BEFAP currently is not suitable. For problems of this type, as well as those involving material nonlinearity, the Finite Element Fracture

Analysis Program, FEFAP,[36,38] has been developed. Again, the two-dimensional version of FEFAP is operational, and the three-dimensional version is being implemented in a high-level interactive computer graphics environment. Examples of problems solved using FEFAP will also be presented later.

5.2. *User-Computer Interface*

Three levels of user-computer interfacing are available for operation of the BEFAP/FEFAP group:

1. Interactive Without Graphics: the standard keyboard entry of data, editing of files, and spooling of output to a printer.
2. Low-Level Interactive Computer Graphics: storage tube graphic display devices. Display of initial mesh, deformed mesh, principal stress vectors, crack increment trajectories, G-curves, and load displacement curves. Interactive programming capability, meaning that the user participates in the real time solution of the problem by, for examples, editing each mesh update, or selecting the length of a crack increment or magnitude of a load increment.
3. High-Level Interactive Computer Graphics: vector refresh graphic display devices. With high-level graphics, all of the capabilities of low-level exist but the display is continually updated. This means that selected regions of the display can be changed nearly instantaneously without the necessity for redraw of the entire display. The mesh can be "zoomed" or panned to highlight detail, and three-dimensional objects can be translated and rotated to enhance the user's perception of a complex object and its mesh.

The operational version of BEFAP currently operates only in the interactive mode, while FEFAP operates in interactive and low-level interactive graphic modes. Two- and three-dimensional versions of these programs are under development for use in the high-level interactive graphics mode.

In the example problems to follow, one will notice an evolution in the user-computer interface towards increasing use of interactive computer graphics. This evolution is still underway but the original objectives are the same. These are:

1. Minimize Manual Generation of Input Data.
   This applies both to the total amount of data necessary to define the problem and the physical act of transferring this information to the computer. The user should communicate geometrical information to his code by way of interactive graphics, e.g., a digitizing tablet and pen in conjunction with a vector refresh display terminal or a cursor and key system such as on some storage tube terminals. Only the geometrical information absolutely required for automatic mesh generation should be input in this manner.
2. Make the Programs Interactive and Adaptive.
   On request, the user should be informed of real time progress of the analysis by way of graphic displays. Moreover, he should be given the freedom to modify the course of the analysis by changing the data base while it is in progress.
3. Results Should be Displayed in a Simple and Effective Way.
   The user should be able to see graphic display of intermediate and final stress and displacement fields, load histories, crack patterns.

5.3. *Automatic Remeshing*

All of the programs developed at Cornell University employ automatic remeshing algorithms. This means that the programs, after computing stress intensities, interaction, angle change, and change in load or fracture length, automatically relocate each crack tip and remesh accordingly. The remeshing algorithms developed by Blandford[35] for BEFAP and Saouma[36] for FEFAP are very versatile. They accommodate mixtures of element types and allow a wide range of crack configurations to be modelled. This versatility will be evident in the examples to follow.

5.4. *Example Solutions*

Example #2: Simulation of the Tests Described in Example #1
Numerical Method: Finite Element and Boundary Element Solutions
User-Computer Interface: Punch Cards and Manual Remeshing for Finite Element Solution; Interactive Without Graphics for Boundary Element Solution

The author simulated[3,5] the behavior shown in Fig.3 for the first, and most rudimentary, example of fracture propagation modelling using the techniques described in this chapter. The initial mesh, shown in Fig.20a, was generated by hand. Each fracture increment required manual remeshing and a job resubmission. Dozens of man-hours were required to produce the typical result of ten primary crack increments shown in Fig.20b.

Note in Fig.20b that a corridor of tensile principal stresses was created as a result of primary crack propagation (Observation #6). The secondary crack nucleation observed in Fig.3b closely corresponds in location to this predicted tension zone. This zone was noticed only when the Gauss-point stresses in each element were plotted manually.

More recently the same problem was analyzed by Blandford[35,37] using BEFAP. A typical mesh is shown in Fig.21. Initial data preparation required about 3 man-hours. The analysis itself, involving 7 primary crack increments and computation of domain stresses, required about 10 CPU minutes on an IBM 370/168. Still, however, the computed stress field had to be manually plotted to mark the area of secondary crack nucleation. A comparison of typical finite element, boundary element, and experimental results is shown in Fig.22.

Example #3: Collapse of Underground Cavity
Numerical Method: Finite Element
User-Computer Interface: High Level Interactive Computer Graphics

The structure shown in Fig.23 represents a cross-section through a deep underground cavity loaded by overburden and horizontal stresses. The model shown was tested by Hoek.[39] An analysis of this structure was performed using the high-level interactive computer graphics facilities

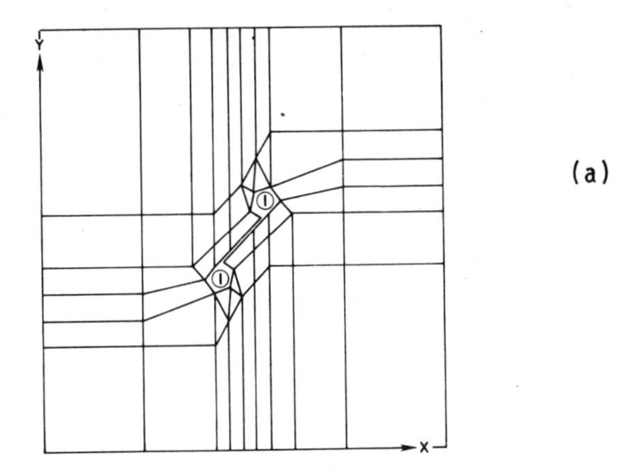

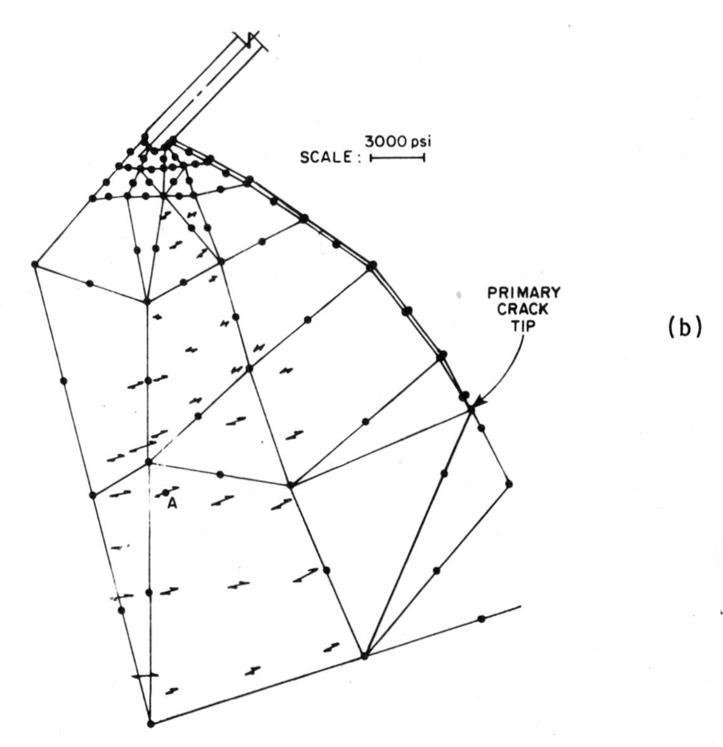

Fig.20 a) Initial finite element mesh for problem of example #1; b) typical detail ' showing primary crack path. (1 psi = 6.9 kPa) From Rf ence 5.

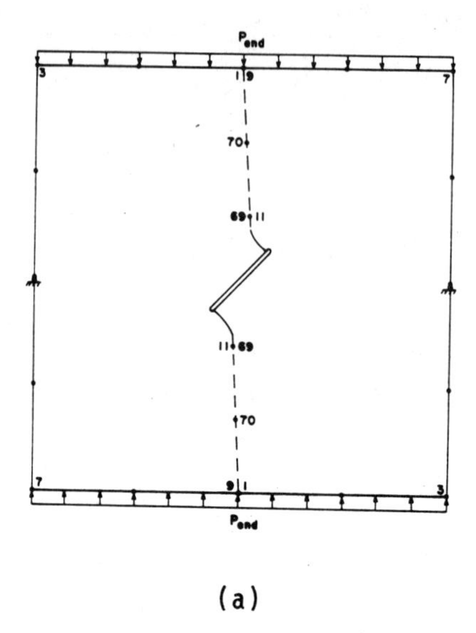

(a)

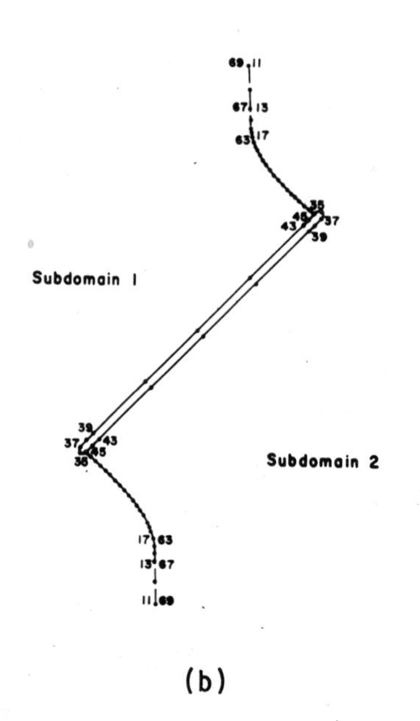

(b)

Fig.21 a) Typical boundary element mesh for problem of example #1; b) typical notch tip detail. From Reference 37.

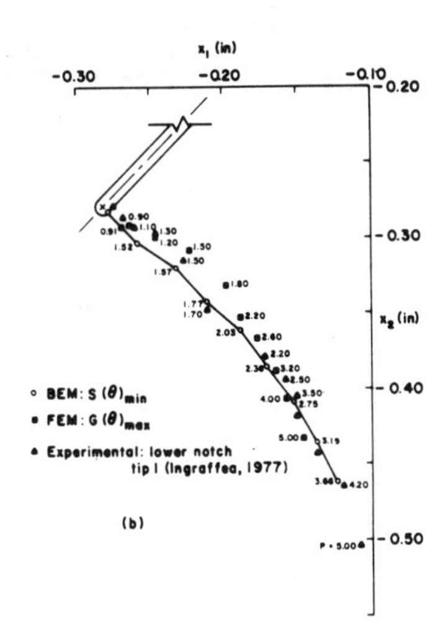

Fig.22 Comparison of predicted and observed primary crack trajectories for problem of example #1. P in ksi. (1 ksi = 6.9 MPa, 1 in = 25.4 mm) From Reference 35.

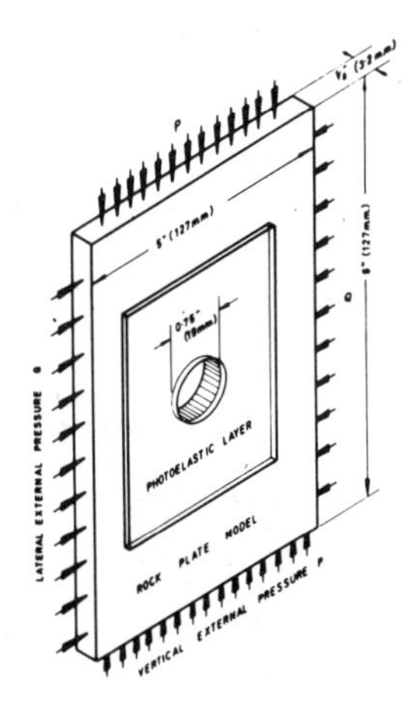

Fig.23 Rock plate model used in experimental study of collapse of underground cavity. From Reference 39.

at the Cornell Program for Computer Graphics. An interactive graphics preprocessor[40] was used to generate the initial finite element mesh, shown in Fig.24a, and its attributes. Generation of all initial data required about 10 man-minutes.

The fracture response of this structure is similar to that of the previous example. Primary cracks initiate at crown and invert and propagate in the manner described previously with respect to Fig.19.

The initial major principal stress field is depicted in Fig.24b. Shown is a photograph taken from a color postprocessor display.[41] Regions in which the stress exceeds the postulated tensile strength are shown here in black. Such postprocessing can be performed at the end of each fracture increment; fields of principal, normal, and shear stress, strain energy density, and displacement can be quickly displayed. Moreover, no additional man-effort is required to generate an image since the postprocessor data base is common with that of the preprocessor.

Next, secondary cracks nucleate in the plate interior in a tension zone developed in response to primary crack propagation. This zone can be seen as the blackened area shown in the postprocessor image of Fig.25a. A "zoomed" detail of the final mesh showing the predicted secondary crack path is shown in Fig.25b. Experimental results for a similar problem[42] are shown in Fig.26.

Example #4: Fracture Propagation Under Indentation Loading
Numerical Method: Finite Element
User-Computer Interface: Low-Level Interactive Computer Graphics

The mechanisms of fracture propagation under a tunnel boring machine roller cutter have been much studied but, in the author's opinion, are not yet completely understood. Paul[43] proposed that at the point of contact with the rock surface high bearing stresses would generate a bulb of very high hydrostatic pressure. He used finite element analysis to prove that such a condition would produce a primary radial crack as shown in Fig.27. FEFAP was used to model this occurrence as well as to predict the trajectories of the secondary radial cracks which Paul[43] surmised would occur after primary crack stabilization.

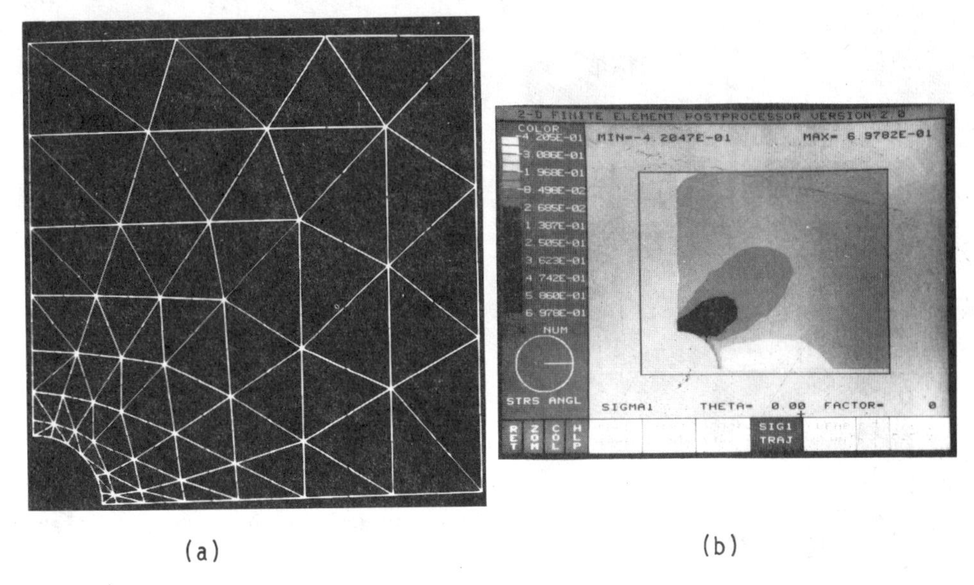

(a) (b)

Fig.24 a) Inital mesh for example #3; b) initial major principal stress field.

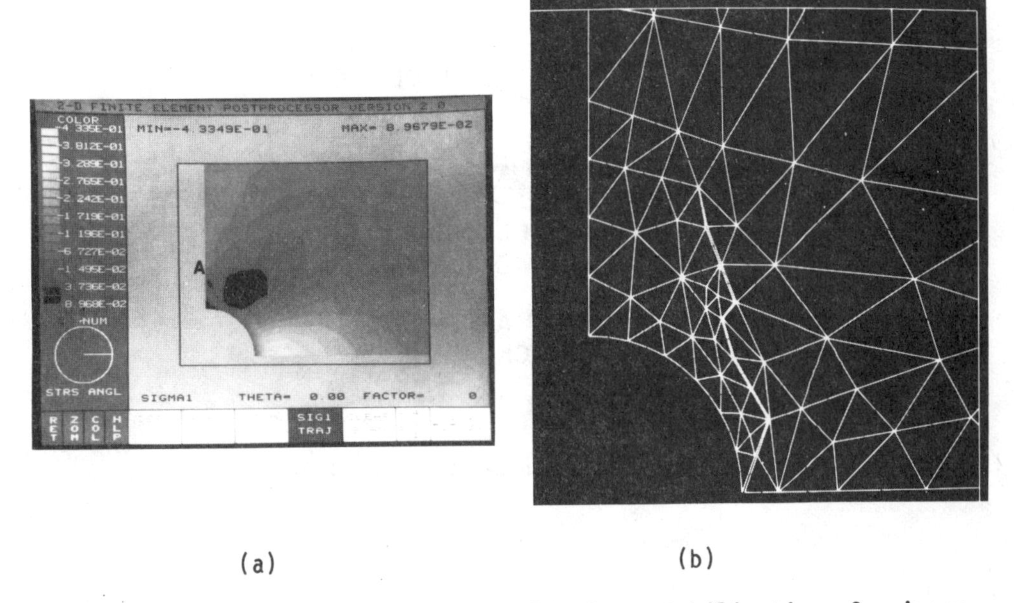

(a) (b)

Fig.25 a) Major principal stress field after stabilization of primary crack at point A; b) detail of final mesh.

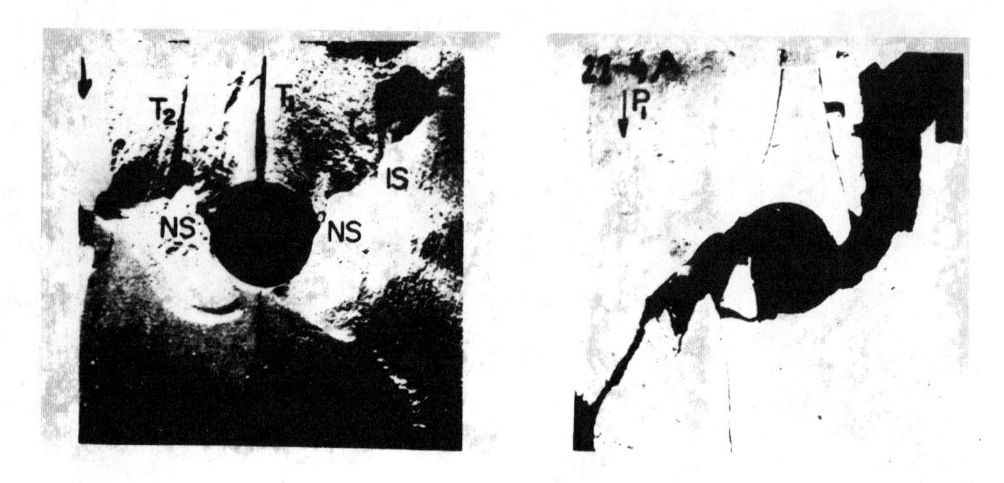

Fig.26 Fracture patterns observed in tests on plaster models similar to figure 23. From Reference 42.

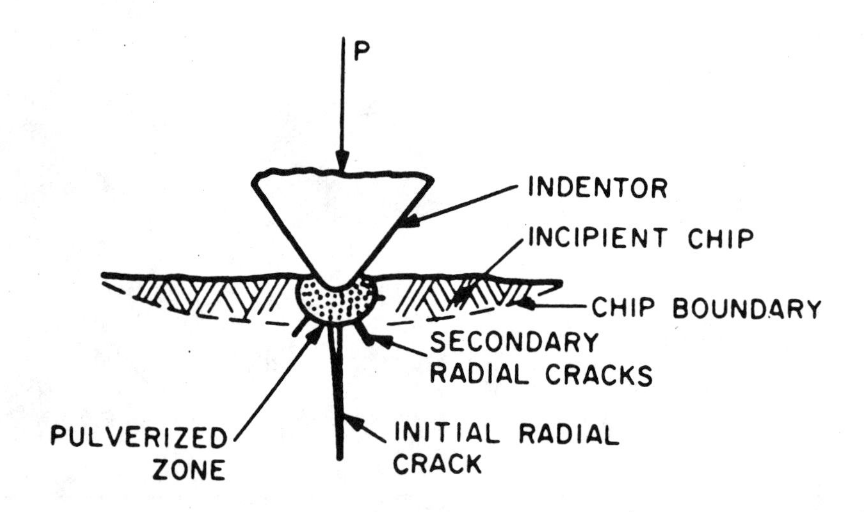

Fig.27 Postulated fracture pattern under indentor loading. From Reference 43.

The initial and final meshes are shown in Fig.28. The predicted fracture pattern closely resembles those observed after TBM roller cutter passage over granite,[44] Fig.29a, and a scratch on a glass plate,[45] Fig.29b.

FEFAP, designed by Saouma and the author,[36,38] is highly interactive and adaptive. The user receives information graphically after each analysis step, and is put in control of each subsequent step. For example, Fig.28a shows a typical fracture increment control page and its question/response dialogue. The automatic mesh modification algorithm in FEFAP can accommodate multiple cracks, a mixture of Q8, LST, and quarter-point singular elements, and interior or symmetry line cracks. The user is given the option of modifying a generated mesh: Figs.30a and 30b show before and after images, respectively, of a typical mesh modification step.

6. *Fracture Propagation Modelling — The Future*

The techniques described in this chapter are certainly not the only ones available for modelling of fracture propagation. Alternative approaches can be based on other numerical methods, theories, and algorithms. However, it is the author's strongly held opinion that, regardless which model is pursued, interactive computer graphics will play a decisive role in determining the viability of and program in the marketplace of "real world" problems. The continuing, rapid revolution in graphics hardware capability and software development and the ever-increasing cost-effectiveness of large, virtual-memory mini-computers are the driving forces in the evolution of sophisticated fracture propagation programs.

Nowhere will this be more evident than in the area of fully three-dimensional modelling. The present high cost of performing analysis of three-dimensional structures is due largely to the human effort required to define and check geometrical data, element topology, boundary conditions, and material properties. In fact, the complexity of error detection, or even slight modification, with three-dimensional meshes can

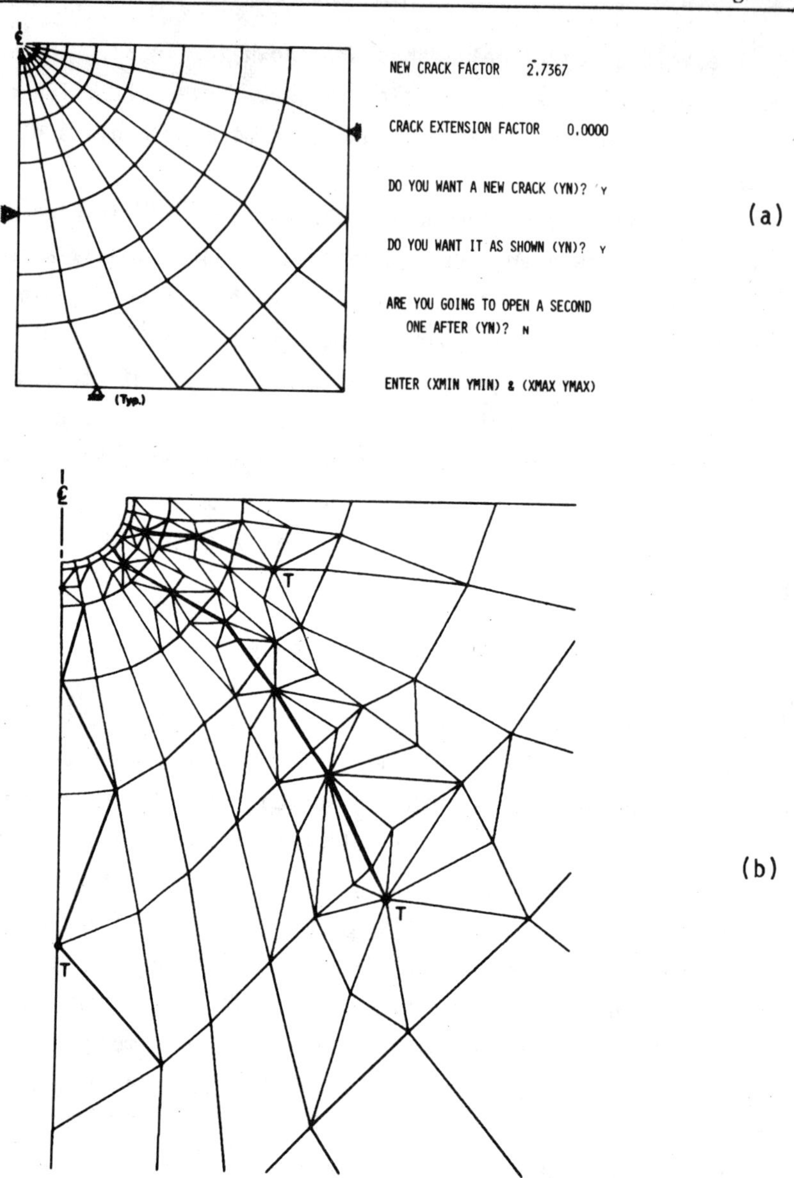

Fig.28 Initial, a), and final, b), meshes for example #4. Crack tips are at points T.

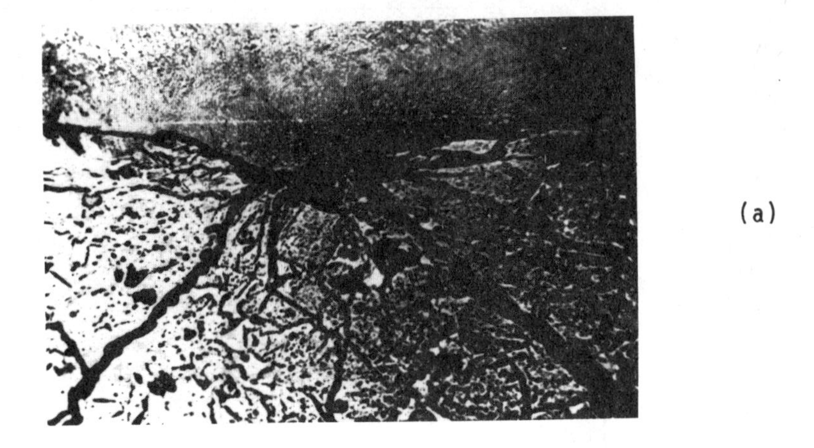

(a)

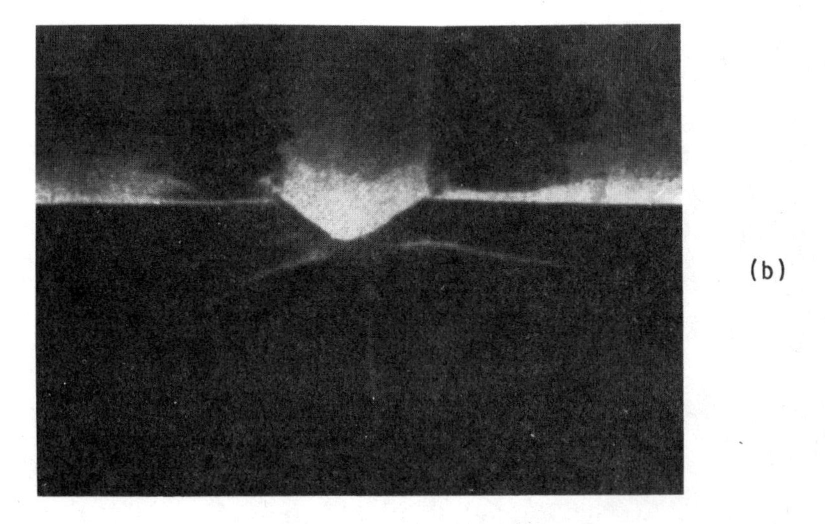

(b)

Fig.29 Fracture patterns on cross section of a) granite plate after TBM cutter pass. From Reference 44; b) glass plate after scratching. From Reference 45.

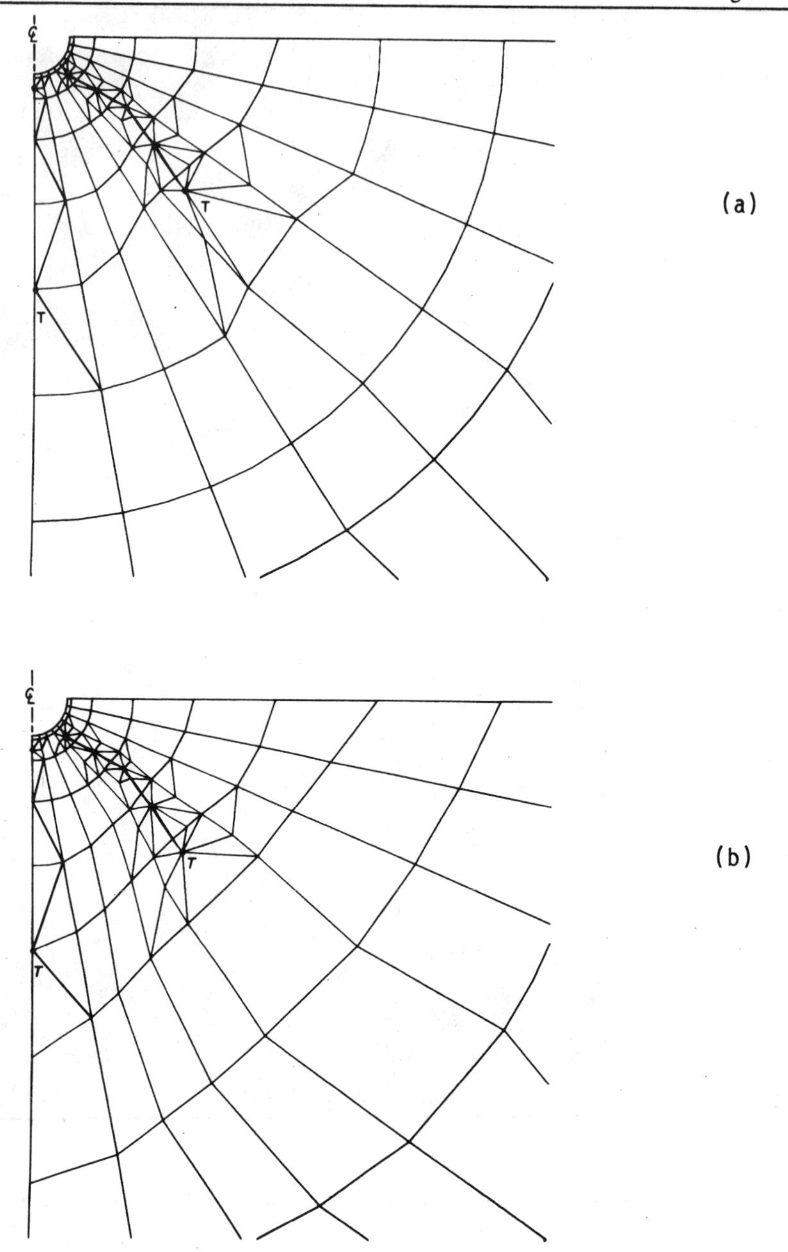

Fig.30 Interactive mesh modification in FEFAP. a) before; b) after.

substantially reduce the cost effectiveness of a program. The user falls back onto a two-dimensional or axisymmetric model that is "good enough," sacrificing the realism of the three-dimensional problem in the face of the reality of tremendous labor cost.

However, interactive/adaptive preprocessing can eliminate a large percentage of such cost while simultaneously placing the engineer back in control of computer analysis. For example Perucchio and the author[46,47] have developed three-dimensional finite and boundary element preprocessors with awesome potential for use with fracture propagation codes. Recall the hydrofracture problem of Fig.1a. During the course of an analysis to predict the size, shape, and orientation of a propagating hydrofracture it will be necessary to generate, display, and modify a mesh many times. The mesh shown in Fig.31 was generated using a boundary element preprocessor[47] in about 15 minutes.

The complete boundary element mesh of Fig.31a can be rotated, translated, zoomed, and depth cued. The user need not sacrifice to the CPU his engineering insight into the physics of the problem. Fig.31b shows

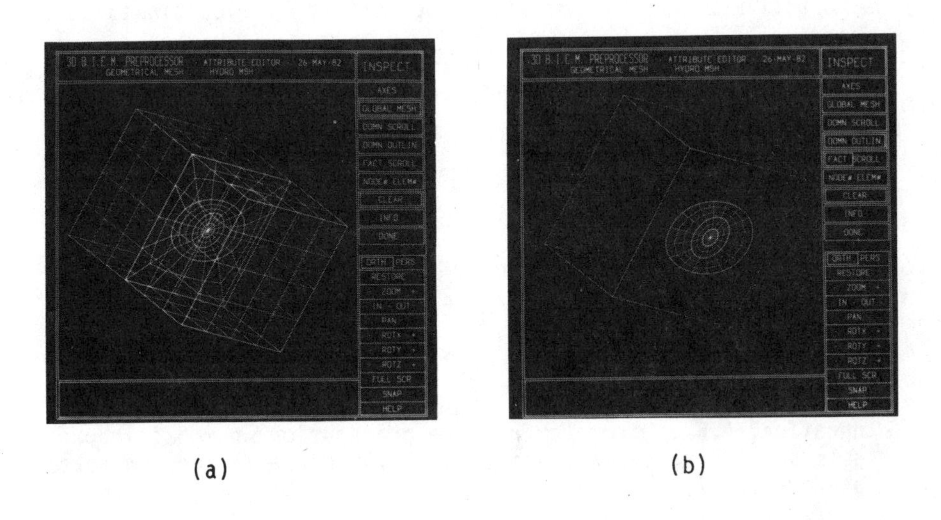

Fig.31 Three-dimensional boundary element mesh for hypothetical hydrofracture problem. a) Entire mesh; b) sub-domain with fracture on a facet.

that the mesh can be taken apart to study cross-sections, sub-domains, or even elements of interest. This capability will be invaluable during the course of a fracture propagation analysis.

Given the changes in the speed and cost of computers and the increasing use of interactive graphics, the author sees the use of truly three-dimensional fracture propagation codes, with the versatility of existing two-dimensional programs, as a certainty in only a few years.

Acknowledgements

The author would like to thank the following former and present students for making the developments reported here possible: Prof. Victor Saouma, Prof. George Blandford, Dr. Corneliu Manu, Renato Perucchio, and Walter Gerstle. Special thanks go to the Mr. Perucchio and Mr. Gerstle for their enthusiastic assistance in preparing this chapter, and to John Beech and Priscilla Nelson for their helpful discussions.

The author would also like to thank Prof. Donald Greenberg, Director of the Cornell Program for Computer Graphics, for the continued use of his facility.

Research grants for the development of the techniques and programs described here came from the National Science Foundation.

References

/1/ Griffith, A.A.: The phenomenon of rupture and flow in solids. Phil. Trans. Roy. Soc. London, Ser. A221,163 (1921)

/2/ Griffith, A.A.: Theory of rupture. Proc.First Int. Congress Appl. Mech., Delft 55 (1924)

/3/ Ingraffea, A.R.: Discrete fracture propagation in rock: laboratory tests and finite element analysis. Ph.D. dissertation, University of Colorado (1977)

/4/ Ingraffea, A.R. and H.-Y. Ko: Determination of fracture parameters for rock. Proceedings of First USA-Greece Symposium Mixed Mode Crack Propagation, G.C. Sih and P.S. Theocaris, eds., Sijthoff & Noordhoff, Alphen aan den Rijn, the Netherlands, 349 (1981)

/5/ Ingraffea, A.R. and F.E. Heuze: Finite element models for rock fracture mechanics. Int.J.Num.Analy.Meths. in Geomechs. 4,25 (1980)

/6/ Cotterell, B.: Brittle fracture in compression. Int.J.Fract.Mech. 8,195(1972)

/7/ Bombalakis, E.G.: Photoelastic study of initial stages of brittle fracture in compression. Tectonophysics 6,461 (1968)

/8/ Hoek, E. and Z.T. Bieniawski: Brittle fracture propagation in rock under compression. Int.J.Fract.Mech. 1,139 (1965)

/9/ Ingraffea, A.R.: The strength-ratio effect in rock fracture. J. of Am.Cer.Soc., in press.

/10/ Parks, D.M.: A stiffness derivative finite element technique for determination of crack tip states intensity factors. Int.J.Fract. 10,4,487 (1974)

/11/ Atluri, S.N., A.S. Kobayashi, and M. Nakagaki: An assumed displacement hybrid element model for linear fracture mechanics. Int.J. Fract. 11,2,257 (1975)

/12/ Chan, S.F., I.S. Tuba, and W.K. Wilson: On the finite element method in linear fracture mechanics. Engr.Fract.Mech. 2,1 (1970)

/13/ Wilson, W.K.: On combined mode fracture mechanics. Ph.D. dissertation, University of Pittsburgh (1969)

/14/ Tracey, D.M.: Finite elements for determination of crack tip elastic stress intensity factors. Engr.Fract.Mech. 3,255 (1971)

/15/ Jordan, W.B.: The plane isoparametric structural element. General Electric Co., Knolls Atomic Power Lab., Report No. KAPL-M-7112 (1970)

/16/ Barsoum, R.S.: On the use of isoparametric finite elements in linear fracture mechanics. Int.J.Num.Meth.Engrg. 10,25 (1976)

/17/ Barsoum, R.S.: Triangular quarter-point elements as elastic and perfectly-plastic crack tip elements. Int.J.Num.Meth.Engrg. 11,85 (1977)

/18/ Freese, C.E. and D.M. Tracey: The natural isoparametric triangle versus collapsed quadrilateral for elastic crack analysis. Int.J. Fract. 12,767 (1976)

/19/ Shih, C.F., H.G. de Lorenzi, and M.D. German: Crack extension modeling with singular quadratic isoparametric elements. Int.J. Fract. 1,647 (1976)

/20/ Ingraffea, A.R. and C. Manu: Stress-intensity factor computation in three dimensions with quarter-point elements. Int.J.Num.Meth.Engrg. 15,10,1427 (1980)

/21/ Cruse, T.A. and R.B. Wilson: Boundary-integral equation method for elastic fracture mechanics. AFOSR-TR-0355 (1977)

/22/ Blandford, G., A.R. Ingraffea, and J.A. Liggett: Two-dimensional stress intensity factor calculations using boundary elements methods. Int.J.Num.Meth.Engrg. 17,387 (1982)

/23/ Brown, W.F. and J.E. Srawley: Plane strain crack toughness testing of high strength metallic materials. ASTM Special Tech. Publication, 410 (1966)

/24/ Erdogan, F. and G.C. Sih: On the crack extension in plates under plane loading and transverse shear. ASME J.Basic Engr. 85,519 (1963)

/25/ Sih, G.C.: Some basic problems in fracture mechanics and new concepts. Engr.Fract.Mech. 5,365 (1973)

/26/ Sih, G.C.: Strain-energy-density factor applied to mixed mode crack problems. Int.J.Fract. 10,305 (1974)

/27/ Sih, G.C. and B. MacDonald: Fracture mechanics applied to engineering problems – strain energy density fracture criterion. Eng.Fract. Mech. 6,361 (1974)

/28/ Sih, G.C.: Surface layer energy and strain energy density for a blunted notch or crack. Proc.Int.Conf.Prosp.Fract.Mech., Delft (1974)

/29/ Sih, G.C.: Mechanics of fracture: linear response. Proc. of the First Int.Conf. on Num.Meth. in Fract.Mech., University College, Swansea 155 (1978)

/30/ Hussain, M.A., S.L. Pu, and J. Underwood: Strain energy release rate for a crack under combined Mode I and Mode II. Frac. Analysis, ASTM-STP 560 2 (1974)

/31/ Kordisch, H. and E.Sommer: Bruchkriterien bei überlagerter Normal- und Scherbeanspruchung von Rissen. Report W6/77, Fraunhofer-Gesellschaft, Institut für Werkstoffmechanik, IWM, Freiburg, West Germany.

/32/ Ingraffea, A.R.: Mixed-mode fracture initiation in Indiana limestone and Westerly granite. Proc. 22nd U.S. Symposium on Rock Mechanics, Cambridge, MA 186 (1981)

/33/ Saouma, V., A.R. Ingraffea, and D. Catalano: Fracture toughness of concrete: K_{Ic} revisited. ASCE J.Engr.Mech.Div., in press.

/34/ Beech, J. and A.R. Ingraffea: Three-dimensional finite element stress intensity factor calibration of the short rod specimen. Int. J.Fract. 18 3 217 (1982)

/35/ Blandford, G.E.: Automatic two-dimensional quasi-static and fatigue crack propagation using the boundary element method. Ph.D. dissertation, Cornell University (1981)

/36/ Saouma, V.E.: Interactive finite element analysis of reinforced concrete: a fracture mechanics approach. Ph.D. dissertation, Cornell University (1981)

/37/ Ingraffea, A.R., G. Blandford, and J.A. Liggett: Automatic modelling of mixed-mode fatigue and quasi-static crack propagation using the boundary element method. Proceedings 14th National Symposium on Fracture Mechanics, ASTM-STP 791 (1981) in press.

/38/ Saouma, V.E. and A.R. Ingraffea: Fracture mechanics analysis of discrete cracking. Proceedings, IABSE Colloquium on Advanced Mechanics of Reinforced Concrete, Delft 393 (1981)

/39/ Hoek, E.: Rock fracture around mining excavations. Proc. Fourth Int. Conf. Strata Control and Rock Mech., Columbia University, NY 334 (1964)

/40/ Haber, R.B., M.S. Shephard, R.H. Gallagher, and D.P. Greenberg: A generalized graphic preprocessor for two-dimensional finite element analysis. Computer Graphics, a quarterly report of SIGGRAPH-ACM 12, 3,323 (1978)

/41/ Schulman, M.A.: The interactive display of parameters on two- and three-dimensional surfaces. M.S. thesis, Cornell University (1981)

/42/ Lajtai, E.Z. and V.N. Lajtai: The collapse of cavities. Int.J. Rock Mech.Min.Sci. and Geomech.Abstr., 12,81 (1975)

/43/ Paul, B. and M.D. Gangal: Why compressive loads on drill bits produce tensile splitting in rock. SPE 2392, Proc. Fourth Conf. on Drilling and Rock Mech., University of Texas at Austin 109 (1969)

/44/ Wan, F.-D., L. Ozdemir, and L. Snyder: Prediction and verification of tunnel boring machine performance. Paper presented at Euro Tunnel, Basel, Switzerland (1978)

/45/ Dick, E. and K. Peter: Generation of deep cracks in glass. J.Amer. Ceram.Soc. 52,338 (1969)

/46/ Perucchio, R., A.R. Ingraffea, and J.F. Abel: Interactive computer graphic preprocessing for three-dimensional finite element analysis. Int.J.Num.Meth.Engrg. 18 909-926 (1982)

/47/ Perucchio, R. and A.R. Ingraffea: Interactive computer graphic preprocessing for three-dimensional boundary integral element analysis. J. of Comp. and Struct. 16 153-166 (1983)

DYNAMIC PHOTOELASTICITY AND HOLOGRAPHY APPLIED TO CRACK AND WAVE PROPAGATION

H.P. Rossmanith
Institute of Mechanics
Technical University Vienna, Austria

W.L. Fourney
Department of Mechanical Engineering
University of Maryland, USA

The field of elastodynamics covers the class of problems in solid mechanics where the inertia term on the right hand side of the equations of motion

$$(\lambda+2\mu)\,\frac{\partial e}{\partial x} - \mu\,\frac{\partial \omega}{\partial y} = \rho\,\frac{\partial^2 u}{\partial t^2} \qquad (\lambda+2\mu)\,\frac{\partial e}{\partial y} + \mu\,\frac{\partial \omega}{\partial x} = \rho\,\frac{\partial^2 v}{\partial t^2} \quad ,$$

where e is the dilatation and ω is the in-plane rotation of an element, may not be neglected because of rapid changes of stress and displacement in time. These variations of the stresses are due to loads or displacements which change in time, or they are due to relatively sudden changes in the geometry of the body. From the broad field of elastodynamics only stress wave propagation, fracture propagation and their interaction, and penetration (high speed impact) problems will be highlighted here.

Collision impact and vibration with possible exception of ultrasonic excitation both involve time-varying forces which, in general, are relatively long compared to the observation period. Because of cyclic force application and the difficulties involved in the proper photoelastic scaling of material damping, dynamic photoelasticity has not been extensively employed in problems of vibration. Although, in collision impact problems the developement of stresses depends upon the velocity of impact and the elastic wave transmission characteristics of the colliding bodies, the stress waves are often of such a low magnitude that fringe multiplication techniques are required to develop a feasable isochromatic pattern. Quasi-static problems, such as transient thermal stress problems, where the inertia term is negligible, can be investigated using normal methods of recording photoelastic fringe patterns.

1. *Photoelasticity*

This section summarizes the basic relationships of the theory of photoelasticity. Many transparent noncrystalline materials that are optically isotropic in a stress free condition become temporarily birefringent when subjected to stress loading. This optical anisotropy persists while the material is stressed but disappears when the stresses are removed. The method of photomechanics is based on this physical phenomenon.

Maxwell formulated the relationship between the change in the indices of refraction of a material exhibiting temporary birefringence and the loads applied (for plane stress)/1,2/:

$$n_1-n_o = \sigma_1 d_1 + \sigma_2 d_2 \quad , \quad n_2-n_o = \sigma_1 d_2 + \sigma_2 d_1 \tag{1}$$

where σ_1, σ_2 = principal stress at point, n_o= refractive index associated with unstressed state, n_1, n_2= refractive indices associated with stressed state and d_1, d_2= stress optic coefficients.

From equ.(1) follows

$$\Delta n = n_2-n_1 = d_o(\sigma_1-\sigma_2) \tag{2}$$

where d_o is the relative stress optic coefficient. During birefringence the two perpendicular components of a light ray propagate with different speeds

$$v_1 = c'/n_1 \quad , \quad v_2 = c'/n_2 \quad , \tag{3}$$

with c' the speed of light in vacuum. Upon passage of a plane layer of thickness h the retardation is

$$\Delta t = t_1 - t_2 = h/v_1 - h/v_2 = h/c'(n_1 - n_2) = h/c'.\Delta n \tag{4}$$

or the relative linear phase shift

$$\delta s = s_1 - s_2 = h\,\Delta n\, c_L'/c' \sim h\,\Delta n \quad . \tag{4'}$$

The relative angular phase shift

$$\Delta = \delta s\, 2\pi/\lambda = \Delta n\, 2\pi h/\lambda \tag{5}$$

and equ.(2) one obtains the basic stress optic equation (λ= wave length)

$$\Delta = (\sigma_1 - \sigma_2)\, 2\pi h d_o/\lambda \tag{6}$$

which is frequently expressed in the form

$$\sigma_1 - \sigma_2 = N\, f_\sigma /h \ . \tag{7}$$

Here, $N = \Delta/2\pi$ is the relative retardation in terms of a complete cycle of retardation, and $f_\sigma = \lambda/d_o$ (N/m) is known as the material fringe value.

Application of Hooke's law (plane stress) yields also

$$N\, f_\sigma /h = (\varepsilon_1 - \varepsilon_2)\, E/(1+\nu) \quad , \tag{8}$$

where $\varepsilon_1, \varepsilon_2$ = principal strains.

Equ.(7) depends on the wave length of light utilized, hence f_σ has to be measured.

The principal stress difference $\sigma_1 - \sigma_2$ can be determined in a two-dimensional model if N is measured at each point in the model. This can be accomplished by means of optical devices termed polariscopes (see e.g. Ref./1/).

In a circular polariscope as shown in Fig.1 an incident monochromatic light wave is resolved by the polarizer into components which vibrate parallel and perpendicular to the axis of the polarizer the parallel component being transmitted only.

The plane polarized light beam emerging from the polarizer enters the first quarter-wave plate where it is resolved into components E_s and E_f with vibrations parallel to the fast and slow axes oriented at 45° with respect to the axis of the polarizer. As the light components pass the quarter-wave plate they develop a relative angular phase shift $\Delta=\pi/2$ which gives rise to a counterclockwise rotation of the light vector. This vector enters the stressed model which, due to transient or sustained loads , acts as a temporary inhomogeneous wave plate. The components of the light vector are again resolved into components E_{s1}, E_{s2} and E_{f1}, E_{f2} which vibrate parallel to the principal stress directions of the model, accumulating an additional relative

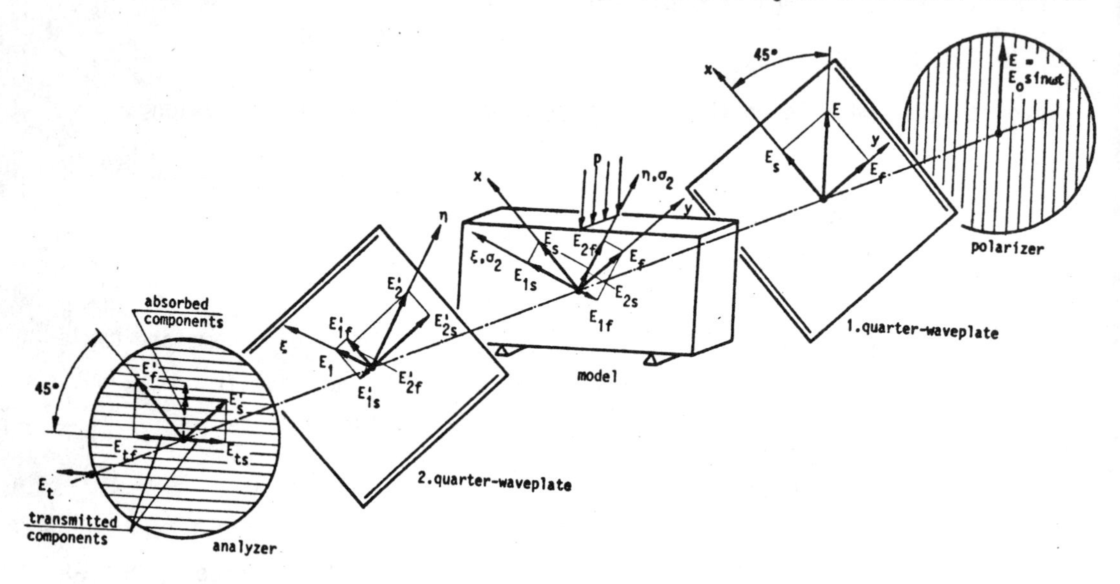

Fig.1 A stressed photoelastic model in a circular polariscope

retardation Δ, equ.(5), during passage through the model. While passing the second quarter-wave plate, a relative phase shift of $\Delta=\pi/2$ develops between the fast and slow components yielding linearly polarized light, the vertical component of which being absorbed by the analyzer. The light emerging from the analyzer (horizontal component) is associated with the intensity distribution

$$I_{\perp} = M \sin^2(\Delta/2) = M \sin^2(\pi N) \quad , \qquad (9)$$

which is a function only of the principal stress difference. From equ.(9) follows that extinction ($I_{\perp}=0$) will occur when $\Delta/2=n\pi$ (n=0,1,2,..) (dark field arrangement). Since $N=\Delta/2\pi=n$, integer order isochromatics appear as black lines.

If the orientation of polarizer and analyzer are parallel (light field arrangement) the intensity distribution is given by

$$I_{\parallel} = M \cos^2(\Delta/2) = M (1 - I_{\perp}/M) . \tag{10}$$

Extinction ($I_{\parallel}=0$) will occur when $\Delta/2 = (n+1/2)\pi$ (n=0,1,2,..). Since $N=\Delta/2\pi = n+1/2$, integer order isochromatics appear as white bands. The first fringe to appear is a black line of the order N=1/2.

In commonplace photoelasticity two photoegraphs (one light-field, the other dark-field) are recorded for determination of fringe orders. Reduction of fringe width and improvement of contrast by increasing the exposure time see Ref./1/. The use of white light results in coloured fringe patterns where any particular coloured line (isochromatics) is associated with a particular level of stress $\tau_m=(\sigma_1-\sigma_2)/2$. Note, that the loci of extinction vary for different colours. In a dark-field arrangement stress free loci ($\sigma_1=\sigma_2=0$) appear as black spots or lines in a coloured isochromatic fringe pattern. Fringe patterns obtained by circular polariscopes are not disturbed by isoclinics which appear in linear polariscope fringe patterns.

2. *High Speed Photography and Requirements of an Optimum Photographic System in Dynamic Photoelasticity*

Perhaps the most difficult aspect of dynamic photoelasticity is recording the isochromatic fringe pattern which represents the transient state of stress. Experimental recording is difficult because the fringes propagate at extremly high velocities which range up to 2 500 m/s for dilatational (P) waves in two-dimensional models fabricated from Homalite 100. The instruments required for recording are often sophisticated and very expensive and, as a result, most laboratories are not equipped to deal with dynamic photoelasticity.

Since research in the field of dynamic photoelasticity was initiated by Tuzi /3/ in 1928, there has been a continuous development of high speed photographic systems. With the advent of newer high speed films, extremely intense light sources, advanced camera designs, and reliable electronic circuitry, continuous improvement has been made in the quality of the photographs of dynamic fringe patterns propagating at high velocity in photoelastic models.

The characteristics which determine the adaptability of a high-speed recording system to dynamic photoelasticity include:

(a) framing rate,

(b) resolution,

(c) exposure,

(d) synchronisation,

(e) size of field and recording image.

The objective of a dynamic photoelastic study is to obtain a prescribed time-sequenced series of negatives exhibiting sharply defined isochromatic fringe patterns /4/. The patterns represent the propagation of several types of stress waves which occur during the dynamic event. The velocities of the fringes being recorded may vary considerable in a high-modulus photoelastic model. For instance, typical velocities of the dilatational (P) wave, shear (S) wave, and Rayleigh (R) wave are 1900, 1075, and 1000 m/s, respectively. In order to adequately record these high-velocity fringes, a camera is required which operates in the range between 5o ooo and 5oo ooo frames/ second. The lower framing rates are used to study Rayleigh wave propagation over reasonably large model distances; the higher rates are used in examining dilatational waves interacting with modelboundaries where changes in the fringe pattern occur very rapidly and fringe velocities exceed those of the incident wave.

Resolution of dynamic fringe patterns is a function of the exposure time t_e, the velocity of the fringe packet (the group velocity), c_f, and the fringe gradient G /5/. The fringe movement during the exposure period, which is approximated by $s = c_f t_e$, must be small relative to 1/G to maintain contrast and to prevent washout or blurring on the negative. To minimize s, exposure times approaching zero are required, which represents a condition impossible to achieve. In practice values of t_e, are held as short as possible, 0.5 to $1.0 \times 1o^{-6}$ sec are considered acceptable, provided fringe gradients do not exceed 4 to 8 fringes per cm.

Proper exposure of film in dynamic photoelasticity is usually a more difficult problem than in other high-speed photographic applications because of the large light losses in the optical system. A typical dynamic polariscope utilizing two sheets of circular polaroid (HNCP-38) will absorb about 84 percent of the incident light. Also, a significant amount of light is absorbed by the notched filter, which must be employed to give a close approximation to monochromatic light. In most cases, the intensity of light transmitted to the film is less than 1 percent of the intensity of the intensitiy of the light source.

In a dynamic event, the time interval of interest ranges from about 50 μsec, as a lower limit for high-velocity waves in small-sized models, to about 300 μsec, as an upper limit for low-velocity waves in relatively large models (model size about 30x30 cm). The dynamic event is often initiated by a small explosive charge which loads the model. The starting of the camera is usually delayed to permit the fringe pattern to propagate from the load source to the region of interest in the model. The preselected delay times usually range from 10 to 100 μsec. The method of introducting delay into the recording system must be variable and the control such that the actual delay period be within ± 2 μsec of the preset delay period.

The required field size (25 - 45 cm ∅) of a dynamic polariscpoe is usually large in order to investigate a wide range of problems which arise in geophysical and in engineering applications. The size of the image on the film must also be large, since negatives on high-speed films do not produce fine quality prints when the amount of enlargement is high. Usually an image-to-field magnification ratio of about one-tenth produces images of 2.5 to 5 cm in diameter which can be subsequently enlarged by a factor of 5 to 10 without encountering objectionable grain-size effects.

Of the many different recording systems employed today only the Cranz-Schardin high-speed camera will be presented here.

2.1. *Cranz-Schardin Camera*

This camera developed originally by Cranz and Schardin in 1929 /6/ has been used extensively in fluid dynamics, ballistics, fracture mechanics, wave propagation and impact studies /7-10/. Recent applications of dynamic photoelasticity are related to problems in geophysics /11-14/, to optimal blasting and fragmentation in mining engineering /15-18/, and to fundamental phenomena of crack-wave interaction /19-23/.

The Cranz-Schardin camera is comprised of three basic subsystems which include: the spark-gap assembly, the optical bench and the control circuits for synchronization /24/.

The Spark-Gap Assembly: A near square-like array of p x q (Maryland: p=q=4, Austria: p=4, q=6) spark gaps provides a time-sequenced series of short-duration, high intensity light flashes used both to illuminate the object and to shutter the camera. The flash pulse duration is in the range of 400-500 nanoseconds where exposure time is measured at 1/3 of the peak intensity. The array of spark gaps is energized by a series connected LC-line. With the capacitors charged to 15 kV DC, the firing sequence of the spark gaps G_i is initiated by applying a 2 kV pulse to the trigger gap. The following sequential discharge through the spark gaps produces short intense flashes of light. The framing rate f can be controlled by varying the inductance L for a given capacitor setting C, $f=\pi^{-1}\sqrt{2/(LC)}$. Framing rates vary from about 20,000 to 800,000 fps and the interframe time can be varied from spark to spark in observing a single dynamic event.

The optical bench associated with the Cranz-Schardin camera is illustrated in Fig.2 and performs three functions which include polarisation of the light, separation of the images and magnification. The light is polarized by placing two circular polarizers between the two field lenses. The images from the n spark gaps are separated by the geometric positioning of the spark gaps, field lenses and camera lenses. The light from fiber optic end F_i is collected and focused on camera lense L_i to establish the image I_i on a sheet of stationary film. The magnification of the optical system depends upon the selection of the lenses; however, a magnification ratio of -0.1 is common which gives an image of 35 to 50 mm in diameter for relatively large-field lenses (350-500 mm ø).

The dynamic resolution which can be obtained from a Cranz-Schardin camera is about 0.6 fringes/mm for a dilatational (plate)-type stress wave

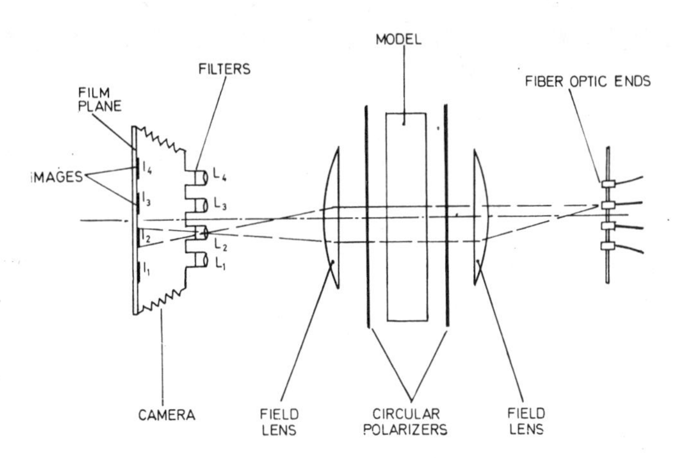

Fig.2 Optical arrangement of a typical Cranz-Schardin multiple-spark gap high-speed recording system adapted for photoelastic experiments

propagating at 2500 m/s. Sufficient intensity levels can be achieved from the sparks at 15 kV to record large-diameter images on a relatively slow (ASA 25/15 DIN = ISO 25/15°) blue sensitive film. The film utilized in recording the image is Agfaortho 25 Professional in conjunction with several different filters, e.g. Kodak K-16, to obtain a transmitted band sufficiently narrow to approximated monochromatic light.

Synchronisation circuits: In dynamic photoelastic applications involving propagating stress waves, event times normally range from about 50 to 300 μs. In this brief interval, the model must be loaded, the camera delayed and then startet, and the time of the individual frames recorded. All of these functions must be initiated on a microsecond time scale of the dynamic event is to be satisfactorily recorded. The synchronisation function is performed by using a number of sequenced electronic circuits. The type of synchronisation circuit depends on the individual type of problem under investigation.

3. *Dynamic Holography*

In dynamic holography the surface of an opaque model is illuminated with a short duration coherent light source. When dealing with photographs taken with this particular light source frequency as well as amplitude information can be recaptured with proper treatment. A typical experimental set up is shown in Fig.3.

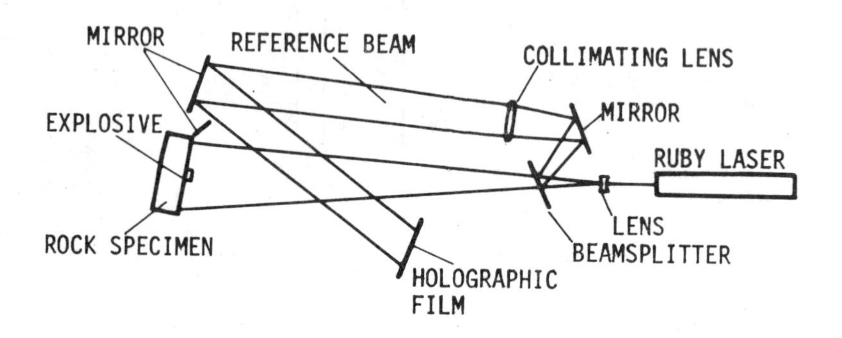

Fig.3 The experimental holographic arrangement.

A pulsed ruby laser (the coherent source) with an exposure time of $50 \cdot 10^{-9}$sec illuminates a model (in this case a rock specimen). The formation of the hologram requires that a portion of the light be diverted by a beam splitter and a series of mirrors so as to strike the holographic plate directly. This is called the reference beam and it will interfere with the light scattered from the object surface. The interference is recorded with a high resolution photographic emulsion. Two exposures are taken of the rock; one in its undeformed state and the other at a controlled time after the initiation of an explosive charge. After processing the exposed film it is re-illuminated with the light from a He-Ne gas laser that duplicates the original reference beam. An observer (or camera) looking through the processed film toward the rock sample views a three-dimensional image of the rock. Superimposed on the rock face are interferometric fringes that are related to the deformation of the surface at the time of the second laser exposure. Holloway et al./25/ has shown that the light intensity distribution of the fringes can be written as

$$I \sim 1 + \cos \frac{2\pi}{\lambda} \{(\hat{r}_s + \hat{r}_v) \cdot \vec{d}\} \tag{11}$$

where λ is the wavelength of the illumination, $\vec{d}$ is the vector surface displacement at a point of interest and $\hat{r}_s$ and $\hat{r}_v$ are unit vectors in the directions of the source and viewing positions, respectively. These are illustrated in Fig.4. A dark band will be present in the image whenever

$$\frac{2\pi}{\lambda}\ \{(\hat{r}_s + \hat{r}_v)\cdot\vec{d}\ \} \quad = \quad n \qquad (\ n \text{ odd}\ ) \tag{12}$$

or

$$(\hat{r}_s + \hat{r}_v)\cdot\vec{d} \quad = \quad N \qquad (\ N = \tfrac{1}{2}, \tfrac{3}{2}, \tfrac{5}{2}, \ldots\ ) \quad ; \tag{13}$$

thus, the dark areas are numbered as half order fringes and light areas as integer fringes.

If α, β, γ, and $\alpha_1, \beta_1, \gamma_1$ are the angles shown in Fig.4 and if $\vec{d}$ is written with respect to the x,y,z axes as $\vec{d} = u\hat{i} + v\hat{j} + w\hat{k}$, then equ.(13) becomes

$$(\cos\alpha + \cos\alpha_1)u + (\cos\beta + \cos\beta_1)v + (\cos\gamma + \cos\gamma_1)w = N\lambda \quad . \tag{14}$$

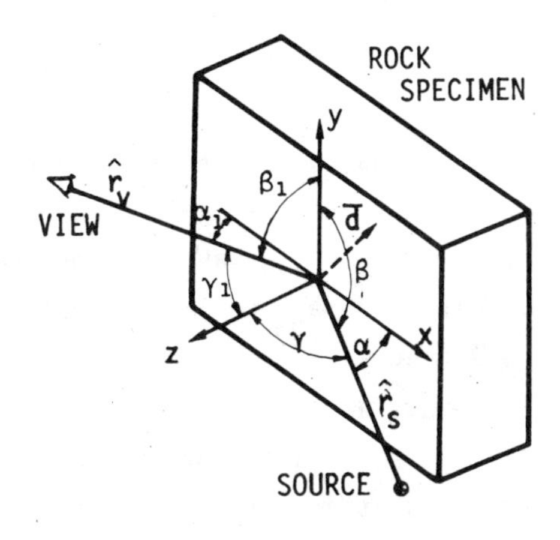

Fig.4 Vector geometry used in the analysis of the holographic fringe pattern

This equation contains the three unknown displacement components, along with the known angles, fringe order, and wave length. Thus at least three views through the same holograms or three separate holograms at different locations are necessary to solve for u,v, and w. Viewing the hologram along the z-axis or taking the hologram along the z-axis yields α or $\beta \to 0$ and restricting the light in the hologram to be in the xz-plane yields

$$2\,w \;=\; N\,\lambda \qquad \text{or} \qquad w \;=\; N\,\lambda/2 \quad . \tag{15}$$

The sensitivity is approximately $\lambda/4$ or 0,152μm for a wavelength $\lambda = 6328$ Å (He-Ne laser) for out-of-plane displacement. The technique is best suited and can be best applied to real materials with no need for fabrication of models.

4. *Applications*

Dynamic photoelasticity and dynamic holography are extremely effective methods to employ in the investigation of a wide variety of transient problems in many fields. To show the versatility of the methods fringe patterns from selected application topics will be briefly demonstrated and discussed.

4.1. *Wave Propagation*

Dynamic photoelasticity is particularly useful in both the qualitative and quantitative evaluation of stress wave problems.

The first example pertains to an explosive wave source on the free edge of a half-plane and subsequent body-wave and surface-wave motions. A typical set of 16 sequential isochromatic fringe patterns is shown in Fig.5. The period of observation varies from 60 to 200 μs, while the small amplitude leading P-wave travels further than across the entire field of view. Presence of a strong S-wave is evident and a Rayleigh-wave (R-wave) the photoelastic trace of which can clearly be observed from frame 6 to 16 propagates along the half-plane boundary to the right. A head-wave (von Schmidt shear wave) generated by gracing incidence of the P-wave at the free boundary links up with the tangent to the shear wave front. The enlargement, shown in Fig.4 depicts one of the transient stress configurations at one particular point in time and demonstrates the individual structure of the particluar wave types.

From a geophysical point of view the buried pulse source is of utmost importance in earthquake engineering and tectonics. When the explosive source ignites at a certain depth below a free surface, a longitudinal wave train with rotational symmetry propagates radially outward from the center of excitation. The free surface conditions give rise to reflected longitudinal (P) and shear waves which can clearly be identified in Fig.6. Further studies

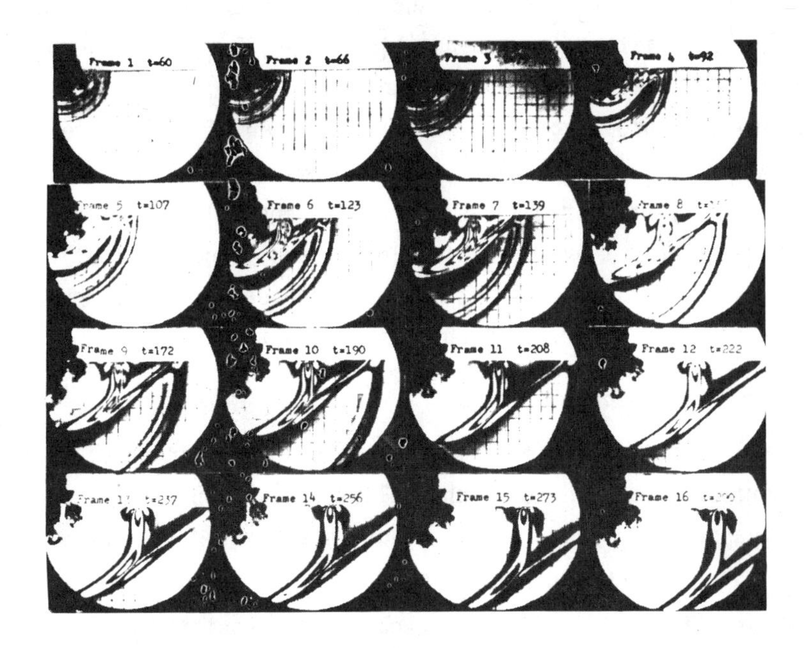

Fig.5 Dynamic isochromatic fringe pattern representing stress-wave propagation recorded with a Cranz-Schardin camera.

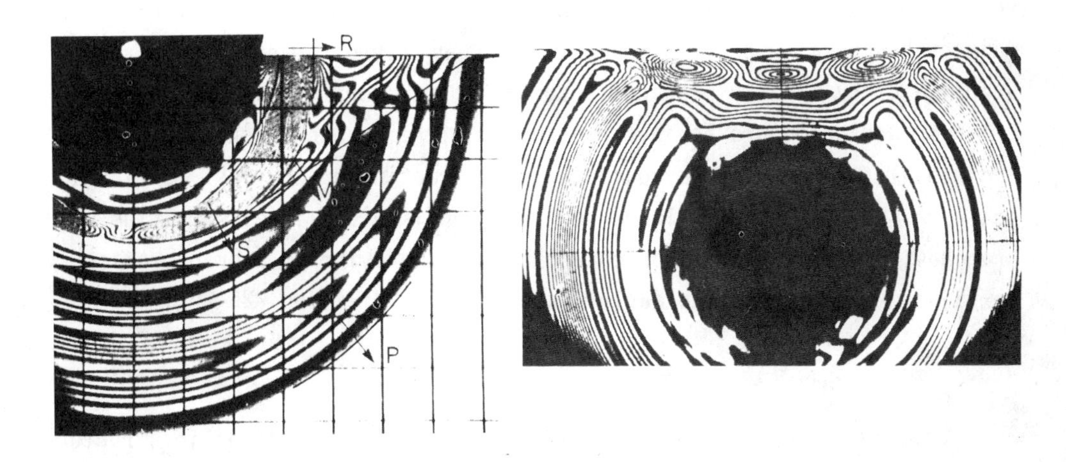

Fig.6 Enlargement of Fig.5 showing longitudinal (P) and shear (S) waves, head wave (V) and Rayeigh (R) surface wave.

Fig.7 Reflection of a P-wave with circular wave front at the free surface of a half-plane.

deal with wave propagation in two- and multilayered similar and dissimilar models with many different impedance mismatches, Rayleigh-wave propagation along curved, cracked, and stepped boundaries.

In many wave propagation problems the number of waves generated by reflection, refraction, and diffraction becomes very large and several of the theoretically predicted waves may not be possibly identified in the resulting pattern because they are of too low an amplitude. Evaluation of the resulting isochromatic fringe pattern may then become very complex and the wave identification a difficult and cumbersome task.

4.2. *Fracture Mechanics*

A large variety of interesting problems in dynamic fracture mechanics can be approached using dynamic photoelasticity. When a structural model containing a stationary or running crack is subjected to static or transient loading, a nested group of fringe loops form above and below the crack path. A typical set of isochromatic fringe patterns associated with high-speed crack propagation traversing a plate specimen is shown in Fig.8. The period of observation is about 440 μs and the crack propagates with nearly constant velocity of about 380 m/s during the process. In the last four frames the main crack has attempted to branch. The unsuccessful branches appear just before a crack finally branches into two or more branches. If successful crack branching occurs as is shown in Fig.9 the branch cracks exhibit individual fringe systems around their interacting crack tips. The large scale fringe pattern with large seemingly undisturbed lobes engulfing both branches reveals a notch type characteristic behavior of a branch crack system pertaining at least to the early phase of crack division /25-27/.

4.3. *Crack - Wave Interaction*

Crack-wave interaction problems arise in many situations in structural engineering, geophysics, and mining engineering. In non-destructive flaw detection methods stress waves are introduced into the component suspected to contain a flaw and the interaction of these waves with flaws are studied in order to detect the flaw and to predict its size, shape and orientation.

Fig.8 Sequence from a sixteen frame Cranz-Schardin camera system of the photoelastic fringe patterns associated with a running crack traversing a plate specimen.

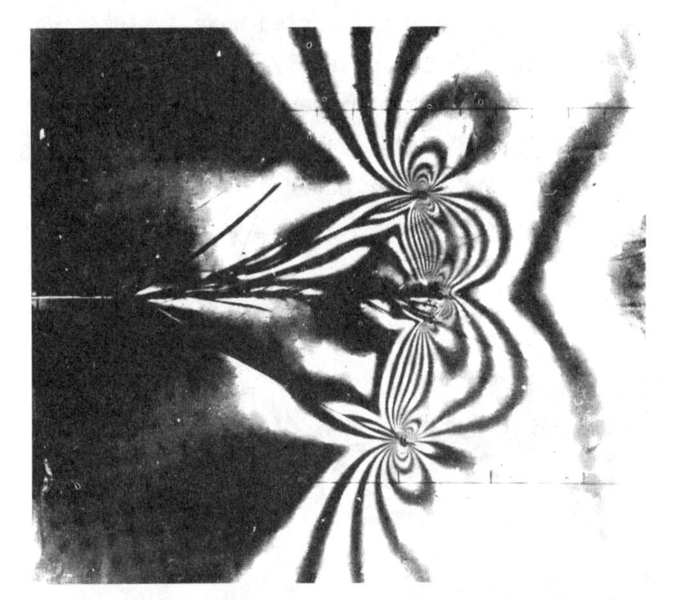

Fig.9 Dynamic multiple crack branching at very high crack speed in a Single-Edge-Notched specimen fabricated from Homalite 100

For embedded flaws and inhomogeneities longitudinal and shear waves impinge on the flaw border generating reflected and diffracted wave systems. Wave intensities and arrival times at several conveniently located receiver positions assist in the identification of the hidden inhomogeneity. Complete identification of embedded crack-type defects is a very difficult task enormously complicated by secondary and higher order diffraction processes at crack tips. Fig.10 shows a dynamic isochromatic fringe pattern associated with fully developed P-wave diffraction about both crack tips of an embedded finite crack and the early phase of interaction of the impinging S-wave about

Fig.10 Primary and secondary P-wave diffraction about the tips of an embedded finite crack.

the left hand side crack tip of a stationary crack-type defect buried at some depth below the free surface. Rayleigh waves associated with diffracted shear waves travel along the free crack surfaces. Notice, that the very complex stress pattern in the vicinity of the stationary crack tips gives rise to mixed-mode stress intensifications. In particular, the left hand side crack tip carries a dominant mode-1 butterfly fringe pattern whereas

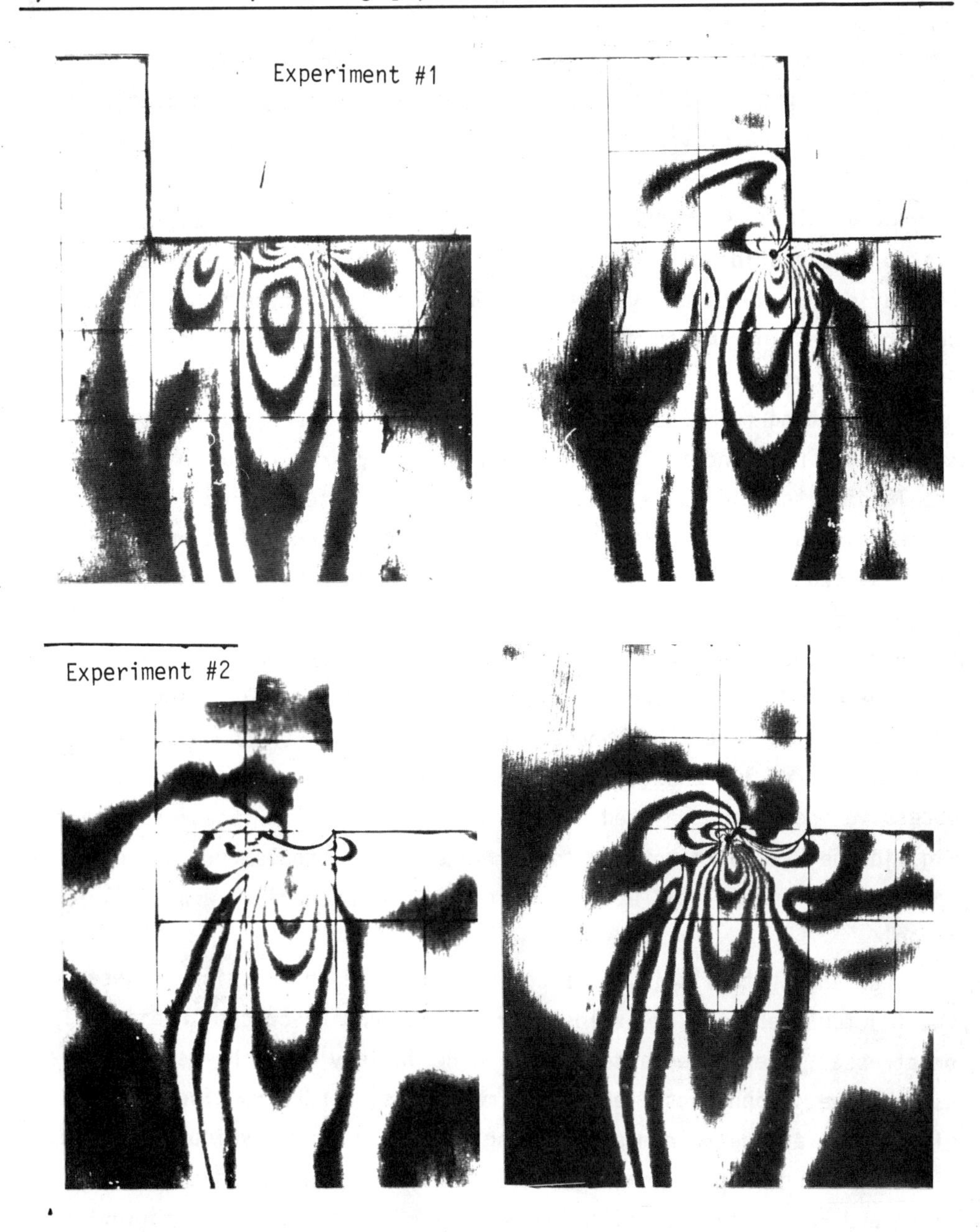

Fig.11 Successive fracture development by Rayleigh-pulse interaction with a sharp re-entrant corner in a stepped half-plane /30/.

the right hand side crack tip shows a dominat mode-2 fringe pattern /22,28/.

Dynamic photoelasticity was utilized in a series of experiments to investigate the interaction of a Rayleigh wave with a crack on the surface of a half-plane /29/. Crack generation due to the build-up of high stress concentrations in a re-entrant corner during the passage of a Rayleigh-wave past a step in elevation is shown in Fig.11/30/. Part of the energy carried by the Rayleigh-wave is being converted into energy to create new fracture surface. Two experiments have been conducted with the same model, experiment #1 with the virgin model (no crack) and experiment #2 with the corner-cracked model resulting from experiment #1. Fig.11a shows the incident Rayleigh-wave traveling along the free specimen boundary from right to left. When the leading tensile pulse of the Rayleigh wave interacts with the re-entrant corner high stress concentrations exceeding the fracture strength of the material cause fracture initiation as shown in Fig.11b. With this fracture emerging from the corner a second experiment identical to the first one is performed. The R-wave now experiences the newly created fracture surface as an extended part of the free surface and, hencefore, propagates along the lower crack face. This phase is illustrated in Fig.11c. Since the R-wave carries most of its energy in a thin layer bordering the free surface, excessive energy can be used to create new fracture surfaces as shown in Fig.11d. The R-wave re-entrant corner interaction problem presented has many applications in reality, e.g. in tectonics, in mechanical and civil engineering.

Current photoelastic research in crack-wave interaction focusses on the fracture behavior of explosively driven running cracks in biaxially prestressed layered media to study a wide variety of problems in fields such as fracture, geophysics, tectonics, mechanics, nondestructive testing, mining, gas and petroleum engineering. Although more development in the field of stress separation and scaling relations is necessary, the quantitative and restricted quantitative results which can be achieved today assist the engineer and scientist in the interpretation and understanding of the phenomena involved and, hence, contribute significantly toward solutions of complicated engineering problems.

References

/1/ Dally J.W. and W.F.Riley: Experimental Stress Analysis, 2nd Edition, McGraw-Hill (1978)

/2/ Föppl L,and E.Mönch: Praktische Spannungsoptik, 3rd Edition, Springer-Verlag, Berlin (1972)

/3/ Tuzi Z.: Photographic and kinematographic study of photoelasticity. Jnl.Soc.Mech.Eng.31,334-339 (1928)

/4/ Riley W.F. and J.W.Dally: Recording dynamic fringe patterns with a Cranz-Schardin camera. Exp.Mech.9,27N-33N (1969)

/5/ Dally J.W.: Dynamic Photoelasticity, CISM Lectures, Udine 1976, Springer-Verlag, Vienna-New York (1976).

/6/ Cranz C. and H. Schardin: Kinematographie auf ruhendem Film und mit extrem hoher Bildfrequenz, Zeits.f.Physik, 56,147 (1929)

/7/ Christie D.G.: A multiple spark camera for dynamic stress analysis. Jnl.Phot.Sci.3, 153-159 (1955)

/8/ Wells A.A. and D.Post: The dynamic stress distribution surrounding a running crack - A photoelastic analysis. Proc.SESA 16,1,69-92 (1957)

/9/ Dally J.W.: An introduction to dynamic photoelasticity. Exp.Mech.20, 409-416 (1980)

/10/ Dally J.W. and W.F.Riley: Stress wave propagation in a half-plane due to a transient point load."Developments in Theoretical and Applied Mechanics",Vol.3, Pergamon Press, New York,357-377 (1967)

/11/ Riley W.F. and Dally J.W.: A photoelastic analysis of stress wave propagation in a layered model, Geophysics 31,881-889 (1966)

/12/ Dally J.W. and S.A.Thau: Observations of stress wave propagation in a half-plane with boundary loading. Int.Jnl.Solids and Structures,3, 293-308 (1967)

/13/ Burger C.P. and Riley W.F.: Effects of impedance mismatch on the strength of waves in layered solids. Exp.Mech.14,129-137 (1974)

/14/ Dally J.W. and D.Lewis: A photoelastic analysis of propagation of Rayleigh waves past a step change in elevation.Bull.Seis.Soc.Amer. 58,539-562 (1968)

/15/ Dally J.W.: Dynamic photoelastic studies of dynamic fracture. Exp. Mech.19,349-361 (1979)

/16/ Reinhardt H.W. and Dally J.W.: Dynamic photoelastic investigation of stress wave interaction with a bench face. Trans.Soc.Min.Eng.AIME, 250,35-42 (1971)

/17/ Fourney W.L. et al: Fracture initiation and propagation from a center of dilatation. Int.J.Fract.11,1011-1029 (1975)

/18/ University of Maryland, Photoemechanics Laboratory Report Series on "Blasting and Fragmentation", 1974-1982.

/19/ Schardin H.: Ergebnisse der kinematographischen Untersuchung des Glasbruchvorganges, Part I,II,III. Glastechnische Berichte, 23 (1950)

/20/ Kolsky H.: Stress Waves in Solids, Dover Publ. New York (1963)

/21/ Kerkhof F.: Report Series of the Institut für Festkörperphysik (IWM) Freiburg, Germany.

/22/ Rossmanith H.P. and A.Shukla: Photoelastic investigation of stress wave diffraction about stationary crack tips. J.Mech.Phys.Solids,29,397-412 (1981)

/23/ Rossmanith H.P. and A.Shukla: Dynamic photoelastic investigation of interaction of stress waves with running cracks.Exp.Mech.21,415-422, (1981)

/24/ Dally J.W. and Brillhart L.V.: Application of multiple spark gap camera to dynamic photoelasticity. J.Soc.Motion Pic.Telev.Sci. 77,116-120 (1968)

/25/ Holloway D.C. et al.: A study of surface wave propagation in rock by holographic interferometry. Exp.Mech.17(8),281-289 (1977)

/26/ Kobayashi A.S. et al: Crack branching in Homalite 100 sheets. Eng.Fract. Mech.6,81 (1974)

/27/ Rossmanith H.P.: Crack branching in brittle materials - Part I: Analytical Aspects; Part II: Dynamic Photoelastic Investigation. NSF-Reports University of Maryland (1980)

/28/ Rossmanith H.P. and W.L.Fourney: Fracture initiation and stress wave diffraction at cracked interfaces in layered media. I-Brittle/brittle transition. Rock Mechanics 14,209-233 (1982)

/29/ Thau S.A. and J.W.Dally: Subsurface characteristics of the Rayleigh wave. Int.Jnl.Sci.7,37 (1969)

/30/ Rossmanith H.P.: The role of Rayleigh waves in rock mechanics. Proc. 18th U.S.Rock Mechanics Conf. Nevada, 138-147 (1978).

ELASTIC WAVE PROPAGATION

H.P. R o s s m a n i t h
Institute of Mechanics
Technical University Vienna
A-1040 Vienna, Austria

When a mechanical disturbance propagates in an isotropic elastic body inertial and elastic properties control the velocity of the advance and the form of the disturbance frequently changes with progression depending upon the initial character of the displacement and the history of propagation. Elastodynamic theory shows that only two types of waves can be propagated through an unbounded elastic solid: a longitudinal (dilatational,P-)wave, and a transverse (rotational, shear, S-)wave. Particle motion in a P-wave (S-wave) is parallel (normal) to the direction of wave propagation. When the solid has a free surface or an interface between two different media surface waves may also be propagated. In layered structures several types of waves may be generated and propagated. In most practical engineering applications plane, cylindrical, and spherical waves form the majority of waves encountered.

1. *Waves in unbounded media*

1.1. *Plane waves*

For a general mechanical disturbance extending in a pre-stressed (S_x, S_y) unbounded isotropic homogeneous elastic medium wave equations may be derived from the equations of motion for the displacement potentials ϕ and ψ in x- and y-directions, respectively:

$$\text{x-direction:} \quad \Delta\phi\, c_{1x}^2 = \phi_{,tt} \quad ; \quad \Delta\psi\, c_{2x}^2 = \psi_{,tt} \tag{1}$$

$$\text{y-direction:} \quad \Delta\phi\, c_{1y}^2 = \phi_{,tt} \quad ; \quad \Delta\psi\, c_{2y}^2 = \psi_{,tt} \tag{2}$$

with

$$\begin{aligned} c_{1x} &= \sqrt{(\lambda+2\mu+S)/\rho} && \text{... P-wave velocity along the x-axis} \\ c_{1y} &= \sqrt{(\lambda+2\mu)/\rho} && \text{... P-wave velocity along the y-axis} \\ c_{2x} &= \sqrt{(\mu-S/2)/\rho} && \text{... S-wave velocity along the x-axis} \\ c_{2y} &= \sqrt{(\mu+S/2)/\rho} && \text{... S-wave velocity along the y-axis} \end{aligned} \tag{3}$$

where λ and μ are Lame's constants, ρ is the material density, $S = S_x - S_y$ is the difference of the pre-stress components, and Δ denotes the Laplace-operator.

The relations between stresses σ_x, σ_y, τ_{xy}, strains ε_x, ε_y, ε_{xy}, displacements u and v and wave potentials ϕ and ψ for incremental isotropy are given in Ref./5/.

The eqs.(1) and (2) differ in form and in number from those of the classical theory pertaining to zero initial stress. For non-pre-stressed or hydrostatic stress situations (S=0) classical theory applies.

When the deformation is a function of only one coordinate, e.g. x, eqs.(1) become

$$X_{,tt} = c_i^2\, X_{,xx} \qquad (X = \phi\ ,\psi\ ;\ i=1,2) \tag{4}$$

with the general solution

$$X(x,t) = f_+(x-c_i t) + f_-(x+c_i t) \quad , \tag{5}$$

where the arbitrary functions f_+(f_-) correspond to a wave propagating in the positive (negative) x-direction. The shape f_+ and f_- of a transient wave depends upon the source of disturbance, the distance of travel and the material behavior. The dilatation wave propagates with speed c_{1x} and the shear wave propagates with speed c_{2x} with $c_1/c_2 > 1$ for $S/E > -3$ with E = Young's modulus. In a P-wave both the bulk modulus K and the shear modulus μ control the velocity of propagation because $\lambda+2\mu=K+4\mu/3$. The stress distribution in a plane P-wave follows from the plane strain condition $\varepsilon_t=0$, $\sigma_y = \sigma_z = \sigma_x \nu/(1-\nu)$.

Representative values for wave velocities in zero pre-stress conditions are given in Table 1.

Table I: P-wave and S-wave speeds for common materials (metals, nonmetals, ceramics and rock)

Material	c_1 m/s	c_2 m/s	Material	c_1 m/s	c_2 m/s
Aluminium	6100	3100	Chalk, Texas	2800	1100
Brass	4300	2000	Granite, Quincy	5900	3000
Copper	4560	2250	Granite, Westerly	5800	3200
Glass (window)	6800	3300	Limestone, Solenhofen	6000	2900
Lead	2200	700	Norite, Sudbury	6200	340
Magnesium	6400	3100	Sandstone, Pennsylvania	2900	--
Plexiglas	2600	1300	Taconite, Minnesota	5300	--
Polystyrene	2300	1200	Rubber	1040	27
Steel	5800	3100	Water (room temp.)	1500	0

In rocks the state of stress, temperature, composition, mechanical history, consolidation, saturation, microfractures, high pressures, etc., cause a wide spread in the values of their P-wave speeds (Fig.1).

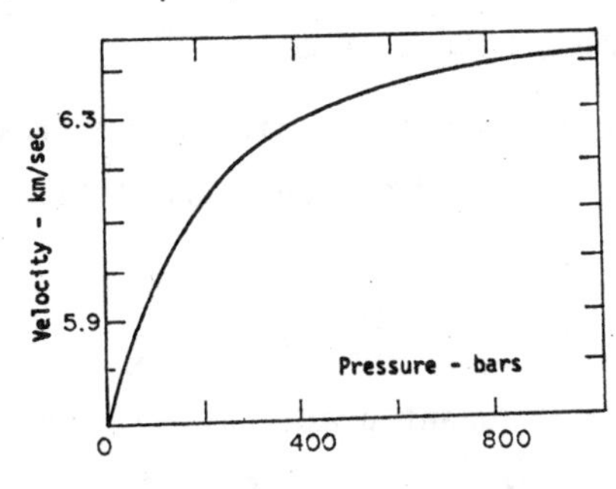

Fig.1 P-wave speed in dolomite at 27°C as a function of confining pressure

During the passage of a transient stress wave material particles are displaced and a relation between the associated particle velocity and the instantaneous stress is obtained from a momentum transfer consideration during progress of a transient stress wave (Fig.2).

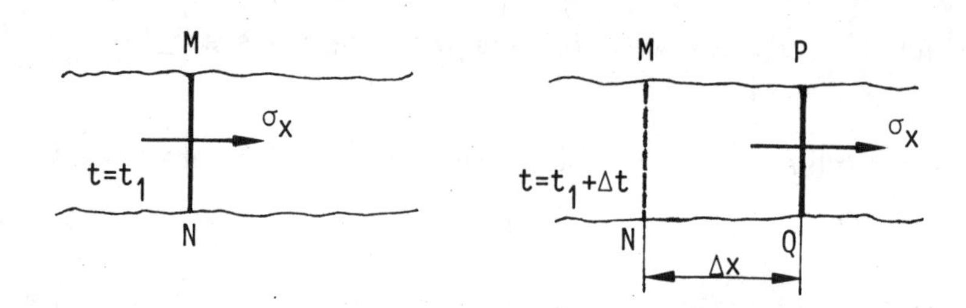

Fig.2 Momentum transfer during passage of a transient stress wave

Newton's second law of motion, impulse is equal to change of momentum, $\sigma_x \, \Delta t = \rho v_x \, \Delta x = \rho v_x \, c_1 \Delta t$, yields

$$\text{P-wave:} \quad \sigma_x = \rho c_1 v_x \qquad \text{S-wave:} \quad \tau_{xy} = \rho c_2 v_y \quad . \tag{6}$$

The product c_1 is called specific acoustic resistance, wave resistance or acoustic impedance of the material. Note, that the linear relationship (6) and the constancy of wave speed enable one to shift from a time to a position description point of view: $(\sigma,t) \leftrightarrow (v,t) \leftrightarrow (v,x) \leftrightarrow (\sigma,x)$.

The displacement of a particle during passage of a P-wave is obtained by integration of equ.(6):

$$d_x(t) = \int_0^t v_x(t)dt = \frac{1}{\rho c_1} \int_0^t \sigma_x(t)dt \quad , \tag{7}$$

with $d_x(T)$ as the resulting displacement upon wave passage.

The momentum dM carried by an infinitesimally thick layer of a plane P-wave is $dM = \rho v_x(x)dx$ and the total momentum per unit cross section is obtained upon integration.

1.2. *Spherical and cylindrical waves*

Propagation of nonplanar waves is of great practical importance because these types of waves are often produced by explosions and highly localized impact. Nonplanar waves change shape as they advance altering considerably the distribution of stress and particle velocity within the wave. A compressional input pulse rapidly develops tensile stresses and becomes oscillatory in the wake of the wave front. Stress or particle velocity at the wave front decay proportional to $1/r$ for a spherical wave and proportional to $r^{-1/2}$ for a cylindrical wave with r being the distance from the source of the disturbance.

1.2.1 *Spherical waves*

If a disturbance spreads out from a point the deformation will depend only on the value of r, the distance from the source. Regarding $r^2=x^2+y^2+z^2$ equ.(4) transforms into

$$\frac{\partial^2(rX)}{\partial t^2} = c^2\frac{\partial^2(rX)}{\partial r^2} \tag{8}$$

with the general solution

$$X(r,t) = \frac{1}{r}\{ f_+(r-ct) + f_-(r+ct) \}, \tag{9}$$

where $f_+(f_-)$ represent a diverging (converging) spherical wave. The amplitude is proportional to 1/r !

If X stands for a scalar displacement potential ϕ, the radial displacement u_r is given by $u_r = \partial\phi/\partial r$.

For practical applications, e.g. explosions in a cavity of radius a the use of the auxiliary quantity $\tau = t-(r-a)/c$ is advantageous, where $\tau = 0$ corresponds to the wave front.

1.2.2 *Cylindrical Waves*

In the case of a cylindrical cavity (2a ø) embedded in a homogeneous, elastic isotropic infinite medium the solution of the wave equation is much more complicated than for the spherically expanding wave. Theory shows, that at the wave front holds

$$\sigma_r = p_0\sqrt{a/r}\,,\quad \sigma_\theta = \sigma_z = p_0\sqrt{a/r}\,(\lambda/(\lambda+2\mu)) \tag{10}$$

i.e. the stresses decrease as $r^{-1/2}$ in contrast to the r^{-2} decrease in static equilibrium. Figs 5a-5c pertain to the cylindrical unit step pressure pulse problem.

1.3. *Superposition of Elastic Waves*

Interference between two or more elastic waves can be analyzed by means of the principle of superposition, first suggested by Lord Rayleigh. The main cases to be considered are interference of the type P-P, P-S, and

S-S waves. Magnitude and directions of the stresses and particle velocities are obtained.

All three characteristic types of wave interaction, P-P, P-S, and S-S, will be encountered in the problem where a detonation is ignited at the center of a square plate. Upon reflection of the outgoing P-wave at the free boundary, the reflected P_rP-waves and S_rP-waves interact. The associated photoelastic fringe pattern to one particular wave interaction configuration is shown in Fig.3, where the individual crossings can easily be detected.

2. *Boundary Effects*

Wave transformations, reflections, refractions and diffractions occur at boundaries and/or interfaces of similar and dissimilar materials with the acoustic impedances playing the major role.

Several different important situations will be mentioned here:

a) normal incidence (at free, rigid, cohesive and noncohesive boundary)

b) oblique incidence (at free, rigid, cohesive and noncohesive boundary).

The following analysis pertains to plane strain conditions.

2.1. *Cohesive Joint*

When a plane elastic wave, σ_1, strikes a plane interface between two dissimilar media ($\rho' c_1'$ versus ρc_1), the conditions at the interface and the initial stress S control the wave reflection/refraction process.

The boundary conditions at the cohesive joint (y=0) are:

continuous displacements

$$u = u' \tag{11}$$

$$v = v' \tag{12}$$

incremental boundary forces

$$\Delta F_x = \Delta F_x' \tag{13}$$

$$\Delta F_y = \Delta F_y' \tag{14}$$

where ΔF_x, ΔF_y ($\Delta F_x'$, $\Delta F_y'$) denote incremental boundary forces per unit initial area in the lower (upper) medium /1/. Primed symbols refer to the upper medium.

Fig.3 Dynamic photoelastic fringe pattern associated with wave reflection from a 90°-corner showing the superposition of P_rP- and S_rP-waves.

The continuity conditions (11)-(14) yield four equations for the amplitude ratios between incident wave and reflected and refracted waves.

An incident P-wave generates reflected P_1P_1 and S_1P_1 waves and refracted P_2P_1 and S_2P_1 waves (Fig.4a), whereas an impinging S-wave with particle motion in the xy-plane (SV-wave) induces reflected P_1S_1 and S_1S_2 waves and refracted P_2S_1 and S_2S_1 waves (Fig.4b).

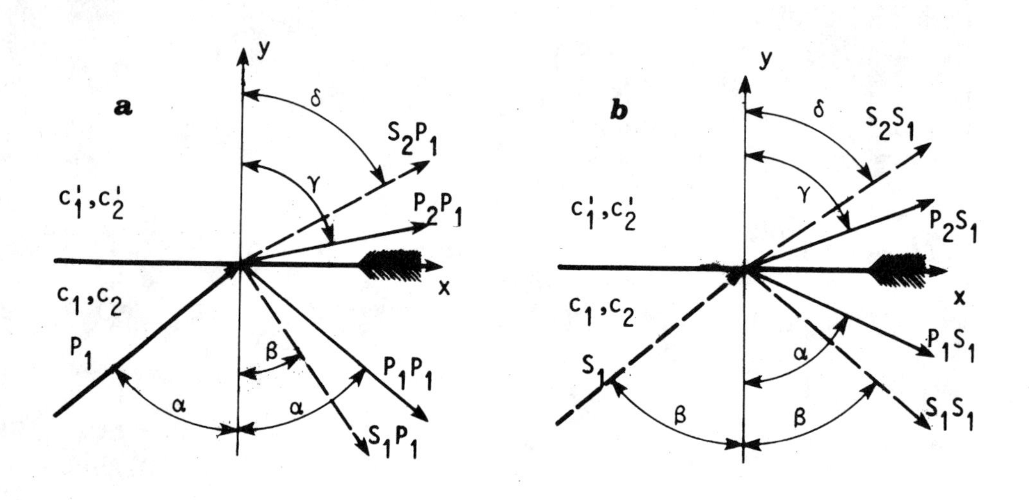

Fig.4a Obliquely incident P-wave Fig.4b Obliquely incident S-wave

Numerical values for reflection and refraction coefficients associated with oblique P-wave incidence, $R_{PP}=\sigma_{P_1P_1}/\sigma_{P_1}$, $R_{SP}=\tau_{S_1P_1}/\sigma_{P_1}$, $T_{SP}=\tau_{S_2P_1}/\sigma_{P_1}$ and $T_{PP}=\sigma_{P_2P_1}/\sigma_{P_1}$, respectively, are shown in Figs.5a and 5b. The angle of P-wave incidence $\alpha = 20°$, $\mu'/\mu=4.16$ and $\rho'/\rho=1.17$; the pre-stress is normalized with respect to the shear modulus: $\eta=S_x/2\mu$ and $\eta'=S_x'/2\mu'$. Positive values of η and η' represent tensile initial stresses and vice versa.

Several special cases are of importance (normal P-wave incidence):

free boundary: $\rho' c_1'=0 \rightarrow T_{PP}=0,\ R_{PP}=-1 \rightarrow$ sgn(inc.wave)=-sgn(refl.wave)
rigid boundary: $\rho' c_1'=\infty \rightarrow T_{PP}=2,\ R_{PP}=1 \rightarrow$ sgn(inc.wave)= sgn(refl.wave)

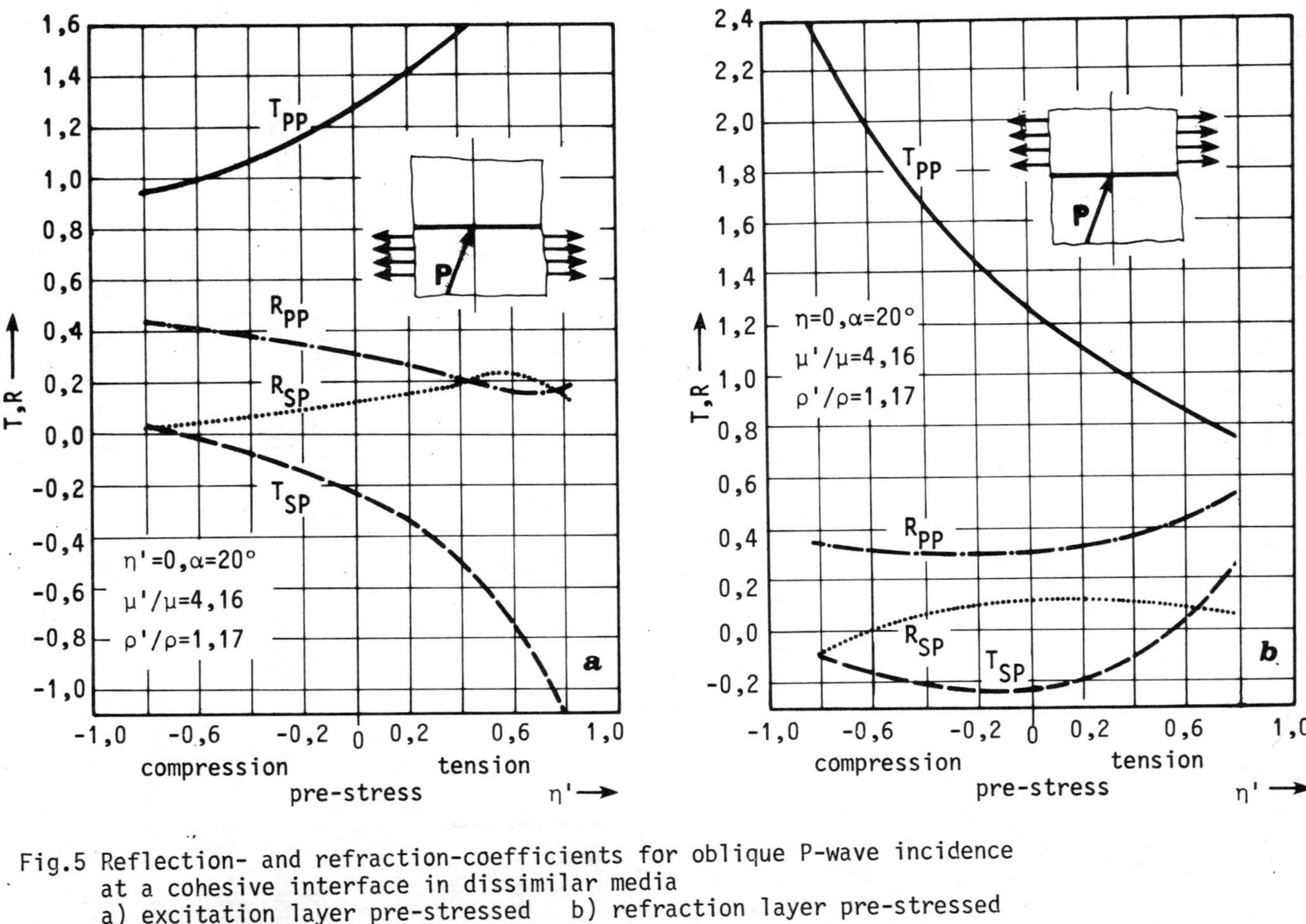

Fig.5 Reflection- and refraction-coefficients for oblique P-wave incidence at a cohesive interface in dissimilar media
a) excitation layer pre-stressed b) refraction layer pre-stressed

Total reflection of SV-waves occurs and the P_2S_1-wave ceases to exist in its classical form when $\beta > \beta_c = \sin^{-1}(c_2/c_1)$. In place of a reflected or refracted plane wave, a disturbance is now generated which decays exponentially with distance from the interface /5/.

When the particle motion in the shear wave is perpendicular to the plane of incidence (xy-plane), SH-wave, displacements and stresses normal to the interface vanish and no P_iS-waves are generated. Although SH-waves play an important role in geophysics and tectonics they will be excluded from consideration here because they do not appear in plane photoelastic recordings.

Explicit expressions for the amplitude ratios for reflection and refraction of P- and SV-waves at perfectly contacting dissimilar interfaces are given in Refs /1,3/.

2.2. *Loose Joints*

For loose joints do not transmit tensile waves, the conditions of continuity of tangential displacement and tangential stress across the interface have to be replaced by the condition of zero shear stress on either side of the joint (Fig.6). Moreover the tangential displacement u is discontinuous for this situation. Thus, boundary conditions (11) and (13) are replaced by the loose joint interface conditions:

$$u \neq u' \quad (11') \qquad \Delta F_x = \Delta F'_x = 0 \quad (13')$$

In the following zero initial stress situations will be considered only.

The ratios of particle velocities of the respective waves are the same as the ratios between the associated amplitudes, e.g. $v_{P_1P_1}/v_{P_1} = R_P$, etc. The distribution of particle velocities for an obliquely incident P-wave is shown in Fig.7.

The relative movement and the total displacement of the face $\overline{ab}$ of the joint is determined by

$$v_{ab,x} = v_{P_1}\sin\alpha + v_{S_1P_1}\cos\beta - v_{P_1P_1}\sin\alpha \quad (15)$$

$$v_{ab,y} = v_{P_2P_1}\cos\alpha + v_{S_2P_1}\cos\beta \quad (16)$$

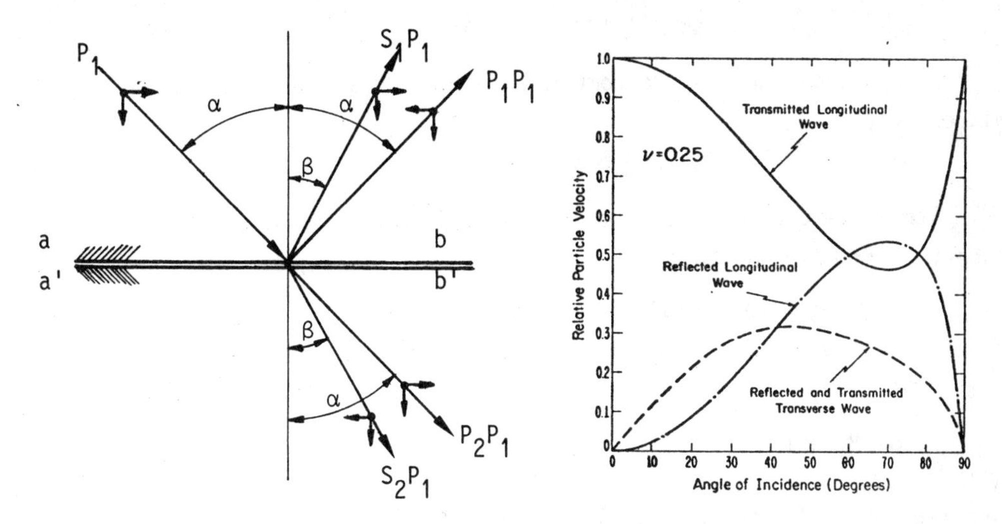

Fig.6 Compressive P-wave transfer across a loose joint

Fig.7 Oblique incidence of P-wave at a loose joint

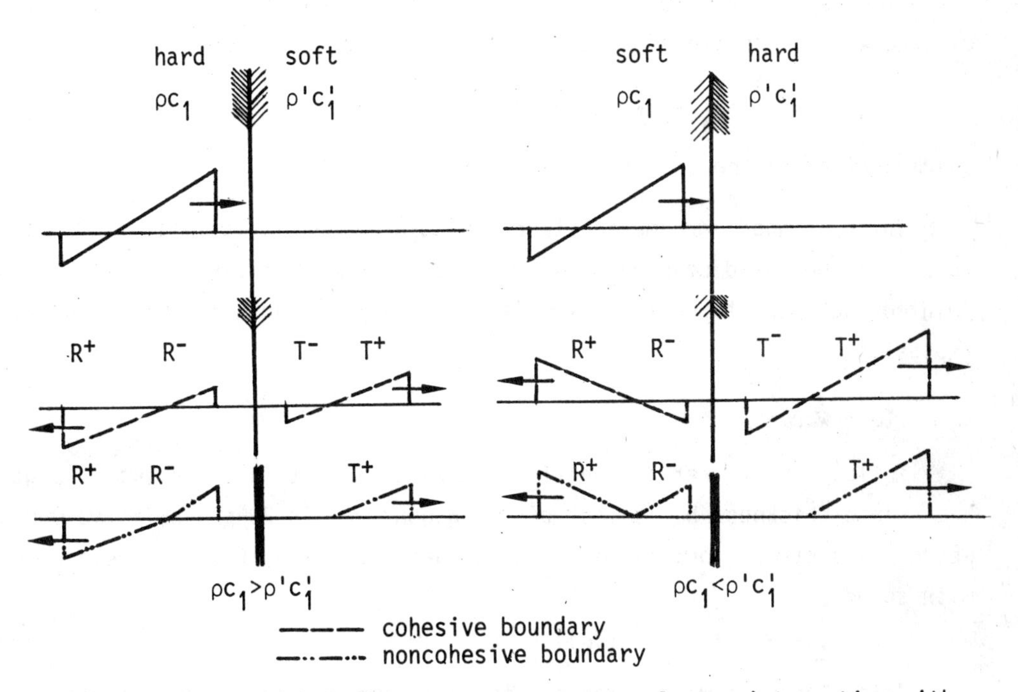

Fig.8 Stress partitioning as a consequence of wave interaction with dissimilar cohesive and noncohesive interfaces

$$d_x = \int v_{ab,x}(t)\, dt \quad . \tag{17}$$

At a loose boundary the compressive part of an incident P-wave is transmitted only:

$$T^+ \geq 0, \quad R^+ \gtreqless 0 \ ; \ T^- = 0 \ ; \ R^- = -1 \tag{18}$$

Stress partitioning as a consequence of wave interaction with dissimilar cohesive and noncohesive interfaces is shown in Fig.8.

If the right hand side half plane is replaced by a layer of thickness D, the relative amount of momentum trapped per unit area depends upon the ratio $\delta = D/\Lambda$:

$$M_T = T \int_0^{2D/c_1'} \sigma_i(t)\, dt \quad , \tag{19}$$

and the throw-off velocity will be $v = M_T/\rho' D$.

No momentum can be transferred across a gap (open crack) of width d until the narrow free space is closed. With respect to a compressive pulse closure will be completed after the time delay t_c given by

$$d = 2 \int_0^{t_c} v_i(t) dt = \frac{2}{\rho c_1} \int_0^{t_c} \sigma_i(t) dt \quad , \tag{20}$$

characterized by the cut-off stress level.

There is good agreement between calculated results utilizing the above euqations and field observations in the development of displacements along jointed rock structures and pre-existing faults during underground nuclear explosions.

2.3. *Plate Waves*

Reflection and refraction of plane waves in a thin plate (wave length plate thickness h) is most often encountered in 2D-photoelastic models where plane stress conditions, $\sigma_z = 0$, prevail anywhere. Plate waves propagate with speed

$$c_L^2 = 4\mu(\lambda+\mu)/(\rho(\lambda+2\mu)) = E/(\rho(1-\nu^2)) \tag{21}$$

This change in wave speed has to be accounted for in the reflection and transmission coefficients.

2.4. *Surface Waves*

2.4.1 *Glancing Angle Diffraction*

If a pulse advances at glancing incidence along the free boundary of e.g. a half-plane, a strong trailing shear wave, socalled von Schmidt or head wave (V) is generated by diffraction controlled by the presence of the free boundary, and the fact that a P-wave cannot satisfy stress free boundary conditions. The motion of particles located at different positions along the wave front is different from the boundary particle which is free to expand in a direction normal to the free surface. The head wave makes an angle of $\sin\beta_V = c_2/c_1$ with the free surface.

2.4.2 *Rayleigh-Waves*

Transient disturbances moving along the free surface of an elastic body leave in their wake surface waves, termed Rayleigh-waves (R), the amplitude of which decreasing rapidly with increasing distance from the surface and propagating with speed $c_R/c_2 = (0.862 + 1.14\nu)/(1+\nu) < 1$.

Plane Rayleigh-waves propagate without dispersion and without change of form. They carry the bulk of their energy confined within a very thin layer just below the surface. The wave decays with increasing depth showing a σ_x-stress reversal and a u-displacement reversal at some finite depths below the surface. The quantity w_o in Fig. 9 is the vertical surface displacement and Λ is the wave length. A representative photoelastic fringe pattern of a Rayleigh-pulse is shown in Fig. 10. The leading tensile pulse of a Rayleigh-pulse has the tendency to open up surface flaws and violent wave diffraction is expected to occur when Rayleigh-pulses interact with ensembles of deep surface cracks.

Since Rayleigh-waves spread only in two dimensions and consequently attenuate more slowly with distance than body waves, they play an important role in seismic and detonic phenomena.

2.4.3 *P-Wave Propagation in Layers*

When a plane-fronted compressive P-pulse propagates in a plate of finite

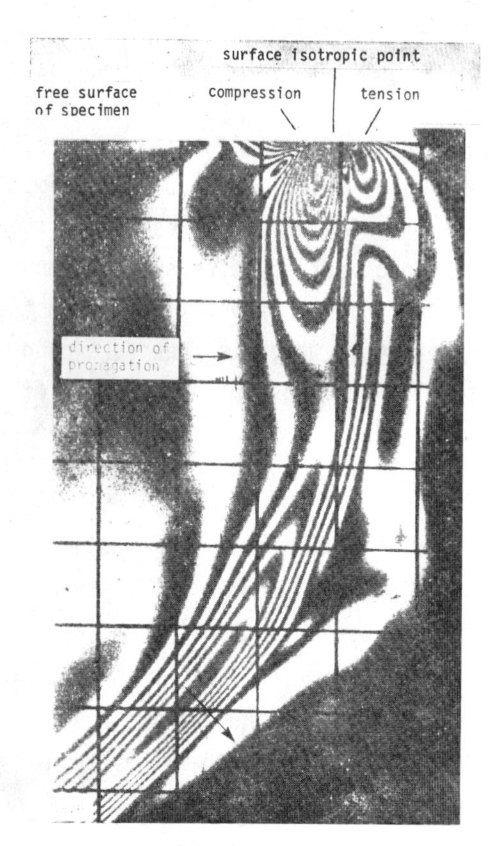

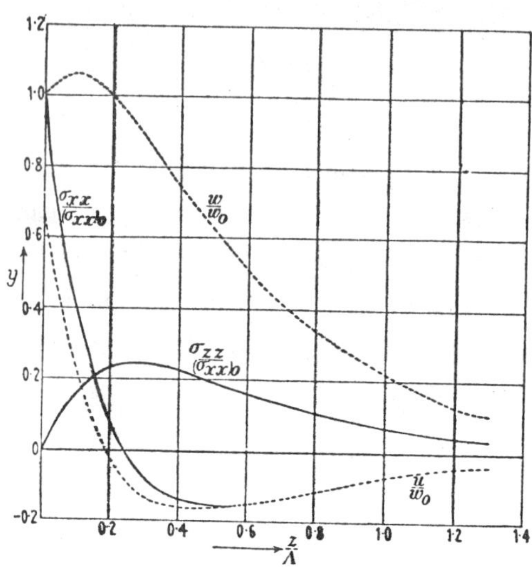

Fig. 9

Amplitudes of the stresses and displacements associated with Rayleigh surface waves in steel. ($\nu = 0{\cdot}29$.)

Fig.10 Photoelastic fringe pattern of a Rayleigh-wave (R)

thickness tensile and shear waves are generated upon reflection of the initial wave front at the boundary surfaces. The resulting complex interaction rapidly degrades the original wave front and sets up oscillations (Fig.11a,b).

After the wave has travelled a distance of about 10 times the plate thickness the speed of energy transportation has slowed down from c_1 to c_L, equ.(21), and the frequency of the oscillations is regulated by the plate thickness and the elastic properties:

initially: $\Lambda_1 = \overline{AF} = 2h.\mathrm{ctg}\beta_1 = 2h/\sqrt{1-2\nu}$ (22)

subsequently: $\Lambda_L = \overline{FG} = 2h\ \mathrm{ctg}\beta_L = 2h\sqrt{(1+\nu)/(1-\nu)}$ (23)

The frequency of oscillations is $f_1 = c_1/\overline{AF} = c_1/\Lambda_1$ or $f_L = c_L/\Lambda_L$. As the wave progresses trough the plate the separation L between the wave

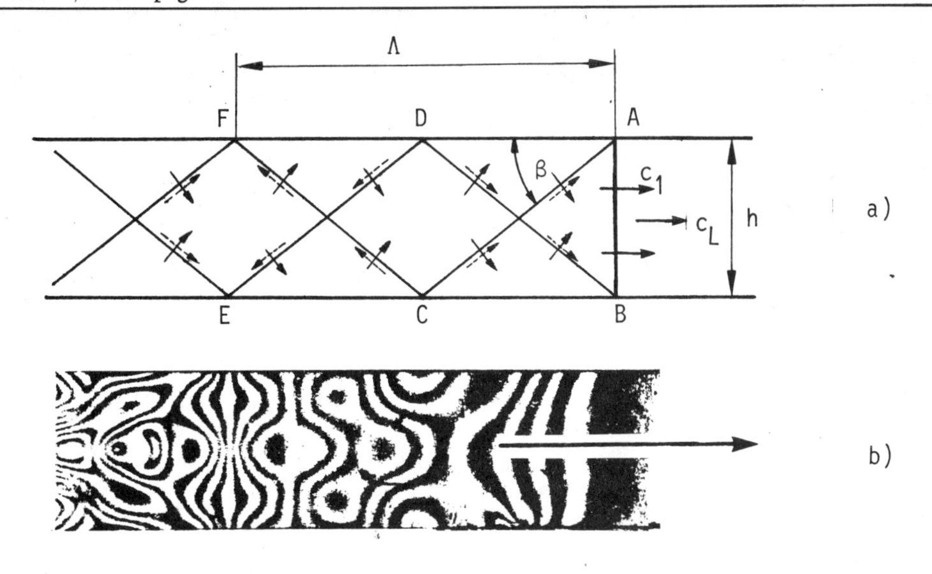

--→ direction of particle movement
—→ direction of wave front propagation

Fig.11 Propagation of plate waves ($\Lambda/h > 1$)

front and a particular shear wave segment is $L = (c_L - c_2 \sin\beta_L)t$ with t the time elapsed since the wave entered the plate.

2.5. *Nonplanar Wave Fronts at Boundaries and Interfaces*

In many practical applications nonplanar wave fronts play a dominant role. In order to locate the regions of high stress concentration the various wave patterns in both space and time have to be established. The principles of ray and physical optics, Snellius' law, Huygen's principle, and Fermat's principle are extremely helpful in reconstructing and analyzing the wave interaction and configurations of the wave fronts. Most often the complexities induced by nonplanar wave fronts may be circumvented by assuming that the wave front is made up of an infinite number of infinitesimally small planar elements. Hence, an arc length element is replaced by its straight tangential element. This represents a good approximation only for large distances from the wave source. Although sometimes simple table computer programs can be developed easily, more frequently geometrical construction using rule and compass is the best and often only approach to obtain an (approximate) engineer solution.

As an example consider the reflection and refraction of a cylindrical detonation wave at the cohesive interface of two dissimilar media. The application of optical ray theory is convenient to use in constructing a series of progressive wave fronts. Each ray of the incident P_1 (or S_1) pulse double-bifurcates at the interface and the emerging rays are associated with the reflected P_1P_1 (P_1S_1) and S_1P_1 (S_1S_1) waves, and the refracted P_2P_1 (P_2S_1) and S_2P_1 (S_2S_1) waves, where Snellius' law holds in the form:

$$\sin\alpha/c_1 = \sin\beta/c_2 = \sin\gamma/c_1' = \sin\delta/c_2' \quad . \tag{24}$$

The geometrical situation associated with an incident P-pulse is shown in Fig.21. The wave fronts in the xy-plane for all reflected and diffracted waves are defined by the parametric form:

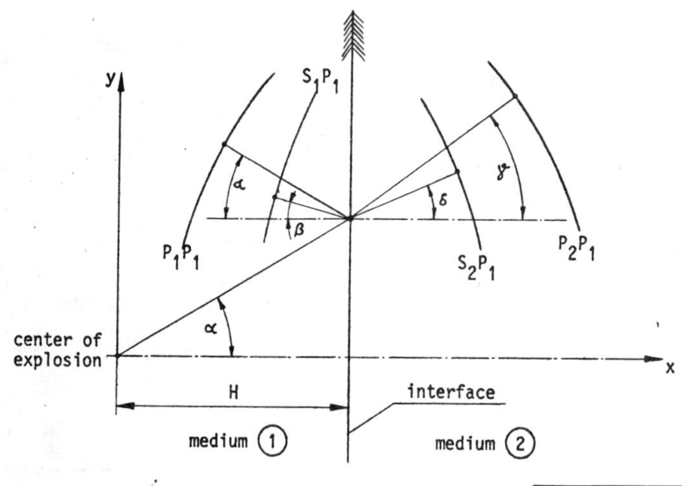

Fig.12 Development of reflected and refracted waves upon incidence of a detonation pulse at an interface

$$\begin{Bmatrix} x_Q \\ y_Q \end{Bmatrix} = \begin{Bmatrix} H \pm (tc_i\cos\alpha_i - H)\frac{c_Q}{c_i}\sqrt{1-(\sin\alpha_i\cdot\frac{c_Q}{c_i})^2}\,/\cos\alpha_i \\ tc_i\sin\alpha_i\cdot(\frac{c_Q}{c_i})^2 + H\tan\alpha_i\{1-(c_Q/c_i)^2\} \end{Bmatrix} \tag{25}$$

For an incident P-pulse Q corresponds to P_1P_1, P_2P_1, S_1P_1, and S_2P_1 for i=1 ($\alpha_1 \equiv \alpha$) and eqs.(25) are valid for $c_1 > c_1'$. If, $c_1 < c_1'$, the limitation is given by $\sin\alpha_c = c_1/c_1'$. Plus and minus signs in the first of eqs.(25) pertain to refracted and reflected waves, respectively.

The character of wave transmission changes completely when $c_1' > c_1$, i.e. when the P_2P_1-wave in medium 2 outdistances the P_1P_1-wave which propagates in medium 1 / 1 /. Several conical and refracted waves are then generated which can be traced in the isochromatic fringe pattern as shown in Fig. 13.

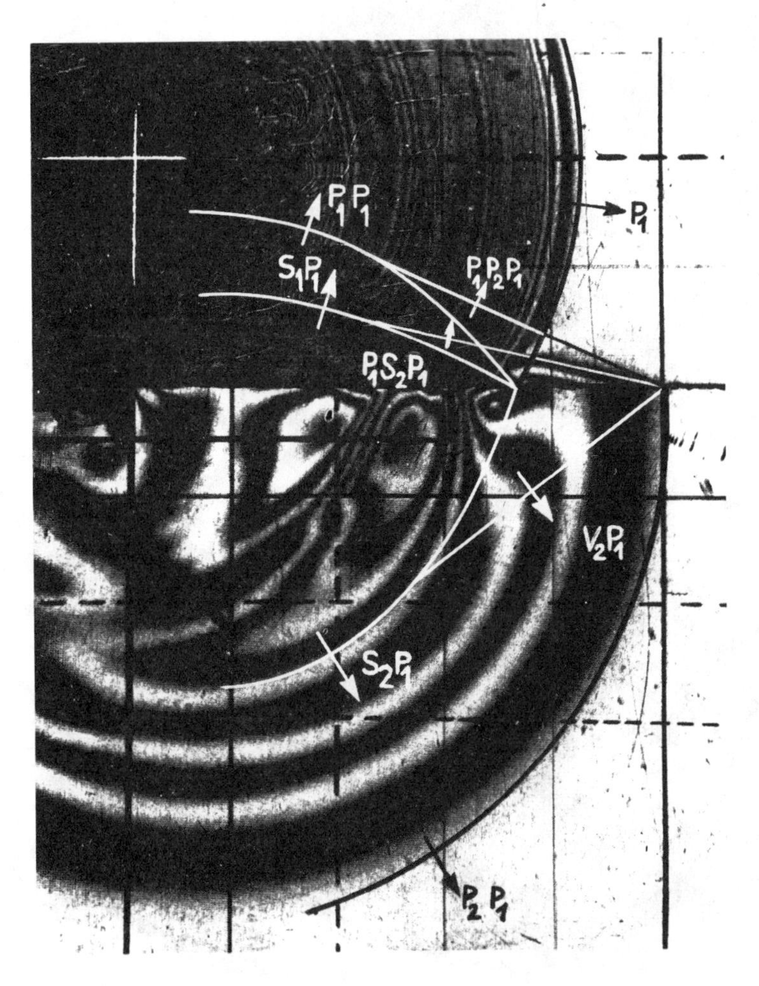

Fig.13 Isochromatic fringe pattern associated with P-wave reflection and refraction at the interface between two dissimilar media

Reflection of plane or nonplanar wave fronts at curved boundaries is treated as a series of interactions between successive infinitesimally small tangent wave front elements with similar elements of the boundary. Of course,

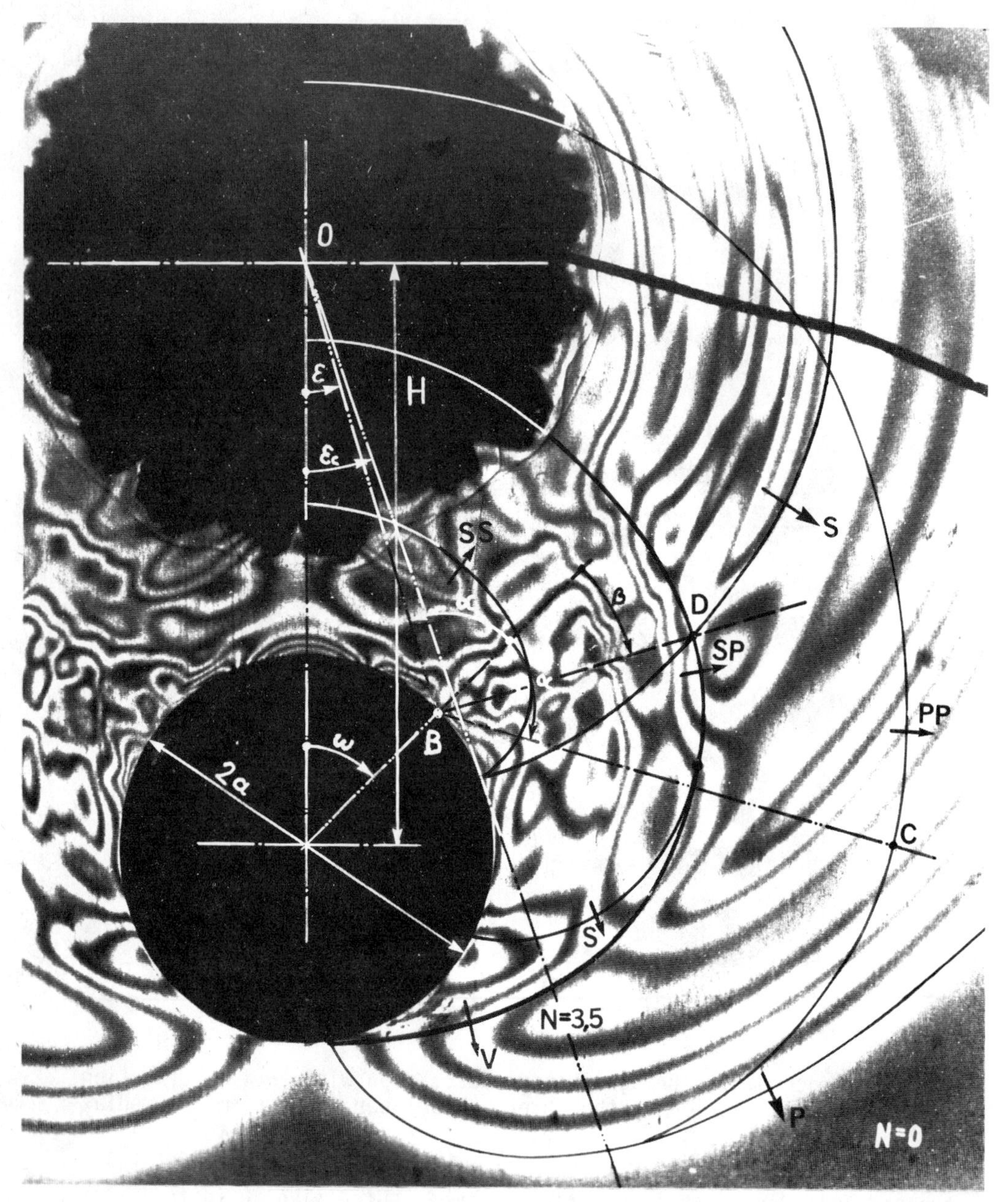

Fig.14 Diffraction and reflection of an incident detonation wave about a cylindrical tunnel

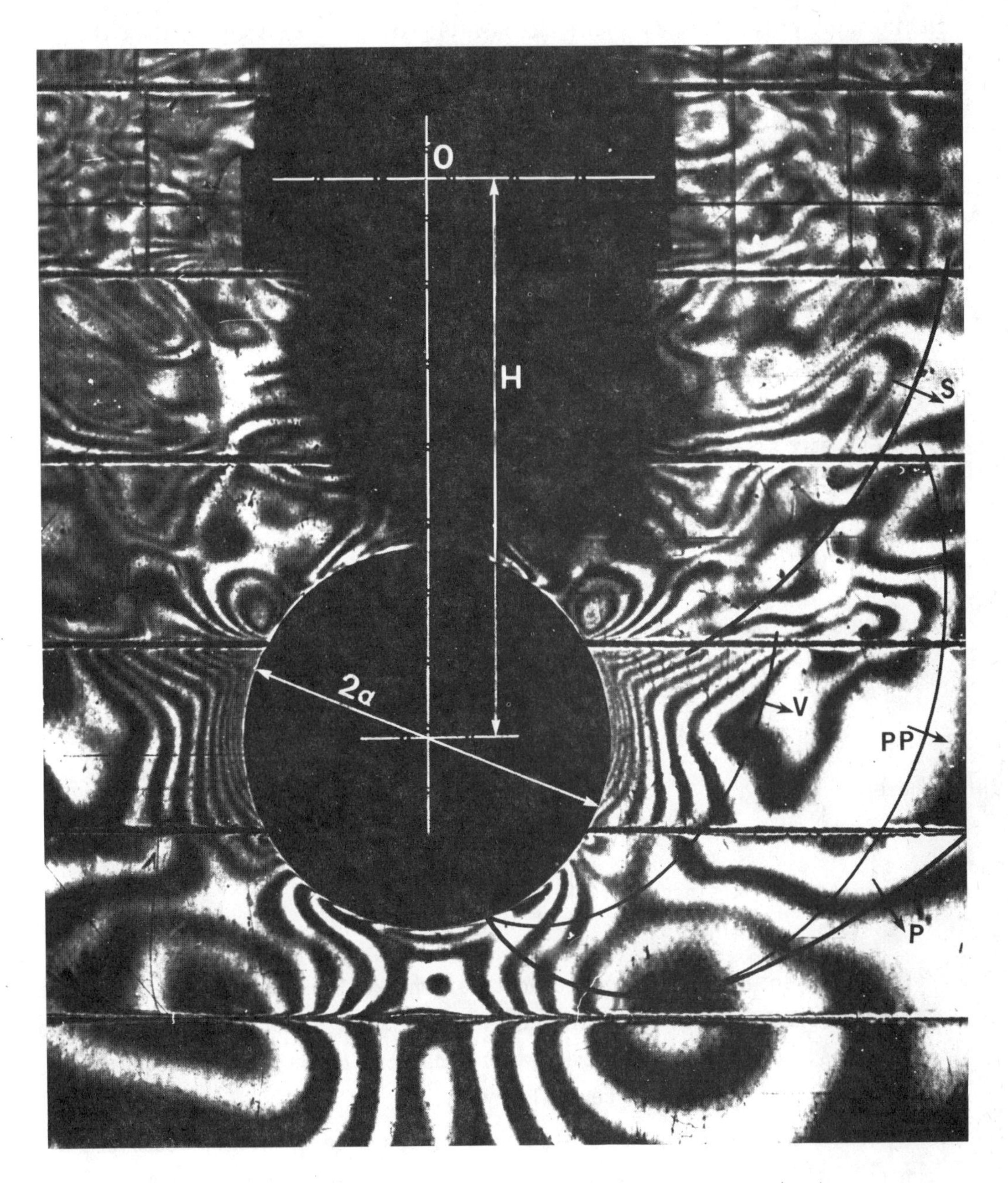

Fig.15 Diffraction and reflection of an incident wave about a cylindrical tunnel in layered media

this holds only in cases where the wave length is much smaller than the characteristic radius of curvature.

A second practical example is taken from tunneling engineering. When excavating a second tunnel in the close proximity of an existing cavity the distribution of stresses induced by impulsive detonation waves from the blasting is of utmost importance. A representative isochromatic fringe pattern is shown in Fig.14. Observe, that geometrical optics is valid only for the range $\varepsilon > \varepsilon_c$. Wave extension into the geometrical shadow zone is a result of wave diffraction about the cavity because of the intrinsic ratio of wave length Λ to cavity diameter 2a of $\Lambda/2a \sim 1$.

The parametric representation of the reflected strong shear SP-wave of order N=3,5 in the range $0 < \varepsilon < \varepsilon_c$ is determined by

$$\begin{matrix} x_D \\ y_D \end{matrix} = \begin{matrix} Hk\cos\varepsilon\,\{\cos\varepsilon + \frac{c_2}{c_1}\cos(\alpha+\beta-\varepsilon)\} - c_2 t\cos(\alpha+\beta-\varepsilon) \\ Hk\cos\varepsilon\,\{\sin\varepsilon - \frac{c_2}{c_1}\sin(\alpha+\beta-\varepsilon)\} + c_2 t\sin(\alpha+\beta-\varepsilon) \end{matrix} \qquad (26)$$

and the very low amplitude reflected PP-wave is given by

$$\begin{matrix} x_C \\ y_C \end{matrix} = \begin{matrix} Hk\cos\varepsilon\,\{\cos\varepsilon + \cos(2\alpha-\varepsilon)\} - c_1 t\cos(2\alpha-\varepsilon) \\ Hk\cos\varepsilon\,\{\sin\varepsilon - \sin(2\alpha-\varepsilon)\} + c_1 t\sin(2\alpha-\varepsilon) \end{matrix} \qquad (27)$$

where

$$k = 1 - \{1 - (1 - (\tfrac{a}{H})^2 (1+\tan^2\varepsilon)\}^{1/2}$$

$$\alpha = \varepsilon + \omega$$

$$\tan\omega = k\sin\varepsilon\cos\varepsilon/(1-k\cos^2\varepsilon)$$

$$\sin\beta = \sin\alpha \cdot c_2/c_1 \quad .$$

Fig.15 shows the same tunnel-detonation problem except for the layered structure of the rock model. In addition the model shows considerable normal pre-stress. Localized high stress amplifications occur at the wedge-shaped tunnel-wall section of the central layers.

A fracture is often generated as a consequence of highly localized stress concentrations when two transient reflected tensile body waves superimpose or when two approaching Rayleigh-waves $R_+(x+x_0-c_Rt)$ and $R_-(x-x_0+c_Rt)$ interact.

3. *Dynamic Layer Detachment and Spallation*

When a (plane) wave strikes a loose interface obliquely, the momentum is partitioned and the transmitted normal component causes the second body (often a layer) to move off in a direction normal to the interface. The remainder of the momentum is trapped in the first body (bulk) causing relative movement along the interface. The relative amount of momentum transferred to the receptor layer follows from

$$M = \int c_2' \, v_{S_2P_1,y} dt + \int c_1' \, v_{P_2P_1,y} dt \tag{28}$$

Separation of the layer from the bulk will occur regressively for a cylindrical detonation wave because at angles of obliquity ranging from 4o° to 75° a fairly large fraction of the momentum will be reflected at the interface and, thus, will never reach the receptor layer. The relative position of the point of separation A with respect to the wave front B depends on the

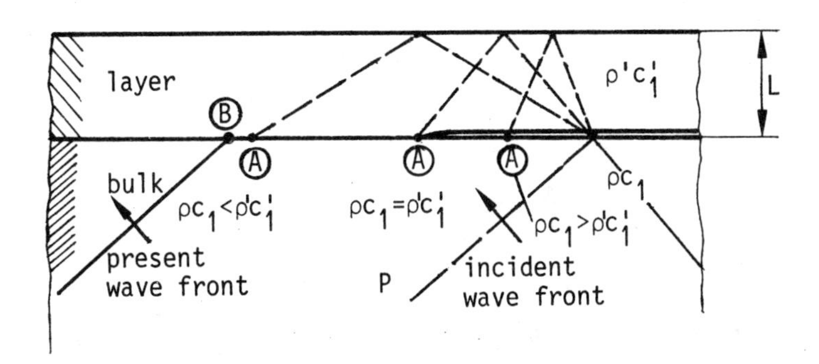

Fig.16 Dynamic layer detachment for dissimilar media junction

ratio c_1'/c_1 and the angle of incidence α (Fig.16). The distance $\overline{BA}$ for a dissimilar media junction is given by

$$\overline{BA} = 2L \tan\gamma \, \{ (c_1/c_1')^2/\sin^2\alpha - 1 \} \tag{29}$$

which reduces to $\overline{BA} = 2L \, \mathrm{ctg}\alpha$ for similar media. Note that the detachment point A anticipates the incident wave front B for $c_1' > c_1$ and angle of incidence $\alpha > \sin^{-1}(c_1/c_1')$.

EXAMPLE: An explosion is ignited at depth H below the interface of a half-plane covered with a top layer of thickness L (Fig.17).

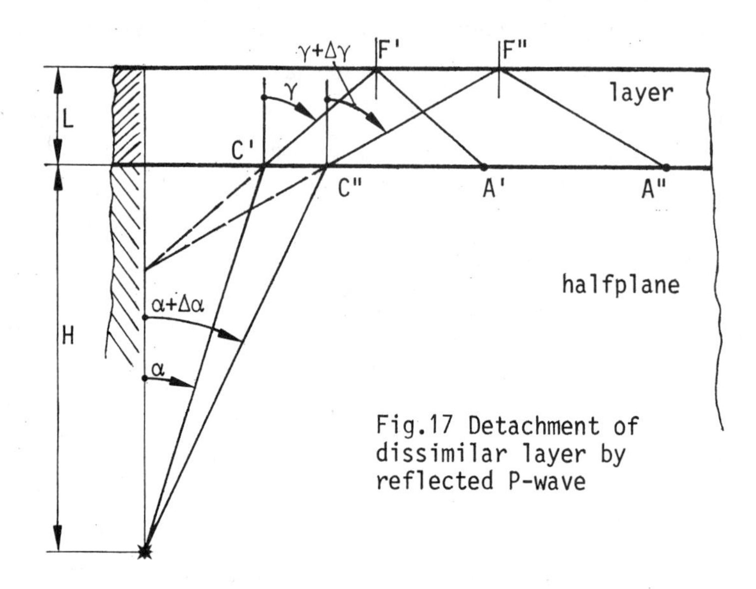

Fig.17 Detachment of dissimilar layer by reflected P-wave

The speed of detachment c_a follows from the differential wave front construction and is given by

$$c_a/c_1 = \frac{1}{\sin\alpha}\ \{1 + \frac{c_1'}{c_1}\ \frac{2L}{H}\ \frac{\cos^3\alpha}{\cos^3\gamma}\}/\ \{1 + \frac{c_1}{c_1}\ \frac{2L}{H}\ \frac{\cos^3\alpha}{\cos^3\gamma}\}\ . \tag{30}$$

For a similar layer, equ.(30) reduces to the known result $c_a = c_1/\sin\alpha$. It should be noted, that this problem involves many more complexities as it represents a generalisation of the buried pulse problem by Pekeris et al./ 6 /. A natural way to generalize wave reflection is to incorporate the actual shape of real wave pulses.

Spalling is the direct consequence of interference near a free surface between the trailing tensile part of the incoming detonation wave and that part of the leading compressive pulse which has been reflected and transformed into a tensile wave. When the intensity of the superimposed tensile waves exceeds the fracture strength, σ_f, of the material a fracture is initiated. Spalling due to obliquely incident pulses is more complex because of the decrease in amplitude. Transient maximum stress and obliquity of wave front control the position of spall fracture.

Multiple spalling, i.e. the development of several parallel, juxtaposed fractures, occurs when the level of the stress wave, σ_0, becomes more than twice the fracture strength, σ_f, of the material. The new spall surface acts as a free surface for the remaining cut-off stress pulse, creating a sequence of spalls each one diminishing the intensity of the rear portion of the trapped stress wave. Spalling becomes a very complicated phenomenon in dissimilar layered rock and/or partly cohesionless junctions. Granite, with a ratio between compressive and tensile strength of as high as 80, fails easily under tensile stress. A tensile stress cannot develop in cohesionless material because of extensive flaking by spalling at the free surface. The fly-off speed v_f of a flake is proportional to the instantaneous stress σ at the front of the chopped wave, $v_f=2\sigma/\rho c_1$. In typical junk rock the flakes appear as a cloud of debris.

The process of spall formation for an obliquely incident wave is rather complex. Micro-fractures open up just in the wake of the point of intersection of the waves leaving behind single micro-cracks that cannot propagate over any appreciable distance because the energy-feeding wave outdistances them.

References

/1/ Achenbach, J.D.: Wave Propagation in Elastic Solids. North-Holland, Publ.Co.(1973)

/2/ Rinehart, J.S.: Stress Transients in Solids. Hyperdynamics, New Mexico, (1975)

/3/ Ewing, W.M., W.S.Jardetzky and F.Press: Elastic Waves in Layered Media. New York, McGraw-Hill (1957)

/4/ Kolsky H.: Stress Waves in Solids. Dover Publ., New York (1963)

/5/ Biot M.A.: Mechanics of Incremental Deformations. John Wiley, New York (1965)

/6/ Pekeris, C.L.: The seismic buried pulse.Proc.Natl.Acad.Sci.41,629-639 (1965).

/7/ Dey S. and Addy S.K.: Reflection and refraction of plane waves under initial stresses at an interface. Int.J.Non-Linear Mech.14,101 (1979).

ANALYSIS OF DYNAMIC PHOTOELASTIC FRINGE PATTERNS

H.P. R o s s m a n i t h
Institute of Mechanics
Technical University Vienna
A-1040 Vienna, Austria

In static photoelastic analysis schemes for identification of isochromatic fringe patterns and stress separation techniques have been developed in the past /1,2/.

The validity of the stress optic law for the dynamic case was established by Thau and Dally /3/ provided the static material fringe value f_σ is replaced by its dynamic counterpart, $f_{\sigma d}$:

$$2\,\tau_m = \sigma_1 - \sigma_2 = N\, f_{\sigma d}/h \qquad (1)$$

where N= fringe order and h= model thickness. Note, that $f_{\sigma d}$ is usually 10-30% higher than f_σ and must be determined by a dynamic calibration. This is frequently accomplished by impacting the end of a long rod with a square cross section by a projectile fired from an airgun and simultaneously recording the signals on the oscilloscope.

Methods for determination of stress in the general dynamic case are not available at present time because the most-common static separation methods which are based on stress equations of equilibrium or on Laplace's equation are not valid for the dynamic problem. Moreover, lack of additional information such as isoclinics or isopachics hampers efficient dynamic stress separation. Only in a few special cases where additional information associated with natural boundary conditions or a special state of stress is available remaining unknown stress components may be determined along certain contours. These include free and rigid boundaries as well as fringes associated with axisymmetric dilatational wave propagation, and pure shear-wave propagation (Ref./5/).

1. *Identification of Isochromatic Fringes*

The first step in fringe identification is the numbering of isochromatic fringes on the basis of the type of optical polarizing arrangement (light field or dark field). Although in simple cases an easy job, in more complex cases this problem may be so difficult that few investigators attempt the analysis. It is only recently that some guidelines for general fringe numbering have been set up regarding the topological character of a fringe pattern /4/. In a Oxyz cartesian coordinate system the maxim shear stress τ_m, equ.(1), describes a topological map characterized by the elements of differential topological geometry, such as peaks, valleys, saddlepoints, nodal points, etc. Although the isochromatic fringe order N per definition is taken its positive absolute value the view point featuring positive and negative fringe orders facilitates the identification and characterization of the topological map in many cases considerably, because valleys and peaks then become associated with valleys and peaks of principal stress difference. Consider e.g. Fig.1 where a nonplanar circular crested P-pulse is being reflected from the free boundary. The outgoing P-pulse has a leading compressive pulse P^+ with peak $N = 6{,}5$ followed by a steep fringe gradient and a trailing tensile pulse P^- with valley of fringe order $N = -6{,}5$. The puls length $\Lambda = 8$ cm. Along the vertical axis the wave fronts of the reflected PP and SP waves are indicated. The outgoing P^- pulse and the reflected P^-P pulse overlap to generate a valley of fringe order $N = -7{,}5$. Notice, that the amplitude of the reflected shear wave is zero at the line of symmetry. A strong isolated island of maximum fringe order

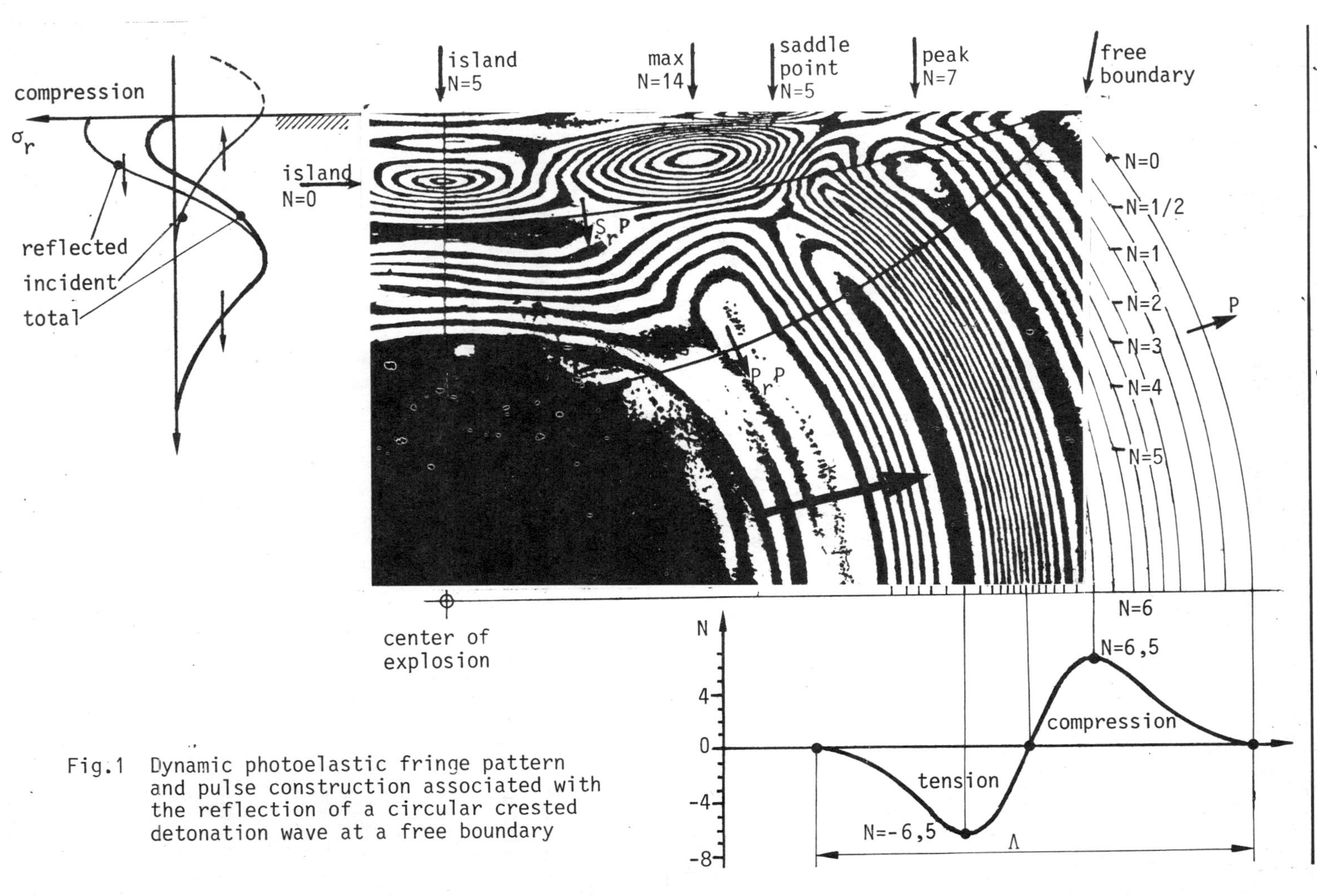

Fig.1 Dynamic photoelastic fringe pattern and pulse construction associated with the reflection of a circular crested detonation wave at a free boundary

N=14 is associated with the superposition of the outgoing P^--pulse and the reflected shear pulse SP^+.

Generally, conditions of continuity and uniqueness of elastodynamic displacement and stresses within the framework of continuum mechanics serve the numbering of fringe orders. Guidelines mention square corners, points of stress reversal at free boundaries etc as loci with N=0 to start from.

In a general case space and time derivatives may be employed for fringe identification. Regarding an isochromatic fringe pattern a topographic map representing the τ_{max}-landscape, the first and second total derivatives

$$d\tau_{max} = 0 \ , \quad d^2\tau_{max} \gtrless 0 \tag{2}$$

determines all peaks (<) and valleys (>) of the map. Saddle points (Fig.1) are characterized by main curvatures of different sign, and they are characteristically bounded by two isochromatic lines with fringe orders $N \pm \Delta N$ with N the fringe order at the saddle point.

The concept of time and space derivative has become an advantageous method in the evaluation of particle velocities and accelerations from a single dynamic photograph. The method of time derivatives as mentioned in Ref./4/ cannot be applied to dynamic photoelasticity. In static photoelastic work the motion of fringes towards the region of lower τ_{max} with changing load is often observed for exact fringe numbering. In contrast to static photoelasticity where conditions of static equilibrium hold for any particular load level, conditions of dynamic equilibrium govern the dynamic phase between two states of static equilibrium.

2. *Data Analysis*

2.1. *Boundary Conditions*

Data analysis of dynamic isochromatic fringe patterns usually employs (a) model characteristics and/or (b) stress wave characteristics.

Free Boundary (specimen boundary, open crack faces):

Boundary conditions: $\sigma_n=0$, $\tau_{nt}=0$

Result:

$$\sigma_t = \pm N f_{\sigma d}/h \quad , \quad u_t = \pm(f_{\sigma d}/Eh) \int N\, ds \tag{3}$$

where N is a function of position along the boundary and s is the arc length.

Rigid boundary (clamped, built-in):

Boundary conditions: $u_t = u_n = 0$

Result: $$\sigma_t = \frac{\nu}{1-\nu} N f_{\sigma d}/h \quad , \quad \sigma_n = \frac{1}{1-\nu} N f_{\sigma d}/h \tag{4}$$

2.2. *Wave Form Conditions*

Plane dilatational (P)-wave:

Direction of progation $\vec{n}$

Conditions: irrotational $\omega_z = 0$, $\varepsilon_t = 0$ (plane strain)

Result:

$$\sigma_n = \frac{1}{1-\nu} f_{\sigma d} N/h \quad , \quad \sigma_t = \nu \sigma_n \tag{5}$$

$$u_n = \frac{1}{1-\nu} (f_{\sigma d}/Eh) \int N\, ds \quad , \quad u_t = 0 \tag{6}$$

Plane shear (S)-wave:

Direction of propagation $\vec{n}$

Conditions: equivoluminal $\sigma_1 + \sigma_2 = 0$

Result:

$$\sigma_1 = -\sigma_2 = N f_{\sigma d}/2h \quad , \quad \tau_{nt} = \frac{1}{2}(\sigma_1 - \sigma_2) = N f_{\sigma d}/2h \tag{7}$$

Notice, that the directions of σ_1 and σ_2 make an angle of 45° to $\vec{n}$.

$$u_t = \frac{(1+\nu)}{Eh} f_{\sigma d} \int N\, ds \ . \tag{8}$$

Cylindrical P-wave:

Condition: rotational symmetry

Result: $$\varepsilon_\theta = -\frac{1+\nu}{Eh} f_{\sigma d} \int N(r)/r\, dr$$
$$\varepsilon_r = -\frac{1+\nu}{Eh} f_{\sigma d} \{ N(r) + \int N(r)/r\, dr \} \tag{9}$$

$$\sigma_r = \frac{E}{1-\nu^2} (\varepsilon_r + \nu\varepsilon_\theta) \quad , \quad \sigma_\theta = \frac{E}{1-\nu^2} (\varepsilon_\theta + \nu\varepsilon_r) \tag{10}$$

General 2D- Wave

Direction of propagation n in a x,y-coordinate system, $\phi = \measuredangle(\sigma_1 x) = \measuredangle(\vec{n},x)$. The abbreviation $A = (1+\nu)f_{\sigma d}/Eh$ will be employed in the following.

P-wave:

$$u_x = A/2 \int N(y)\sin 2\phi \, dy$$
$$u_y = A/2 \int N(x)\sin 2\phi \, dx$$

S-wave:

$$u_x = A/2 \int N(x)\cos 2\phi \, dx$$
$$u_y = -A/2 \int N(y)\cos 2\phi \, dy \quad (12)$$

for both wave types:

$$\sigma_1+\sigma_2 = \frac{E}{1-\nu}\{\partial u_x/\partial x + \partial u_y/\partial y\} \quad , \quad \sigma_1-\sigma_2 = N\, f_{\sigma d}/h \quad (13)$$

$$\sigma_x = (\sigma_1+\sigma_2)/2 + (\sigma_1-\sigma_2)/2.\cos 2\beta$$
$$\sigma_y = (\sigma_1+\sigma_2)/2 - (\sigma_1-\sigma_2)/2.\cos 2\beta \quad (14)$$
$$\tau_{xy} = (\sigma_2-\sigma_1)/2.\sin 2\beta$$

For data analysis employing stress wave characteristics in three dimensions see Ref./5/. The above-mentioned data reduction methods are valid for single waves only; if several wave interact more elaborate data reduction schemes based on superpostion principles have to be developed.

3. *Fracture Analysis using Photoelastic Data*

A great amount of research work was performed in the late 50's and early 60's to establish stress analysis methods, to develop material testing techniques, and to demonstrate the applicability of fracture mechanics principles to many significant and important engineering problems/6,7/. Among experimental studies of fracture mechanics the method of photo-elasticity is one of the most suitable means for determining the stress intensity factor K at the crack tip.

During the period 1952 and 1954, the inovative photo-elastic research studies by Post /8/ drew attention to use of photo-elasticity for the investigation of both static and dynamic crack stress fields. At the same laboratory a subsequent photo-elastic investigation of running cracks was made by Wells and Post /9/.

3.1. *Mixed-Mode Fracture Problem*

The stresses in the local neighborhood of a crack tip which is subjected to combined loading (mode-1 = opening mode plus mode-2 = in-plane shearing mode ≡ mixed-mode) are given by

$$\begin{aligned}
\sigma_x &= \frac{K_1}{\sqrt{2\pi r}}\cos\frac{\theta}{2}\left(1-\sin\frac{\theta}{2}\sin\frac{3\theta}{2}\right) - \frac{K_2}{\sqrt{2\pi r}}\sin\frac{\theta}{2}\left(2+\cos\frac{\theta}{2}\cos\frac{3\theta}{2}\right) + \sigma_{xo} \\
\sigma_y &= \frac{K_1}{\sqrt{2\pi r}}\cos\frac{\theta}{2}\left(1+\sin\frac{\theta}{2}\sin\frac{3\theta}{2}\right) + \frac{K_2}{\sqrt{2\pi r}}\sin\frac{\theta}{2}\cos\frac{\theta}{2}\cos\frac{3\theta}{2} \\
\tau_{xy} &= \frac{K_1}{\sqrt{2\pi r}}\sin\frac{\theta}{2}\cos\frac{\theta}{2}\cos\frac{3\theta}{2} + \frac{K_2}{\sqrt{2\pi r}}\cos\frac{\theta}{2}\left(1-\sin\frac{\theta}{2}\sin\frac{3\theta}{2}\right)
\end{aligned} \qquad (15)$$

Upon substitution of equ.(15) into

$$\tau_m^2 = (\sigma_y - \sigma_x)^2/4 + \tau_{xy}^2 \quad , \qquad (16)$$

one obtains a relationship which defines the isochromatic fringe pattern in the close proximity of the crack tip:

$$\begin{aligned}
8\pi r\tau_m^2/K_1 &= \sin^2\theta + m^2(4-\sin^2\theta) + 2m\{\sin2\theta - \sigma_{xo}\sqrt{2\pi r}(2\sin\frac{\theta}{2}+\sin\theta\cos\frac{3\theta}{2})\} \\
&+ \sigma_{xo}^2\, 2\pi r - 2\,\sigma_{xo}\sqrt{2\pi r}\,\sin\theta\sin\frac{3\theta}{2}
\end{aligned} \qquad (17)$$

with $m = K_2/K_1$.

With the three variables K_1, K_2, and σ_{xo} as the control parameters, mixed-mode isochromatic crack-tip fringe patterns can be characterized and classified. Classification schemes have been worked out by Dally and Sanford /10,11/ and Rossmanith /12/. The most prominent feature of a mixed-mode fringe pattern is the loss of symmetry with respect to the plane of the crack. If $m^{-1} \to 0$ for pure mode-2, the fringe pattern regains a symmetrical shape. An illustrative example for the change of the fringe pattern surrounding a static crack tip when the mixed-mode index m increases from zero (Fig.2a) to m=0.1 (Fig.2c) is presented in Fig.2 for a DCB-type fracture specimen where $\sigma_{xo} > 0$.

In many crack problems the presence of a small in-plane shearing load can be considered a small disturbance of the "undisturbed" mode-1 situation. The crack responds to these "load imperfections" during quasi-static crack growth by changing its crack path to attain a new "undisturbed" travel path.

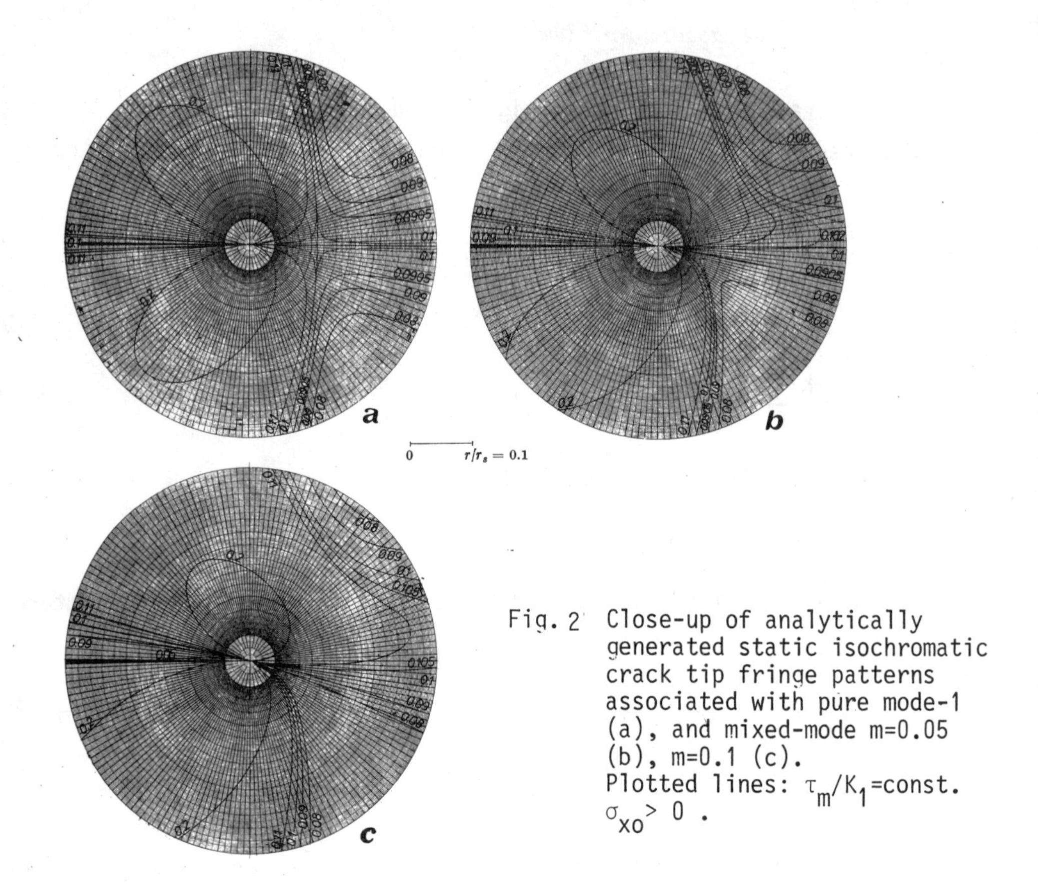

Fig. 2 Close-up of analytically generated static isochromatic crack tip fringe patterns associated with pure mode-1 (a), and mixed-mode m=0.05 (b), m=0.1 (c).
Plotted lines: τ_m/K_1=const.
$\sigma_{xo} > 0$.

This type of crack path stability is essentially influenced by at least the first non-singular terms in the stress function expansion, i.e. it is influenced by the parameter σ_{ox}.

Irwin /13/, in a discussion to Ref. /6/ suggested a simple engineering approach for determination of K-values from mode-1 crack-tip fringe patterns. Irwin's analysis will be generalized here to include mixed-mode problems.

Observing the fact that for a given mixed-mode fringe loop (τ_m=const.), $\partial\tau_m/\partial\theta$ is zero at the apogee points A_1 and A_2 (Fig.3), a measurement of the apogee distances, r_{mj}, and fringe loop tilt angles, θ_{mj}, can then be used to obtain values of the stress intensity factors K_1 and K_2, and the

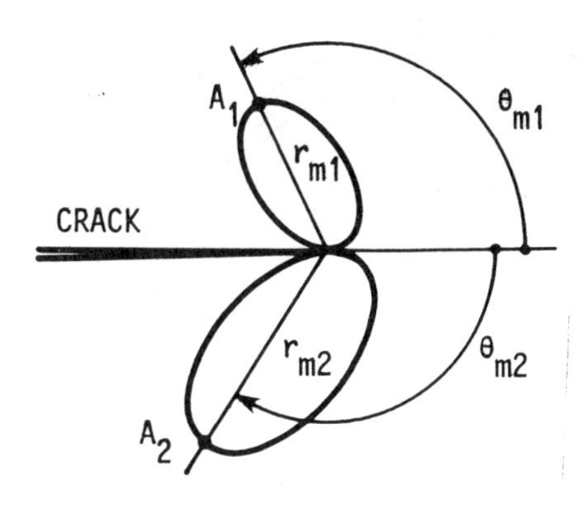

Fig.3 Geometrical configuration of asymmetric isochromatic crack-tip fringe loops

stress field parameter σ_{ox}. For each one of the two apogee points holds

$$\partial\tau/\partial\theta \; (\theta=\theta_{mj};r=r_{mj}) = 0 \quad , \tag{18}$$

which leads to a relation for determination of σ_{ox} as a function of the mixed-mode index $m = K_2/K_1$. Because the regular stress field σ_{ox}, acting parallel to the crack line, is the same for the upper and the lower lobe of the loop, equating the governing relation for σ_{ox} for j=1 and j=2 yields a cubic equation for the mixed-mode index m with the solution:

$$K_2 = m\,K_1 \quad , \qquad m = H_m(\,r_{m1},r_{m2};\theta_{m1},\theta_{m2}) \quad , \tag{19}$$

where H_m is a function of four geometrical parameters.

For the case of pure mode-1 - associated with a symmetrical crack-tip fringe pattern - the H-function can be tabulated or represented in graphical form (Fig.4). Then, the stress intensity factor K_1 takes the form:

$$K_1 = (Nf_\sigma/h)\sqrt{2\pi r_m}\;H(\theta_m) \quad . \tag{20}$$

The dynamic counterpart of equ.(20) for the dynamic stress intensity factor is given by

$$K_{1d} = (Nf_\sigma/h)\;\sqrt{2\pi r}\;\;H(\theta_m,c) \tag{21}$$

where the function $H(\theta_m,c)$ is presented in Fig.4.

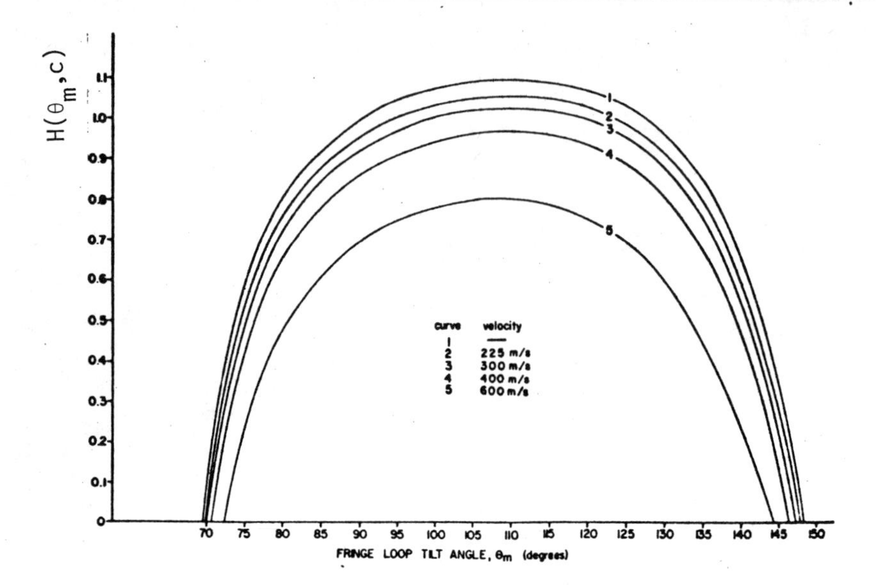

Fig.4 Normalized stress intensity factor $H(\theta_m,c)$ versus fringe loop tilt angle θ_m

The effect of crack tip velocity is to increase the size of the isochromatic loops and to increase the fringe loop tilt angle. Below a speed limit $c/c_2 < 0.1$ a much simpler static analysis may replace the dynamic fracture analysis for many practical engineering applications.

3.2. *A Multiparameter Approach*

The study of stress and strain fields around the base of stationary and running cracks is fundamental to fracture mechanics. Under linear-elastic conditions, both stresses and strains display a singular behavior at the crack tip, with the stress intensity factor, K, providing a convenient measure of the magnitude of the singularity. A single parameter characterization of the crack tip stress field in terms of K is generally recognized as adequate for engineering practice and is valid if attention is restricted to a very small region around the crack tip (the singularity-dominated zone).

Frequently, however, crack tip stress field regions that are too large for a one-parameter representation become of interest, either due to a growth in size of the fracture process zone /15/ or as a practical consequence of experimental methods for the determination of K. Reasons such as crack front curvature, fringe clarity, an unknown degree of plane strain constraint close to the crack tip, pseudo-caustic formation around the crack tip etc , require data to be taken from fringes further away from the crack tip in order to eliminate or at least reduce experimental uncertainties.

It is generally recognized that equ.(15) adequately describes the state of stress in the immediate neighborhood of the crack tip, excluding a very small region around the crack tip itself. The region of validity of this equation is however quite restricted and the problem is complicated by the transition in the local neighborhood of the crack tip, from a state of plane strain to generalized plane stress. To avoid this uncertainty the stress analyst is constrained to take measurements in regions which may lie outside the range of validity of the two-parameter stress field representation. It thus becomes necessary to use parameters additional to K and α to provide an adequate representation of the stress field in an enlarged region around the crack tip and to better match experimentally recorded and analytically generated fringe patterns. Considerable work in this area has taken place in the past decade. A summary of the historical development of K-determination procedures is given in Table I.

It has been shown /16/, that the stress state associated with two-dimensional cracks under static opening-mode (m=0) and edge-sliding mode ($m=\infty$) can be described by a generalized form of the original Westergaard stress equations /10,17/. By combining the governing optical equations for isochromatic fringe patterns with these stress function series representations, $Z_j(z)$ and $Y(z)$, the isochromatic fringe loop is determined by

$$\tau_m = F\{ Z_j(z), Y(z) \} \quad . \tag{22}$$

and the analysis of isochromatic fringe patterns reduces to the problem of determining the coefficients of the two series Z and Y that produce the best match to the experimental pattern, over the region relected for data acquisition. Particular attention has been paid to the effects of including a fist higher order term proportional to $z^{1/2}$, which is the lowest order

TABLE I Summary Showing the Historical Development of K-Determination Procedures

		Number of Data Points Used		
		1	2	large
Number of Parameters Retained in the Analysis	1			Static (0,-) C.W. Smith, 1972
	2	Static (0,0) Irwin, 1958 Dynamic (0,0) Irwin et al, 1975	Static (0,0) Bradley and A.S. Kobayashi, 1970	Static (0,0) Sanford and Dally, 1979 Dynamic (0,0) Irwin et al, 1979
	3	Static and Dynamic (1,0) Dally and Etheridge, 1978; Rossmanith and Irwin, 1979	Static and Dynamic (1,0) Rossmanith and Irwin, 1979	Static and Dynamic (1,0) University of Maryland, Photo-mechanics Lab., 1979-1980
	Arbitrary			Static and Dynamic (N,M) University of Maryland, Photo-mechanics Lab., 1979-1980

non-singular term in the function Z(z).

The evaluation of the desired parameters can be accomplished in a number of different ways. A logical extension of an analysis based on one or two fringes is to take advantage of the full field data available from isochromatic fringes, and develop a method of K-determination based on a finite set of measurement points. These points may lie on different loops (and hence may belong to different levels and states of stress) and could be either randomly distributed or selected in some particular fashion, and can be conveniently used for analysis based on an n-parameter approach to the problem.

The procedure based on an n-parameter least-squares method can briefly be summarized as follows. A data acquistion region is selected for a given experimental pattern using the guidelines suggested in Refs./10,17/. Data points are taken over the entire region in a distributed fashion, and this set is input to the least-squares algorithm (details of which can be found in Refs./11,18,19/) to obtain a best-fit set of coefficients. The number of coefficients necessary for an adequate representation of the stress field over the data acquisition region can be estimated by examining, as a function of the number of coefficients, the value of the average fringe order error. The computed set of best-fit parameters can then be used to reconstruct an isochromatic fringe pattern which, when compared with the experimental fringe pattern being analyzed, serves as a visual check on the adequacy of the model assumed. Illustrative examples and detailed discussions of the three standard fracture test specimens are given in Ref./17/. Results for a MCT-specimen (Fig.5) are presented in Figs.6a-6d. The fringe pattern (Fig.6a) was analyzed to obtain the values of the associated parameters using several different methods of K-determination. The methods used range from a 2-parameter apogee point technique to a 6-parameter least squares solution. The fringe patterns corresponding to the parameter values obtained by different methods have been plotted and are shown in Figs.6b-6d. The MCT-example shown in Fig.5 represents a rather deep crack (a/w=0.8), and the fringe pattern of Fig.6a has obvious boundary influences present. The 2-parameter apogee point and least squares solution completely fails to predict the salient features of the actual fringe pattern. The 3-parameter least squares solution, although capable of matching medium-sized main loops (N=3.5 to N=5.5) quite well, does not correctly predict the fringe distribution ahead of the crack tip. The fringes produced by the steep gradient field ahead of the crack require the use of at least a 6-parameter solution (M=N=2, Fig.6d) before the global appearance of the predicted fringe pattern begins to match that seen experimentally. However, differences still exist between Figs.6a and 6d, indicating that for the crack location and data acquisition region used, a model of order higher than six may well be required.

In summation, it can be seen from the example presented, that adequate modelling of the stress state present can require the retention of as many as six terms in equ.(22) and that reliable values of both K and σ_{ox} can only be obtained when the solution includes at least the lowest order higher order terms.

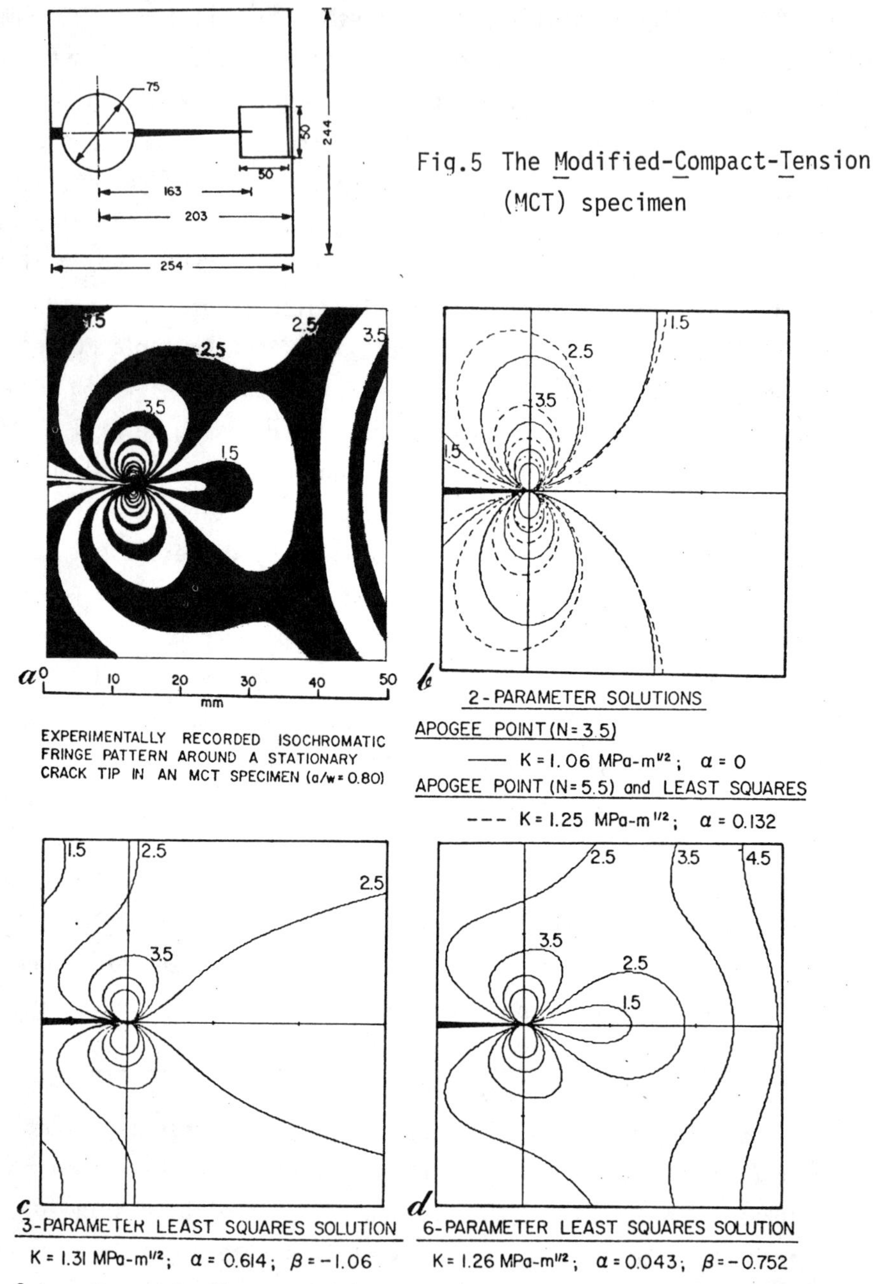

Fig.5 The Modified-Compact-Tension (MCT) specimen

Fig.6 Experimentally recorded isochromatic fringe pattern from an MCT specimen and the predicted fringe pattern and parameter values corresponding to the different K-determination methods used.

It is worth mentioning that the least-squares method, by utilizing data form a global field is preferable to a single-point, apogee-point method, and is likely to give more reliable results. It is also the only method available which permits the simultaneous determination of more than three parameters.

Recently, dynamic mixed-mode crack propagation has become a subject of increasing interest, particularly in conjunction with dynamic crack path stability in fragmentation blasting and fracture control and wave-crack-interaction. Employing a Irwin-procedure for K_1,K_2, and σ_{xo} determination from progressively smaller-sized (higher order) loops, results obtained indicate that the "apparent mixed-mode index" $m_{Ni}=K_2(N_i)/K_1(N_i)$ vanishes upon extrapolation to the crack tip. For obvious reasons the extremely dense fringe gradient at the crack tip cannot be resolved in dynamics. Dynamic isochromatic fringe patterns associated with a fast propagating crack in a DCB specimen fabricated from Homalite 100 are shown in Fig.7, where Fig.7a pertains to the initially path-stable crack propagation phase (mode-1) and Fig.7b shows the crack 's response to unfavorable global stress field conditions ($\sigma_{xo}> 0$) by deviating from the line of symmetry /20/.

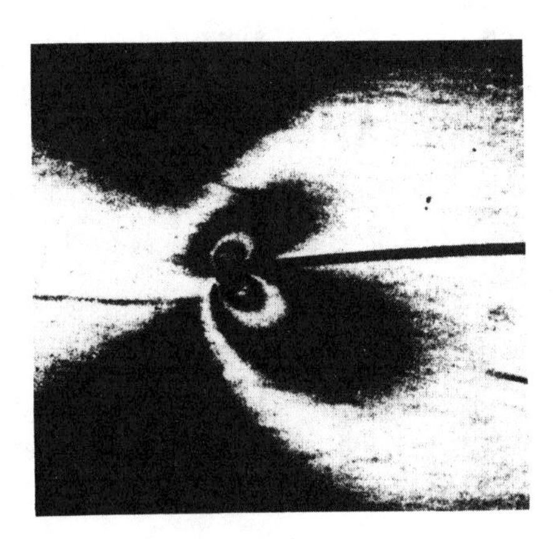

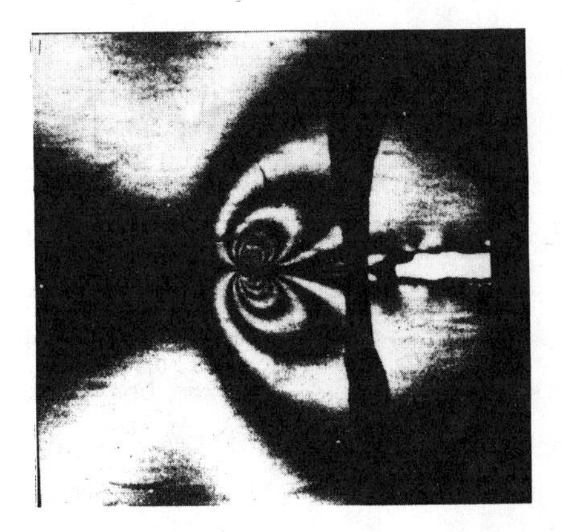

Fig.7 Dynamic isochromatic crack tip fringe patterns associated with a DCB-type specimen: (a) initially path-stable phase, (b) later path-unstable phase.

A further delicate field of application for K-determination procedures is crack branching. At the onset of branching the situation is governed by the K-factor (G-factor) of the butterfly-pattern of the main crack. When the individual branches emerge from the branching site they carry individual local butterfly-pattern which show distortion due to branch tip interaction which decreases with increasing distance of individual travel. During the initial post-branching phase a global mode-1 butterfly-pattern still engulfes the branching.

4. *Crack Speed versus Stress Intensity Factor Characterisation of Dynamic Fracture*

The dynamic stress intensity factor K_d is an important material property in the study of dynamic fracture. The dynamic behavior of propagating cracks can be characterized over the entire range from crack arrest to crack branching from the relationship between the stress intensity factor, K, and the crack tip speed, c. For brittle materials (e.g. rock subjected to dynamic loading conditions) three values of the stress intensity factor are important:

- K_c... the value at which the crack will initiate,
- K_b... the value at which the crack will branch, and
- K_a... the value at which propagating cracks will arrest.

In many rock fracture applications a sufficient amount of (explosive) energy must be supplied to the rock in order to raise the stress intensity to initiate cracks. If, in addition, unwanted fragmentation is required to be kept to a minimum the energy level during crack extension must be controlled. In order to accomplish this goal knowledge of the relationship between stress intensity factor (or energy release rate) and crack speed is imperative. Numerous experiments have been conducted in many laboratories with various types of fracture test specimens (SEN,MCT,DCB,3-PB,..) cut from different kinds of rock and other materials in order to obtain crack velocity versus stress intensity curves /21,22/.

References

/1/ Dally,J.W. and W.F. Riley: Experimental Stress Analysis. McGraw-Hill, New York (1978)

/2/ Kuske, A. and G.Robertson: Photoelastic Stress Analysis. Wiley, New-York (1974)

/3/ Thau, S.A. and J.W.Dally: Subsurface characteristics of the Rayleigh-wave. Int.Jnl of Engng.Sci.7,37 (1969)

/4/ Durelli, A.J. and A.Shukla: Identification of isochromatic fringes (to appear; 1981)

/5/ Dally, J.W. et al.: Dynamic Photoelasticity and Photoplasticity. CISM-Course, October 1976)

/6/ Fracture Toughness Testing and its Applications.ASTM STP-381, 67th Annual Meeting ASTM, Chicago, Ill., 1964 (1965)

/7/ Irwin, G.R. and A.A.Wells: A continuum mechanics view of crack propagation, Metallurgical Reviews, 10,259-266 (1965)

/8/ Post, D.: Photoelastic stress analysis for an edge crack in a tensile field. Proc.SESA, XII,99-116 (1955)

/9/ Wells, A.A. and D.Post: The dynamic stress distribution surrounding a running crack - A photoelastic analysis, Proc.SESA 16(1), 69-93 (1958)

/10/ Sanford, R.J. and J.W.Dally: A general method for determining mixed-mode stress intensity factors from isochromatic fringe patterns. Eng.Fract. Mech.11,621-633 (1979)

/11/ Dally, J.W. and Sanford R.J.: Classification of stress intensity factors from isochromatic fringe patterns. Exp.Mech.18(12), 441-448 (1978)

/12/ Rossmanith,H.P.: Analysis of mixed-mode isochromatic crack-tip fringe patterns, Acta Mechanica 34,1-38 (1979)

/13/ Irwin, G.R.: Discussion to Ref./9/, Proc.SESA 16(1),93-96 (1958)

/14/ Westergaard, H.M.: Bearing pressures and cracks. Trans-ASME, 61,A49-A53 (1939)

/15/ Rossmanith H.P.: Modelling of fracture process zones and singularity dominated zones, Eng.Fract.Mech.(to appear, 1982)

/16/ Sanford, J.R.: A critical re-examination of the Westergaard method for solving opening-mode crack problems. Mech.Res.Comm.6,289-294 (1979)

/17/ Rossmanith H.P. and R.Chona: A survey of recent developments in the evaluation of stress intensity factors from isochromatic crack-tip fringe patterns. ICF V, Vol.5,2507-2516 (1981)

/18/ Sanford, R.J. and R.Chona: Analysis of photoelastic fringe patterns with a sampled least squares method. Proc.SESA Spring Meeting, Dearborn, 273-276 (1981)

/19/ Dally, J.W.: Dynamic photoelastic studies of fracture. Exp.Mech.19, 349-367 (1979)

/20/ Rossmanith,H.P.: Crack Propagation and Branching. Proc.of Symp.on Absorbed Spec.Energy/Strain Energy Density; Guillemot Symp., Budapest, September 1980; Proc.283-294 (1981)

/21/ Irwin,G.R. et al.: Annual Reports 1975-1981 of the Photomechanics Lab. of the University of Maryland to the U.S.Nuclear Regulatory Commission.

/22/ Der, V.K., D.C.Holloway and T.Kobayashi: Techniques for dynamic fracture toughness measurements, NSF-Report, University of Maryland, (1978).

DYNAMIC CRACK ANALYSIS AND THE INTERACTION BETWEEN CRACKS AND WAVES

H.P. Rossmanith
Institute of Mechanics
Technical University Vienna
A-1040 Vienna, Austria

The distribution of stress and strain near the tip of a running crack in an elastic material will be influenced by dynamic effects of inertia. Rapid unloading occurs in material regions behind the crack tip and the newly created crack surfaces quickly attain new conditions of dynamic equilibrium. This is accompanied by elastic wave emission into the elastic body as the crack rapidly advances in order to compensate for continuous change of crack configuration.

The term "dynamic fracture" will be used to denote the effects of inertia resulting from the "fast propagation" of a crack. Inertia effects in conjunction with fracture phenomena are also important in problems where external loading gives rise to mechanical disturbances which interact with the crack and cause fracture.

Dynamic crack propagation in finite bodies ranks among the most difficult problems of the theory of elasticity and, hence, analytical solutions are available only for a number of basic problems. Finite elements and finite difference techniques form a powerful numerical tool for dynamic fracture studies.

1. The Moving Crack

The stress distribution near the tip of an opening-mode crack running at (constant) velocity through an elastic material takes the form

$$\sigma_{ij} = \frac{K(a,\dot{a})}{\sqrt{2\pi r}} \{ f_{ij}(c;\theta) + F_{ij}(c;r,\theta) \} + O(1) \qquad (1)$$

where $K(a,\dot{a})$ is the dynamic stress intensity factor for the instantaneous crack length $a(t)$ and crack speed $\dot{a}(t) = c(t)$. Along the ligament $(y=0, x>0)$ the bracketed term is equal to unity, so that the dynamic stress intensity factor is related to the stress component σ_y ahead of the crack tip in the same way as in the static case. Neglecting terms of order O(1), i.e. regular stress fields that dominate in the far field only, equ.(63), although derived originally for cracks extending at constant speed, is valid also for non-uniform motions of the crack tip. The equation indicates that the dynamic stress intensity factor $K(a,\dot{a})$ is a measure of the severity of loading in the region close to the moving crack tip. The evaluation of $K(a,\dot{a})$-distributions for elastodynamic crack problems is a fundamental necessity in solving dynamic engineering crack problems.

The first problem solved in dynamic crack propagation was the constant speed moving central crack of finite length /1/(Fig.1). The problem may be solved in various ways, such as Westergaard stress function methods, integral transform techniques etc.

The boundary conditions of the problem considered require: $\sigma_y = \tau_{xy} = 0$ on the cracked section $-a < x < a$, and $\sigma_x = \tau_{xy} = 0$, $\sigma_y = \sigma$ for $R = \sqrt{(x^2+y^2)} \to \infty$ at infinity.

Carrying out the analysis one obtains for the circumferential stress component σ_θ in a local polar coordinate system (r,θ) centered at the crack tip the expression:

$$\sigma_\theta = \frac{\sigma\sqrt{\pi a}}{\sqrt{2\pi r}} \{ f_\theta^{(1)}(r_1;\theta) + f_\theta^{(2)}(r_2;\theta) \} \qquad (2)$$

where the functions $f_\theta^{(j)}(r_j;\theta)$ depend on the ratios c/c_j and the circumferential angle θ . Notice, that the finite crack length 2a does not enter the bracketed factor in equ.(61) and therefore, it has no influence on the distrib-

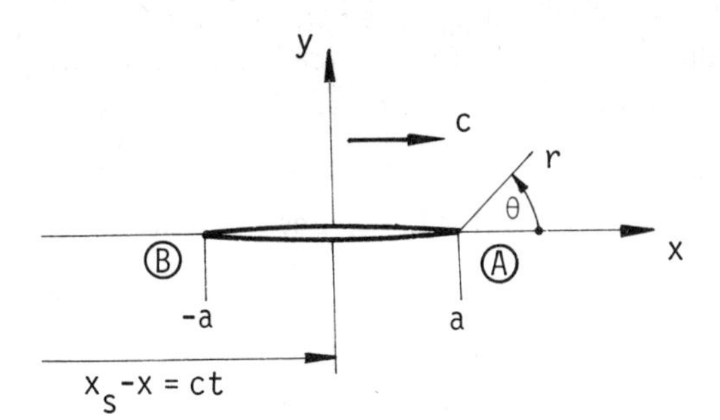

Fig.1 The moving central crack (Yoffe-problem) /1/

ution of σ_θ . In the limit, as $c \to 0$, one recovers the static σ_θ -distribution

$$\sigma_\theta = \frac{\sigma\sqrt{\pi a}}{\sqrt{2\pi r}} \cos^3\frac{\theta}{2} \quad , \tag{3}$$

in agreement with results of Inglis and Westergaard. Yoffe's results for $\nu=0.25$ are presented in Fig.2 and also in tabulated form in Table I.

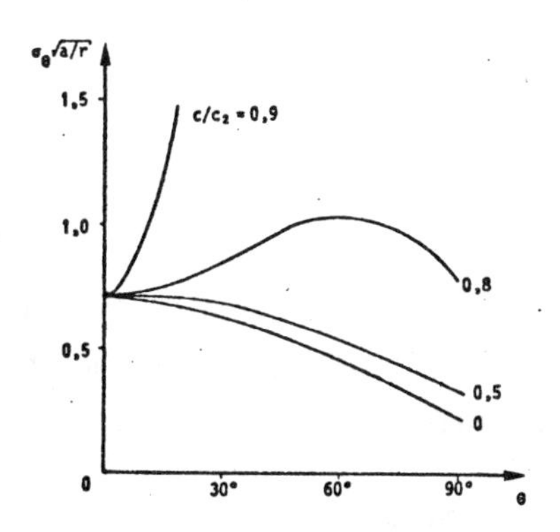

Fig.2 σ_θ vs θ for Yoffe-problem

Table I :

$\sigma\sqrt{\frac{a}{r}}$		θ			
		0^0	30^0	60^0	90^0
c/c_2	0	0,707	0,64	0,46	0,25
	0,5	0,707	0,70	0,55	0,35
	0,8	0,707	0,85	1,04	0,78
	0,9	0,707	2,3	5,0	4,6

The results show a distinctive maximum off the centerline for σ_θ with crack speeds $c > 0.6\ c_2$. Adopting a maximum tensile stress criterion as a fracture criterion Yoffe concluded from a redistribution of maximum circumferential tensile stress that the crack would change its orientation for sufficiently high crack speeds. Yoffe did not consider the magnitude and locus of the maximum principal tensile stress which is located at a different place and slightly larger than the maximum circumferential tensile stress. This path instability was assumed to indicate a necessary condition for successful crack branching. Unfortunately, the pathological condition of perfect healing behind the trailing crack tip B yields identical stress intensity factors for the static and dynamic crack and, therefore, does not provide a relationship between the crack velocity c and the external loading applied.

The first transient fracture problem has been solved by Maue /2/ by computing the stress distribution around the tip of a semi-infinite crack which appears suddenly at time t=0 and remains stationary. The same generation problem but with the crack propagating at constant speed for t>0 has been studied by Baker /3/. Broberg /4/ gave the solution for the fundamental problem of a crack in an uniaxial tension sheet extending at t=0 from a microflaw of length "0^+" in both directions with constant and equal speeds. Despite the complexity and difficulty of Broberg's crack problem results obtained for the close-field approach show several distinct features.

Defining the dynamic stress intensity factor the same way as in the static case, one obtains

$$K_d = \sigma\sqrt{\pi a}\, k(c) = K_s\, k(c) \tag{4}$$

with the velocity-dependent function k(c) to be evaluated later in this chapter. Similarly, employing Irwin's "crack sewing method" for the running crack, one obtains for the strain energy release rate, G_d,

$$G_d = \frac{1}{\Delta a} \int_a^{a+\Delta a} \sigma_y(x) v(a+\Delta a-x)\, dx = \frac{1}{E'} \pi\sigma^2 a\, \gamma(c) \tag{5}$$

where E' = E for plane stress and E' = E/(1-ν^2) for plane strain. The universal dynamic correction function $\gamma(c)$ is given later in this chapter.

From an engineering applications point of view a thorough mathematical analysis by Freund /5/ of the general dynamic crack problem is of importance. Freund considers an infinitely homogeneous, isotropic, linear-elastic solid weakened by a straight-crested half-plane crack defined by ($x < ct, y=0$) in a fixed cartesian coordinate system. The crack front is parallel to the z-axis. The crack extends with uniform velocity passing the origin 0 at time t=0. For $t < 0$ then are no body forces or surface tractions acting on the body. At time $t=0$ a pair of splitting forces of unit magnitude appears at the tip of the crack acting in the direction of the outward normal on each face. For $t>0$, the crack continues to move in the positive x-direction and the crack faces begin to separate, these forces do work, and stress waves are generated. The resulting dynamic stress field is called the 'fundamental solution'.

The mathematical analysis yields an important relation for the determination of the dynamic stress intensity factor, K_d:

$$K_d = \lim_{\xi \to 0^+} \{ \sqrt{2\pi\xi}\, \sigma_y(\xi,o;t) \} = \sqrt{2/\pi}\, \frac{k(c)}{c}\, t^{-1/2} \qquad (t>0) \quad . \tag{6}$$

Consider the mode-1 extension of a half-plane crack subjected to time-independent varying loading. Let the resulting normal stress in the y-direction along $x>0, y=0$ be $p(x)$. During constant rate crack extension (after $t=0$) new traction free crack surface has been created along $0 < x < ct$, $y=0^{\pm}$. The stress wave field associated with crack extension may be obtained by superimposing the solution of a dynamic problem on the static field (solution for $t<0$). Application of the fundamental solution, a superposition integral may be defined with respect to the stress $p(x)$ to be removed.

Regarding the transformation of variables $x-ct=\xi$ and $ct-x_o=\xi_o$ (Fig.3) the limit $\xi \to 0^+$ yields the dynamic stress intensity factor

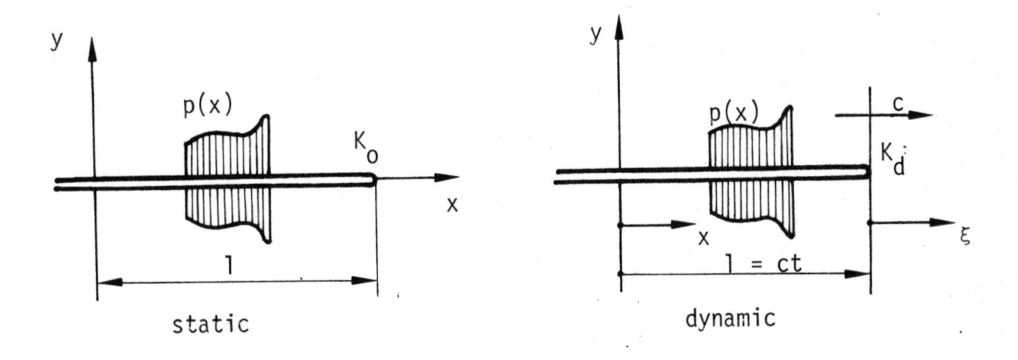

Fig.3 Dynamic and equivalent length static crack problem

$$K_d = \lim_{\xi\to 0^+} \{\sqrt{2\pi\xi}\ \sigma_y^*(x,o;t)\} = \sqrt{2/\pi}\ k(c) \int_0^{ct} \frac{p(ct-\xi_o)}{\sqrt{\xi_o}}\ d\xi_o \quad , \tag{7}$$

in the form of a product of a function of velocity c and of a functional of the loading applied, p(x). This form is well suited for comparison with the static value of the stress intensity factor

$$K_s = \sqrt{2/\pi} \int_0^1 \frac{p(1-\xi_o)}{\sqrt{\xi_o}}\ d\xi_o \quad , \tag{8}$$

for a static crack of equivalent length $l = ct$, Fig.3 .

The dynamic correction $k(c) = K_d/K_s$ is very similar to k(c) defined in equ.(4) , and it is independent of both crack length and loading p(x) and is only a function of the crack speed and material constants.

Because wave reflections from other tips or boundaries do not occur here, an expression for the energy-release rate or crack extension force, G_d, for the extending semi-infinite crack may be derived

$$G_d = \frac{1-\nu^2}{E} K_d^2\ g(c) = G_s\ \gamma(c) \quad , \tag{9}$$

with $g(0)=1$ and $g(c)\to\infty$ when c approaches the Rayleigh-wave speed, c_R. Graphs of the functions k(c) and $\gamma(c)$ are shown in Figs. 4 and 5 , respectively. The tendency of the function $\gamma(c)$ to decrease with increasing crack speed allows for the possibility of crack branching when G_d drops below 0.5 G_s, thus defining an estimate of the upper limit for possible crack speeds. From Fig.5 , $\nu=2/7$, $c/c_R=0.45$ for $\gamma=0.5$; many experimental records show the limiting crack speed to be well below this limit.

The results of the constant speed running crack problem can be utilized to determine the stress intensity factor and the energy-release rate as functionals of the crack tip motion for the case of a non-uniformly propagating crack.

By solving the dynamic problems of a starting or a stopping crack these elementary start-move-stop crack extension sequence can successfully be applied to the problem of a crack extending at a non-uniform rate c(t), as is shown in Fig.6.

The velocity profile l(t) of the moving crack is approximated by a

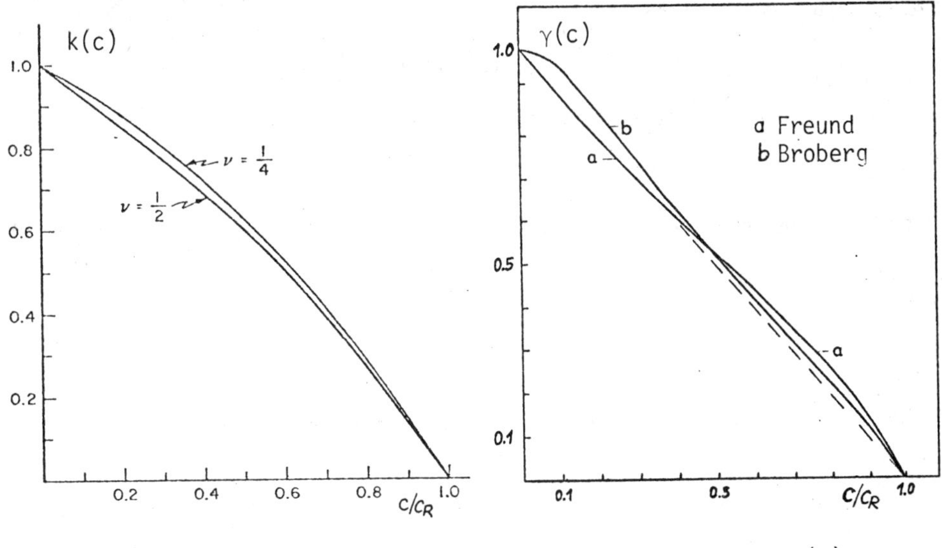

Fig.4 The function k(c)

Fig.5 The function γ(c)

polygonal curve L(t) with segments of constant crack speeds, $c_i = \Delta L_{i,i+1}/\Delta t_{i,i+1}$, during time intervals $\Delta t_{i,i+1} = t_{i+1} - t_i$, respectively. The preceding analysis allows for an expression for the stress intensity factor as a functional of L(t).

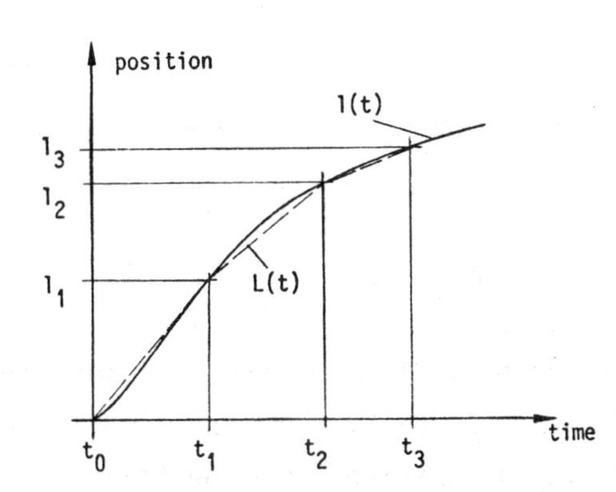

Fig.6 Position versus time graph for non-uniform crack propagation /5/

For $\Delta t_{i,i+1} \to 0$, the polygonal curve L(t) approaches l(t) for all i. In the limit it follows

$$K_d(l,c) = k(c)\, K_s\{l(t)\} \quad . \tag{10}$$

Utilizing a form of the energy-release rate G_d proposed by Atkinson and Eshelby one obtains

$$G_d(1,c) = \frac{1-\nu^2}{E} K_d^2(1,c)\, g(c) \quad . \tag{11}$$

Assuming a fracture criterion of the form

$$G_d = 2\,\Gamma(c) \quad , \tag{12}$$

with Γ the generalized specific surface energy, the actual motion of the crack is given by a non-linear ordinary differential equation:

$$\frac{2E\Gamma(c)}{1-\nu^2}\,\frac{1}{K_S^2(1)} = k^2(c)\, g(c) = \gamma(c) \quad . \tag{13}$$

EXAMPLE: Infinite plate with uniformly extending central crack (Broberg-crack problem /4/).

The energy released from the stress field around a constant speed bilaterally extending crack subjected to a uniform external traction σ_∞ acting in y-direction is given by:

$$G_d = \pi(1-\nu^2)/(2E)\, \sigma^2 a\, \gamma(c) \qquad \text{(plane strain)} \tag{14}$$

with $\gamma(c)$ given in Fig.5 . For plane stress, $\nu \to \nu/(1+\nu)$ and $E \to E(1-2\nu)/(1+\nu$

The fracture criterion (12) yields

$$\pi(1-\nu^2)/(2E)\, \gamma(c)\sigma^2 a = 2\Gamma(c) \quad . \tag{15}$$

Suppose the crack extends from a pre-existing flaw of length $2a_o$ and at crack initiation, σ reaches a critical value σ_c. Then, it holds

$$\pi(1-\nu^2)/(2E)\, \gamma(o)\sigma_c^2 a_o = 2\Gamma(o) \tag{16}$$

with $\gamma(o) = 1$.

Keeping the load at level σ_c, the subsequent crack motion is determined by equ.(15) with $\sigma = \sigma_c$. From eqs(15) and (16) follows an equation for c/c_R.

The shape of the $\Gamma(c)$ curves for various materials may be found in the literature /6/. Dynamic photoelastic investigations lead to the assumption that acceleration effects should have negligible influence up to velocity ratios $c/c_R \sim 0.4$.

Employing the principle of superposition this crack problem can be converted into one associated with constant pressure hydraulic fracturing /7/.

1.1. *Discontinuous Change of Crack Speed*

It is of considerable practical interest to investigate the conditions which may cause a sudden change in crack speed during propagation. Suppose that at time $t=t_1$ the moving semi-infinite crack with tip located at $x=x_1$ changes tip speed from $c^-=c^-(t_1^-)$ to $c^+=c^+(t_1^+)$ and the crack continues to propagate at speed c^+ for $t>t_1$.

The stress intensity factor at the new speed c^+ is obtained /9/

$$K_I(t_1^+,c^+) = h(c^-,c^+)\, K_I(t_1^-,c^-) \tag{17}$$

with the jump function $h(c^-,c^+)$ given in Fig.7.

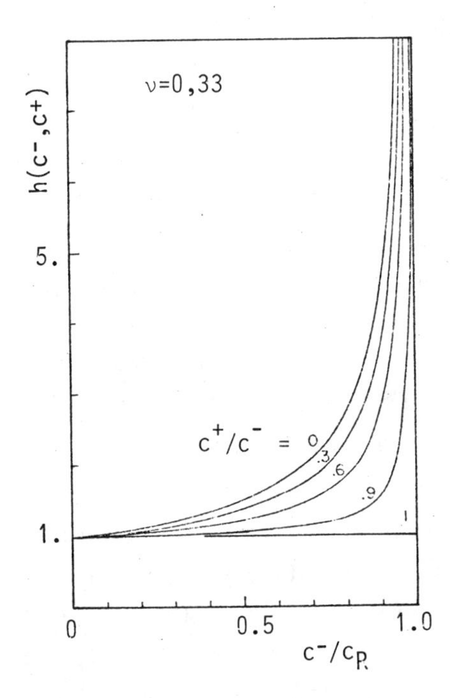

Fig.7 Jump function $h(c^-,c^+)$

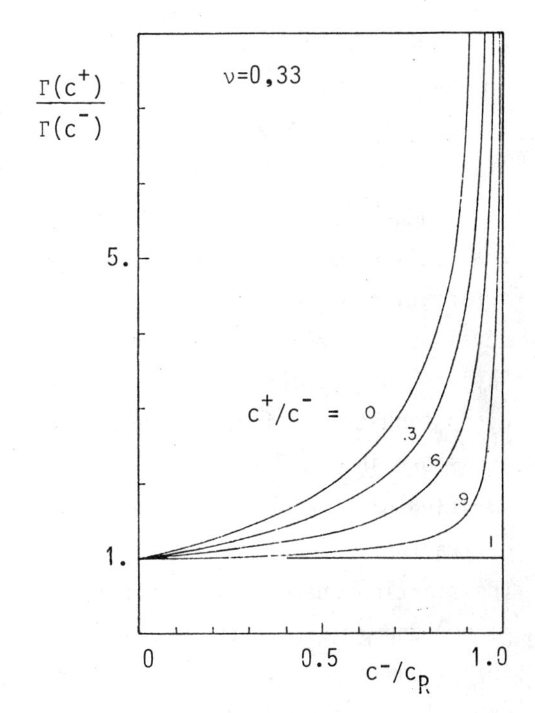

Fig.8 Jump function for $\Gamma(c)$

The ratio of specific crack extension energies required for a sudden velocity jump $[c] = c^+-c^-$ follows from equations (11),(12), and (17):

$$\Gamma(c^+)/\Gamma(c^-) = h^2(c^+,c^-)\, g(c^+)/g(c^-) \quad . \tag{18}$$

For crack arrest, $c^+=0$, one obtains

$$K_I(t_a^+,o) = h(c^-,o)\, K_I(t_1^-,c^-) \tag{19}$$

The results for crack arrest, $c^+=0$, are plotted in Fig.8. The graph shows that larger ratios of specific crack extension energies are required for cracks suddenly arresting from higher crack speeds. This fact has important consequences in crack arrest design methodologies where running cracks are to be arrested by special arrest segments. In rock fracture mechanics sudden changes of fracture toughness in a system of rock layers induces discontinuous changes in crack speed. A practical application for the discontinuous extension of a Broberg-crack is given when a central crack centered at a borehole is propagated due to the detonation stress waves and extends on its own when the elastic stress field has outdistanced the crack. Another application of the constant velocity central crack is found in layer separation along a pre-existing weak interface plane such as a bond line, etc.

Sudden changes in crack speed give rise to stress discontinuities in the radiated body wave pattern. When a crack initiates, arrests or bifurcates a radiation pattern of wave fronts associated with this discontinuity phase develops and expands which consists of a P-wave, S-wave, head wave, and Rayleigh-waves, the latter ones propagating along the crack faces. Stress discontinuity amplitudes have been computed and are reported in Ref./10/ and shown in Fig.9 for crack initiation. Note that the jump of $\{\sigma_{rr}\}_P$ is negative because a compressive stress release wave is radiated from the site of fracture initiation. Fig.9a shows the wave front pattern associated with the starting phase; radial plots of the step discontinuity $\{\sigma_{rr}\}_P$ (upper half-plane) and $\{\tau_{r\theta}\}_S$ (lower half-plane) are shown in Fig.9b .The photoelastic fringe pattern recording associated with the starting phase (Fig.10) clearly exhibits the expanding stress wave jumps by the fringe loop kinks. A detailed analysis /10/ shows that $\max\{\tau_{r\theta}\}_S / \max\{\sigma_{rr}\}_P < 0.25$. During constant or slowly varying speed crack propagation the crack tip acts as an energy sink with the energy flux given by $G_d \cdot c$. The dynamic crack tip fringe pattern of a mode-1 crack represents the stress field associated with steady state or slowly transient kinetic energy radiation from the crack tip.

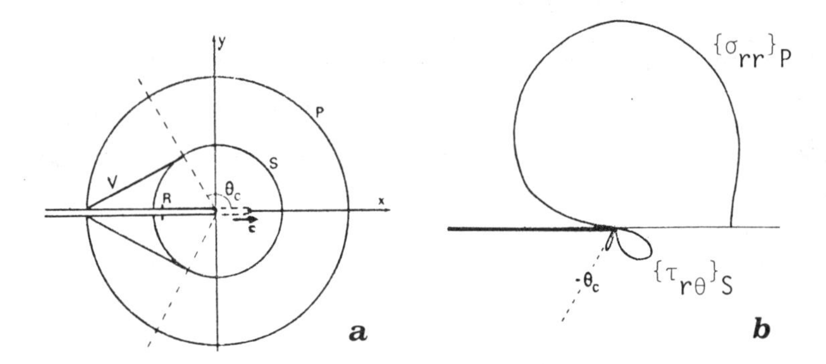

Fig.9 Sudden crack initiation; a) wave fronts
b) radial plots of stress discontinuities

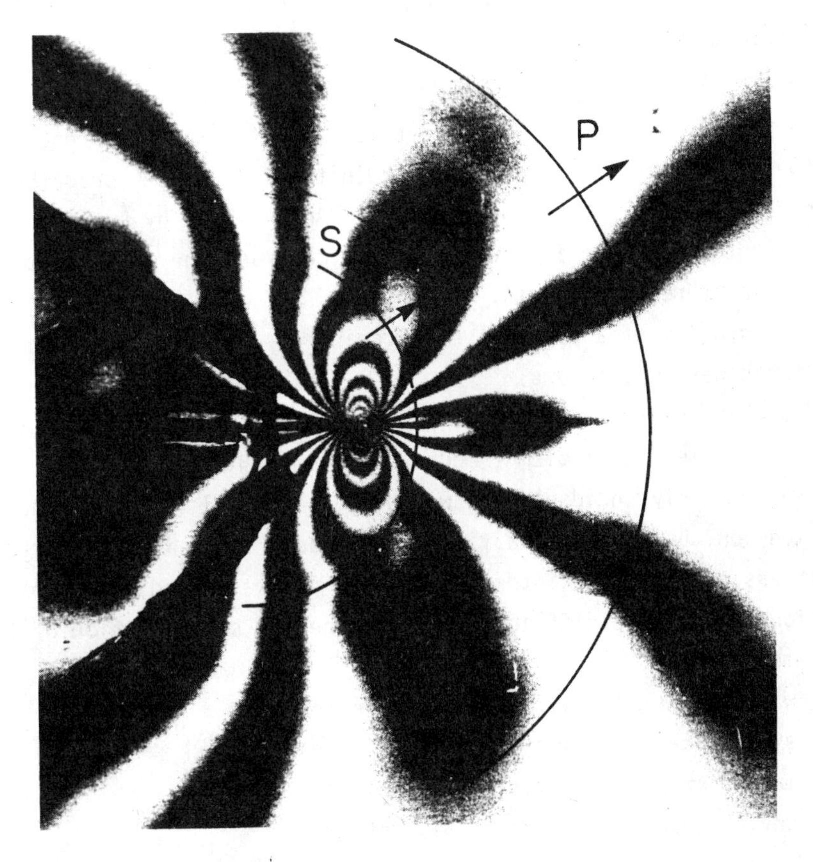

Fig.10 Dynamic photoelastic recording of stress release waves at crack initiation in Homalite 100

2. *Crack - Wave Interaction*

This chapter is concerned with the generalisation of earlier dynamic crack propagation problems to the case where the static or running crack is subjected to time-dependent stress loading. The crack may represent an interface crack located at the (slipless, frictional, wavy etc.) interface between two similar or dissimilar media.

2.1. *Cracks Subjected to Stress Wave Loading*

Consider a half-plane crack in an unbounded isotropic, homogeneous, linear elastic solid. The material in the vicinity of the crack is initially at rest and the crack tip is located at x=0. A plane longitudinal (tension) wave propagates toward the crack, the wavefront being obliquely oriented with respect to the plane of the crack (Fig.11a).

At time t=0 the wave strikes the crack tip and it is partially reflected and partically diffracted. At some time after the wave has reached the crack tip the energy input allows for crack initiation and the crack begins to extend at an arbitrary non-uniform rate. An analysis by Freund /10/ assumes the crack to extend in its own plane, in contrast to results obtained from dynamic fracture experiments.

The problem of a plane wave impinging on a planar crack can be decomposed into two separate problems:

a) a symmetric problem (mode-1 problem, Fig.11b).
b) a antisymmetric problem (mode-2 problem, Fig.11c).

An obliquely incident wave would induce normal (σ_θ) and shear (τ_θ) tractions on the planes of a fictitious crack. Since the real crack faces are stress free (or the internal pressure is prescribed), the complete solution to the wave diffraction is obtained by superposition of several problems:

A : wave propagation with no crack present
B : symmetrically or antisymmetrically loaded crack
C : moving crack → negating radiation field
D : moving crack → negating wave stresses

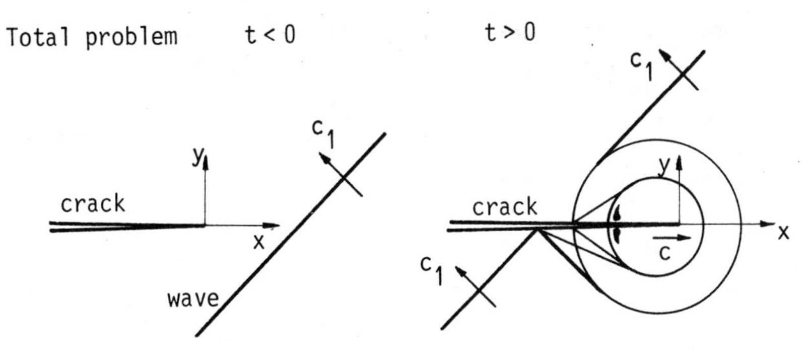

A: wave propagation problem (no crack)

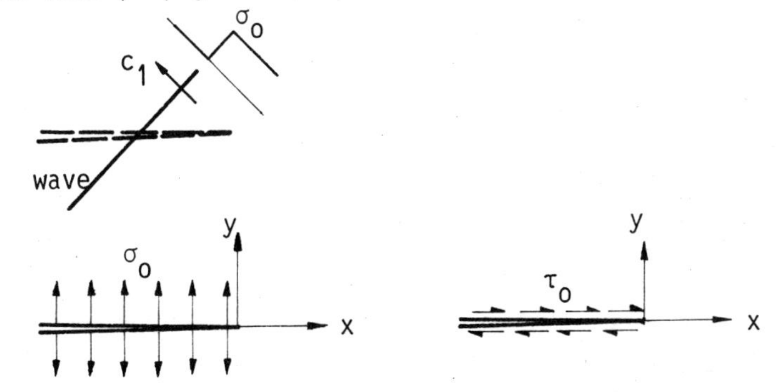

B_I: symmetric crack loading B_{II}: antisymmetric crack loading

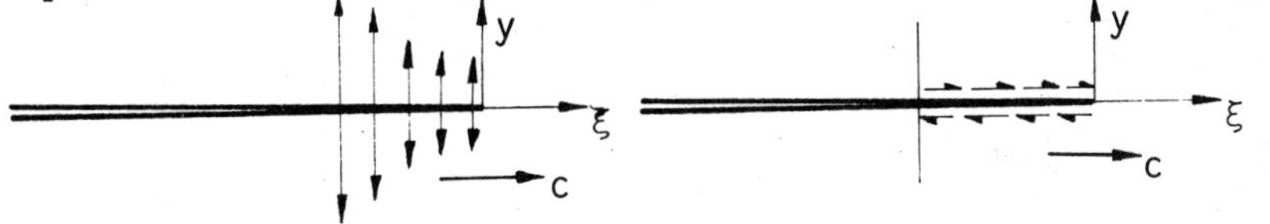

C_I: moving crack (negating radiation field stresses) : C_{II}

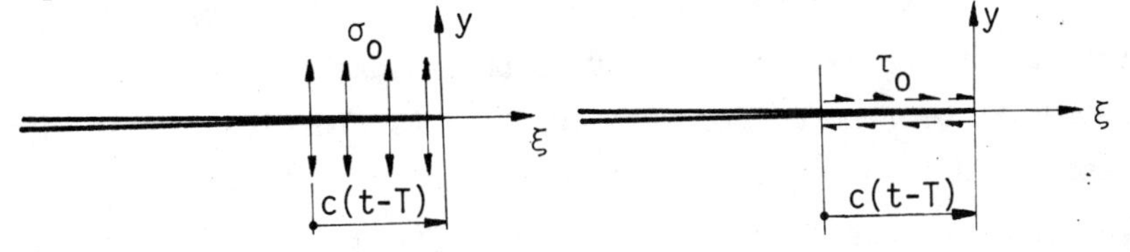

D_I: moving crack (negating wave stresses) : D_{II}

Fig.11 Superposition scheme for oblique incident stress waves

Problems B,C, and D can be treated by means of generalized forms of Freund's fundamental solutions /5,11/.

For a step-stress incident pulse of magnitude σ_0 the amplitudes of the normal traction (σ_θ) and the shear traction (τ_θ) are given by

$$\sigma_\theta = \sigma_0 \{1 - 2 \sin^2\theta \, (c_2/c_1)^2\} \tag{20}$$

$$\tau_\theta = \sigma_0 \, \sin 2\theta \, (c_2/c_1)^2 \quad , \tag{21}$$

where c_1= longitudinal wave speed and c_2= shear wave speed.

For isotropic material behavior the two modes are uncoupled and independen and the two problems can be treated separately, superposition playing a significant role in obtaining the results.

Wave diffraction about a crack tip gives rise to some stress distributions radiated out on the prospective fracture plane ahead of the crack. If the crack now begins to extend at a constant rate at some finite delay time after diffraction has started, the crack must negate this stress distribution. During crack extension a traction distribution appears through the moving tip equal but opposite to the traction on the prospective fracture plane. This problem is denoted by D. The sum of the solutions of these four problems (see superposition schemes) is the solution for the initially stated general wave-crack interaction problem.

2.1.1. *Diffraction by a Stationary Crack*

In the plane strain problem of diffraction of a pulse by a stationary crack a (tensile) stress of magnitude σ_0 is induced on the portion of the plane y=0 for which $t + x\sin\theta / c_1 > 0$ upon passage of the incident wave front. The stresses to be removed from the crack faces are given by $-\sigma H(t + x\sin\theta/c_1)$. The stress is zero on y=0 and the normal displacement v=0 ahead of the crack tip. The normal stress distribution $\sigma_y(x)$ and shear stress distribution $\tau_{xy}(x)$ on y=0, x<0, determine the stress intensity factors $K_{I,d}$ and $K_{II,d}$, respectively, at time t for zero crack tip speed:

$$K_{I,d}(t,o) = \lim_{x\to 0^+} \{\sqrt{2\pi x}\, \sigma_y(x,t)\} = 2\sqrt{2/\pi}\, \sigma_\theta \omega_I \sqrt{t} \tag{22}$$

$$K_{II,d}(t,o) = \lim_{x\to 0+} \{\sqrt{2\pi x}\tau_{xy}(x,t)\} = 2\sqrt{2/\pi}\ \tau_\theta \omega_2 \sqrt{t} \tag{23}$$

where

$$\omega_2 = \omega_1 \sqrt{\frac{c_1/c_2 + \sin\theta}{1 + \sin\theta}} . \tag{24}$$

The $\sqrt{t}$ - increase of $K_{I,d}$ and $K_{II,d}$ for a step pulse has been obtained earlier by various authors.

Assume that a rectified square pulse of the form

$$\sigma(t) = \begin{cases} 0 & t < 0 \\ \sigma_o & 0 < t < t_1 \\ \lambda\sigma_o & t_1 < t < t_2 \\ 0 & t < t_2 \end{cases} \tag{25}$$

impinges on the stationary crack tip located at x=0 at t=0. The associated stress intensity factor is obtained by superposition of three Heaviside-step pulses of magnitudes $+\sigma_o$, $-(1+\lambda)\sigma_o$, and $\lambda\sigma_o$, and

$$K_i(t) = \begin{cases} 0 & t < 0 \\ 2\sqrt{2/\pi}\sigma_\theta\ \omega_i\sqrt{t} & o < t < t_1 \\ 2\sqrt{2/\pi}\ \omega_i\ \{\sigma_\theta\sqrt{t} - (1+\lambda)\sigma_\theta\sqrt{t-t_1}\} & t_1 < t < t_2 \\ 2\sqrt{2/\pi}\ \omega_i\ \{\sigma_\theta\sqrt{t} - (1+\lambda)\sigma_\theta\sqrt{t-t_1} + \lambda\sigma_\theta\sqrt{t-t_2}\} & t_2 < t < t_3 \\ 0 & t < t_3 \end{cases} \tag{26}$$

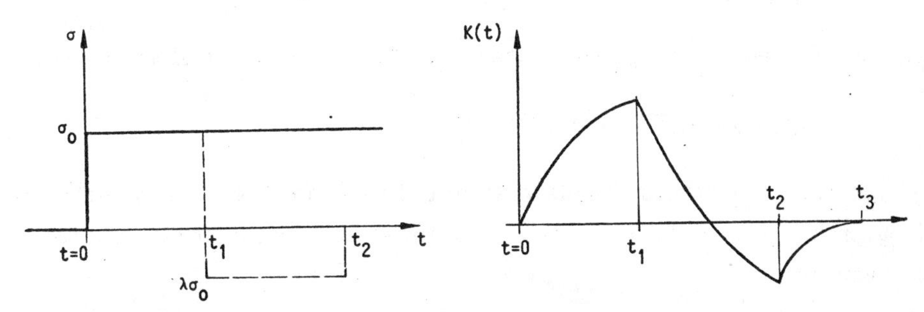

Fig.12 Rectified stress pulse and associated K-distribution

The stress intensity factors $K_i(t)$ change sign at $t=t^*$

$$t^* = t_1\ (1+\lambda)^2 / \{\lambda(2+\lambda)\} , \tag{27}$$

and the time $t=t_3$ for having reestablished stress free conditions at the crack tip is obtained by equating to zero the last line of equ. (26).

In dynamic photoelasticity a plane detonation wave is characterized by its fringe order distribution $N(t) = N(x,c_j)$, and its discretized form $\{N(t_i),t_i\}$ or $\{N(x_i/c_j),x_i/c_j\}$ can directly be utilized in conjunction with eqs.(25)-(27) for numerical evaluation of stress intensity factors for P(j=1) or S(j=2) wave interaction with a stationary crack tip.

2.1.2. *Crack Extension Following Wave Diffraction*

If K(t) exceeds a critical value, K_{dc}, the dynamic fracture toughness, crack initiation under dynamic loading conditions occurs, and the crack extends at a (non)uniform rate. Hence, this section refers to problems C and D of the superposition scheme, Fig.11.

Suppose at first that the crack extends in mode-1 (normal incidence) at constant speed c for $t \geqq T$. The incident plane wave (normal incidence), a simple tension step pulse of magnitude σ_0, is diffracted about the moving crack tip and partically reflected at the crack face. In order to extend, the crack must negate the tractions on y=0, $o \leqq x < c(t-T)$ found in problems A and B.

The mode-1 stress intensity factor for problem C_I is given by /1/

$$K^C = 2\sigma_0 \sqrt{2/\pi}\ \{\omega_0\sqrt{t} - \sqrt{c(t-T)}\}\ k(c) \tag{28}$$

and for problem D_I the stress intensity factor is obtained from equ. (7)

$$K^D = 2\sigma_0 \sqrt{2/\pi}\ k(c)\ \sqrt{c(t-T)} \quad . \tag{29}$$

Because the stress fields associated with problems A and B are non-singular at y=0, x=c(t-T), the stress intensity factor for the total problem is given by

$$K(t,c) = K^C + K^D = 2\sqrt{2/\pi}\ \sigma_0\omega_0\ k(c)\ \sqrt{t} \qquad (t < T) \quad . \tag{30}$$

Note, that the stress intensity factor K(t,c) for constant speed crack extension with delay time T after the crack-wave front interaction is independent of the delay time T, except for the condition $t > T$. The delay time between step-wave impingement and crack initiation follows from equ.(22) upon specializing for a fracture criterion, e.g. $K_{dc} = K_I(t=T,o)$,

$$T = \frac{\pi}{8}\ K_{dc}^2/(\sigma_\theta\omega_1)^2 \quad . \tag{31}$$

An expression for the stress intensity factor for an arbitrary nonuniformly moving crack may be constructed on the basis of the equations of earlier sections associated with run-stop crack extension behavior and polygonal approximation of the time-position graph of the crack tip. In this approximation the crack tip moves with constant speed $c_i=(l_{i+1}-l_i)/(t_{i+1}-t_i)$ during the time interval $t_i < t < t_{i+1}$. Because the stress intensity factor of the constant speed problem is independent of the delay time, the stress intensity factor for any interval $t_i < t < t_{i+1}$ is given by

$$K(t,c_i) = k(c_i)\ K(t,o) \quad , \tag{32}$$

which, in the limit as L(t) approaches l(t) yields

$$K(t,c) = k(c)\ K(t,o) \quad , \tag{33}$$

where K(t,o) is the stress intensity factor of the associated stationary crack under the action of the given incident stress wave.

The stress intensity factor for a step stress pulse of magnitude σ_o arriving at t=0 is given by equ.(7)

$$K(t_1,c) = 2\sqrt{2/\pi}\sigma_o\omega_o\ \sqrt{t}\ k(c) \quad , \tag{34}$$

and the result for a differential step stress pulse of magnitude $\Delta\sigma(s)$, arriving at time t=s, is

$$2\sqrt{2/\pi}\ \Delta\sigma(s)\ k(c)\ \omega_o\ \sqrt{t-s} \quad . \tag{35}$$

Superposition of a large number of such incremental jumps $\Delta\sigma(s)=\partial s/\partial s\ \Delta s$ in the range $0 < s < t$ yields, in the limit as $\Delta s \to 0$, the stress intensity factor

$$K^*(t^*,c) = 2\sqrt{2/\pi}\ k(c)\ \omega_o \int_{o^-}^{t^*} \frac{\partial\sigma}{\partial s}\sqrt{t-s}\ ds \quad , \tag{36}$$

for any motion of the crack tip l(t) due to the interaction of an arbitrarily shaped tensile stress pulse of profile $\sigma(t)$ arriving at the crack tip at t=0. The equation of motion, equ.(13), may be generalized by exchanging $K_s(l)$ by $K^*(t,o)$.

The fracture capability of stress pulses and the reciprocity between pulse and flaw spectrum has been investigated in Ref./12/. It was found that pulse attenuation by fracture depends not only on pulse energy but also on the energy density of the pulse (see also Ref./13/).

For obliquely incident stress waves, the stress intensity factors for problem C and problem D are given by

$$K_I^C(t,c) = 2\sqrt{2/\pi}\,\sigma_\theta\,k(c)\,\{\omega_1\sqrt{t+\frac{c}{c_1}(t-T)\sin\theta}-\sqrt{c(t-T)}\} \tag{37}$$

and

$$K_I^D(t,c) = 2\sqrt{2/\pi}\,\sigma_\theta\,k(c)\,\sqrt{c(t-T)} \quad , \tag{38}$$

respectively.

Again, the stress fields associated with problems A and B are nonsingular and therefore the total stress intensity factors $K_I(t,c)$ and $K_{II}(t,c)$ are obtained in the form

$$K_I(t',c) = 2\sqrt{2/\pi}\,\sigma_\theta\,\omega_1\,k_I(c)\,\sqrt{t'} \tag{39}$$

$$K_{II}(t',c) = 2\sqrt{2/\pi}\,\tau_\theta\,\omega_2\,k_{II}(c)\,\sqrt{t'} \quad , \tag{40}$$

with

$$t' = t + \frac{c}{c_1}\cdot(t-T)\sin\theta. \tag{41}$$

The mode-2 dynamic correction function

$$k_{II}(c) = K_{II}(t',c)/K_{II}(t',o) = k_I(c)\,\frac{\sqrt{1-c/c_1}}{\sqrt{1-c/c_2}} \tag{42}$$

is due to Fossum /14/. It should be mentioned that theoretical calculations associated with the mode-2 contribution assume the crack to extend in its plane.

Eqs.(39) and (40) show that the stress intensity factors are not independent of the delay time T for constant speed crack extension under oblique incidence stress wave loading. However, the formal form and character of the equations is preserved, the real time being replaced by the generalized time quantity t', which is a measure of the time that would have elapsed since the incident wave struck the crack tip if the crack had always been at its instantaneous position. For normal incidence (θ=0) holds t=t' (Fig.13).

For oblique incidence the parameter t' plays the same role as t for normal incidence. The dynamic stress intensity factors for nonuniformly moving cracks subjected to arbitrary stress wave loading may now be constructed on the basis that for any motion of the crack they depend on time and on crack motion only through the parameter t' and crack tip speed c(t), where t' is generalized for non-uniform motion as

$$t' = t + \sin\theta \; l(t)/c_1 \quad . \tag{43}$$

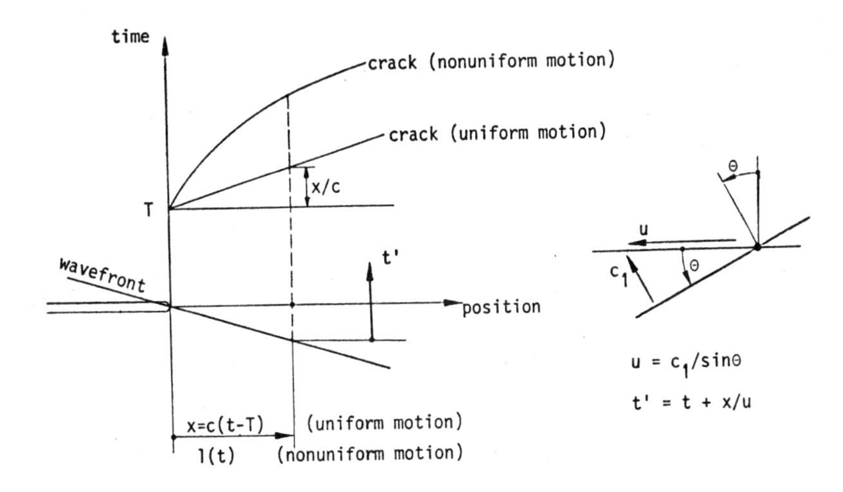

Fig.13 Time vs position diagram for obliquely incident P-wave

The stress intensity factors take the form

$$K_I(t',c) = k_I(c)\, K_I(t',o) \tag{44}$$

$$K_{II}(t',c) = k_{II}(c)\, K_{II}(t',o) \quad , \tag{45}$$

and the dynamic energy release rate is given by

$$G(t',c) = \frac{1-\nu^2}{E} \{g_I(c)K_I^2(t',c) + g_{II}(c)K_{II}^2(t',c)\} \tag{46}$$

where $g_I(c)$ follows from equ.(9) and $g_{II}(c)$ is defined by

$$g_{II}(c) = g_I(c) \frac{\sqrt{1-(c/c_2)^2}}{\sqrt{1-(c/c_1)^2}} = g_I(c)\; r_2/r_1 \quad . \tag{47}$$

Employing static quantities, equ.(46) may be written in the form

$$G(t',c) = \gamma(c)\; G(t',o) \tag{48}$$

with the universal function $\gamma(c)$ given in Fig.5. For a tensile step stress pulse the analysis has been carried out by Freund /11/. The result for an arbitrary stress pulse is obtained by superposition integrals.

2.2. *Photoelastic Investigation of Crack-Wave Interaction*

This section describes the application of dynamic photoelasticity and fracture mechanics to the interaction of stress waves with stationary and moving cracks /13,15-19/. Fracture initiation due to various wave types is considered and stress magnitudes and stress intensity factor determinations were made in some cases.

General wave diffraction at crack tips gives rise to a mixed-mode transient fracture problem which for purposes of analysis is conveniently divided into a symmetric part (mode-1) and an antisymmetric part (mode-2). When a P-wave impinges onto a stationary (or moving) crack tip the incident wave is diffracted and scattered about the crack tip (Fig.14). The total stress field $\sigma^t_{ij}(x,y;t)$ will be composed of the stress field of the incident wave $\sigma^i_{ij}(x,y;t)$ and the interaction (scattered) field $\sigma^s_{ij}(x,y;t)$:

$$\sigma^t_{ij}(x,y;t) = \sigma^s_{ij}(x,y;t) + \sigma^i_{ij}(x,y;t) \tag{49}$$

where the scattered field must satisfy the radiation condition.

As a result of wave diffraction, the following zones can be identified during crack-(P)wave interaction (Fig.14):

I ... undisturbed zone
II ... incident wave zone
III ... longitudinal wave scattering zone
IV ... shear wave scattering zone
V ... longitudinal wave reflection zone
VI ... shear wave reflection zone
VII ... head wave zone
VIII ... longitudinal wave diffraction zone
IX ... shadow zone

The general crack-wave interaction process is best discussed in terms of the following elementary phases:

1 ... incident P-wave
2 ... P-wave reflection
3 ... primary P-wave diffraction at crack tip A
4 ... secondary P-wave diffraction at crack tip B
5 ... incident S-wave
6 ... S-wave reflection
7 ... primary S-wave diffraction at crack tip A
8 ... secondary S-wave diffraction at crack tip B

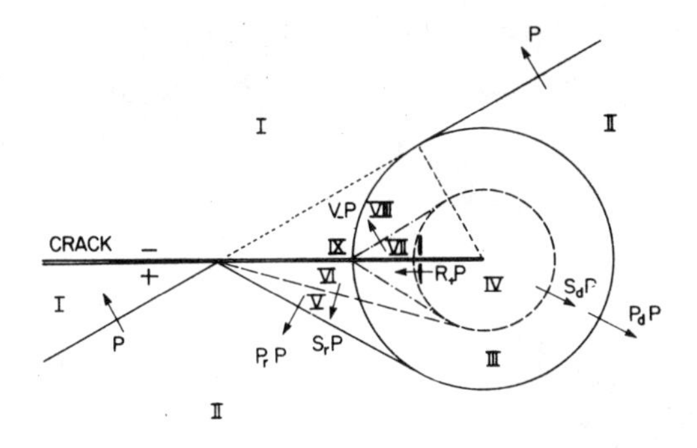

Fig.14 Diffraction of a plane elastic P-wave about a semi-infinite crack

9 ... Rayleigh wave generation
10 ... higher order diffraction.

Because the scattered wave systems will contain both, P- and S-wave contributions both stress intensity factors, K_I and K_{II} will be present in the general case.

In the experimental model studies the cracked specimen was fabricated from a 9.5 mm thick sheet of brittle polyester, Araldite B. This polyester becomes temporarily birefringent when subjected to a state of stress and this gives rise to isochromatic fringes. A crack of length 85 mm and width 0.5mm buried parallel to the free edge at a depth 110mm was saw-cut and its tips sharpened to yield a crack length of 85mm. The explosive excitation source is located at the epicentre of the crack tip A (Fig.15). A Cranz-Schardin camera provided 16 discrete configurations of the dynamic interaction process /13/.

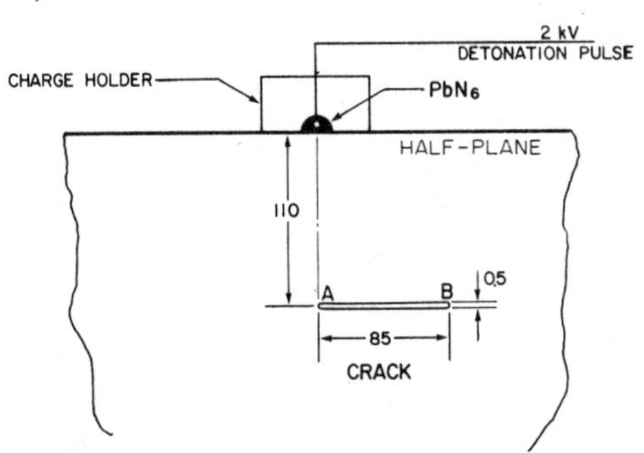

Fig.15 Specimen geometry and dynamic loading conditions

The dynamic isochromatic recording of the reflection and diffraction of a P-wave about the crack-tip A and the associated wave front constructions are shown in Figs.16a,16b, and 16c. Fig.16a is the isochromatic counterpart to the wave front construction, Fig.16b, whereas Fig.16c corresponds to a later phase of the process where secondary P-wave diffraction at crack tip B has occurred. A variety of reflected and diffracted wave systems can be identified.

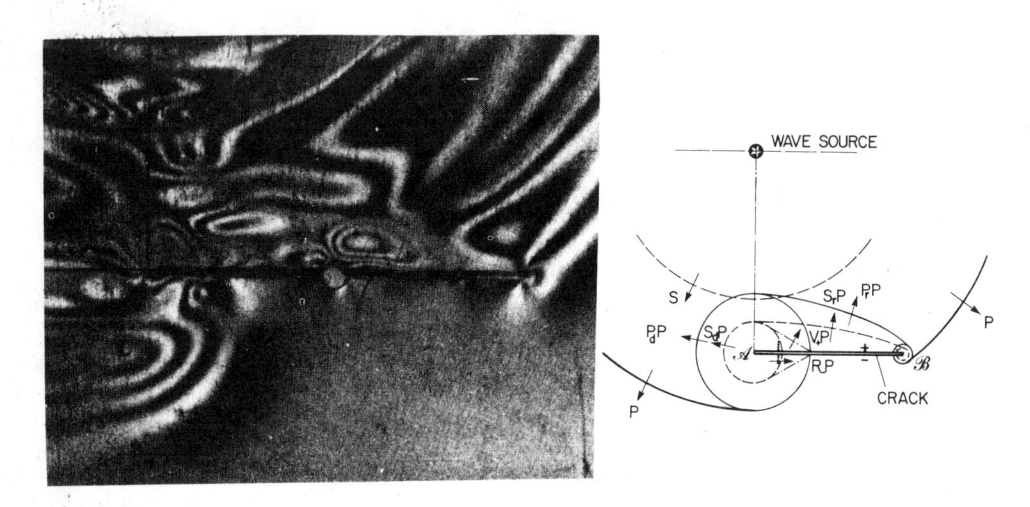

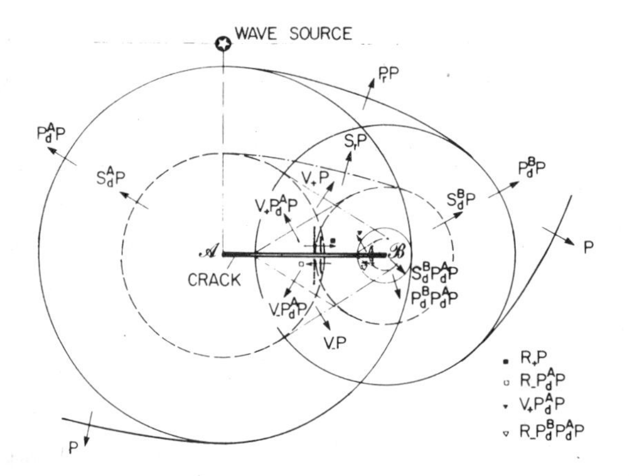

Fig.16 Diffraction of explosively generated P-wave about a static crack-tip
(a) dynamic isochromatic fringe pattern (test #2, frame 4, t=72 μsec)
(b) stress wave front construction
(c) stress wave front construction for secondary P-wave diffraction at crack-tip B

Oblique incidence of the leading compressional pulse of the P-wave and the secondary P_dP-wave diffracted at crack-tip B induce a state of mixed-mode crack wall deformation where the shearing mode contribution is effective only leading to a pure mode-2 isochromatic near crack-tip fringe pattern. A close-up of the experimentally-recorded isochromatic pattern in the near crack-tip zone about B is compared in Fig.17 with an analytically-generated mode-2 fringe pattern /20/.

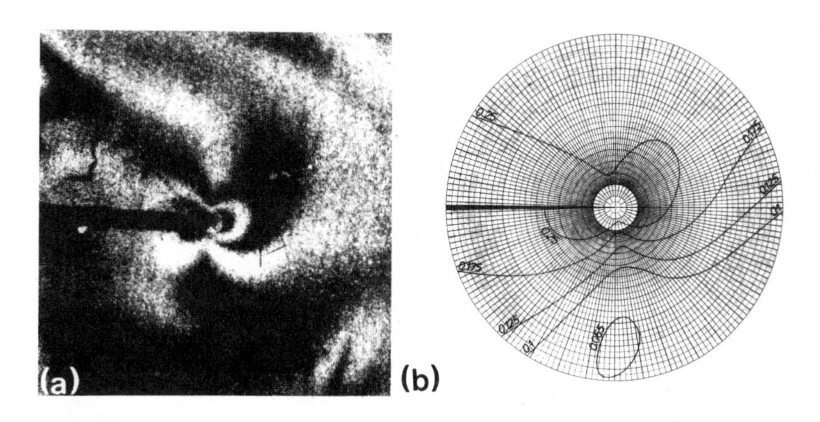

Fig.17 Comparison of dynamically-recorded with analytically-generated isochromatic crack-tip fringe pattern associated with oblique diffraction of cylindrical P-wave /13/

Consider a tensile P-wave impinging at a crack tip as shown in Fig.14. The normal (σ_n) and tangential (σ_t) stress component of the P-wave are given by $\sigma_n = \sigma_0$ and $\sigma_t = \sigma_0\, \nu/(1-\nu)$, where $\nu/(1-\nu) = 1-2(c_2/c_1)^2$. The stresses induced along the crack faces are then given by eqs.(20) and (21) with

$$\begin{aligned}
\sigma_x^i &= \sigma_0\{1-2\cos^2\theta\ (c_2/c_1)^2\} = \sigma_t \cos^2\theta + \sigma_n \sin^2\theta \\
\sigma_y^i &= \sigma_0\{1-2\sin^2\theta\ (c_2/c_1)^2\} = \sigma_n \cos^2\theta + \sigma_t \sin^2\theta \qquad (50)\\
\tau_{xy}^i &= \sigma_0 \sin2\theta\ (c_2/c_1)^2 = (\sigma_n-\sigma_t)\,\frac{1}{2}\sin2\theta \quad .
\end{aligned}$$

The scattered near tip stress field associated with combined opening and shearing mode crack deformation is given by

$$\sigma_x^s = \frac{K_1}{\sqrt{2\pi r}}\cos\frac{\theta}{2}\left(1-\sin\frac{\theta}{2}\sin\frac{3\theta}{2}\right) - \frac{K_2}{\sqrt{2\pi r}}\sin\frac{\theta}{2}\left(2+\cos\frac{\theta}{2}\cos\frac{3\theta}{2}\right) + \sigma_{ox} \qquad (51)$$

$$\sigma_y^s = \frac{K_1}{\sqrt{2\pi r}} \cos\frac{\theta}{2}(1+\sin\frac{\theta}{2}\sin\frac{3\theta}{2}) + \frac{K_2}{\sqrt{2\pi r}} \sin\frac{\theta}{2}\cos\frac{\theta}{2}\cos\frac{3\theta}{2}$$
$$\sigma_{xy}^s = \frac{K_1}{\sqrt{2\pi r}} \sin\frac{\theta}{2}\cos\frac{\theta}{2}\cos\frac{3\theta}{2} + \frac{K_2}{\sqrt{2\pi r}} \cos\frac{\theta}{2}(1-\sin\frac{\theta}{2}\sin\frac{3\theta}{2}). \tag{51}$$

with K_1 and K_2 the stress intensity factors which depend on the time and the geometrical configuration as well as on the wave profile.

The total stress field may be written in the form:

$$\sigma_x^t = \sigma_x^s + \sigma_x^i + O(1) \quad , \quad \text{etc.} \tag{52}$$

Considering the photoelastic relation for the incident wave

$$N^i f_o/h = \sigma_1-\sigma_2 = \sigma_n-\sigma_t = \sigma_n(1-\nu/(1-\nu)) = \sigma_n 2(c_2/c_1)^2 \tag{53}$$

$$\sigma_n = \frac{1}{2}(c_1/c_2)^2 N^i f_o/h$$
$$\sigma_t = \frac{1}{2}((c_1/c_2)^2-2) N^i f_o/h \, , \tag{54}$$

and regarding the relation $\tau_m^2=(\sigma_y-\sigma_x)^2+\tau_{xy}^2$, one obtains from eqs. (50)-(54) the basic relation for the isochromatic field in the vicinity of the crack tip:

$$(\tau_m^t)^2 = F(K_1,K_2;\sigma_n,\sigma_t;N^i) = (f_o N^t/2h)^2 \tag{55}$$

This is a nonlinear relation in the variables K_1, K_2, and N^t. The quantity N^t can be measured near the crack and N^i can be measured at a point on the incident wave front remote from the crack tip but corresponding to the same radial distance. If the stress intensity factors are normalized with respect to a crack loaded statically with $\bar{\sigma}_o^i$ and $\bar{\tau}_o^i$ at infinity,

$$K_{1,o} = \bar{\sigma}_o^i \sqrt{\pi a}\, Y_1(2a/\Lambda;t) \quad , \quad K_{2,o} = \bar{\tau}_o^i \sqrt{\pi a}\, Y_2(2a/\Lambda;t) \tag{56}$$

with $\bar{\sigma}_o^i$ and $\bar{\tau}_o^i$ determined photoelastically by loading a crack of length 2a in pure mode-1 and pure mode-2, respectively.

Finally, putting together all the former equations and rearranging, one obtains the useful relations for determination of the stress amplifications as a function of time and wave pulse:

$$G(Y_1,Y_2;N^i,N^t,N) = 0 \tag{57}$$

This equation can be utilized in conjunction with the method of least squares for Y-determination as outlined in the Lecture "Analysis of Dynamic Photoelastic Fringe Patterns". For mode-1 problems equ. (57) reduces to a simple relation $Y_1(2a/\ ;t) = G(N^i, N^t, N)$.

3. *Interface Cracks and Joints*

Wave propagation and wave induced fracture in a layered media is of importance in the area of seismology as well as in removal of natural resources by mining engineering and hence has been the subject of many analytical and experimental investigations. The phenomenon of wave-crack interaction plays a basic role in geothermal heat exploitation, oil shale retort formation and typical bench fragmentation blasting. When this fragmentation is done inefficiently in a quarry operation additional expense is incurred to the mine operator. This added expense is seen as lost time in handling oversize fragments, time required in secondary breakage, and more frequent repairs to the primary crushers. Knowledge of stress wave interaction with layered interfaces would help one to optimize fragmentation and consequently reduce cost in mining operations or increase success rate in retort formation.

Formations encountered at quarry sites and oil shale deposits form stacks of layered rock with bedding planes and sets of joints present. Upon detonation of an explosive the wave pattern generated in layered media is extremely complicated where incident and reflected wave systems not only interact with running cracks but also with pre-existing cracks, flaws, fissures, inclusions or other inhomogeneities within the formation. Depending on the nature of the waves the cracks may accelerate or decelerate during the interaction phase, and in some situations crack branching may be induced.

In spite of an enormous body of theoretical background literature it is difficult to obtain quantitative information in terms of stress magnitudes for real wave pulses and real layer geometries with differing acoustical impedances. Experimental work has concentrated on point by point observations on the surface of the body and has consisted in measuring arrival times and wave velocities. More recently, dynamic photoelastic methods have been employed to study wave and crack propagation /13,15-24/.

Consider a crack lying at the interface between two different isotropic elastic media and the interface is welded or strongly bonded as shown in Fig. 18a. If the plane problem is considered the analysis shows that the stress ahead of the crack tip has the form

$$\begin{aligned} \sigma_y &= \frac{1}{\sqrt{r}} K_o \cos(\eta \ln r - \phi) \\ \tau_{xy} &= \frac{-1}{\sqrt{r}} K_o \sin(\eta \ln r - \phi) \end{aligned} \tag{58}$$

where $\phi = \tan^{-1}(K_2/K_1)$ and $K_o^2 = K_1^2 + K_2^2$. The factor η depends on the material properties of the dissimilar layers (for similar layers holds $\eta = 1$). The oscillatory character of the stress field implies that the displacement discontinuity behind the crack tip changes sign infinitely often and, hence, interpenetration of the crack faces occurs. The region of interpenetration however, is confined to a very small distance behind the crack tip and for practical purposes it can be neglected. A more refined analysis allowing for slip at the crack faces removes these oscillatory behavior. In addition, when the strain energy release rate is computed by a local work argument these oscillatory terms disappear. Notice that the singularity is the same as for the similar media case.

If, however, the crack is perpendicular to the interface, the stresses at the crack tip behave like $\sigma \sim r^{-\lambda}$, where λ lies between zero and one depending on the ratio of the moduli of the two layers. There is no oscillatory behavior in this case but the fact that the stress singularity is, in general, no longer $r^{-1/2}$ means that the fracture mechanics arguments cannot be applied to the point where the crack meets the interface. Numerical calculations have been made for various different geometrical crack-interface configurations (Fig.19) and stress intensity factors have been defined as the coefficients of $r^{-\lambda}$, however, there is no clear connection between the stress intensity factor and the strain energy release rate. If the crack crosses the interface there is an additional singularity at the point of crack face / interface intersection and again pseudo-stress intensity factors may be defined. For a crack approaching a free surface at an angle of 90° in an infinite medium the stress intensity factor tends to infinity like $1/(\sqrt{w} \ln w)$ with w the unbroken ligament tending to zero. If the crack approaches a layer at an angle other than 90° the singularity may have an oscillatory component.

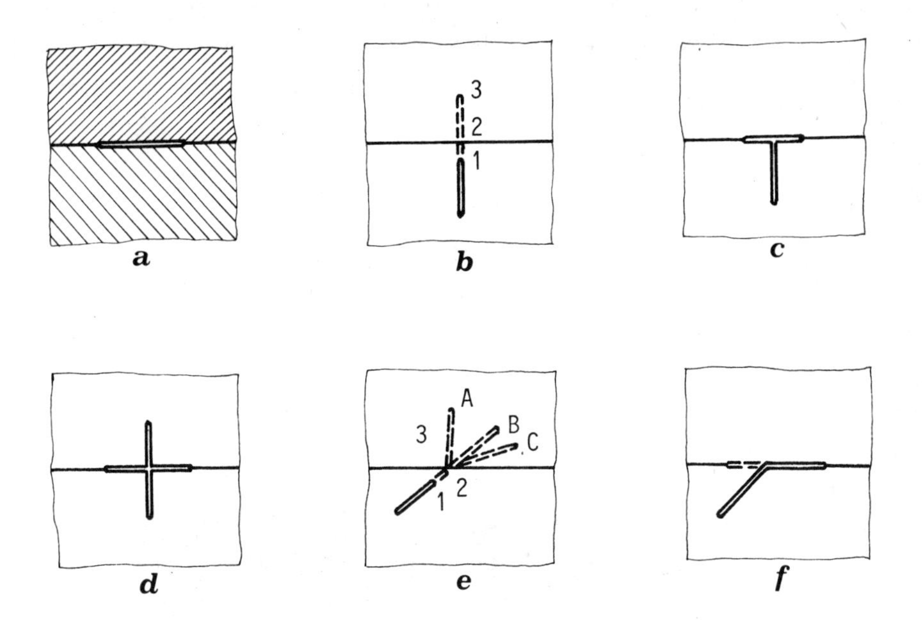

Fig. 18 Various crack-interface configurations encountered in rock fracture mechanics applications

Consider crack-type bond imperfections located at the interface between two similar or dissimilar materials as shown in Fig.19. The cracks are loaded by a detonation pulse ignited at some distance h from the interface.

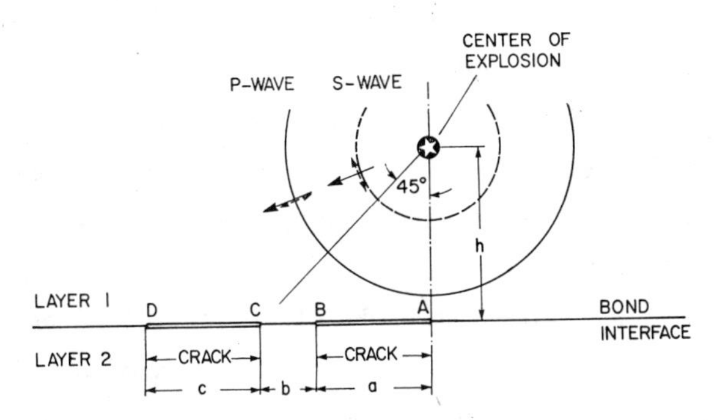

Fig.19 Geometry and dynamic loading condition of cracked layer model. Diffraction of cylindrical P- and S-waves about the tips of a stationary double crack at a bonded interface.

Two fracture mechanisms have been observed in the course of these experiments; shear cracking initiated by the obliquely incident leading compressive part of the P-pulse and mixed-mode cracking as a result of the interaction of the trailing tensile part of the P-pulse possibly in conjunction with the leading shear wave front.

Fig.20 depicts a progressive phase of the wave-crack interaction process. The incident P-wave has just outdistanced crack tip D and the reflected and diffracted waves are clearly visible. Layer cracking starts from crack tips A,B, and C, such, that the direction of the emerging cracks is at an angle of slightly less than 90° clockwise with respect to the inverse crack growth direction.

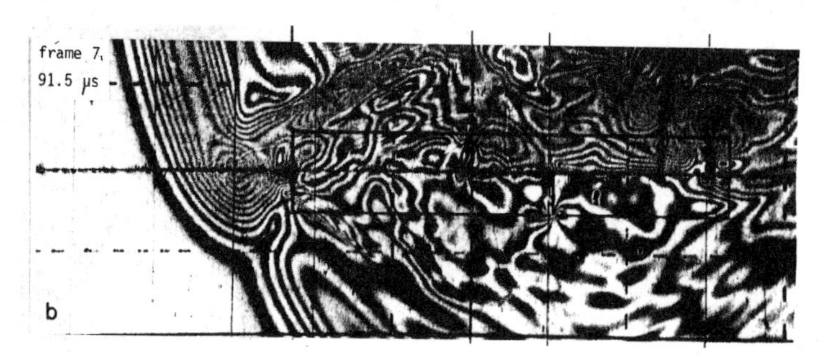

Fig.20 P-wave induced shear layer cracking from bond imperfections

More complex interactions will occur when several materials of different acoustic impedances are juxtaposed, as in layered media, such as layered rock. The several transmitted and reflected stresses can be calculated on the basis of former sections. The fact that layers thin with respect to the wave length will have substantially no effect on the wave is of utmost importance for the passage of energy across a multi-layered composite. Thus, layered rock consisting of an assembly of like blocks cemented together by quarzite etc. will transmit a wave nearly as effective as a monolothic block of rock. Likewise, the cement layers between dissimilar blocks will not effect the wave interactions taking place at the joints provided the bonding is perfect. The two-dimensional layered models were an attempt at modelling rock blasting and fragmentation in jointed formations. Observation obtained from production quarry blasting and oil shale exploitation sites seem to support the mechanisms of initiation that have been identified with this type of model testing.

References

/1/ Yoffe,E.: The moving Griffith crack. Phil.Mag.42,739 (1951)

/2/ Maue, A.W.: Die Beugung elastischer Wellen an der Halbebene. ZAMM 33, 1-10 (1953)

/3/ Baker, B.R.: Dynamic stresses created by a moving crack.J.Appl.Mech. 29,449-458 (1962)

/4/ Broberg, B.K.: The propagation of a brittle crack. Arkiv för Fysik, 18,159-192 (1960)

/5/ Freund, L.B.: Crack propagation in an elastic solid subjected to general loading -I. Constant rate of extension.(J.Mech Phys.Solids 20,129-140-1972), -II.Non-uniform rate of extension.(J.Mech.Phys.Solids 20,141-152,1972).

/6/ Bergkvist, H.: The motion of a brittle crack. J.Mech.Phys.Solids 21, 229-239 (1973)

/7/ Cleary, M.P.: Modelling and Development of Hydraulic Fracturing Technology (this Volume)

/8/ Irwin, G.R.: Comments on dynamic fracturing; in "Fast Fracture and Crack Arrest", ASTM-STP 627,7-18 (1977)

/9/ Achenbach, J.D. and P.K. Tolikas: Elastodynamic effects on crack arrest, ASTM-STP on "Fast Fracture and Crack Arrest", 59-76 (1977)

/10/ Rose, L.R.F.: Recent theoretical and experimental results on fast brittle fracture. Int.J.Fract. 12,6,799-813 (1976)

/11/ Freund, L.B.: Crack propagation in an elastic solid subjected to general loading-III. Stress wave loading.J.Mech.Phys.Solids,21,47-61, 1973); IV. Obliq-ely incident stress pulse (J.Mech.Phys.Solids, 22,137-146, 1974).

/12/ Steverding, B. and S.H.Lehnigk: The fracture penetration depth of stress pulses. Int.J.RockMech.Min.Sci.&Geomech.Abstr.13,75-80(1976)

/13/ Rossmanith, H.P. and A. Shukla: Photoelastic investigation of stress wave diffraction about stationary crack tips. J.Mech.Phys.Solids, 29, 397-412 (1981).

/14/ Fossum, A.F.: PhD Thesis, Brown University, 1973.

/15/ Rossmanith, H.P. and A. Shukla: Dynamic photoelastic investigation of interaction of stress waves with running cracks. Exp.Mech.21,11, 415-422 (1981).

/16/ Fourney, L.W. et al.: University of Maryland Reports to the National Science Foundation, USA, Reports 1977-1982.

/17/ Rossmanith, H.P. and W.L.Fourney: Fracture initiation and stress wave diffraction at cracked interfaces in layered media. I.- Brittle-brittle transition. Rock Mechanics, 14,209-233 (1982)

/18/ Smith, D.G.: A photoelastic investigation of stress wave loading of a crack. Paper presented at 1971 SESA Spring Meeting, Salt Lake City, Utah (1971)

/19/ Fourney, W.L. et al.: Explosive fragmentation of a jointed brittle media. Submitted to Int.J.Rock Mech.and Ming.Sci.(1981)

/20/ Rossmanith, H.P.: Analysis of mixed-mode isochromatic crack-tip fringe patterns. Acta Mechanica, 34, 1-38 (1979).

/21/ University of Maryland, Photomechanics Laboratory: Private Communication, (1977-1982)

/22/ Fourney, W.L. and H.P.Rossmanith: Crack tip position and speed as determi from Rayleigh-wave patterns.Mech.Res.Comm./,5,277-281 (1981)

/23/ Rossmanith H.P. and W.L. Fourney: Determination of crack speed history and tip locations for cracks moving with non-uniform velocity. Exp.Mech.22,111-116 (1982).

/24/ Rossmanith, H.P. and G.R.Irwin: Analysis of Dynamic Crack-Tip Stress Patterns. University of Maryland Report, Department of Mechanical Engineering (1979)

Acknowledgement

This research work was conducted under the Contract Numbers # 3864 and # 4532 under the sponsorship of the Fonds zur Förderung der wissenschaftlichen Forschung in Austria.

FRACTURE CONTROL BLASTING

William L. Fourney
University of Maryland
Mechanical Engineering
College of Engineering
College Park, Maryland 20742

INTRODUCTION

The information that I will present in this publication will describe for the most part results obtained from research projects conducted in the Photomechanics Laboratory at the University of Maryland over the past twelve years. The work has been accomplished by the author in conjunction with Drs. J.W. Dally, D.B. Barker, D.C. Holloway and Mr. Anders Ladegaard-Pedersen (Swedish Detonics Research Foundation). The research work was dynamic in nature and involved the application of dynamic photoelasticity, dynamic holography, or high speed photography.

The particular topics to be discussed involve fracture control blasting, fragmentation, well stimulation, and vibrational analysis. This first chapter describes model testing and field testing conducted to improve smooth wall blasting techniques. Other topics will follow in subsequent chapters.

Excavation in hard rock is usually accomplished with a drill and blast procedure where a hole is drilled in the rock, packed with high explosives, stemmed and the explosive detonated. The detonation pressures are extremely high and an extensive amount of energy is dissipated at the borehole by crushing the adjacent rock and in producing a dense radial crack pattern about the hole. These radial cracks arrest quickly and only about 8 to 12 randomly oriented cracks extend for any significant distance from the borehole.

Since this process dissipates much of the energy in crushing the rock at the borehole and the resulting crack pattern is randomly oriented, very little control of the fracture plane is achieved. Where control of the fracture process is important, the conventional drill and blast procedure is modified. Pre-splitting, post-splitting, and smooth blasting procedures have been developed which, to some degree, control the fracture process.

In pre-splitting, a row of closely spaced and highly charged holes are detonated simultaneously. The resulting stress waves interact to produce cracking in the region between the holes where the stress waves overlap and double the dynamic stresses. These highly charged holes, of course, produce extensive cracking at the borehole and weaken the wall of an excavation. Also, simultaneous detonation results in excessively high ground shocks when excavations are made in populated urban areas. Post-splitting is almost identical except that the holes are fired after the central core of the excavation has been fragmented.

In smooth blasting, the holes are drilled on very close centers and cushioned charges are used. As control is obtained by spacing the holes, delays can be used and the ground shock reduced. Smooth blasting gives satisfactory results when enough holes are drilled and when the charge is properly cushioned; however, the number of holes which must be drilled and loaded increases the cost of the excavation.

EFFECTS OF NOTCHES IN A BOREHOLE

In early work conducted with Dally and Ladegaard-Pedersen [1] using dynamic photoelasticity it soon became obvious that to truly achieve fracture control, one had to alter either the load as applied to the borehole or else alter the borehole so that an arbitrarily applied load created the desired results. It was also evident that any alteration had to be effective immediately at the time of detonation before the fracture process began. Dummy (unloaded) holes for example cannot affect the fracture pattern until the outgoing P and S waves reach them and by that time radial cracks have been initiated in all directions at the borehole.

Several workable options for altering the load and or the borehole have been determined but the technique of grooving the borehole will be discussed here.

The concept of using notches on the side of a borehole is not new - Foster [2] as well as Langefors and Kihlstrom [3] have indicated that notching could be used to initiate fractures at desired locations and hence to control the fracture plane. What was not well understood is that unless care is taken when deciding upon the amount of charge that is to be placed in the notched hole that cracks will start at other locations in addition to the notch location and the beneficial effects of the notch will be lost. A fracture mechanics analysis will be used to define the pressure range which can be applied to a notched borehole which will result in successful fracture control.

CONTROL OF CRACK INITIATION

Consider the borehole and the surrounding rock as a two dimensional body as shown in Figure 1. The borehole with a radius R is loaded by a pressure p which is considered to be constant over the time required for crack initiation.* Consider next a number of radial cracks of length a extending from the wall into the media.

---

* It is recognized that the pressure in blasting does vary markedly with time. However, a static analysis will give a close indication of the pressure required for crack initiation in terms of the fracture properties of the rock.

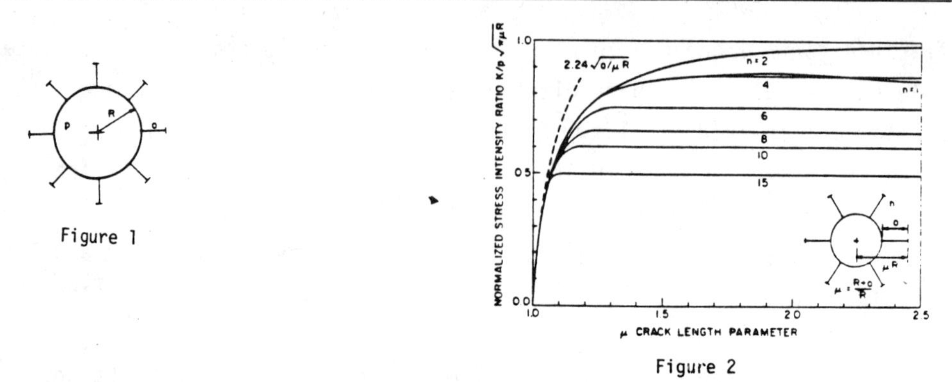

Figure 1 Pressurized borehole with radial cracks.

Figure 2 Stress intensity factor for a pressurized circular cavity as a function of crack length parameter.

This problem was first treated by Bowie [4] who showed the methods of mapping and treated one and two radial cracks. Kutter [5] extended Bowie's treatment to include k cracks at the hole and introduced the mapping function for the star-crack without a circular cavity. Ouchterlony [6,7] combined the theoretical findings of Bowie and Kutter with Irwin's [8] linear fracture mechanics and treated radial crack growth from a pressurized borehole. Ouchterlony showed that the stress intensity factor K for a pressurized circular hole with n radial cracks is given by the relation shown in Figure 2.

It is important to note from Figure 2 that the critical pressure required to initiate cracks is independent of the number of cracks when the cracks are sufficiently short (i.e. a = 0.05 R). In this case

$$p_c = K_{Ic}/2.24\sqrt{\pi a} \tag{1}$$

where K_{Ic} is the fracture toughness of the rock for crack initiation.

The results from Equation 1 can be used to compute the pressure required to initiate cracks at the tip of sharp notches on the side of the borehole as shown in Figure 3. Two facts are evident from this figure. First, the borehole pressures required to initiate cracks at the notches are quite low even for very shallow notches. This implies that high explosives with detonation pressures above 200 k bars give over-pressures which are too large by a factor of 30.

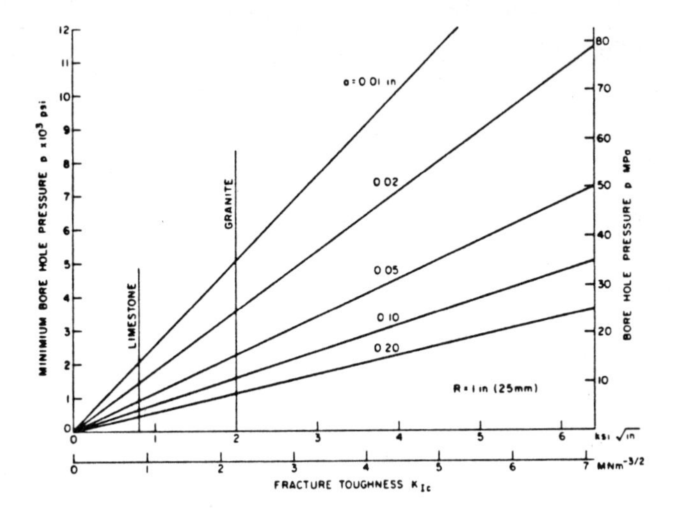

Figure 3 Pressure required to initiate cracks at the borehole.

The second result from Figure 3 is that cracks can be initiated at notches with low pressures for all types of rock. The rock property which is important is the fracture toughness, K_{Ic}. Measurements of the fracture toughness of rock is relatively new; however, some limited data exists which indicates that the value of K_{Ic} is between about .18 to 15 MPa $m^{1/2}$ (0.2 - 17 ksi-$in^{1/2}$) for most of the commonly excavated rock. A complete listing of K_{Ic} and G_{Ic}* is presented in Table 1.

TABLE 1.

Fracture Toughness of Several Rock Types

	K_{Ic}		G_c	
	MPa-$m^{1/2}$	(psi-$in^{1/2}$)	N-m/m^2	(in-lb/in^2)
Limestone	0.657-0.992	(598-903)	7.71-42.03	(0.044-0.24)
Sandstone	0.180-0.198	(164-180)	21.89-26.27	(0.125-0.15)
Gray Granite	2.224	(2024)	67.25	(0.384)
Red Granite	2.221	(2021)	89.49	(0.511)
Salem Limestone	0.947-2.626	(862-2390)	29.95-229.9	(0.171-1.31)
Barre Granite	5.121-7.593	(4660-6910)	149.9-329.2	(0.856-1.88)
Sioux Quartzite	1.407-5.736	(1280-5220)	70.93-1177	(0.405-6.72)
Dresser Basalt	5.363-18.97	(4880-17260)	239.9-2998	(1.37-17.12)

Using limestone as an example and referencing Figure 3 shows that a crack can be initiated from a notch 0.5 mm (0.020 in) deep if the pressure exceeds 10.3 MPa (1500 psi). Similarly, the same notch in granite would require 24.8 MPa (3600 psi).

The next factor which must be considered in control of crack initiation is over-pressure in the borehole. The pressures given in Figure 3 are minimum pressures which should be exceeded to overcome the effect of notch end bluntness and to provide additional energy to drive the crack. The amount of over-pressure which can be tolerated can also be determined from Eq. 1. In this instance, the flaw length, a_f, is equated to the natural flaw size. These natural flaws occur in large numbers randomly distributed about the periphery of the wall of the borehole. Little is

* Note that $K_{Ic} = \sqrt{G_{IC}E}$ for plane stress

known about the distribution of the size of these natural flaws; however, photomicrographs by Bienawski [9] suggest that many grain boundaries represent small cracks. Thus, it appears reasonable as a first approximation to equate the natural flaw length, a_f, to the grain size of the rock.

Taking a_f ranging from 0.025 mm (0.001 in) to 0.25 mm (0.01 in) and using Eq. 1 gives the maximum pressure which can be tolerated in the borehole before producing a large number of random cracks. This pressure as a function of flaw size and fracture toughness is shown in Figure 4. These results show that the very fine grain rock materials with a_f 0.025 mm (0.001 in) support much higher pressures prior to random crack initiation than the coarse grain materials where large size natural flaws are encountered.

The pressure range in which crack initiation can be controlled will depend on three factors - the fracture toughness, K_{Ic}, of the rock, the natural flaw size, a_f, and the depth, a, of the side notches. By employing the results presented in Figures 3 and 4, the operating range can be determined. A typical example for a granite with K_{Ic} = 0.18 MPa-m$^{1/2}$ (2 ksi - in$^{1/2}$) showing the allowable pressure range for several sets of operating conditions is presented in Table 2.

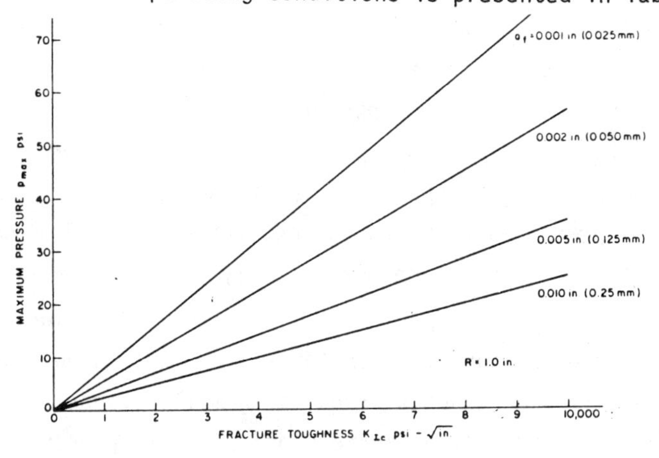

Figure 4 Pressure required to initiate small natural flaws at the walls of the borehole.

TABLE 2.
Pressure Range for Controlling Crack Initiation with Side Grooving

Operating Conditions										
Rock Grain Size	Notch Size	a_f mm	a_f in	a mm	a in	p_{max} MPa	p_{max} psi	p_{min} MPa	p_{min} psi	$\frac{p_{max}}{p_{min}}$
very fine	deep	0.025	0.001	5.00	0.20	110	16000	7.6	1100	14.5
fine	medium	0.050	0.002	2.50	0.10	76	11000	11.0	1600	6.9
medium	medium	0.125	0.005	2.50	0.10	48	7000	11.0	1600	4.4
coarse	shallow	0.250	0.010	1.25	0.05	34	5000	15.9	2300	2.2

for granite with K_{Ic} = 1.8 MPa-m$^{1/2}$ (2 ksi-in$^{1/2}$)

Inspection of these results shows a very wide range of pressure, 7.6 to 110 MPa (1100 to 16000 psi), which gives satisfactory control of crack initiation when deep grooves are used in very fine grain material. However, when relatively shallow grooves are used in coarse grain rock the range of pressure 15.9 to 34 MPa (2300 to 5000 psi) is much narrower.

It appears that crack initiation can be controlled by using side grooving. The rock must be characterized by determining K_{Ic} and the natural flaw size, a_f. Then the depth of the side groove can be selected to give a satisfactory ratio of p_{max}/p_{min}. Finally, the type of explosive and the size of the charge is determined.

The very low pressures required suggest the use of a low explosive where pressure could be controlled more closely and where hole stemming can be maintained for a larger period of time after detonation. Black powder, smokeless powder, and nitro cotton all detonate slowly and complete the conversion of gaseous products in milliseconds rather than the microseconds associated with high explosives. The pressures generated are usually less than 34.5 MPa (5000 psi) and can be controlled by adjusting the weight of the charge per unit length of borehole and the diameter of the hole.

In laboratory and field work it has been convenient to use a very highly cushioned charge of PETN. In this case, the charge employed is in the form of a length of prima-cord spaced centrally relative to the borehole. The approximate weight of the charge required for crack initiation can be computed from the pressure versus specific-volume curve presented in Figure 5.

For example, consider a fine grain limestone where a borehole pressure of 27.6 MPa (4000 psi) due to the gaseous products of the explosive charge is specified. Reference to Figure 5 at p = 0.27 kbars indicates a specific volume 15,500 cm^3/kg. For a 38 mm (1.5 in) diameter borehole 3 m (9.8 ft) long, the total volume is 3400 cm^3 which requires a distributed explosive weighing 0.22 kg (0.483 lbs). The distributed charge is 0.073 kg/m which is equivalent to 0.049 lbs/ft or 343 grains/ft.

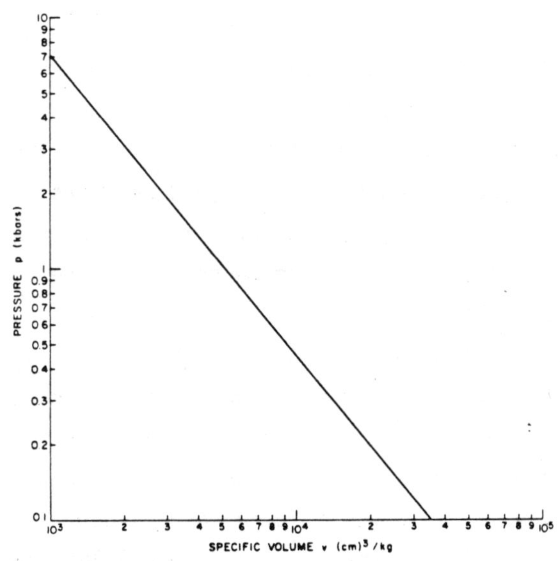

Figure 5 Pressure as a function of specific volume for nitroglycerine detonation gas.

As the data in Figure 5 is for isentropic expansion of nitroglycerine detonation gas, the computed charge weight should be considered approximate. Comparison with charges recommended by Langefors and Kihlstrom [3] for smooth blasting or presplitting indicates that the notching reduces the amount of explosive by a factor of at least two.

PHOTOELASTIC STUDIES OF FRACTURE CONTROL

Typical high speed photographs which illustrate the mechanism of fracture with a grooved borehole are shown in Figure 6. Eight of the sixteen frames have been shown which cover a period of 677 μs after detonation. The unsymmetrical loading due to unequal packing of the explosive about the through-bolt is evident in Figure 6a taken 47 μs after detonation. The dilatational or P type stress wave has propagated far into the model while the S wave is visible near the pressure containment device. The horizontal line along the diagonal indicates the specified fracture control plane. The unsymmetrical loading is not important in this test since the stress wave does not contribute significantly to the propagation of the cracks.

The P wave has propagated to the model boundaries in Figure 6b and the S wave has extended well into the model. The cracks are evident on the diagonal about 10 mm (0.4 in) beyond the pressure cap.

The cracks are more easily observed in Figure 6c at t = 102 μs and are propagating along the fracture control plane. The reflected PS wave, generated by the reflection of the P wave from the model boundary, is propagating back towards the borehole. It is interesting to note that the cracks are propagating in the region behind the stress waves and are being driven by the residual gases in the borehole. The fact that the cracks propagated at high velocity in the low stressed region behind the stress waves is even more evident in Figure 6d where the higher amplitude stress waves are all located near the boundary. Several fringe loops at the crack tips are evident indicating that there is a significant amount of energy available to drive the crack.

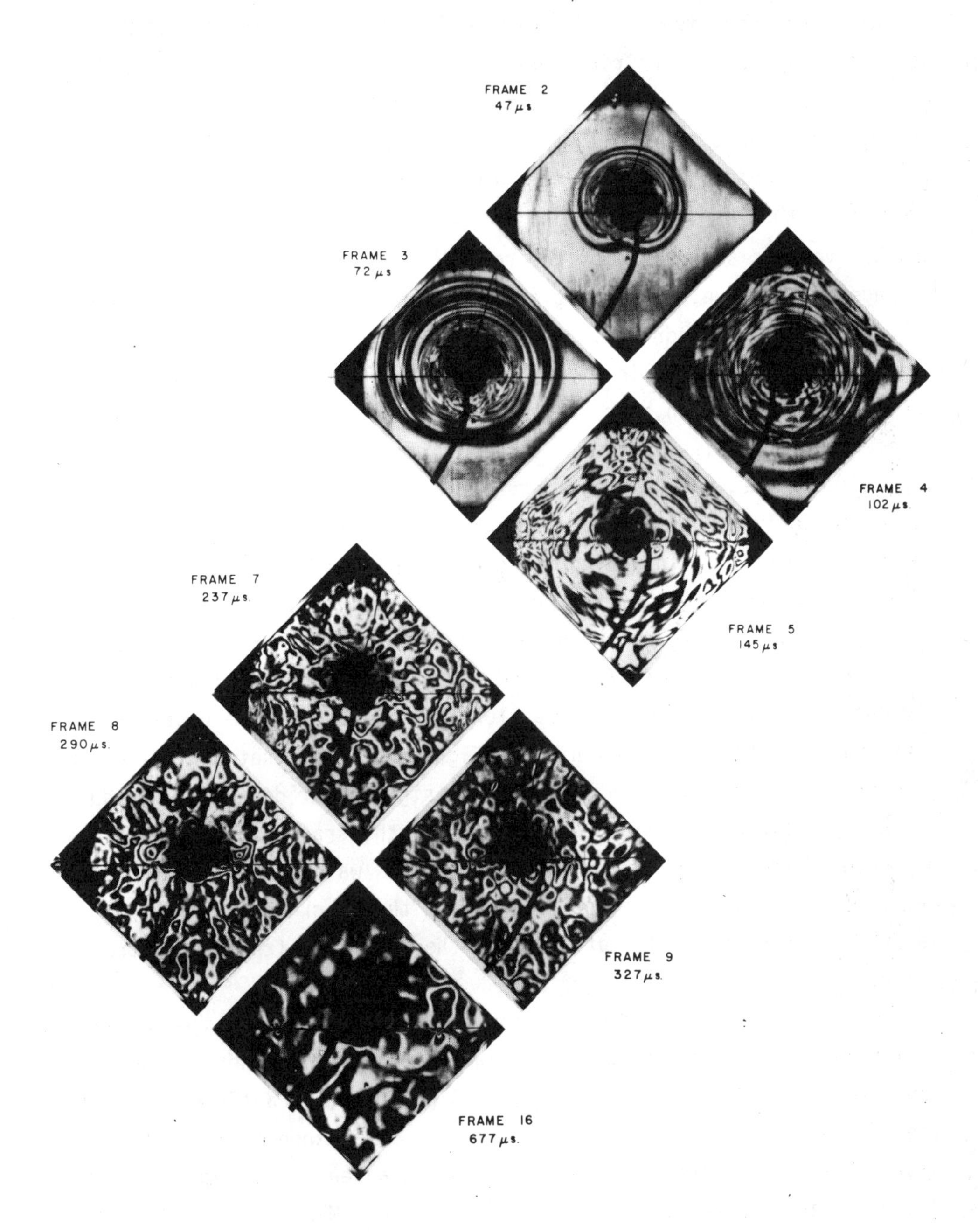

Figure 6 Dynamic crack propagation during a fracture control event.

The final four frames presented in Figures 6e-h shows the complex but relatively low order fringe patterns associated with the mixing of the reflected PP, PS, SS and SP waves. No single wave can be clearly distinguished and the overall fringe order decreases as time increases. These stress waves will affect the crack behavior. For instance in Figure 6e, the fringe loops which indicate the driving force on the crack have been suppressed by the reflected shear wave (SS). After passage of the wave, Figure 6g and h, the fringe loops recover and the crack is again being driven.

The fringe pattern in Figure 6h is particularly significant. The very low order pattern over the field indicates that the stress waves have essentially dissipated. Yet the multiple fringe loops at the crack tip indicate that the driving force on the crack (stress intensity factor) is sufficiently large to keep the crack from arresting. The pressure in the borehole is maintaining the stress intensity factor by a flow of gas along the length of the crack.

While the dynamic recording of fracture propagation was limited to an observation period of 677 μs, the cracks continued to propagate and divided the model into two halves along the diagonal. The post-test photograph of the model presented in Figure 7 shows that control of the fracture plane was excellent and that no significant damage was produced on the borehole wall. Slight deviations in the fracture path are produced by the action of the reflected stress waves; however, the deviations are not significantly large and the crack on the average maintains its radial orientation along the specified fracture control plane.

The pressure in the borehole for the early part of this event, as monitored with a pressure bar, is shown in Figure 8. The pressure exhibited a rapid rise time peaking in 5 μs at 53.5 MPa (7756 psi) and then decayed exponentially until t = 20 μs. The exponential decrease was followed by a minor oscillation and then a second much more gradual decay. The pressure at t = 45 μsec is about 4.1 MPa (600 psi). It is

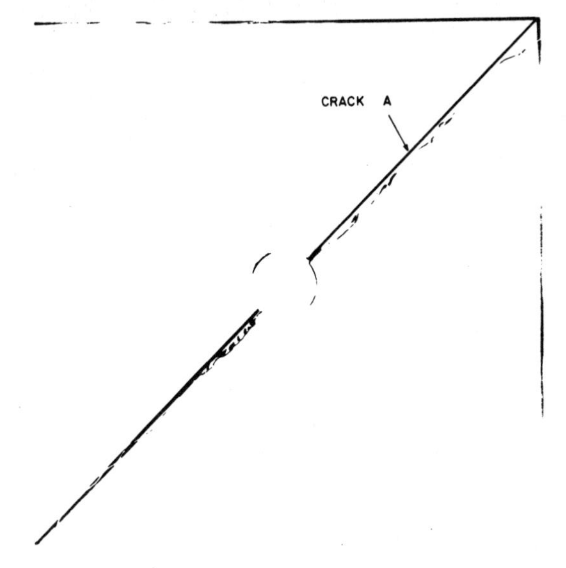

Figure 7 Development of the fracture control plane. Note minimal damage at the borehole.

believed that this value represents the pressure of the residual gases which will decrease only slightly during the remaining time in the dynamic event.

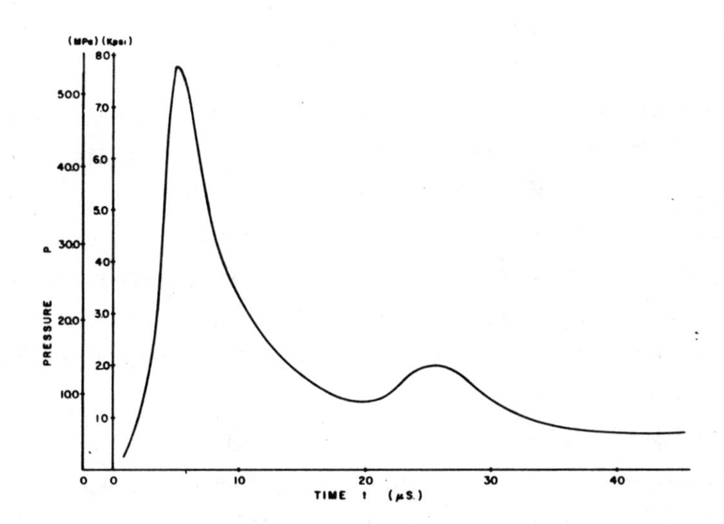

Figure 8 Pressure in the borehole as a function of time.

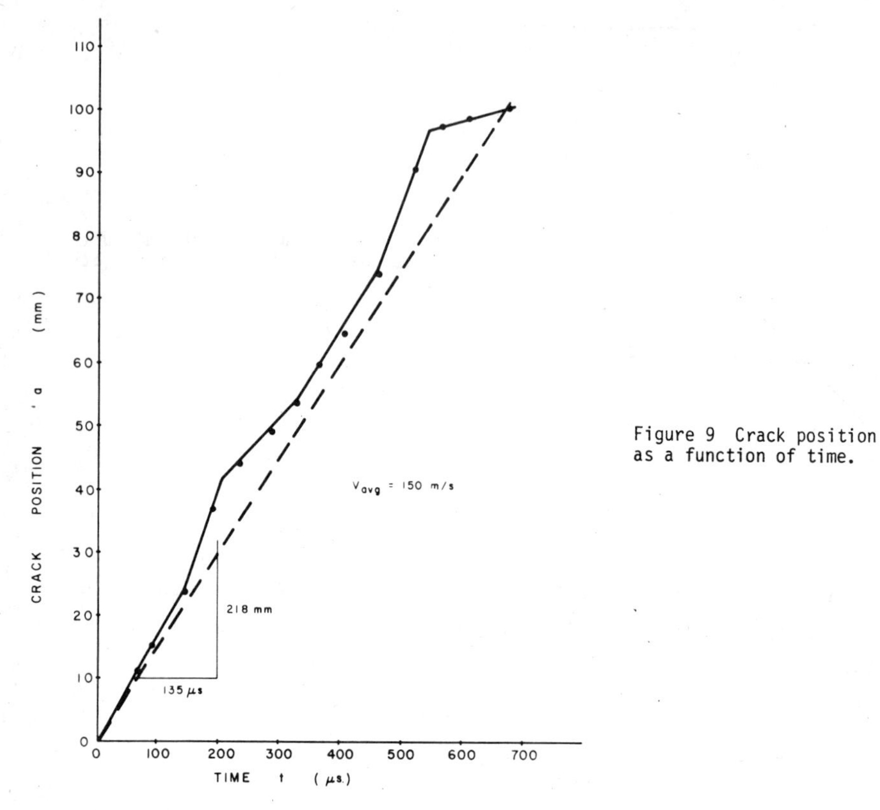

Figure 9 Crack position as a function of time.

A graph of the crack position as a function of time is shown in Figure 9. It is evident that there are significant deviations of the data points from a straight line and that the velocity of crack propagation is not a constant. The average velocity is about 150 m/s (6000 in/sec) over the entire observation interval; however, the velocity probably varies from 50 m/sec (2000 in/sec) to 250 m/sec (10000 in/sec) depending upon the effects of the stress waves which interact with the crack.

An enlargement of the fringe pattern at the tip of crack A for this model is shown in Figure 10. A two-fringe, three-parameter technique developed by Etheridge and Dally [10] was utilized to determine the stress intensity factor K from the geometry of the fringe patterns at the crack tips. The results for the stress intensity factor K as a function of time are presented in Figure 11, and indicate that K varies appreciably over the event. The sharp decay of K in the time period from 200 to 300 μs is due to reflected shear waves (SS) suppressing the stress field in the local neighborhood of the crack tips. The cause of the positive oscillation can not be isolated; however, it is believed that a combination of reflected stress waves reinforced the stress field at the crack tip. Very late in the dynamic event with $t > 600$ μs, the stress waves have decayed and K is shown increasing monotonically. It is believed that K is increasing with time (and crack length) after stress wave decay, because of the action of gas in the borehole.

The Homalite 100 material used in these experiments has been well characterized [11,12] and the K versus velocity relationship given in Figure 12 has been shown to be independent of specimen geometry. It is evident from Figure 12 that the crack will arrest when $K < K_{Im} =$ 0.37 MPa-$m^{1/2}$ (340 psi-$in^{1/2}$). It appears probable that the crack forming the fracture plane arrested momentarily in the time period between $t = 200$ and 300 μs and then re-initiated after passage of the reflected SS wave.

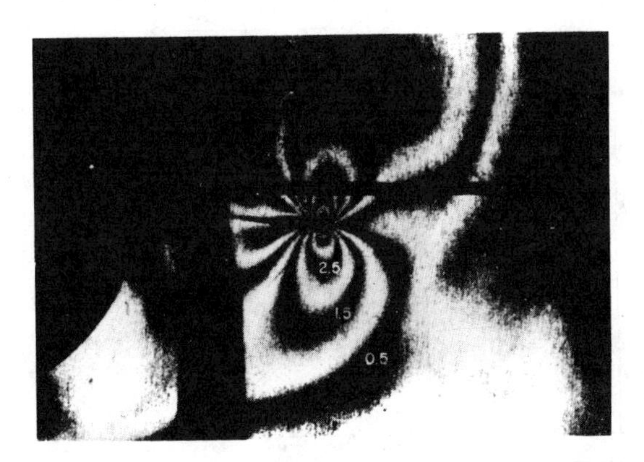

Figure 10 Enlargement showing the fringe pattern at the tip of a propagating crack.

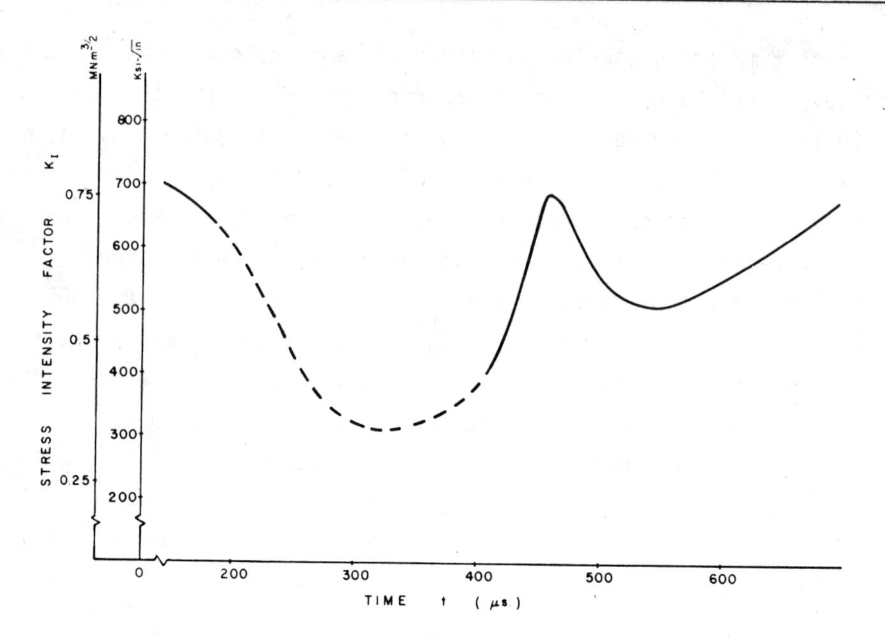

Figure 11 Stress intensity factor as a function of time.

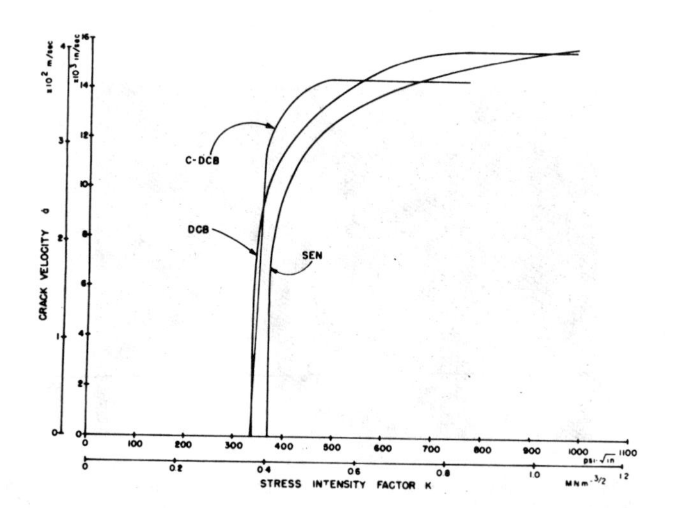

Figure 12 Stress intensity factor versus ȧ for CDCB, RDCB & SEN specimens of Homolite 100.

The effect of gas flow into cracks producing an increase in K with increases in the length of the propagating crack has been noted previously by Dally and Fourney [13]. For open cracks, the increase in K with crack length is shown in Figure 13. It is evident that pressures modestly above 0.110 MPa (100 psi) will produce $K > K_{Im}$ where K_{Im} is the fracture toughness associated with arrest in Homalite 100. Providing $K > K_{Im}$ the crack will continue to propagate forming the fracture control plane. Free gas flow into the crack will then extend the crack over very large distances if the gas does not vent and lower the pressure in the borehole below the level required to maintain $K > K_{Im}$.

It is apparent then that the cracks are open and do not fill with crushed materials and this permits large distances of crack propagation with light charges. The cracks in this test propagated to the free boundaries of the model as a result of the detonation which represents a propagation distance to borehole radius ratio (s/R) of 11.3.

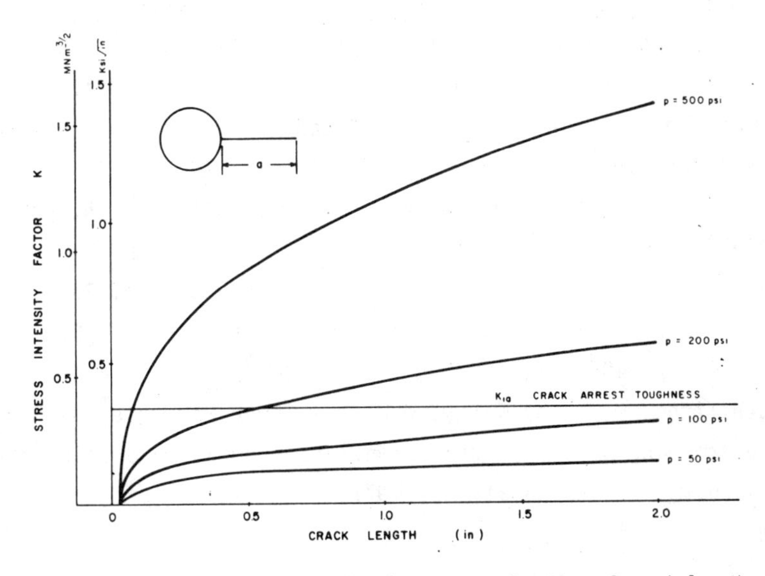

Figure 13 Stress intensity factor as a function of crack length for different borehole pressures.

REFERENCES

[1] Ladegaard-Pedersen, A., Fourney, W.L., and Dally, J.W., "Investigation of Pre-splitting and Smooth Blasting Techniques in Construction Blasting", Report to NSF, Dec. 1974, NSF-RA-T-75-015.

[2] Foster, Clement LeNene, "A Treatise of Ore and Stone Minning", Charles Griffin & Company, 6th Edition, 1905.

[3] Langefors, U. and Kihlstrom, B., "Rock Blasting", John Wiley and Son, pp. 300-301, 1963.

[4] Bowie, O.L., "Analysis of an Infinite Plate Containing Radial Cracks Originating at the Boundary of a Circular Hole", J. Math Physics, Vol. 35, p. 60, 1956.

[5] Kutter, H.K., "Stress Analysis of a Pressurized Circular Hole with Radial Cracks in an Infinite Elastic Plate", Int. J. of Fracture Mechanics, Vol. 6, pp. 233-247, 1970

[6] Ouchterlony, Finn, "Fracture Mechanics Applied to Rock Blasting", 3rd Int. Congress of the Int. Soc. for Rock Mechanics.

[7] Ouchterlony, Finn, "Analys av Spanningstillstandet Kruig Nogra Olika Geometrier Med Radiellt Ricktade Sprickor 1 Ett Oandlight Plant Medium Under Inverkan av Expansionkrafter", Swedish Detonic Research Foundation Report, DS 1972: 11, 1972.

[8] Irwin, G.R., "Fracture", Handbuck der Physik, Vol. 6, pp. 551-590, 1958.

[9] Bienawski, Z.T., "Mechanism of Brittle Fracture of Rock, Part II. Experimental Studies", Int. J. of Rock Mech. & Min. Sci., Vol. 14, No. 4, pp. 407-423, 1967.

[10] Etheridge, M.J. and Dally, J.W., "A Three Parameter Method for Determining Stress Intensity Factors from Isochromatic Fringe Loops", Journal of Strain Analysis, 1977.

REFERENCES (continued)

[11] Irwin, G.R. et.al., "A Photoelastic Characterization of Dynamic Fracture", NRC Report NUREG 0072, Dec. 1976.

[12] Irwin, G.R. et.al., "A Photoelastic Study of the Dynamic Fracture Behavior of Homalite 100", NRC Report NUREG 751107, Sept. 1975.

[13] Dally, J.W. and Fourney, W.L., "Fracture Control in Construction Blasting", University of Maryland, Report to NSF, November 1976.

FRAGMENTATION STUDIES WITH SMALL FLAWS

William L. Fourney
University of Maryland
Mechanical Engineering
College of Engineering
College Park, Maryland 20742

INTRODUCTION

There presently exists disagreement with regard to the mechanism that is reponsible for the fragmentation that occurs in quarry blasting operations. The particular contribution of gas pressurization as opposed to stress wave initiation has not been clearly established. The various schools of thought range from the notion that gas pressurization plays the dominant role [1-5], to the idea that combinations of the stress waves and gas pressurization are responsible for the major amount of damage [6-8] to the thought that stress waves alone [9-11] are responsible for the resulting fragmentation.

Nearly all previous model studies (laboratory as well as small boulder tests) have been conducted primarily in non-flawed plastics or fault free rock. If the mechanism of fragmentation is stress wave related then testing in flaw free models would skew the experimental results towards the theory that gas pressurization played the dominant role in fragmentation.

I am going to report on a study [12] in which flaws found in a typical limestone quarry were artifically simulated in birefringent models made of Homalite 100. By using birefringent models and the technique of dynamic photoelasticity it was possible to determine how the various stress waves generated by the detonation of an explosive charge interacted with the various artificial flaws. It was also possible to view the initiation and growth of both pressurized and non-pressurized cracks. In this manner the mechanisms responsible for crack initiation and propagation were determined.

The experimental program involved the testing of many small two-dimensional polymeric models. The tests were conducted with Homalite 100 which is available from SGL Industries of Wilmington, Delaware. The initiation toughness (K_{Ic}) is around 0.40 MPa-m$^{1/2}$ (364 psi-in$^{1/2}$) and is less than that found for most rock. It is therefore more brittle in behavior than rock. At high crack driving energy levels, crack propagation behavior is felt to be similar to that in rock. Since crack velocities expected in a typical fragmentation round are large, it is felt that the details of the fragmentation process can be studied in Homalite 100 and applied qualitatively to fragmentation in rock.

All models were 6.4 mm (0.25 in) thick and were typically 300 mm (11.8 in) square or 300 × 380 mm (11.8 × 15 in) depending upon the type of test being conducted. Two hundred to two hundred fifty grams of PETN were tightly packed into 6.4 mm (0.25 in) diameter boreholes. An attempt was made to contain the detonation gases by tightly capping the boreholes with O-ring seals and steel caps.

As stated earlier, dynamic photoelasticity [13] is an optical method of stress analysis which permits a full-field visualization of the state of stress in a transparent birefringent material. The dynamic fringe patterns obtained provide a way of simultaneously observing the interaction between the propagating cracks and the local stresses that drive these cracks. Each fringe order N is related to the difference in the principal stresses, σ_1 and σ_2, according to the stress optic law

$$\sigma_1 - \sigma_2 = \frac{Nf_\sigma}{t} \tag{1}$$

where f_σ is the material fringe value and t is the model thickness.

FRAGMENTATION IN A HOMOGENEOUS MODEL

When an explosive source is detonated at an interior point in a half plane, two elastic body waves and four reflected elastic wave systems are generated as shown schematically in Figure 1. The faster of the two waves which propagate within an elastic body is called the primary or P wave. It is also referred to as the dilatational wave because the deformation associated with this wave includes a volume change. Detailed examination of the stress state in this wave from a typical detonation has indicated that in the leading edge both σ_r and σ_θ are compressive with σ_r being about 2.5 times larger than σ_θ. There is a small tensile tail for the radial component of stress but usually very little tensile tail for σ_θ. It is felt that the tensile tail in σ_θ is mostly consumed in borehole crushing. Typical speed of the P wave in Homalite 100 is 2,000 mps (78,740 ips).

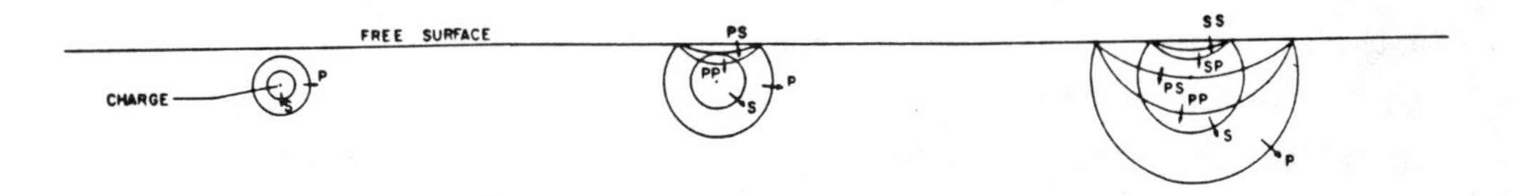

Figure 1 Schematic representation of the stress waves generated by an explosive detonated near a free surface.

The slower of the two elastic body waves is called the secondary or shear wave. A theoretic analysis of a point charge will not predict the presence of a shear wave due to the axisymmetry of the problem. In practice, crushing and crack initiation at the borehole walls destroys the axisymmetry and thus accounts for the existence of a shear wave. The disturbance caused by this shear wave involves no dilatation and the typical speed of propagation in Homalite 100 is found to be 1,100 mps (43,307 ips). The stress state for the shear wave is pure shear. The magnitude of $\sigma_{r\theta}$ varies with angular position around the borehole.

Reflected wave systems are similarly named. When striking a material interface the P wave generates a reflected primary wave, the PP wave, and a reflected secondary, PS, wave. Similarly the S wave generates an SP and SS wave system.

When the difference in acoustical impedance between two materials meeting at an interface is not too great, waves are also transmitted through the interface as well as reflected.

Upon detonation of the explosive, an intense pressure pulse is applied to the borehole wall. A quasi-static solution to the stresses about the borehole can be derived from the classic solution of a pressurized thick-walled cylinder [14]. This solution shows on the borehole wall that the tangential stress, σ_θ, is tensile and equal in magnitude to the compressive radial stress σ_r and that σ_r equals the applied pressure.

Using numerical techniques Bligh [15] has developed a dynamic solution. The quasi-static and dynamic solutions are schematically shown in Figure 2 for an assumed pressure pulse, P(t). Of primary importance in the explosive process is the tangential stress since this is the stress that will initiate borehole cracks. In the dynamic analysis the tangential component starts as a compressive stress and after a few microseconds changes to tension. The tangential component ultimately reaches a maximum value typically 50% greater than that predicted from a quasi-static analysis. This knowledge of the stress field about the borehole helps explain the physically observed crack initiation behavior.

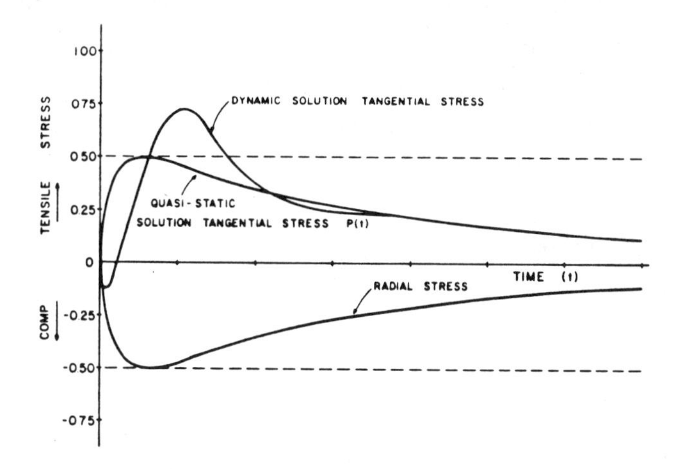

Figure 2 Quasi-static and dynamic borehole stresses for an assumed pressure pulse P(t).

After detonation, crushing begins to occur in the immediate vicinity of the borehole. This crushing is caused by the intense shear stresses generated. Before the tangential stress becomes tensile, the maximum shear stress is small since all three principal stresses are compressive and similar in magnitude and very little crushing will occur. When the tangential stress changes sign the maximum shear stress becomes significant. The resulting crushed zone propagates radially outward until the maximum shear stress decreases below some critical value required to cause fracture. Outside the crushed zone, cracks initiate due to tensile loading and not shear.

In this case existing flaws are opened by the tensile stress component normal to the crack faces. When this stress component produces a stress intensity factor (K) greater than the materials fracture toughness (K_{Ic}), the crack will start to propagate. Once moving, the crack will align itself perpendicular to the maximum tensile stress so as to maximize instantaneous dynamic stress-intensity factor, (K_D). This leads to the general conclusion that cracks always propagate perpendicular to the maximum tensile stress.

Initiated cracks in the borehole vicinity will propagate in a radial direction perpendicular to the maximum tensile stress. These cracks will continue to propagate in a radial direction until the local stress field at the crack tip is altered. When the stress field is altered, the crack will adjust its propagation direction so that it is perpendicular to the maximum tensile stress.

The two stress waves generated by the pressure loading on the borehole wall quickly outdistance the propagating cracks that were initiated at the crushed zone boundary. With the outdistancing of the crack tips by both the P wave and the S wave, a quasi-static situation exists about the borehole. The dense radial crack pattern is driven by borehole pressurization and gases penetrating into the cracks and pressurizing their faces. Ouchterlony [16] has shown that uniform growth of all cracks in a radial system is virtually impossible. Dominant cracks will emerge from the dense radial crack pattern and grow at the expense of the shorter ones. From experimental results Persson and coworkers [2] have found the number of dominant cracks to be about six. Langefors and Kihlstrom [1] put the number of dominant cracks between eight and twelve, a number we tend to confirm in our experiments.

A typical explosively generated radial crack pattern in a homogeneous material is shown in Figure 3. This figure is a post mortem photograph of the results of a small contained charge in Homalite 100. Eleven dominant cracks labeled B through M are clearly evident emerging out of the dense radial borehole crack pattern. These eleven cracks propagated an average distance of 19 times the borehole diameter. These long radial cracks succeed in breaking the material into large pie-shaped segments, instead of fragments. More fragmentation would occur if these dominant radial cracks could be made to bifurcate so as to increase the number of active radial cracks.

If the dominant radial cracks are going to branch, they will do so as soon as they emerge from the dense radial borehole crack pattern. Ouchterlony [16] has shown that for the case of partial pressurization of radial cracks, the crack network is propagating into a decreasing

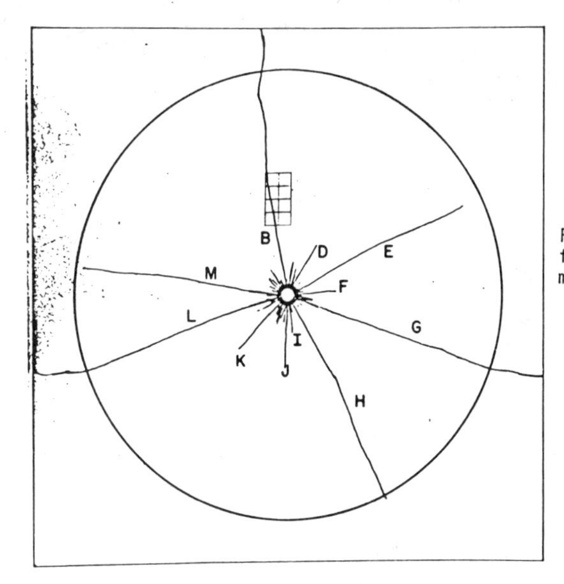

Figure 3 Postmortem radial crack pattern from a light contained charge in a square model.

K field after an extension of less than 1/2 the borehole radius. For the case of complete crack pressurization, K reaches a constant maximum. Thus, borehole cracks reach a maximum K while within the region of dense radial cracking. Thus, higher than normal crack velocities can be expected as stress fields from the neighboring cracks inhibit early branching. As the dominant cracks emerge from the dense radial pattern, the influence of shorter neighboring cracks decreases and branching will occur. Thus, branching due to explosive gas pressurization alone will not occur at distances far outside the dense radial cracking zone.

An example of dominant radial borehole crack branching is shown in the post mortem crack network photograph in Figure 4. In this model and in all cases except when a very small charge is used, the dense radial crack pattern in the immediate vicinity of the borehole is so severe that it is impossible to piece the fragments back together. Five dominant crack systems are labeled A through E in the photograph. Note that successful branching occurs shortly after emerging from the dense

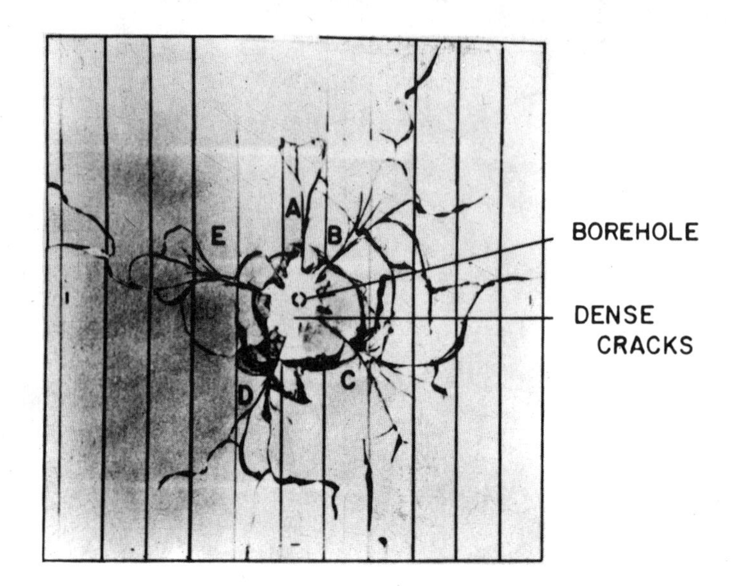

Figure 4 Postmortem showing branching in dominant radial cracks.

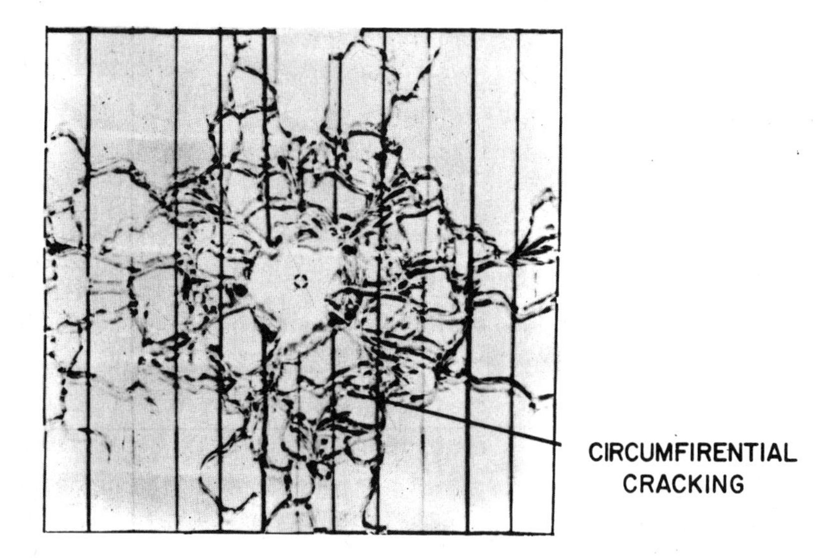

Figure 5 Postmortem showing circumferential cracking due to reflected stress waves.

radial crack pattern except for crack E. The feathery appearance of the cracks as they propagate out of the dense radial cracked region indicates that the K_D is only marginal for successful branching. Crack E finally branched with the additional energy input from a reflected stress wave.

Compare the degree of fragmentation in the model shown in Figure 5 with that of Figure 4. The added fragmentation is due to the development of a circumferential crack pattern. The only difference between the two models is that in Figure 5 a fully coupled charge of PETN was used. In Figure 4 the fully coupled charge consisted of a ratio of 1 part PETN to 5 parts smokeless pistol powder. This mixture of explosives reduced the detonation violence and thus the magnitude of the P wave. The more developed circumferential crack pattern seen in Figure 5 is due to the interaction of the reflected stress waves with the outwardly traveling radial cracks.

To more clearly demonstrate the mechanism responsible for this circumferential pattern, a circular model was fragmented. With this model only two reflected wave systems are present (no PS or SP waves are generated due to the circular boundary). The wave fronts will intersect the radial cracks at a normal angle. The shadows of eleven dominant radial cracks are labeled A through K in Figure 6a. This is a dynamic photoelastic photograph taken by the Cranz-Schardin camera 85 μs after detonation. The dense radial crack pattern is hidden from view by the black circular shadow of the pressure caps. The eleven dominant cracks are branching and, in most cases, now consist of more than one crack. This bifurcation is due to the high K_D from the borehole and crack face pressurization. The stress waves are well removed from the region of the cracks as shown by the location of the wave fronts in the figure. The P wave has already reflected from the free boundary and the S wave has yet to reach the boundary.

Twenty-nine μs later, the PP wave front has almost reached the radial cracks. Figure 6b shows an enlargement of cracks B through G. Note the many arrested cracks due to unsuccessful branching between the

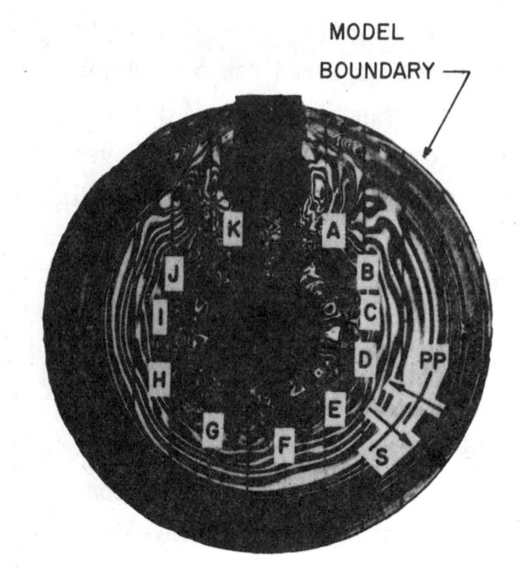

Figure 6(a) Dynamic photoelastic photograph of a circular fragmentation model showing eleven dominant radial cracks.

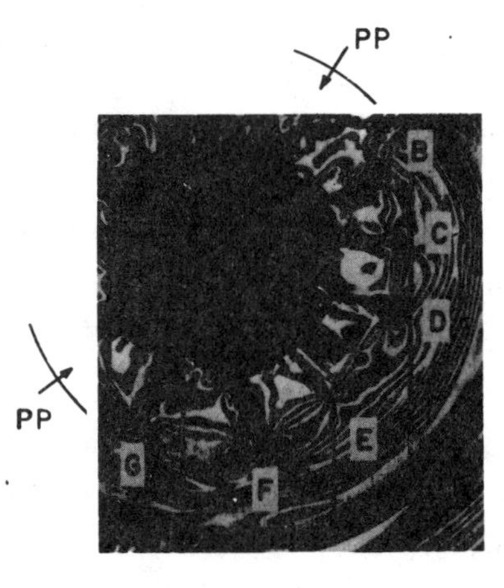

Figure 6(b)

Figure 6(c)

Figure 6(b) Enlarged view of sucessful and unsucessful branching in the dominant radial cracks.

Figure 6(c) Barrier Branching - influence of the PP wave on the dominant radial cracks.

crack tips and the borehole (feathery-like appearance). In Figure 6c taken 153 μs after detonation, the PP wave has interacted with the radial crack tips and created circumferential crack growth. The SS wave has yet to come into view.

The reflected PP wave in this model is a mirror image of the outgoing P wave. The wave consists primarily of biaxial tension with the radial tensile component approximately 3 times the tangential tensile component. When this stress field is superimposed on the local stress field at the crack tips, the local maximum tensile stress changes from a tangential direction to a radial direction. The propagating cracks correspondingly adjust their direction and run in the circumferential direction. Multiple bifurcation will also take place due to the increase in K_D, but it will take place in the new preferred circumferential direction. As the tensile leading edge of the PP wave passes, the maximum tensile stress returns to the tangential direction and the cracks turn and travel once again in the radial direction. This entire process can be followed by tracing the dominant borehole crack paths in Figure 6c.

This intersection of the outgoing radial cracks with the PP wave and the resultant change in crack direction and multiple bifurcation has been termed "barrier branching". The outgoing radial cracks appear as if they run into a barrier, branch, and then continue in the radial direction. This phenomenon of barrier branching is very important for overall fragmentation. First, it initiates circumferential cracks that tend to break up the large pie-shaped segments between the dominant cracks. Second, with the extensive bifurcation under the influence of the reflected stress wave, more cracks are created from the dominant cracks propagating from the borehole. These cracks turn and start to propagate in the radial direction upon passage of the stress wave. Many of these newly initiated cracks shortly arrest due to the limited driving ability of the explosive gas pressure, but they can be easily reinitiated with the passage of the next reflected stress wave.

EFFECTS OF SMALL FLAWS

In order to study the effect of small flaws on fragmentation, models were used which had flaws artifically routed into one surface. These flaws were made with a sharp 60° router bit and were 6 mm (0.24 in) in length and 1 mm (0.04 in) deep. The flaw depth as well as charge size were selected so as to have crack initiation occur primarily at the artificial flaw locations and not in "flaw free" regions of the model.

The model depicted schematically in Figure 7 was one of many used to study flaw initiation. The specimen was fabricated from 6.4 mm (0.25 in) thick Homalite 100. Eight flaws were routed into the front surface of the model as shown in the figure. Flaws labeled A through F were placed 133 mm (5.24 in) from the borehole with flaws D and C also being parallel to and 19 mm (0.75 in) from the lower free surface. Two additional flaws, G and H, were placed parallel to and 19 mm (0.75 in) from the left free surface.

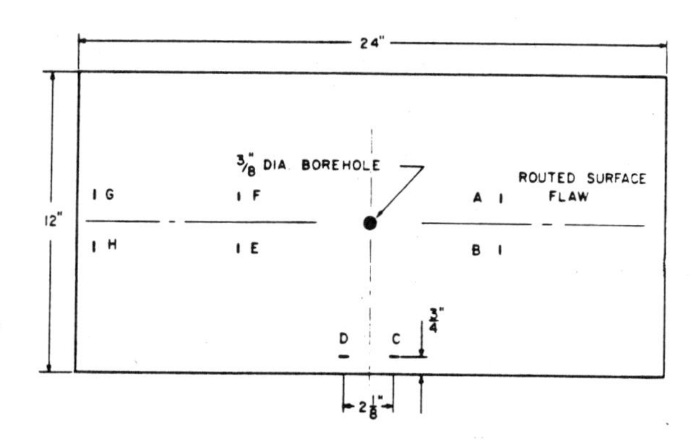

Figure 7 Rectangular fragmentation model.

In the experiment, the P wave tail was found to initiate cracks only at flaw E. Close examination revealed that this flaw had been routed slightly deeper than the others. Due to the increased depth, a critical value of stress-intensity factor occurred as the tail of the P wave passed over the flaw.

The early phases of crack initiation are seen in the sequence of pictures in Figure 8. Each picture is an enlargement of the dynamic isochromatic fringe pattern around flaws E and F. The borehole is outside of the picture frame on the right as marked by the black cross. Flaw E is the lower of the two flaws shown.

Eighty microseconds after detonation, the compressive front of the P wave has just passed over the two flaws as shown in Figure 8a, The compressive peak can be seen as the broad light fringe passing through the vertical grid line, just to the left of the flaws. The tensile region of the P wave is just approaching the flaws.

In Figure 8b, taken 103 μs after detonation, a building stress intensity factor is visible at the lower flaw due to the tensile radial component in the P wave tail. Since the flaw is not oriented in an exact borehole circumferential direction, mixed mode loading exists as confirmed by careful observation of the isochromatics. This mixed mode loading is more obvious in Figure 8c where the isochromatic loops are seen skewed in a counter-clockwise direction. In Figure 8c cracks have already initiated at the upper and lower corner of the flaw. This initiation occurred somewhere between 103 and 116 μs after detonation. In Figure 8c the S wave front has just reached the flaw but its full effect will not be noticed until later when the peak of the S wave arrives. Note that no initiation occurred at the upper flaw. As mentioned previously, the lower flaw was routed slightly deeper and since the stress intensity factor is proportional to the square root of the crack depth, a greater stress intensity factor occurred at the lower flaw.

Figure 8d, e, and f are enlargements of the region around this P wave initiated flaw (flaw E) showing its later propagation behavior. The two cracks that were initiated at the top and bottom of the flaw immediately start to propagate in a circumferential direction, perpendicular to the σ_r tensile component of the P wave tail. Currently in Figure 8d, the cracks are propagating in a direction approximately 22° off of circumferential. This new propagation direction occurred abruptly

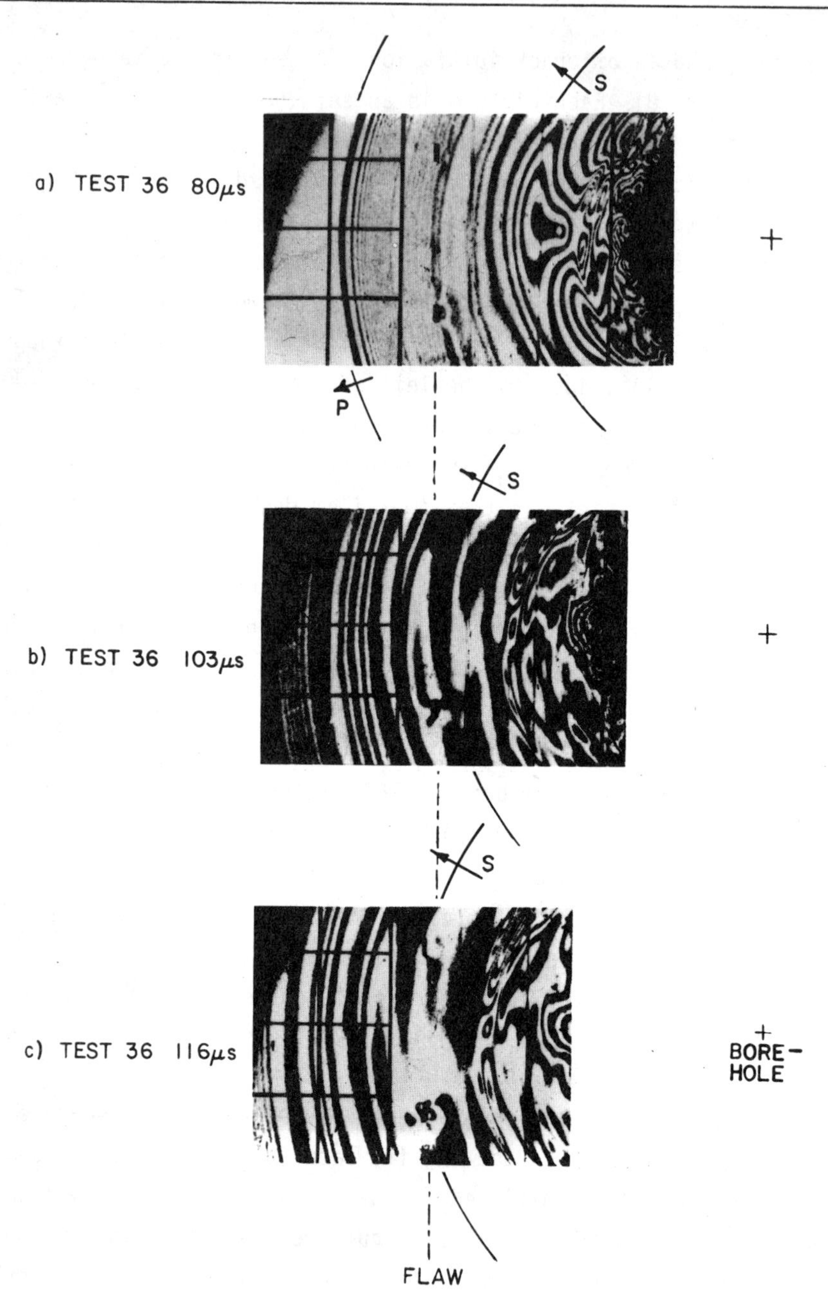

Figure 8 Dynamic photoelastic fringes showing the initiation of a flaw by the tensile tail of the P-wave.

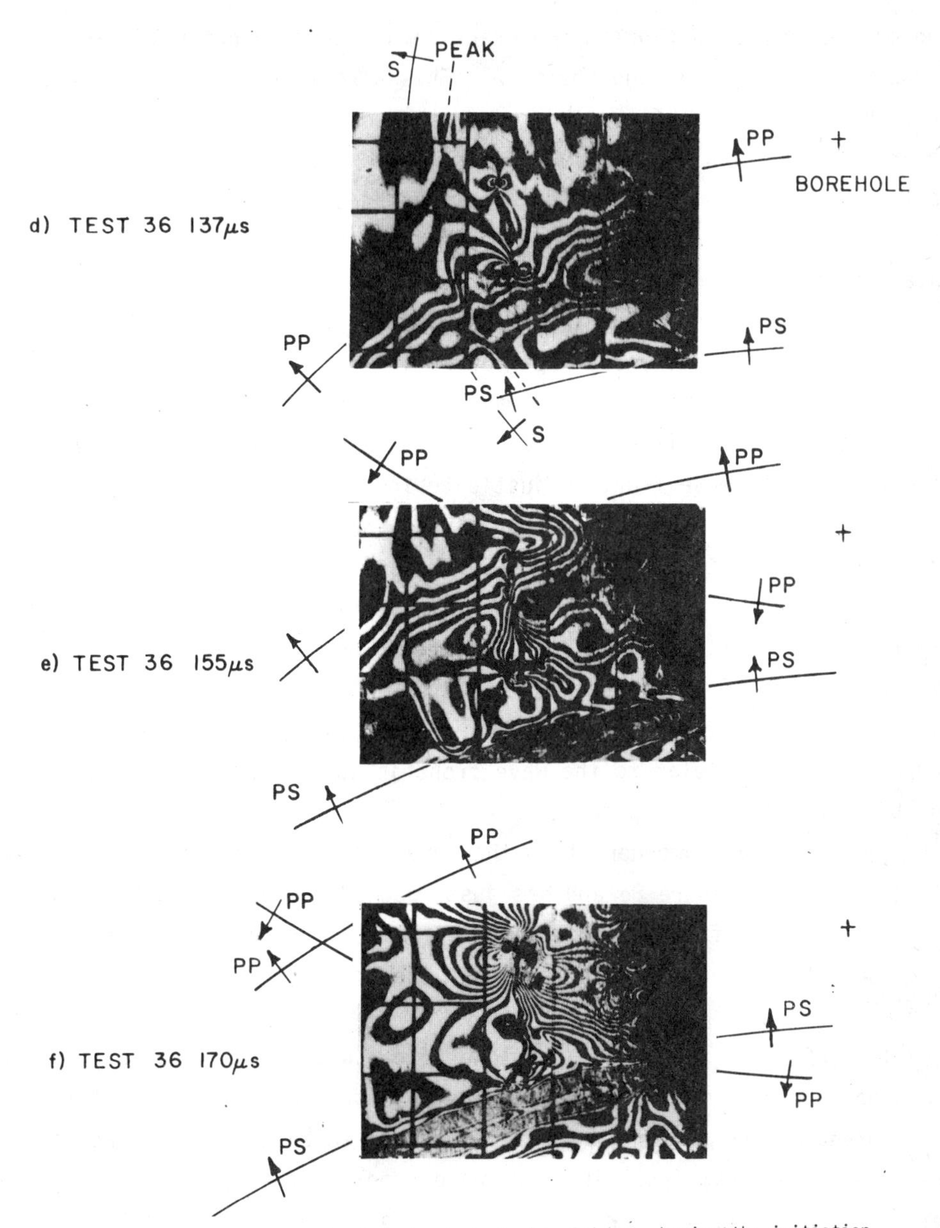

Figure 8(cont.) Dynamic photoelastic fringes showing the initiation of a flaw by the tensile tail of the P-wave.

and can be seen as a kink in the crack path located about half way between the crack tips and the flaw. This abrupt change in direction occurred as the peak of the shear wave passed over the crack. The shear wave peak can be seen as the circumferential fringe located halfway between the flaw and the S wave front. Thus, the current propagation direction is due to the combined influence of decaying P wave tail and the building S wave.

Initiation velocity of the cracks was about 280 m/s (11,024 ips) growing to approximately 355 m/s (13,976 ips) before branching occurred under the influence of reflected waves. In Figure 8d, the PP wave front from the lower free surface has just passed over the lower crack tip and the PS wave front is just entering the field of view in the lower right corner. Since K_D was already nearly equal to K_b, the added energy contained in the PP wave front produced successful multiple branching at the lower crack tip. This is evident in the figure by the multiple lobed fringe pattern, but more obvious in Figure 8e, taken after passage of the wave. The preferential branching direction is parallel to the PP wave front since the tensile stress component perpendicular to the wave front is approximately three times the parallel tensile component.

The PP wave component from the upper free surface is entering the field of view in Figure 8e and has just reached the upper crack tip as has the PP wave from the lower free surface. The combined stress state from these two waves tends to "explode" the crack tip as seen in Figure 8f.

The PS wave front from the lower free surface has also reached the branched tips from the lower main crack in Figure 8f. Only two of these crack branches are currently active as evidenced by the small black dots at two of the crack tips. The PP wave supplied the energy to branch the original crack but with its passage, the majority of the resulting branches arrested. The only additional source of driving energy available is the outgoing S wave. This is further evidenced by noting that the two branches that are still propagating are oriented at approximately 45 degrees to a radial line (the direction perpendicular to the maximum tensile stress in the S wave). The PS wave is a pure shear wave containing a tensile stress of significant magnitude as evidenced by the closely spaced isochromatic fringes shown approaching the lower branches in Figure 8f. Unfortunately, the dynamic recording ended at this time. It is possible to determine further crack extension by comparison of Figure 8f with the final post mortem photograph, Figure 9. The PS wave apparently caused reinitiation and branching of these lower branches as indicated in Figure 9. Further crack extension occurred with the passage of the SP and SS waves which had yet to come into the field of view in Figure 8.

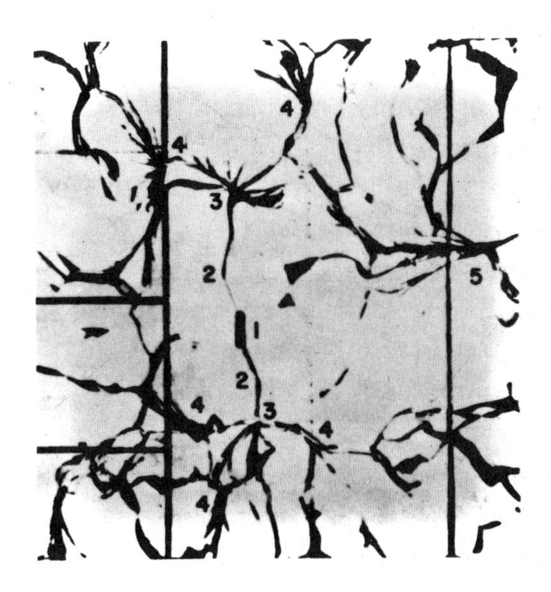

1. FLAW INITIATION BY P WAVE TAIL
2. REORIENTATION WITH S WAVE
3. PP BRANCHING
4. PS BRANCHING
5. PP BRANCHING OF RADIAL CRACK

Figure 9 Postmortem photograph showing the development of the P-wave initiated initiated crack.

Comparison of Figure 8 and 9 also shows that most of the branches in the upper main crack arrested with the passage of the two PP waves. Additional branching did occur on the left and right main branches due to the combined influence of the two PS waves. Again the SP and SS waves served as additional sources of energy to extend the existing cracks.

In Figure 8e and 8f a borehole radial crack can be seen entering the field of view from the right. Branching of the crack can be seen to occur with the passage of the two PP waves. In the postmortem photograph, Figure 9, this radial crack and its branches are more evident.

SUMMARY

The detonation of an explosive in a homogeneous "flaw free" infinite plate produces a very simple radial fragmentation pattern. Branching of the dominant cracks only produces long thin slivers of material along the crack's path and does not succeed in greatly breaking up the material between the dominant cracks. This fragmentation pattern was found to be quite different when flaws were present. The greatest effect of small flaws on fragmentation is caused by cracks initiated at flaw sites remote from the borehole region by the combined action of the P wave tail and the shear wave front. Flaws initiated in the immediate borehole vicinity by these waves have only a minor effect since they generally propagate in a radial direction and rapidly coalesce with the borehole cracks.

Remotely initiated flaws located further from the borehole generally do not propagate in a radial direction. This non-radial propagation succeeds in breaking up the large pie shaped segments created by the dominant radial cracks.

When an explosive is detonated near a free face, the fragmentation pattern becomes quite complex. Additional fragmentation is caused by the reflection of the P and S wave components back into the medium. These reflected wave components drastically influence the propagation behavior of the borehole radial cracks as well as initiate cracks at various flaw sites between the free surface and the borehole.

The strongest flaw initiating mechanism is the combined action of the PP and PS wave near a free surface. This mechanism is not necessarily the most important for overall fragmentation, but it has been demonstrated to initiate flaws near free surfaces at extreme distances from the borehole. Due to the natural attenuation of the reflected P wave components, the majority of flaws seem to be initiated near the shadow zone boundary.

REFERENCES

[1] Langefors, U. and Kihlstrom, B., The Modern Technique of Rock Blasting, Wiley, New York, 1963.

[2] Persson, P.A., Lunborg, N. and Johansson, C.H., "The Basic Mechanism in Rock Blasting", Proc. 2nd International Society for Rock Mechanics, Belgrade, 1970.

[3] Porter, D.D. and Fairhurst, C., "A Study of Crack Propagation Produced by the Sustained Borehole Pressure in Blasting", Proc. 12th Symp. Rock Mechanics, University of Missouri-Rolla, November 1970.

[4] Ash, R.L., The Influence of Geological Discontinuities on Rock Blasting, Ph.D. Dissertation, University of Minnesota, 1973.

[5] Hagen, T.N. and Just, G.D., "Rock Breakage by Explosives - Theory, Practice, and Optimization", Proc. 3rd Congress International Society Rock Mechanics, Denver, Colorado, September 1974.

[6] Kutter, H.K. and Fairhurst, C., "On the Fracture Process in Blasting", Int. J. Rock Mech, Min. Sci. and Geomech. Abstr., Vol. 8, 1971.

[7] Johansson, C.H. and Persson, P.A., "Fragmentation Systems", Proc. 3rd Congress International Society for Rock Mechanics, Denver, Colorado, Sept. 1974.

[8] Coursen, D.L., "Cavities and Gas Penetrations from Blasts in Stressed Rock with Flooded Joints", Acta Astronautica, Vol. 6, 1979.

[9] Duvall, W.I. and Atchison, T.C., "Rock Breakage by Explosives", U. S. Bureau of Mines, R.E. 5356, 1957.

[10] Starfield, A.M., "Strain Wave Theory in Rock Blasting", Proc. 8th Symposium on Rock Mechanics, University of Minnesota, Sept. 1966.

[11] Bhandari, S., "On the Role of Quasi-static Gas Pressurized Stress Waves in Rock Fragmentation by Explosives", Proc. 6th International Colloquium on Gasdynamics of Explosives and Reactive Systems, Stockholm, Sweden, August 1977.

[12] Barker, D.B., Fourney, W.L. and Dally, J.W., "Photoelastic Investigation of Fragmentation Mechanisms, Part I - Borehole Crack Network", NSF Report from University of Maryland, March 1978.

Barker, D.B., Fourney, W.L. and Dally, J.W., "Photoelastic Investigation of Fragmentation Mechanisms, Part II - Flaw Initiated Network", NSF Report from University of Maryland, August 1978.

[13] Dally, J.W., "Applications of Photoelasticity to Elastodynamics", Proc. Symposium on Dynamic Response of Solids and Structures, Stanford University, 1971.

[14] Timoshenko, S. and Goodier, J.N., Theory of Elasticity, McGraw-Hill, New York, 1951.

[15] Bligh, T.P., Gaseous Detonations of Very High Pressures and Their Applications to a Rock Breaking Device, Ph.D. Dissertation, University of Whitwatersrand, Johannesburg, South Africa, 1972.

[16] Ouchterlony, F., "Fracture Mechanics Applied to Rock Blasting", Proc. 3rd Congress International Society for Rock Mechanics, Denver, Colorado, Sept. 1974.

FRAGMENTATION STUDIES WITH LARGE FLAWS

William L. Fourney
University of Maryland
Mechanical Engineering
College of Engineering
College Park, Maryland 20742

INTRODUCTION

This lecture describes results obtained when investigating fragmentation achieved in models which contained large flaws. Figure 1 is a photograph of a working bench face in the Pinesburg limestone quarry located near Hagerstown, Maryland. The segmentation seen in Figure 1 is typical of rock found in quarry locations. Much of the segmentation in a quarry bench is blast damage incurred in fragmenting the material that was removed in previous shots. These detonations loosen up the joints and faults that are present in the formation, but not to such an extent that the material can be removed.

The spacing of these joint sets and bedding planes vary from rock type to rock type and even from point to point in the same rock formation. The orientation of these fault sets also varies, but the assumption is that the joint sets and the bedding planes make up an orthogonal triad of planes. Rock joints in limestone range from open sets filled with mud to very tightly bonded calcite joints with a strength nearly equal to the tensile value found in the adjoining rock masses.

Figure 1 Bench face showing bedding and joint sets at Pinesburg, Maryland quarry.

Jointed models were constructed of Homalite 100. All models were 6.4 mm (0.25 in) thick and were made by bonding together 50 mm (1.97 in) wide strips of Homalite 100. The strips were rough cut on a band saw and then routed to final size to provide smooth edges. One very important parameter in the model is the bond formed between the two photoelastic strips. During the experimental program variations in bonding strengths were utilized from grease filled joints to very tough epoxy glues. The particular bonding agent used throughout the test series to be described here is a product sold under the trade name "M Bond 200". M Bond 200 is marketed by Eastman Chemical Products, Incorporated, and contains cyanoacrylate ester which sets rapidly under a slight pressure. The adhesive was spread uniformly with a resulting bond that was approximately 0.075 mm (0.003 in) in thickness. The joint could be classed as medium in strength in that it normally could be handled without separating, but in some instances did fail to remain intact when the samples were being prepared for testing. Tests conducted to determine bond strength showed a tensile failure in the bond at about 6.9 MPa (6.28 ksi) and a shear strength that exceeded the shear strength of the Homalite 100. This is to say when a model was loaded in tension at 45° to the bond line, the Homalite 100 failed before the bond did. Most models which separated during handling did so as a result of bending loads being applied.

The 6.4 mm (0.25 in) thick 2-D models were either 300 mm × 300 mm (11.8 × 11.8 in) with a single borehole or were 300 mm × 380 mm (11.8 × 15 in) with two boreholes. The geometry of the two different models tested is given in Figure 2. Each borehole was loaded with 200 mg. of PETN. An attempt was made to contain the detonation gases by tightly capping the boreholes with O-ring seals and steel caps.

The models were explosively loaded while being viewed in the multiple spark gap camera equipped to function as a dynamic light field polariscope.

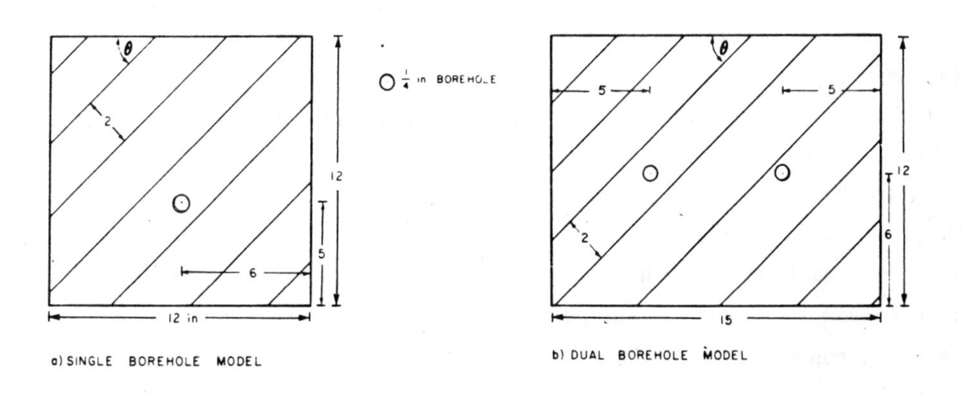

Figure 2 Geometry of the test models.

JOINT INITIATED FRACTURE

The first four of sixteen frames taken during test DL-19 are shown in Figure 3. Figure 3a was taken 34 μs after detonation of the first borehole. The fronts of the outgoing dilitational (P) and distortional (S) waves have been clearly marked. The shadows of the fractures seen in this frame are radial in direction. Although it cannot be proven conclusively, some appear to have originated at the borehole wall and some at the bond lines between layers. All of these fractures were most likely produced by the outgoing P wave. (The stress state in the P wave tail promotes radial cracking.) The second frame, Figure 3b, was taken 62 μs into the dynamic event. The long radial fractures are due to bond-line initiated fractures that coalesce with the borehole radial crack pattern. A large number of non-radial cracks are evident along the joint line. These fractures were initiated at the bond lines adjacent to the borehole and are traveling away from the borehole layer at angles of 60 to 80 degrees to the bond line. These joint initiated cracks are the dominate fragmentation mechanism in a jointed media and have been identified and described earlier [1]. These fractures are felt to be the result of a high shear loading on the bonded interface. In previous testing the initiation was identified with the passage of the compressive peak of the P wave. In this particular test there is also evidence that these cracks were initiated by the compressive peak of the outgoing P wave. Observe from Figure 3a the area just in front of the S wave front and behind the P wave. Small dark areas along the joint lines can be barely discerned. These are initiation sites of the intense joint initiated cracking seen in Figure 3b. Once these sites are initiated by the resolved shear stress in the outgoing P wave, they grow in a direction roughly perpendicular to the bond line as a result of the tensile stress in the outgoing shear wave. The stress states of the P and S waves are responsible for this mechanism of fragmentation. Notice that the maximum fragmentation due to this mechanism should occur where a borehole radial line intersects a joint line at 45°. This observation is confirmed from the photograph in Figure 3d.

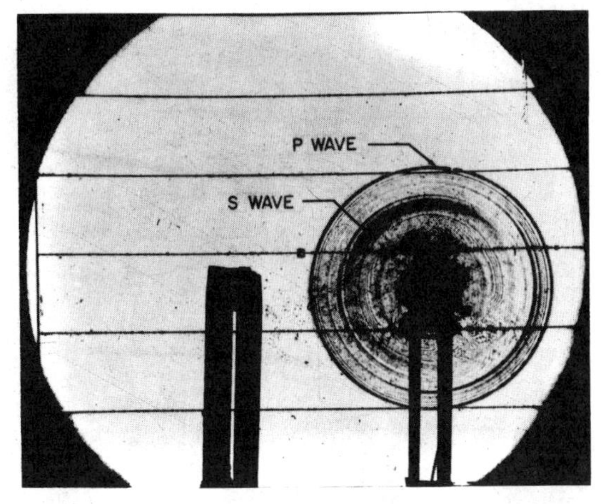

a

34 μs AFTER DETONATION.

62 μs AFTER DETONATION.

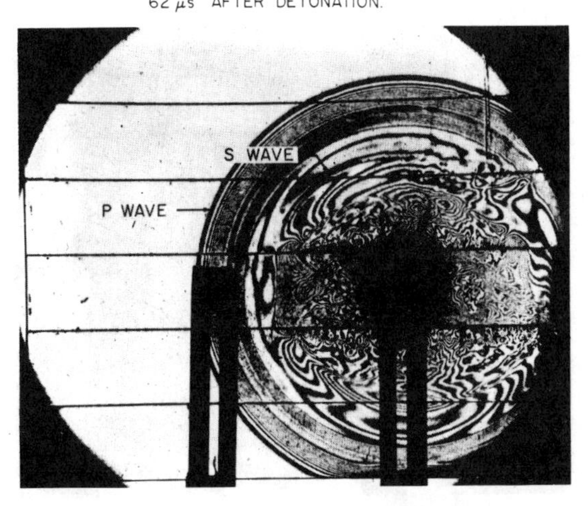

b

Figure 3 Early frames from DL-19. 400 μs delay.

c

86 μs AFTER DETONATION.

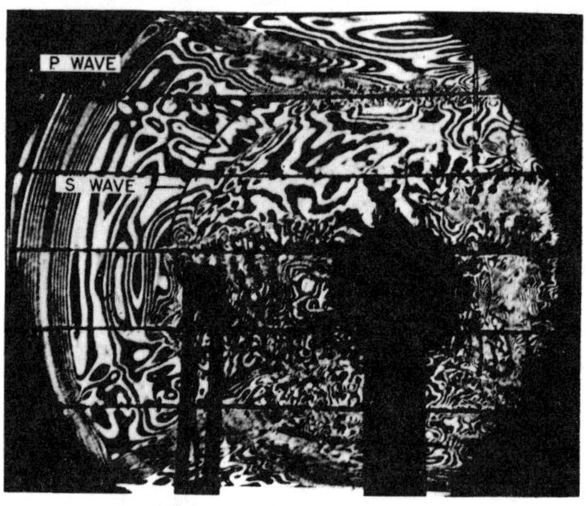

6 μs AFTER DETONATION

d

Figure 3 Early frames from DL-19. 400 μs delay.

Figure 3c taken at 86 μs shows the further development of the fragmentation due to the first detonation. Notice the relatively strong shear wave that is traveling along the bond line above and to the left of the borehole in Figure 3c. Note also that the radial cracks in the borehole layer have turned and are propagating parallel to the bond line. A number of cracks have also been initiated at the second interface directly above the borehole. Unlike the earlier joint initiated cracks these propagate both toward and away from the borehole. These cracks have been initiated and driven by the shear wave since the region of maximum shear in that wave system is directly above and below the borehole.

Cracks initiated from a slight imperfection on the bond line (whether the inperfection is due to small debonded areas or to small areas debonded by the passage of the P wave) by shear or mode II loading should initiate in a direction of 70° from the bond line [2]. All of the joint initiated cracking seems to follow this behavior.

At 116 μs (Figure 3d) the fragmentation in the borehole layer is beginning to develop due to the arrival from the right of the reflected P wave which is now predominately biaxial tension. Some initiations have occurred near the second borehole due to reflections from the undetonated hole. The reflected waves from the right free surface have caused extensive branching when passing over the propagating joint initiated cracks. This is very obvious in the layer just above and to the right of the first borehole.

From close examination of the results from approximately 20 tests similar to the one shown, it appears that the large number of cracks initiated at the bondlines are the result of a shear loading on the layer interface. This conclusion is based upon several different observations:

A) The speed at which the initiation front appears to travel along bondlines agrees with experimental error to the speed of the P wave.

B) The appearance of the caustics which result in successful initiation corresponds in time to the arrival of a high shear loading. Directly above the borehole this shear loading is associated with the arrival of the shear wave while along the layer interfaces at other locations with the arrival of the compressive peak of the P wave.

C) The most intense cracking and the cracks which appear to accelerate the fastest are in locations where the highest shear loading is found.

D) Finally, the direction of propagation of the initiated cracks is between 70 and 90 degrees to the layer interface and in the case of P wave initiation are directed away from the borehole layer. This agrees well with an analytical solution by Erdogan and Sih [2] for mode II (shear) loading of an existing crack. This would imply that the layer interfaces appear to the outgoing stress wave to be a very long partially bonded crack. When the shear loading is applied by the stress waves cracks at about 70 to 90 degrees to the interfaces are initiated wherever the shear stress in the outgoing wave exceeds the level necessary for cracking to occur.

For lack of a more descriptive title we have named this type of fracture "joint initiated" fracture or JI cracking. The typical speed of these cracks during the early part of the dynamic event is around 400 m/s (15,748 in/s) and in many cases the cracks are being driven with sufficient energy to result in branching.

In the early part of the event, the observed results for the 20 or so layered model tests conducted can be summarized as follows:

1. No "joint initiated" cracks (JI) initiated by the P wave form in the borehole layer.
2. The most intense fragmentation originates at the joints which adjoin the borehole layer.

3. The most intense JI cracking when initiated by the P wave occurs on the bondline where a borehole radial line makes a 35 to 45 degree angle with the layer direction

4. The initiation of JI cracks on bondlines adjacent to the borehole layer corresponds to the arrival of the P wave peak.
5. On bondlines directly above or below the borehole and on bondlines at least one joint spacing apart from the borehole layer JI cracks propagate towards (as well as away from the borehole) and in this case appear to be initiated and grown by the S wave.

TIME DELAYS BETWEEN BOREHOLES

Additional testing was conducted with dual borehole models to determine enhancement in fragmentation due to delays between detonations. The delay time between the first and second borehole detonation was varied from zero to 600 microseconds. Fourteen tests were conducted with delay times of approximately 0, 50, 200, 400 and 600 μs. It was found that a very long time delay was implssible to achieve since the second borehole never remained intact long enough to permit a firing.

After the tests were conducted the fragments were screened through sizing screens of 4.8, 9.5, 19, 25 and 38 mm (3/16, 3/8, 3/4, 1 and 1.5 in) openings. The screens used were typical of the type used for sorting sedimentary rocks and other soil samples in a laboratory situation.

The weight distribution of particles was used to construct histograms for graphically illustrating the distribution of fragments for each delay. From the histograms a weighted average fragment size was computed. This average size gives an idea as to the size of a typical fragment. The following formula was utilized for calculating F_{av}.

$$F_{av} = \frac{\sum_{i=1}^{n} X_i W_i}{\sum_{i=1}^{n} W_i}$$

F_{av} is the weighted average X_i is the mean of the ith and $(i-1)^{th}$ sieve dimension. W_i is the total weight of fragments retained by the ith sieve. When more than one test per delay was available an average of the weights retained by a given sieve was used; n is the total number of sieve sizes used.

This same calculation was used by Singh [3] for homogeneous dual borehole models made from Homalite 100. The results from Singh's tests are presented in Figure 4a. The average fragment size for homogeneous models was still decreasing at a 600 μs delay. Unfortunately, Singh had no data between 600 μs and an infinite delay. The dashed portion of the curve therefore is speculative but the decrease in average fragment size between 520 and 600 μs was not as great as the decrease from 380 to 520 μs. The minimum fragment size of about 30.5 to 33 mm (1.2 to 1.3 in) is felt to be accurate in spite of the large extrapolation made. Fragmentation results for one hole detonation only are also shown in the figure.

Similar results for the layered models are shown in Figure 4b. The model size of the homogeneous and the layered models were the same except for thickness. The charge sizes were scaled in the same ratio as the thickness. Single hole detonation resulted in an average fragment size of a little under 25 mm (1 in) for the layered media and over 76 mm (3 in) for the homogeneous models. The minimum average fragment size for the layered media was about 20.3 mm (0.8 in) and occurred at a delay of 360 μs. For the homogeneous models the value extrapolated from Figure 4a is a minimum average fragment size of 31.8 mm (1.25 in) at about 775 μs delay.

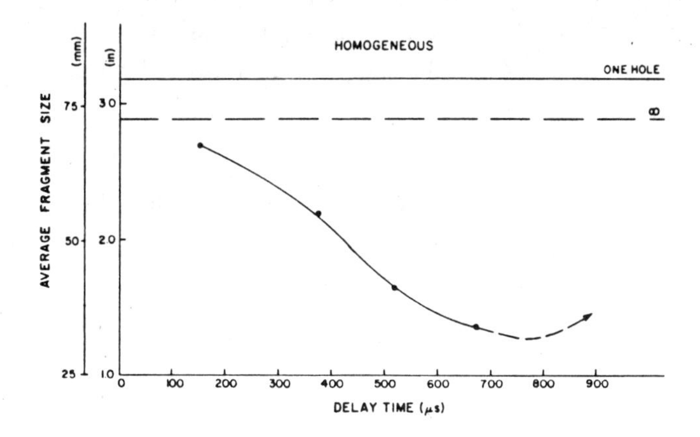

Figure 4(a) Homogeneous models.

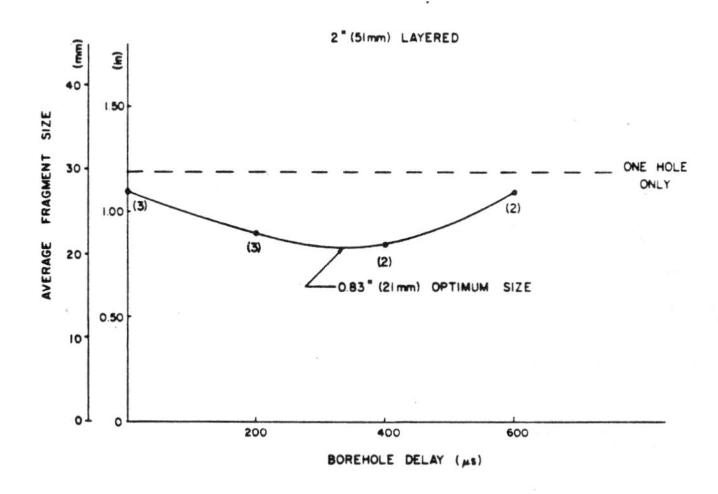

Figure 4(b) Jointed models.

For a layered media the fragmentation mechanism of joint initiated cracking yields a much smaller average fragment size than would be obtained in a homogeneous media. This reduction in fragment size is at least 1.5 times. For some delays it can be as high as 2 times.

The delay needed to optimize the fragment size is much more critical with a jointed media than for a homogeneous one. The optimum delay time is much shorter for the layered media (about 1/2 in these tests). The increase in fragment size with increased delay time once the optimum is exceeded is more drastic with a layered media. This observation is not surprising. After a given amount of time the layers begin to move as rigid bodies. The strips begin to separate to such an extent that the only additional damage done by the second detonation is restricted to the area in the immediate vicinity of the second borehole. Additional details on the effect of time delays in a layered media are available in [4].

REFERENCES

[1] Fourney, W.L., Barker, D.B. and Holloway, D.C., "Mechanism of Fragmentation in a Jointed Formation", NSF report by University of Maryland, July 1979.

[2] Erdogan, F. and Sih, G.C., "On the Crack Extension in Plates Under Plane Loading and Transverse Shear", J. Bas. Engng. (ASME), Dec. 1963.

[3] Singh, Bhupinder, "Experimental Investigation of the Effect of Time Delays on Fragmentation in Homogeneous Models", Report to NSF Photomechanics Lab, Jan. 1979.

[4]. Fourney, W.L. and Barker, D.B., "Effect of Time Delay on Fragmentation in a Jointed Model", NSF report by University of Maryland, Aug. 1979.

GAS WELL STIMULATION STUDIES

William L. Fourney
University of Maryland
Mechanical Engineering
College of Engineering
College Park, Maryland 20742

INTRODUCTION

The research described here is part of a program funded by Morgantown Energy Technology Center under the Eastern Gas Shale Program to determine if tailored pulse loading (TPL) offers a viable means for stimulating Devonian Shale Wells. This technique (TPL) relies on the detonation so deflagration of a chemical substance to create borehole pressures which are low in magnitude, have a relatively slow rise time, and last for an extended period of time so as to pressurize the created fractures. The pressure loading of the cracks is thought to result in fractures that are driven much farther from the borehole wall than cracks which result from detonation of an ordinary high explosive in the wellbore. Schmidt et al., [1] report that too fast a rise time in the pressure loading results in a plastic deformation in the vicinity of the charge and an ultimate residual compressive stress surrounding that borehole. This compressive stress shuts down not only gas flow into the propagating cracks but gas flow from the shale back into the wellbore.

It is suspected on the other hand that too slow a rise time due to a deflagrating charge [2] results in a near quasi-static loading and has the same drawback as hydraulic fracturing -- namely an inability to create new crack sites other than existing joints and an inability to control crack growth direction.

At Maryland we have conducted a large number of model tests over the past two years to help understand the basic mechanism involved in fracture propagation from a wellbore. Results from some of these tests will be described in the following sections.

GROWTH OF FRACTURES FROM A WELLBORE AND GAS FLOW INTO THEM

Rectangular models 51-76 mm (2-3 in) thick, with a geometry similar to that shown in Figure 1, were used in this series. The 12.7 mm (0.5 in) diameter borehole was drilled parallel to the faces of the model as shown in the figure. The borehole was grooved to produce a controlled fracture that would cleave the model in half in the thickness direction. A decoupled cylindrical explosive charge, approximately 3mm (0.125 in) in diameter and 114 mm (4.5 in) long, was placed in the bottom of the borehole. The borehole was then tightly stemmed with 54 mm (2 in) of modeling clay and a 6.4 mm (0.25 in) cap plate of Lexan bonded over the end. The explosive charge (300 mg of PETN) was detonated while dynamic photoelastic photographs were taken with a Cranz-Schardin multiple spark gap camera.

Two modifications to the PMMA model were made for one test and these are indicated in Figure 1. A 3.2 mm (0.125 in) hole was drilled three quarters of the way through the model at Station A. This hole was filled with water and capped with a 6.4 mm (0.25 in) Kistler model 603A pressure transducer. The transducer was used to read the pressure in the fracture surface as it intersected the water filled hole. Station B, on the other side of the borehole, was a 9.5 mm (0.375 in) diameter hole drilled completely through the model and was used to visually determine when the opaque detonation products intersected the hole.

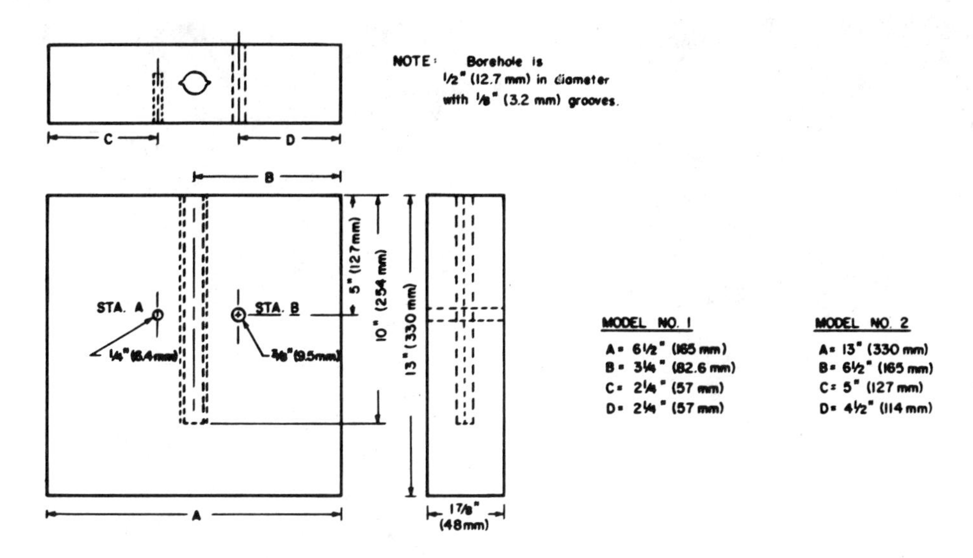

Figure 1 Model used to determine crack pressurization.

Four frames from Test PG5 are presented in Figure 2. Figure 2a taken 8 μs after detonation clearly shows the shadow of the vertical borehole and the shelf illuminated explosive charge in the lower part of the borehole. The pressure transducer is on the left of the borehole and the viewing hole on the right. The viewing hole is not perfectly circular due to the parallax of the camera set-up, but it was possible to see through the hole in all photos prior to detonation. At 53 μs after detonation, Figure 2b shows the well developed P and S waves and the reflected PP and PS wave systems. The fracture front has just reached the pressure transducer and has intersected the viewing hole. Note that no opaque detonation products are visible in the hole even though it appears as if the fracture front has passed completely through the hole. This means that the shadow of the fracture surface is caused by optical interference and reflection of the light trying to pass through the model and is not due to opaque detonation products.

At 172 μs, shown in Figure 2c, the viewing hole has become blocked by detonation products. In all frames up to this time the viewing hole was clear. In all subsequent frames it was blocked.

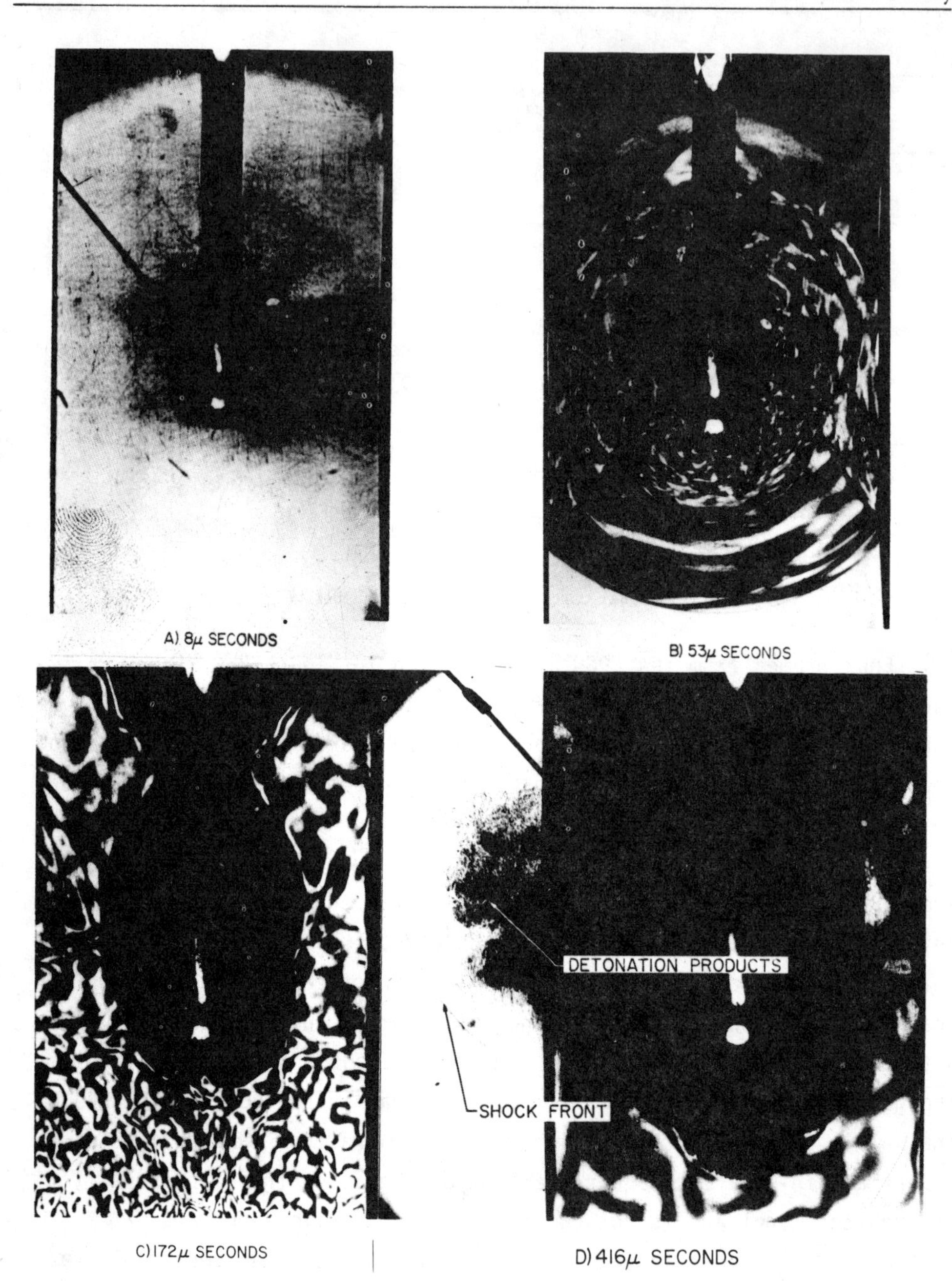

Figure 2 Four frames from Test PG5.

In this particular test the grooves placed in the borehole were misaligned. As a result the cracks propagated toward the front surface and one large fragment was blown off the front face of the model. The final shape of this fragment is shown by the black shadow in Figure 2d taken 416 μs after detonation. Notice in this frame the separation of the shock front and the opaque gaseous products. In this test both are moving towards the camera as well as having a major component of motion parallel to the plane of the model. The approximate separation of the shock front from the gaseous products in a plane parallel to the model front surface is about 25 mm (1 in).

Figure 3 gives some typical crack position-time values determined from the tests. The slope of this plot yields an average crack velocity in the PMMA of 394 m/s (15,500 ips). For the same test the P wave velocity was found to be 2044 m/s (80,500 ips) and the shear wave, 1336 m/s (52,600 ips). From photographs such as Figure 2 from Test PG5 the shock front in air was observed to increase from 422 m/s (16,600 ips) immediately after exiting from the model to around 498 m/s (19,600 ips) in later frames. This would certainly indicate that the shock front was capable of keeping up with the crack front but in fact appears to have been held back by it.

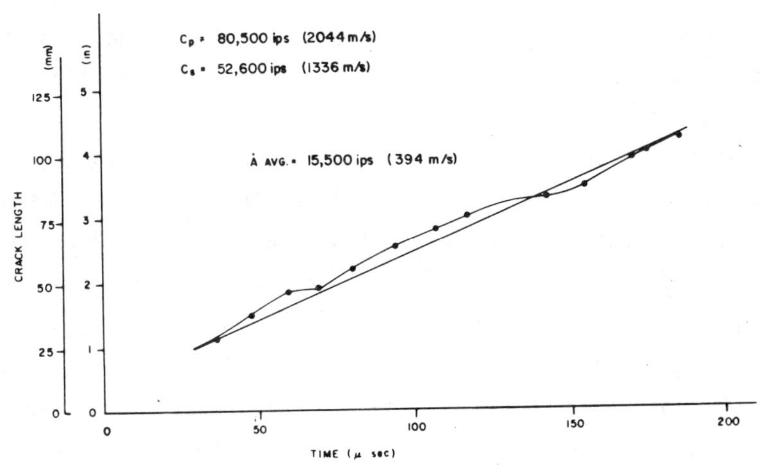

Figure 3 Crack length as a function of time.

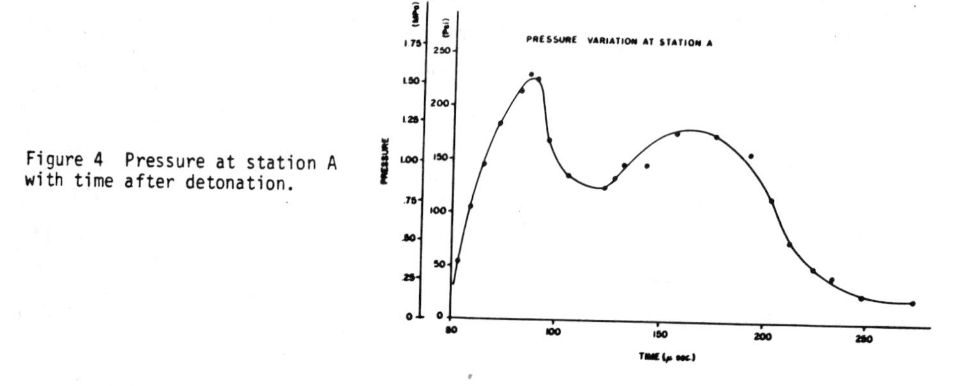

Figure 4 Pressure at station A with time after detonation.

Although no borehole pressure measurements were made for these tests, some earlier measurements made by Stecher [3] by pressure bar techniques indicate pressures in the range of 14 to 21 MPa (2,000 to 3,000 psi) under similar loading conditions.

The data from the pressure transducer located just to the left of the borehole, Station A, in Figure 1, is presented in Figure 4. By 50 μs the pressure is beginning to increase and it peaks at about 75 μs. This corresponds, in time, to the frame taken at 53 μs that shows that the fracture front has reached the capped pressure transducer hole. The second peak in the transducer data occurs at 170 μs. This was the time at which the viewing hole could no longer be seen through. The maximum measured pressure of 1.6 MPa (230 psi) is about an order of magnitude less than the assumed borehole pressure.

Homalite 100 models as shown in Figure 5 were used to determine crack opening as a function of time after detonation. The models were 6.4 mm (0.25 in) thick, 304.8 mm (12 in) square, and had 38.1 mm (1.5 in) borehole routed in the center. 6.4 mm (0.25 in) long grooves were sawn into the sides at the borehole to direct the fractures [4] along one diagonal of the square model. Kaman Eddy Current transducers were situated as shown in Figure 5 so that the crack ran between the proximity transducer and its target. The decoupled explosive charge consisted of 90 mg of PETN.

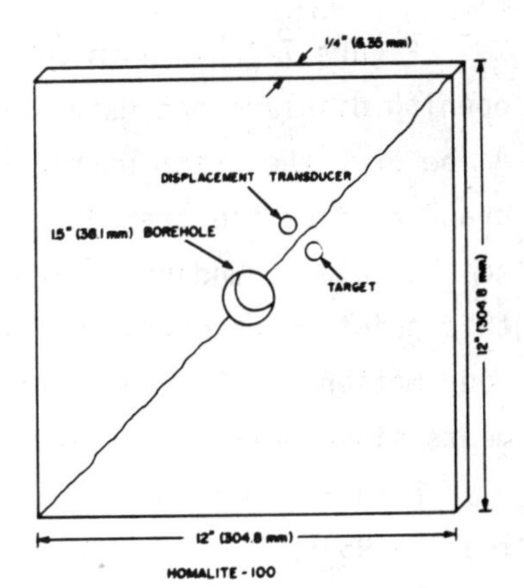

Figure 5 Model used to determine crack opening.

These models were photographed during the dynamic event with the Cranz-Schardin camera so crack position and stress wave location could be correlated with the transducer displacement data.

Figure 6 shows the displacements recorded from two transducers on both sides of a borehole in a single model. Both cracks behaved similarly. If the slopes of the curves shown in Figure 6 are taken as crack opening velocities, one crack opened at a rate of 1.12 m/s (46 ips) and the other at 1.45 m/s (57 ips). The crack velocity in the direction of propagation for this test was about 33 m/s (13,000 ips).

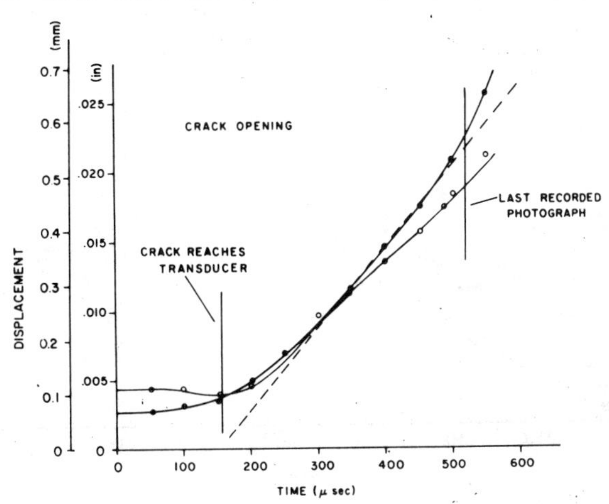

Figure 6 Crack opening as a function of time for two cracks from Test CO-2.

Figure 7 shows similar results from another test, C07, in which opening displacement data was recorded over a much longer period of time. As before, the crack displacement is fairly linear from the time the crack reaches the transducer at about 130 μs, until the crack propagates to the outer boundary of the model (about 600 μs in Figure 7). After this point is reached, the remaining record applies primarily to rigid body motions of the model halves. (In this case due to bending displacements since only one crack propagated.)

Both Figures 6 and 7 indicate crack openings on the order of at least 0.95 to 1.25 mm (20 to 50 mils) can be expected to accompany crack propagation. It should be remembered that these results were obtained for thin models. In addition to the standard plane stress to plane strain displacement correction, bear in mind that pressure was being lost through the front and back faces of the model during propagation. Hence, even though plane strain would indicate a lesser displacement,

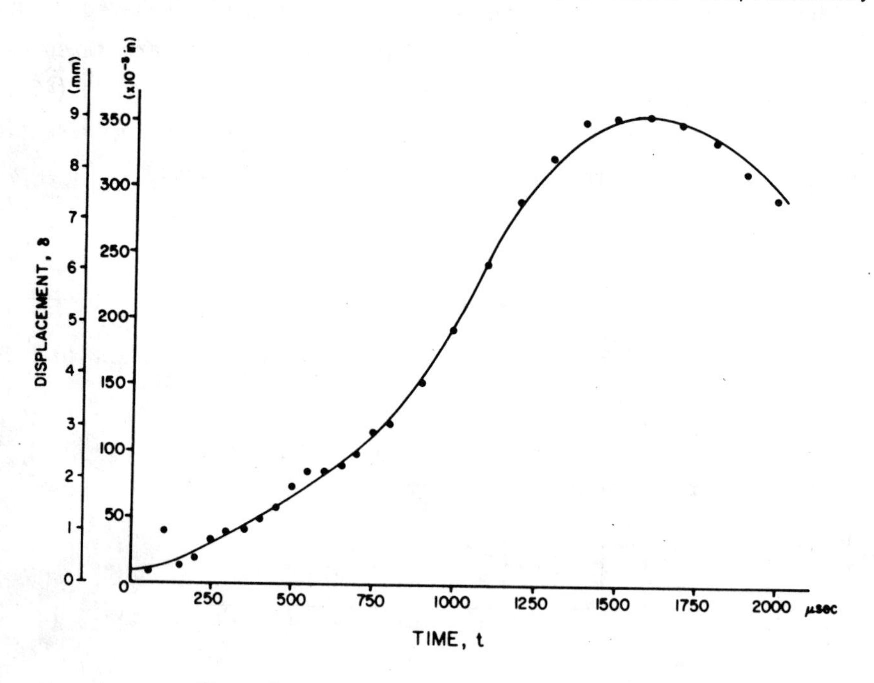

Figure 7 Crack opening over a longer period of time.

it is expected that larger pressures would be acting on the crack faces and even larger displacements would be seen in a thick block situation.

EFFECT OF LOADING RATE

In tests with thick plexiglas blocks we noted a very definite dependence of borehole shattering on loading rate within the borehole. Figures 8 and 9 demonstrate this effect. Results from two tests (Numbers 804 and 807) are described in those figures. Both tests were loaded in a similar fashion with around 300 mg of a propellant (supplied and manufactured by Kinetec Corporation of San Francisco (USA)) in a water filled borehole. The pressure profile recorded in the borehole for Test 807 is presented in Figure 8a. Note that the 300 mg of propellant detonated rather than deflagrated and a very large pressure and pressure rise rate resulted. The resulting fracture patterns produced are shown in Figure 8b. Notice the shattering of the borehole wall that resulted.

Figure 9a shows the pressure pulse recorded in Test 804 where 375 mg of the propellant deflagrated as planned producing only half the maximum pressure and a rise rate of only one sixth as with Test 807. Figure 9b shows the resulting fracturing for the test. Note that there was no shattering of the wellbore and the resulting fractures were extremely smooth.

Our model results appear therefore to support the concepts of tailored pulse loading.

STEM INDUCED FRACTURE

In the course of our model testing, we have discovered what we feel is an important mechanism of fracturing in well stimulation. We have called this mechanism stem induced fracturing and it is best described by the results shown in Figure 10. In this case the model was similar to that shown in Figure 1 but was 152 × 305 × 102 mm (6 × 12 × 4 inches) in size and the borehole was notched to promote fracture in a plane parallel to the camera plane as before. 250 mg of PETN was placed

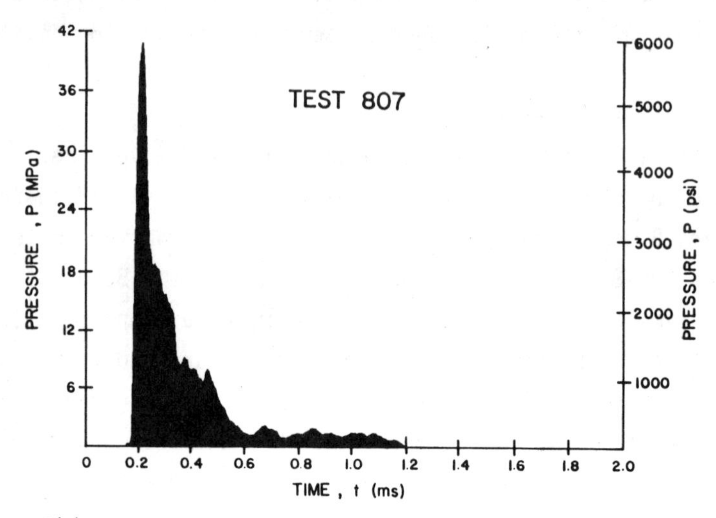

Figure 8(a) Pressure-time record obtained when a propellant detonated rather than rather than deflagrated.

Model Reassembled.

Model Folded Open To Show Borehole Crushing That Occurred.

Figure 8(b) Borehole crushing produced by propellant detonation.

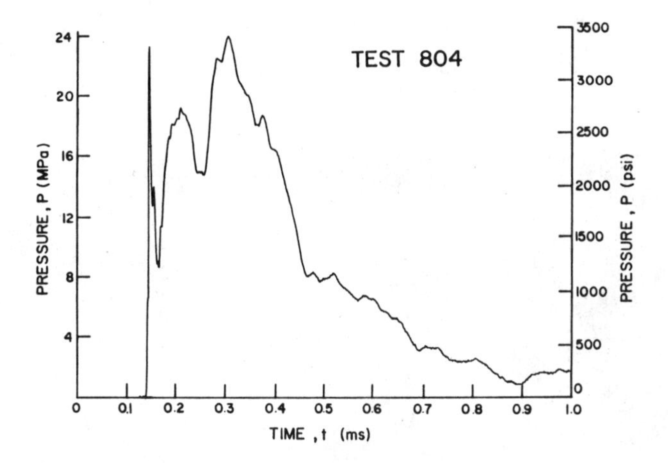

Figure 9(a) Pressure-time record from Test 804. Fluid filled propellant charged.

Model Reassembled.

Model Folded Open To Reveal Undamaged Borehole.

Figure 9(b) Fracture produced by propellant deflagration in a fluid filled borehole.

near the bottom of the 230 mm (9 in) deep 12.7 mm (0.5 in) diameter air filled borehole which was stemmed to a depth of two inches and the charge detonated.

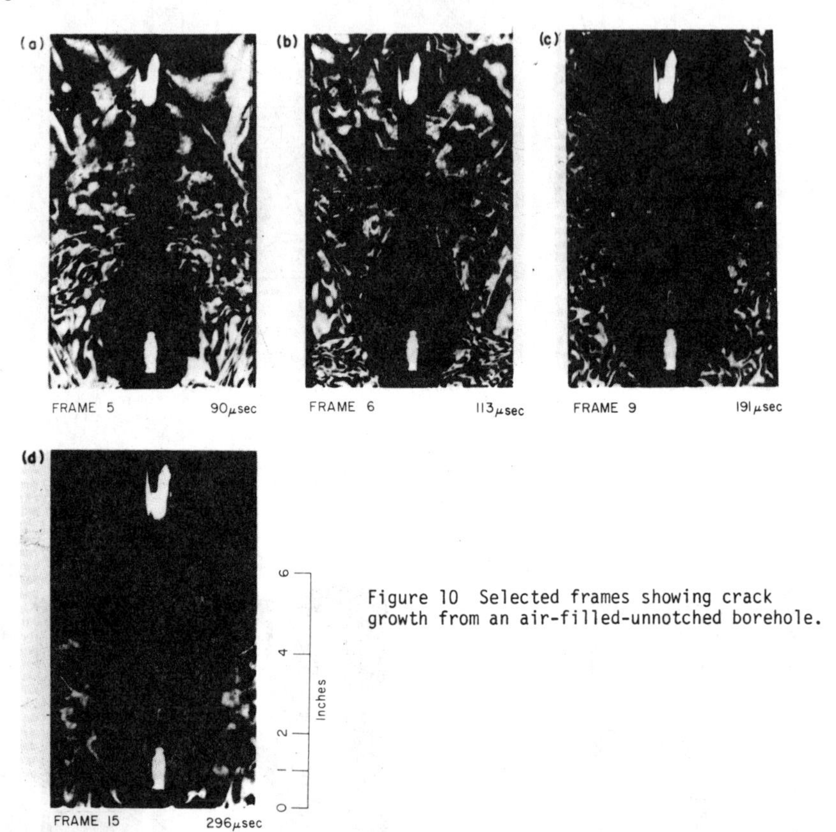

Figure 10 Selected frames showing crack growth from an air-filled-unnotched borehole.

Figure 10 shows four frames from the test as recorded by the multiple spark gap camera. Frame 5, Figure 10a, shows the location at 90 μsec of the fracture (the fracture is the dark bladder shaded area near the bottom of the model) that initiated due to the detonation of the charge. Notice the two long straight fringes located in the upper third of the model. These are due to a shock wave which has travelled up the borehole from the top of the charge at detonation. The fringe makes an angle of 40° with the borehole wall indicating that the ratio of shock wave speed to P-wave speed in the plexiglas is 1.19, or that the shock wave speed is 236 m/s (93,000 ips). Note that there are two bright areas in the borehole. One near the bottom, where the charge was detonated, and one at the stem. The bright area near the stem is due to the increase in pressure and temperature which is caused by the shock wave reflecting from the stem. The light emitted is due to the ionization of the air at that location. The camera used to record the photographs is such that light emitted from any area within the field of view will be visible in all 16 frames and it is not possible to determine at what point in time the light flash occurred.

By the time Frame 6 (Figure 10b) was recorded 23 μsec later, at 113 μsec, a fracture is seen to have initiated at the stemming area and has grown considerably. At this time the charge area fracture is also seen to be growing, but at a much slower rate. At 191 μsec (Figure 10c) the fracture that originated at the stem is larger than the charge area fracture and is beginning to engulf the arrested charge area fracture at 296 μsec after detonation.

Figure 11 is a sketch showing how the fractures grew from frame to frame. Included in the figure are the times at which each frame was photographed, as well as the velocities computed from the crack front positions measured from the photographs. The numbers on the sketch represent the frame numbers and the straight lines indicate directions along which the velocity measurement were made. Notice how quickly the charge area fracture slows and arrests, while the stemming area fracture continues to propagate at a somewhat erratic but high speed. The large velocity obtained between Frames 10 and 12 is due to the fact that the fracture reached the free edge of the model during that time interval but the reason for the low velocity between Frames 12 and 14 is not clear.

The observed ionization of the gas in the stem region was investigated further by conducting the following test. A 51 mm (2 in) thick plexiglas model, with geometry as shown in Figure 12 was used to measure pressure at three locations within the borehole. A small charge was

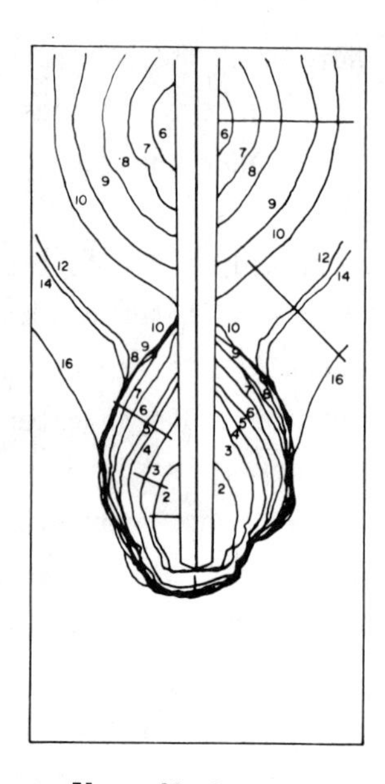

Frame no.	Time (μsec)	Velocities between frames (ips)	
		Charge area	Stem area
1	12.5		
2	34.0	20,854*	
3	57.0	13,760	
4	80.5	11,784	
5	89.5	12,493	
6	112.5	12,613	10,320*
7	140.0	7193	17,742
8	163.5	5050	17,395
9	190.5		21,978
10	214.5		12,363
12	246.5		28,846
14	271.5		4220
16	322.5		20,168

* Velocity probably low since fracture was not likely to initiate at the same instant the time count started.

FIgure 11 Sketch showing fracture front positions at different times and the interframe velocities of the two fracture fronts.

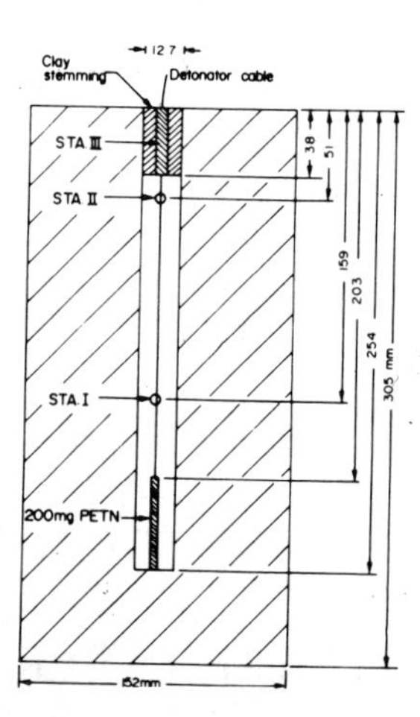

Figure 12 Cross-sectional view showing the geometry used to record borehole pressures at different borehole locations.

used with no grooves, as it was desired only to compare pressures at different points in the borehole and not to fracture the model. Pressures were measured at three locations: (1) Station I -- midway between the top of the charge and the stem; (2) Station II -- 13 mm (0.5 in) down from the stem; and (3) Station III -- in the clay stem. Kistler 603A transducers were mounted flush with the borehole wall for Stations I and II and flush with the bottom of the clay stem for Station III.

Pressure records from the three transducers for a single test are shown in Figure 13. The first curve is the pressure at Station I and shows a peak of 4.1 MPa (600 psi) and a duration of 260 μsec before reflection starts to occur. The second curve shows the pressure recorded at Station II, located below the stem-borehole interface. Notice that a reinforcement has occurred and the magnitude of 8.5 MPa (1,240 psi) and duration of 600 μsec are both about twice the values measured at Station I. The third curve shows the measurements made at the stem borehole interface where the pressure reached 12.0 MPa (1,750 psi) and remained above 1.0 MPa (150 psi) for longer than 600 μsec. Since the pressure is higher at the stem-borehole interface than at any other location (except possibly adjacent to the charge), fracture initiation in that region is to be expected. This fracture does not arrest as quickly as the charge area fracture possibly because no "stress cage" is formed or no "fines" plug the cracks. It is felt that increased pulse duration and longer rise time also aids in driving the fracture for longer distances before arrest occurs.

The mechanism for initiation and growth of the major fractures in an explosively loaded air-filled borehole is attributed to shock wave generation. Propagation of shock waves in glass tubes and the resulting ionization of gases due to pressure build-up upon reflection have been studied earlier by Johannson & Persson [5]. When the explosive detonates at the bottom of the borehole a shock wave is created and travels up the borehole. Upon reaching the stem a reflection occurs that results in increased pressures and temperatures. This local pressure increase is

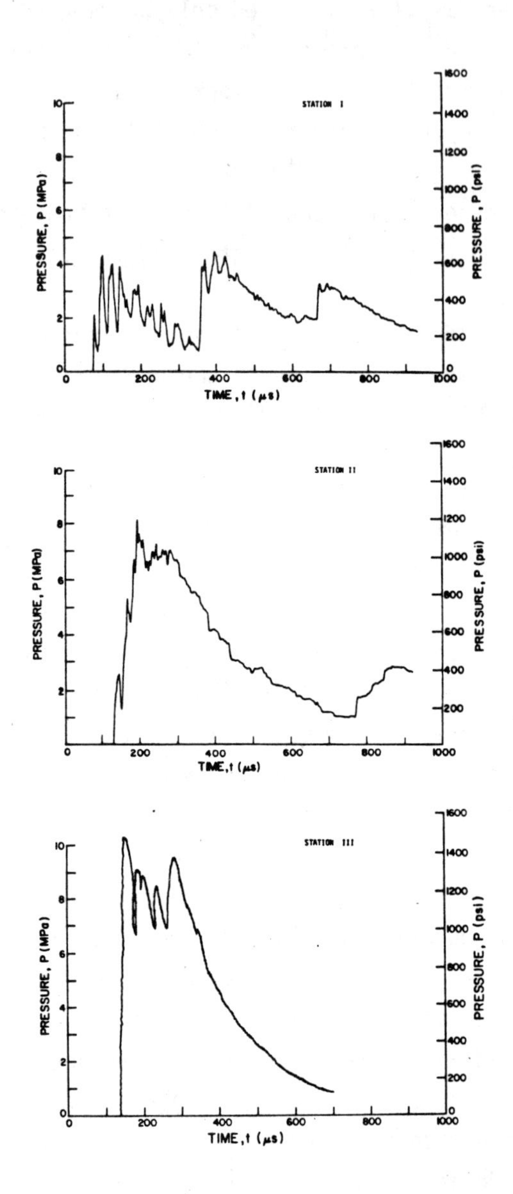

Figure 13 Pressure profiles recorded at different stations.
(a) Borehole Station I (b) Borehole Station II (c) Borehole Station III.

large enough to initiate fractures at the stem. The rise time is longer than that seen at the charge location and the pressure duration is also longer. This condition appears to be more favorable with regard to crack propagation than conditions occurring at the charge area. Hence the total propagation length of fractures originating at the stem is greater than for the fractures which originate at the charge area. This mechanism has been observed regardless of model size, stem location with respect to the top of the model, and model thickness and has been observed in large boulder tests.

REFERENCES

[1] Schmidt, R.A., Boade, R.R., and Bass, R.C., "A New Perspective on Well Shooting--The Behavior of Contained Explosions and Deflagrations", 54th Annual Conference SPE of AIME, September 1979, Las Vegas, Nevada.

[2] Warpinski, N.R., Schmidt, R.A., Cooper, P.W., Walling, H.C., and Northrop, D.A., "High Energy Gas Frac: Multiple Fracturing in a Wellbore", 20th U.S. Symposium on Rock Mechanics, June 1979, Austin, Texas.

[3] Stecher, F.P., "A Numerical Method for Simulating Controlled Blasting Techniques", Ph.D. Thesis, University of Maryland, December 1978.

[4] Gupta, In.N. and Kisslinger, C., "Model Study of Explosion-Generated Rayleigh Waves in a Half Space", Bulletin of the Seismological Society of America, Vol. 54, No. 2, pp. 475-484, 1964.

[5] Johannson, C.H. and Persson, P.A., Detonics of High Explosives, Academic Press, New York (1970).

GROUND VIBRATION STUDIES

William L. Fourney
University of Maryland
Mechanical Engineering
College of Engineering
College Park, Maryland 20742

INTRODUCTION

Many attempts have been made to quantify the propagation and interaction of elastic waves. Woods [1], for example, examined the role of trenches in isolating a building from vibration. Dally and Lewis [2] studied the effects of a slit on a propagating Rayleigh wave through the use of dynamic photoelasticity. Gupta and Kisslinger [3] observed the surface displacements in an explosively generated Rayleigh wave, by modeling a half space with a plane and employing capacitor probes to measure the vertical and radial displacement. DeBremaecker [4] studied the transmission and reflection of Rayleigh waves at corners. All of the above have their shortcomings.

Some of these include modeling a three dimensional problem in two dimensions, employing "model" materials, obtaining single station data, using a sinusoidal source, or obtaining only one component displacement.

I am going to present results of an experimental study of ground motions resulting from an explosive source in a three dimensional rock model. This work was conducted by Dr. D. C. Holloway and myself using dynamic holography.

In dynamic holography the surface of an opaque model is illuminated with a short duration coherent light source. When dealing with photographs taken with this particular light source frequency as well as amplitude information can be recaptured with proper treatment. A typical experimental set up is shown in Figure 1.

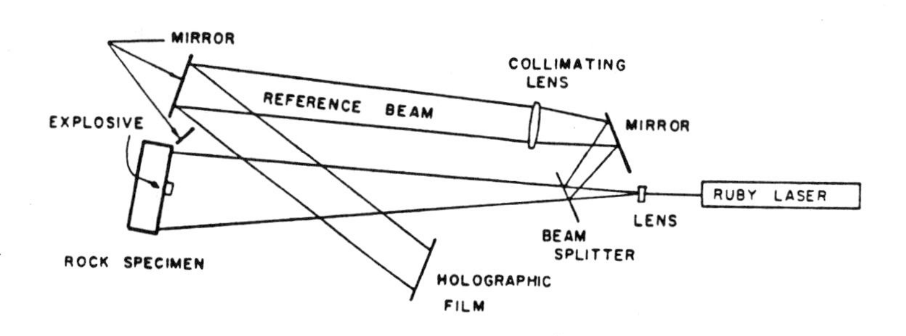

Figure 1 The experimental holographic arrangement.

A pulsed ruby laser (the coherent source) with an exposure time of 50×10^{-9} sec. illuminates a model (in this case a rock specimen). The formation of the hologram requires that a portion of this light be diverted by a beam splitter and a series of mirrors so as to strike the holographic plate directly. This is called the reference beam, and it will interfere with the light scattered from the object surface. The interference is recorded with a high resolution photographic emulsion. Two exposures are taken of the rock; one in its undeformed state and the other at a controlled time after the initiation of an explosive charge. After processing the exposure film it is reilluminated with the light from a He-Ne gas laser that duplicates the original reference beam. An observer (or camera) looking through the processed film toward the rock sample views a three dimensional image of the rock. Superimposed on the rock face are interferometric fringes that are related to the deformation of the surface at the time of the second laser exposure. At least three views through the same hologram or three separate holograms at different locations are necessary to solve for u, v, and w. In many cases, the first alternative does not prove practical due to an insufficient fringe shift over the 102 × 128 mm (4 in × 5 in) window of the holographic film. Here the method of three holograms will be discussed. The views are arranged so that fringes in the first view are predominantly from w, in the second view from u and w, and in the third view from v and w.

This is accomplished by locating the rock and laser such that the view into the hologram yields primarily the w displacement. In addition, a mirror is placed along one side of the rock and another mirror above the rock. By photographing the fringes through the mirror on the side a pattern results that is predominantly influenced by the u and w components of displacement. Similarly, a photograph from the top mirror is mostly a function of the v and w displacement. At each point of interest on the rock face, the fringe orders are determined from the photographs and three simultaneous equations are solved for u, v and w. The operation is then repeated for other points of interest.

The sign of the fringe orders, whether positive or negative corresponding to an upward or downward motion of the surface cannot be determined from the above-described analysis. Apriori knowledge or another technique must be utilized to correctly number the fringes.

The results presented in terms of displacements may be converted to particle velocities by numerical differentiation. The tests described are ones where the explosive was detonated on the surface and then at various depths below the surface. The rock surface was observed at selected times after the initiation of the explosive charge.

THEORY

The solution for surface displacements arising from a buried explosive charge is complex even for a homogeneous isotropic material.

Pekeris [5] examined the problem for a source that behaves as the Heaviside unit function and for a material with equal Lamé constants. His solution has discontinuities at the arrival time of the shear wave. When the source is at the surface it also has discontinuities at the arrival of the Rayleigh wave. Nevertheless, comparisons will be made here with Pekeris' solution.

The arrival and appearance times for the wave systems can be predicted through elementary analysis of wave theory. When a dilatational (P) wave is incident on a free boundary at an angle e, upon reflection it produces both a reflected P wave and a shear wave SV at an angle f. The relationship between the angles e and f and the dilatational wave speed C_1, and the distortional wave speed C_2 is

$$C_1 \cos f = C_2 \cos e \tag{1}$$

When a distortional (SV) wave is incident upon a free boundary upon reflection it produces an SV wave and an SP wave that travels with velocity C_1. Equation 1 also describes this phenomenon. However, a real SP wave is generated when e is > 0. Thus when $e = 0$ the critical angle f becomes

$$f_{crit} = \cos^{-1} \left(\frac{C_2}{C_1}\right) \tag{2}$$

thus whenever $r < \frac{h}{\tan(f_{crit})}$ no SP wave will exist on the surface.

For a half space, a Rayleigh wave forms whenever the curved wave fronts from a P or SV wave reach the free surface. However, a Rayleigh wave will not appear at radial distances less than

$$r < \frac{c_r h}{\sqrt{C_1^2 - c_r^2}} \tag{3a}$$

or

$$r < \frac{c_r h}{\sqrt{C_2 - c_r^2}} \tag{3b}$$

where c_r is the Rayleigh wave velocity. For a source emitting both P and SV waves, the minimum appearance point for the Rayleigh wave is thus given by Equation 3a.

The arrival times of the P and SV waves at the free surface can be calculated knowing h, r, and their respective wave velocities C_1 and C_2. The arrival time for the SP wave for a radial position greater than r_{crit} is given by

$$t_{SP} = \frac{\sqrt{r^2_{crit} + h^2}}{C_2} + \frac{(r - r_{crit})}{C_1} \tag{4}$$

thus the time has components related to shear wave and to dilatational wave travel.

Ewing [6] states that the arrival time for the Rayleigh wave can be computed as

$$t_R = r/c_r \qquad r > \frac{c_r h}{\sqrt{C_1^2 - c_r^2}} \tag{5}$$

For radial distances greater than the minimums above, the order of arrival is P, SP, SV and then Rayleigh.

RESULTS

Figures 2a through 2d illustrate the nature of the holographic fringe pattern at selected times and depths of burial in a granite block. In Figure 2a the charge has been detonated on the free surface and the time after initiation of the explosive is 50 μsec. Fringe orders have been indicated at several locations on the photograph. The observed disturbance is entirely due to the Rayleigh wave, as the P and

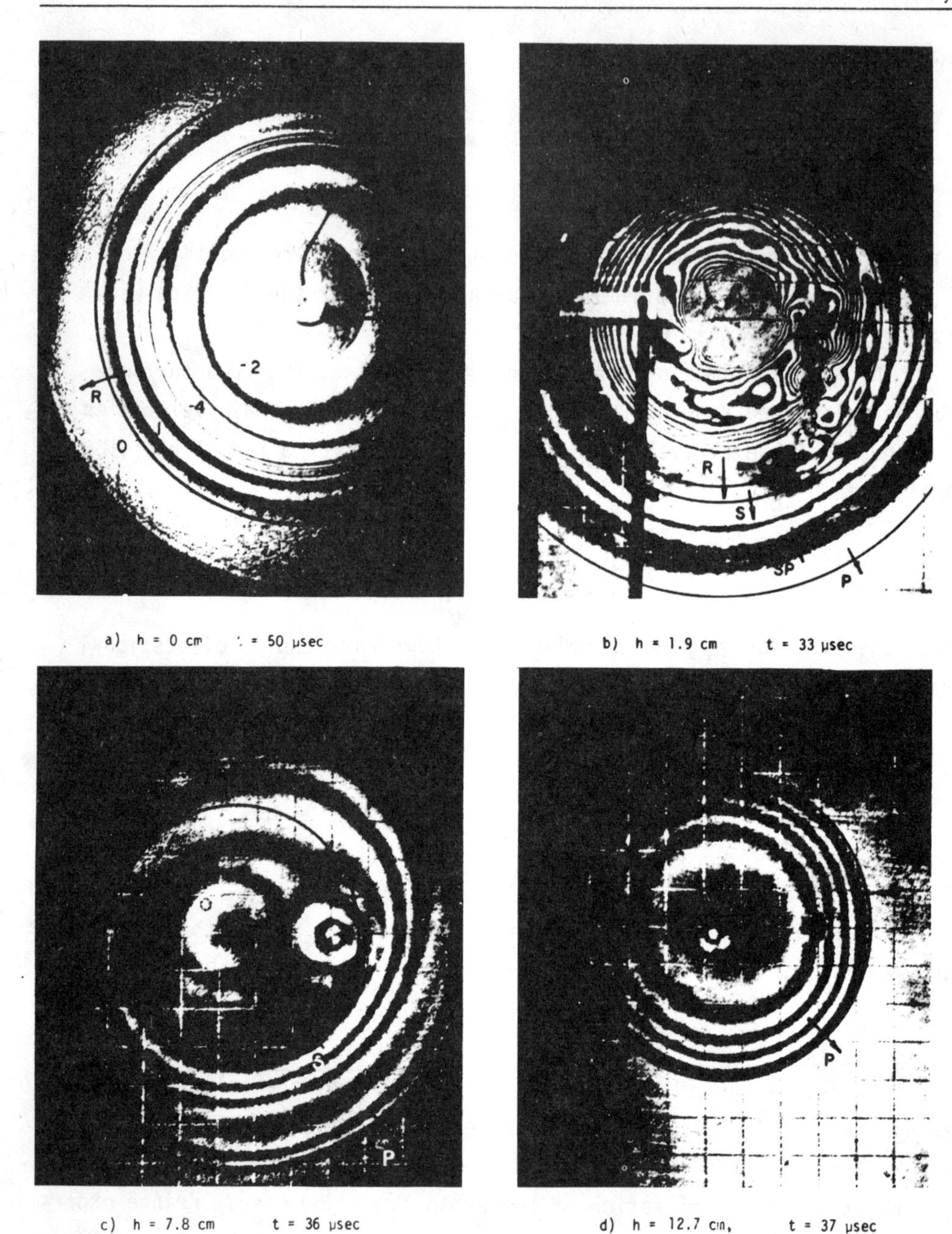

a) h = 0 cm t = 50 μsec

b) h = 1.9 cm t = 33 μsec

c) h = 7.8 cm t = 36 μsec

d) h = 12.7 cm, t = 37 μsec

Figure 2 Holographic fringe patterns of vertical surface motions in a rock half-space resulting from an explosive charge of 200 mg. PETN. All photographs are to the same scale with depths of burial and times indicated below the photographs.

S waves have attenuated to a level below the resolution of the holographic method. It is noted that the distribution is asymmetrical, with the larger activity occurring in the lower left quadrant. This can be attributed to the explosive being concentrated in a similar section of its holder. In all tests, however, the fringe pattern is asymmetrical due to a combination of varying elastic properties of the rock and the slight damage done to the rock from previous tests.

Figure 2b shows the case where the depth of burial is 1.9 cm (3/4 in). Here the influence of all the wave systems is clearly visible. The P wave has nearly reached the edges of the block. The position of the SP wave corresponds to an island in the fringe pattern signifying a relative minimum. The lead positions of the S and Rayleigh waves are in a region of low fringe density indicating a plateau in the displacement. Directly behind the Rayleigh wave is a cluster of fringes with a very high spatial gradient indicating a rapid decrease in displacement. There is then another region of relative minimum followed by an intense area of negative displacement directly above the epicenter. The latter displacement exceeds the resolution capabilities of the holographic method. An anomalous area in the rock appears in the lower right quadrant at about the 4 o'clock position. This particular region had been damaged by some previous testing with surface charges, and the effects of the small cracks sustained earlier are clearly visible.

Figure 2c illustrated the surface motion when the depth of burial is 7.8 cm (3 in). The P wave has reached a distance of 13.4 cm (5.3 in), and the S wave a distance of 6.7 cm (2.6 in). A previously damaged area located 5 cm (2.0 in) to the right of the epicenter is seen to act as a secondary source of disturbance. At this time, there is no evidence of a Rayleigh wave.

Figure 2d shows the fringe patterns for a burial depth of 12.7 cm (0.5 in) and a time of 37 μsec. For this case, the only wave system present is the P wave and its leading edge is located 8 cm (3.1 in) from the epicenter. The relative minimum is located at approximately 3 cm (1.2 in).

Analysis of these and other photographs yielded a P wave velocity of 4 km/sec (15,750 ips), a S wave velocity of 2.66 km/sec (10,472 ips), and a Rayleigh wave velocity of 2.38 km/sec (9,370 ips). These are average values since the elastic properties varied throughout the rock.

A displacement analysis for Figures 2a and 2b along the positive y axis was made using two additional top and side photographs of the fringe patterns. The results are presented in Figure 3a and 3b. For the surface blast in 3a, the displacements are those of a Rayleigh wave. The sign convention employed is consistent with the x, y, z coordinate system located above the source. A positive value of w is upward, a positive value of v is outward and a positive value of u is to the right for an observer facing in the direction of wave propagation. The motion is retrograde eliptical with the peak inward motion occurring at a time of zero upward motion. The radial displacement returns to zero much more rapidly than the vertical motion. The peak ratio of displacement amplitudes (w/v) was found to be 1.3 in comparison to a theoretical value of 1.35. The transverse displacement is a result of the asymmetrical energy distribution in the explosive charge.

In Figure 3b the surface displacements have been determined for the positive y axis of Figure 2b. The behavior is quite complex but the contributions of the various waves can be identified. The vertical motion in the P wave is initially downward with the radial motion being inward until the arrival of the SP wave. At this time, the radial displacement increases outward, and continues to do so through the arrival of the S and Rayleigh waves, however, never reaching its zero position. In the mean time, the vertical motion is now upward. Upon arrival of the Rayleigh wave, the motion of both the vertical and radial directions is like that of Figure 3a. The transverse component of motion (u) is felt to be a result of the asymmetrical explosive charge and an SP wave system producing transverse motion. The area directly over the source could not be analyzed due to a loss of resolution, but other tests conducted at a later time reveal that vertical displacement decays very slowly to its zero position. For this time and depth, the peak amplitudes occur during the Rayleigh wave phase. The trends described for the vertical and radial motion agree with those developed by Pekeris. By comparing Figures 3a and 3b, it can be seen that the effects of containment have approximately doubled the peak particle displacements.

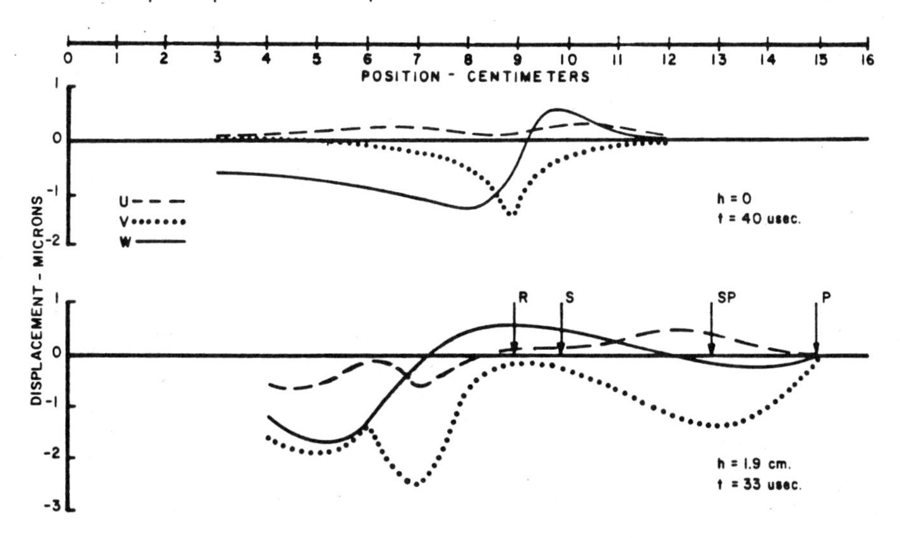

Figure 3 Surface displacements along the positive y-axis for two depths of burial. Source:200 mg. of PETN.

Figure 4 illustrates the ground motion at various times for a fixed depth of burial of 7.8 cm (3.1 in). The analysis line is a positive y axis, thus the u motion is transverse, the v motion radial and the w motion is vertical. The sign conventions used above also hold. The location of the wave systems have been marked on the figure. The transverse displacement is the smallest of the three and is consistently to the right in the leading portion of the P wave and then is to the left in the tail of the P wave. The radial component increases with increasing time and the vertical component decreases with increasing time, according to a 1/r relationship for the times of 31 and 40 μsec.

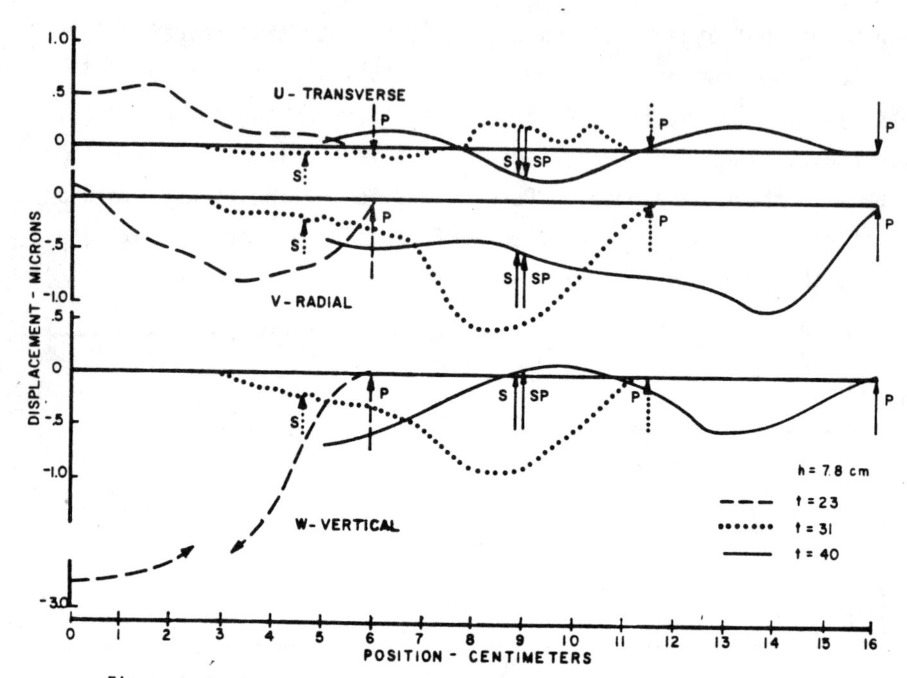

Figure 4 Surface displacements along the positive y-axis for various times at a fixed depth of burial equal to 7.8 cm. Times in μs after the detonation of 200 mg of PETN.

The increase in v and the decrease in w, and the depth of burial and distance r are related through (see Ewing, pg. 27)

$$2 \cos^2 e = \frac{C_1^{\,2}}{C_2^{\,2}} (1 - \sin \bar{e})$$

$$\bar{e} = \tan^{-1}(\frac{V}{W}) \tag{6}$$

Using this formula gave an e of 34° versus an actual value of 29° at 31 μsec and an e of 48° versus an actual value of 42° at 40 μsec. The limited sample size prevented longer times from being examined. In applying Equation 6 there are difficulties in evaluating v and w. For example, at 40 μsec it is seen that the peak v displacement is leading the peak w displacement, whereas at 31 μsec the two deformations appear to be in phase. Therefore the uncertainties in the application point of Equation 6 contribute to the differences.

CONCLUSIONS

These tests were conducted in pink westerly granite blocks 30.5 × 30.5 × 15.25 cm (12 × 12 × 6 in). The three orthogonal components of displacement were determined along selected radial lines at various times after the initiation of 200 mg of PETN. For the charge on the surface, the predominate effects are from the Rayleigh wave. For the buried tests when the radial distance r divided by the depth of burial h is < 1 ($r/h < 1$) the P wave predominates. Between $1 < r/h < 2.5$, the maximum effects from both the P wave and Rayleigh wave are about the same but occur at different times. For $r/h > 2.5$ the Rayleigh wave is the dominating system. There is not a strong contribution from the S or SP waves. The reason for this is the fact that rock adjacent to the borehole did not fracture extensively which thus produced a very weak shear wave system. In all tests the major components of displacement were radial and vertical.

The relative importance of each component varied with the wave system and observation point. That is, for the P wave, the vertical motion is most important close in, as is borne out by comparisons with theory. For the Rayleigh wave, both the vertical and radial components are about equal close in, with the emphasis going to the vertical motion at greater distances. The trends described above agree with the theory of Pekeris.

Overall, the method of holographic interferometry appears to be well suited to the study of ground vibrations from blasting. Here the concentration was on a problem with a known solution. Other problems have been examined where much less is theoretically known, and the results look promising.

REFERENCES

[1] Woods, R., "Screening of Elastic Surface Waves by Trenches", Ph.D. Thesis, Civil Engineering Department, University of Michigan, 1967.

[2] Dally, J.W. and Lewis, III, D., "Photoelastic Analysis of Propagation of Rayleigh Waves Past Step Shcnge in Elevation", Seisomological Society of America Bulletin, Vol. 53, No. 2, pp. 539-563, 1968).

[3] Gupta, In.N. and Kisslinger, C., "Model Study of Explosion-Generated Rayleigh Waves in a Half-Space", Bulletin of the Seismological Society of America, Vol. 54, No. 2, pp. 475-484, 1964.

[4] DeBremaecker, J.C., "Transmission and Reflection of Rayleigh Waves at Corners", Geophysics, Vol. 23, No. 2, pp. 253-266, 1958.

[5] Pekeris, C.L. and Lifson, H., "Motion of the Surface of a Uniform Elastic Half-Space Produced by a Buried Pulse", Journal of the Acoustical Society of America, Vol. 29, No. 11, pp. 1233, 1238, 1957.

[6] Ewing, W.M., Jordetzky, W.S. and Press, F., Elastic Waves in Layered Media, McGraw-Hill, New York, 1957.

Modelling and Development of Hydraulic Fracturing Technology

by Michael P. Cleary
Associate Professor of Mechanical Engineering
Massachusetts Institute of Technology
Cambridge, Mass. 01239

SUMMARY

The technology of underground fracturing has come to a central position in the areas of oil, gas, other minerals and heat extraction from the earth. In the Resource Extraction Laboratory at MIT, we are conducting a comprehensive theoretical and laboratory investigation of fracturing schemes which hold promise for providing or improving access to underground reserves of energy and other natural resources. Our main focus has been on quasi-static methods, particularly on ramifications of the central hydraulic fracturing technique (singled out in this paper) and on enhancing complements such as thermal cracking or induction of high pore pressures (e.g., due to expasion tendencies of highly energetic trapped second phases). Consideration is also being given to various controlled explosive methodologies (e.g., as preparation for hydrafrac) and to possibilities for constructively dispersing electrical or chemical energy (e.g., with pumped fluids), both for permeability induction and mapping.

1. Four major aspects of the area are outlined here: Identification of (e.g., mechanical, thermal, chemical) mechanisms which have potential for creating suitable fracture patterns in underground reservoirs or other target rock.

2. Detailed analytical and numerical modelling of fracture evolution incorporating each potentially important field phenomenon into simulators for use on pocket-calculators, micro-computers and mainframe machines.

3. Experimental verification of analytical predictions at all levels of sophistication, using both laboratory and field data.

4. Provision of insight and readily intelligible formulae (plus computer routines) for immediate application toward improvement of current operations in the field and development of new technology.

INTRODUCTION

The need to generate underground fractures with suitable extent, orientation and distribution is a problem well appreciated by the community of engineers concerned with extraction of energy and natural resources from pay zones which must be exploited under increasingly adverse conditions of inaccessibility, low transmissivity and tightening regulation. Research in this general area has mushroomed over the past decade especially, both among industrial concerns and in government sponsored laboratories; fewer university groups have been involved, perhaps because the problems were not considered to be sufficiently fundamental in nature. However, there is now a growing recognition that some very basic issues have to be resolved quickly if there is to be proper guidance for and understanding of expensive field trials which have yielded all too little insight in the past.

A complete survey of the literature and industrial activity in this area would probably be superfluous for most readers of this report, so we limit the discussion to a few brief remarks on foregoing work. Adequate amount of review may be found in our published papers (Cleary, 1978, 1979, 1980), but a cross-section of relevant papers may be found cited at the end. These reveal, among other things, the great variety of uses that have been found for the technique of hydraulically breaking down a formation by producing excess pressure in a borehole and then continuing to pump as the fracture propagates; other mechanical, thermal and chemical fracturing processes (many common in nature) are also beginning to find technological application in the resource extraction area. As well, it seems that, perhaps predictably, limited success has attached to the use of (chemical or nuclear) explosives for stimulation purposes; some attention has been paid to coupling of controlled detonations or deflagrations with hydrafracing. The primary foci of our work (in our project MIT UFRAC) are, indeed, these twin topics of quasi-static fracturing and suitable methods for creating regimes of dynamic crack growth, whenever these might enhance the success-ratio of stimulation efforts or other reservoir-access activities; we concentrate our discussion here on hydraulic fracturing, central to most practically viable techniques.

The literature on "hydrafracing" abounds with (often inconsistent) assumptions which yield operative models for design of field operations. Two primary groups of working idealisations have evolved among industrial groups (see figure I.1), the one (called PKN after its authors) assuming a vertically-closed laterally-similar elliptical crack opening (Perkins et al, 1961; Nordgren, 1972) while the other (called CGDD after its authors), also vertically bounded, allows enough slippage at interfaces that the crack is vertically-similar (e.g. Christianovich and Zheltov, 1955; Geertsma and De Klerk, 1969; Daneshy, 1973); obviously, neither quite fits the real situation where the crack may partially blunt -- or, especially, may even spread vertically appreciably further than assumed. The CGDD model is much more pessimistic about effective lateral length achieved for a given volume of fluid pumped; it may therefore give a better length, with reference to a model which more realistically allows vertical extension of the fracture; on the other hand, such a more realistic model (e.g. P3DH in Section 2B) is typically closer to the PKN assumption for crack-width! We have performed computations (e.g. based on formulae in Cleary, 1980) for a much broader group of assumptions about shape, formation and fluid parameters; we substantially conclude that predictions typically made for the extent of fractures generated by conventional operational procedures in the field are often too optimistic -- a conclusion that many of our industrial colleagues have apparently reached as a result of various correlation studies in the field.

Although one finds astute empirical realisation (e.g. Pugh et. al, 1978, White and Daniel 1981) of procedures which might enhance the tendency for containment of fracturing within the stratum of interest ("pay zone"), there seems to have been only limited recognition of the mechanics and materials features which govern this aspect. Indeed, some insights have been displayed in the identification of elasticity barriers (e.g. Simonson et al., 1978), in-situ stress contrast (e.g.

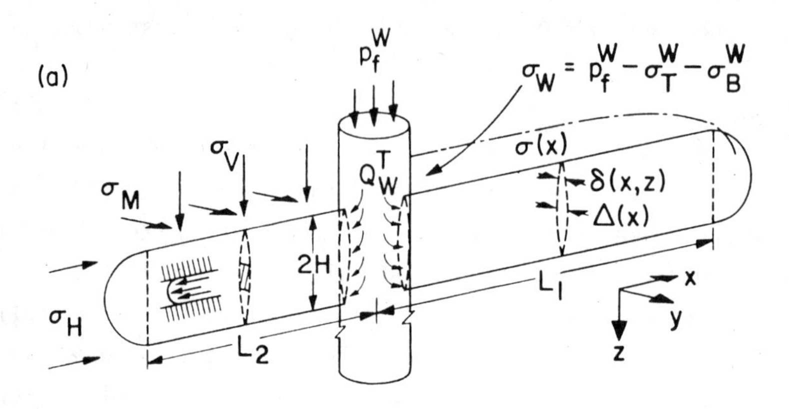

Fig. I.1(a). Schematic of conventional PKN geometry used in making sample calculations for field applications.

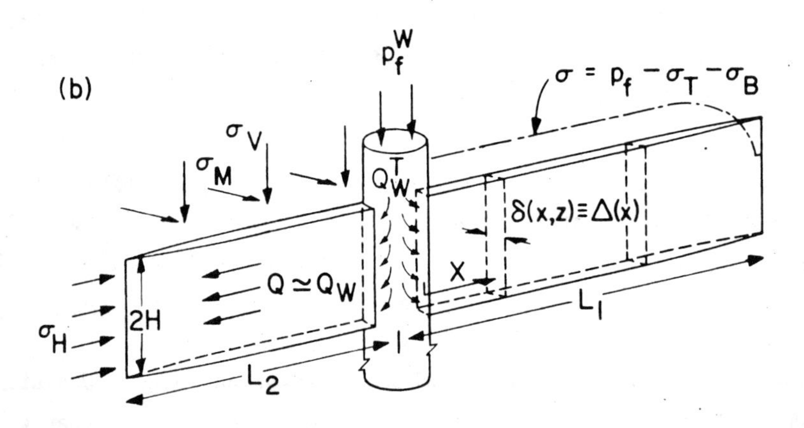

Fig. I.1(b). Demonstrates conventional implementation of CGDD model for computation of lateral extent in field operations.

Abou-Sayed et al. 1978) and other mechanisms (e.g. Daneshy, 1978) of retardation or acceleration of crack growth across interfaces between strata, but some confusion of the issues seems to prevail (e.g. Warpinski et al., 1980); some clarifications and more detailed consideration (Cleary, 1978, SPE 1980) shows that these aspirations are usually over-optimistic, unless some attempt is made to render the formation response more amenable to operation of such mechanisms. We have established a number of possibilities (Cleary, SPE 9260, 1980) and we have detailed various methods whereby this can be achieved (Cleary et al., 1981). It is perhaps interesting that these and numerous other deductions of ours (Cleary, 1980) are based on calculations that we might classify as "back-of-the-envelope"; one tends to miss such phenomena in large computer analyses (e.g. using finite elements), when the mechanisms involved are not model-intrinsic.

However, we do not advocate over-simplification of the fluid-solid structural interaction, which is at the heart of the hydrafrac methodology: indeed, we quickly felt the need for more precise quantification of our estimates for crack shapes and extent at successive increments of time after initiation. To achieve this detailed simulation, while preserving the correct character of discontinuities on crack surfaces and slippy interfaces, we gradually developed a preference for the effective and very insightful method of force and dipole distributions, from which there arises only the need to solve a set of surface integral equations (Cleary, 1978, SPE 9259, 1980) on each of the surface sub-domains which is active at that instant. The development of a general purpose computer scheme, based on this formulation, for simulation of fracture evolution under arbitrary driving stress distributions (e.g. internal hydraulic pressure), has become one of our major undertakings in MIT UFRAC.

That care is needed, to avoid over-simplification of relevant structural calculations, is also well illustrated by the interpretations that have been assigned to critical pressures measured in mini-frac efforts to deduce tectonic stress conditions (e.g. Haimson et al, 1970,

1975, Voegele and Jones, 1981): the problem here is that simple solutions (e.g. for a round hole in isotropically linear elastic material) tend to be employed, to the exclusion of important nonlinear, anisotropic and mechanistic effects. Considerable rationalisation has since been offered (e.g. Abou-Sayed et al, 1978; Zoback et al, 1978) and we have given a thorough discussion of the matter (Cleary, 1979). Simple laboratory experiments (Haimson et al, 1970; Medlin et al, 1976; Cleary, 1979; Morita et al., 1981) quickly illustrate the extent of deviation from the idealisation, and they emphasize the importance of mechanistic considerations.

We are now convinced that mini-frac has enormous potential both for initial breakdown of the well-bore (hence avoiding detrimental initially unstable growth of subsequent massive fractures) and for determination of reservoir parameters (especially in-situ stress and material deformation/transport properties, a potential hitherto unrealised); however, our lab work shows that great care is needed to achieve reliable interpretation of (pressure and flow) measurements made. In this respect, laboratory simulation is indispensable: it may not rigorously represent the field environment but any analytical model of worth should certainly be capable of predicting sample response under these controlled conditions. This feeling has prompted us to pursue the development of a laboratory facility which we can use to test the reliability of our theoretical calculations: our aim is to make this effort increasingly sophisticated, quite feasibly to the point where we eventually have both analytical and experimental analogues of reasonably realistic field circumstances. In beginning this work, we noted that previous lab arrangements did not really take proper account of scaling laws which must govern any reduced model of field operations: great attention was typically focused on use of feed-back control, stiff fluids and other artifacts to produce stable crack growth, whereas the field operation is innately stable. By using suitably viscous fluids, and modulus/toughness of the sample being fractured, properly scaled to the reduced length, and especially by recognizing the dominant stabilizing influence of the confining stress on the exterior, we have been able (Papadopoulos et al, 1980, 1981, 1982; Crichton, 1980) to generate innately stable quasi-static spreading of fractures easily observable visually or deductively: these may well be the first true laboratory simulations of the hydrafrac operation and we now plan to carry the capability through a series of stages -- until we have exploited its full potential. Thus, a second major part of our project (MIT UFRAC) is to design various test configurations and material micro-structures, which will provide invaluable insight into the character of fracture evolution within increasingly complex (e.g. porous, stratified, naturally fractured) structural formations of the kind produced by geological processes.

Conjugate to all this analysis and experimentation has been the desire to evaluate what response is actually elicited in the real-life circumstances of field operations, which are being performed everyday and could serve as the best test of all. Toward this, numerous monitoring methodologies have been proposed (Evans et al, 1980; Keck et al, 1978; Pierce et al, 1974; Power et al, 1975; Greenfield et al, 1977; Komar et al, 1973 Dobkins, 1979, Lester et al. 1979) some have proven out to some extent in the field (Smith et al., 1978) and we have been seeking alternative more comprehensive schemes. Related to this is our desire to work with opaque porous specimens in the laboratory, so that we need a method for delineating the crack perimeter within the sample. Ultrasonic scanning tools were already available but we have also explored the potential of various electromechanical, thermal and deformation field mapping techniques. Meanwhile, we have developed some effective but lab-specific marker techniques for delineating successive perimeters from induced surface markings, corresponding to timed pressure pulses.

A last group of problems--which seem to have received little attention, but which are central to our objectives at MIT, and have foremost importance for a great variety of resource extraction endeavors, are those of sustaining multiple crack propagation and achievement of linkage, in order to generate underground fracture networks. Our early experiments (Crichton, 1980) have shown us that it is quite feasible to make fractures from adjacent boreholes link up, by picking the sequence of borehole pumping in a reasonably careful fashion -- so that the interacting stress fields always favour curving toward each other, rather than away. The broader fundamental relevance of experimentation on this phenomenon, from the viewpoint of three-dimensional crack evolution in solid materials at all size-scales, should also be emphasised. On the other hand, the generation and sustenance of multiple cracks from a single bore-hole, especially in a biased tectonic field, seems also to be in the realm of practical achievability and we are studying a number of possible artifacts to do this, even quasi-statically: the most promising approach may be a rapid (or laboratory jacketed) pressurisation technique for initiation, followed by successive selective pumping and propping of each fracture thereby created. Multiple crack initiation can certainly be achieved by suitable explosive pressure rise times (e.g. Young, 1978, Schmidt et al., 1979) but, even if the obvious drawbacks of explosives are overcome, this must be followed by conventional hydrafrac; we will refer to this combination as the "hybrid fracturing technique", and note that it may alone (among dynamic techniques) have potential for reservoir stimulation. Assistance for such multiple cracking can also be had from the induction of thermal stresses e.g. created by steam injection -- or by cooling near the borehole on the main fracture surface, phenomena which we have recently examined in some depth (Barr, 1980).

These and many other mechanisms are being studied further (in MIT UFRAC) However, our presentation here will concentrate on just one major technique, hydraulic fracturing, which seems to be central to most reasonable plans for (enhanced) resource extraction. As well, we will limit ourselves to some basic models, which adequately illustrate the principal features of the process and of our other more complex simulations; these allow clear verification in the laboratory and subsequent confident application to design of field operations.

1. POTENTIAL AND STATUS OF FRACTURING TECHNOLOGY

1A. Mechanisms of fracture creation in rock

There are many ways to create fractures, above and below ground. A short list of examples is provided here.

1). Mechanical application of stress to the rock.

a. Direct contact with solid machine elements. This approach is typical of conventional excavation and rock breakage equipment (e.g. drill-bits, indentation or drag cutters, even water-jets and the common pick-axe); it has seen improvement, mainly in recognition of rock's low tensile (vs. compressive) strength, but most devices still crush rock.

b. Use of fluid pressure to exploit the tensile weakness. This approach has many advantages and much unexplored potential; it really should be divided into four main types of process:

i) Hydraulic fracturing by external injection of fluid into a propagating crack, one already existing or one created for the purpose. This technique has seen extensive use by industry.

ii) Pore-pressure-induced cracking due to creation of excess internal pressure p in the rock pores, exploiting the effective stress law so that exceeds the tensile strength of the material, causing a crack to propagate under the pressure of fluid inflowing from the pores. The fascinating patterns of this process may be classified as hydrafracs but we distinguish them by the source of the fluid driving the cracks.

iii) Fluid impact and penetration into surface cracks, e.g. in water-jets which have been used both for direct cutting and assist of mechanical indenters.

iv) Explosive generation of high fluid pressures in a confined cavity (e.g. borehole), which induce transient tensile stresses high enough to open cracks in the rock as stress-waves pass by. The main drawbacks of this methodology, apart from complexity of analysis, are the (delaying) need for great safety precautions during the operation and the re-closure of the cracks (under the ever-present confining earth stresses) when the dynamic event has passed; indeed, there are often deleterious "compression cages" which can wipe out permeability in critical regions (e.g. around a borehole). It is surprising that many field trials, based on calculations which predict generated/reflected high tension waves, have been carried through to inevitable fruitlessness: massive nuclear detonations under-ground have apparently lost their purely superficial appeal but misconceptions seem to linger. A classic example is the concept of slurried explosive detonation all along a hydrafrac surface; this does have potential for self-propping of rough fracture surfaces, but may not be effective for the extensive spalling often suggested.

Still, the use of (controlled explosive?) dynamic fracturing may play a helpful role in future stimulation technology, by hybridisation with the preceding hydraulic fracturing, which will follow the initial dynamic breakdown process if the pressure is sustained: thus, for example, it may serve as a useful means of breaking down the borehole, without employing a long over-loaded column of fluid (which is detrimental to containment).

2). <u>Induction of Thermal stresses by temperature alteration</u>

Since reservoirs are hot, there is always the possibility of exploiting the tensile stresses induced by cooling around cavities (e.g. wellbores); most simply, this requires the injection of cold fluid -- either into the body of pore-space or along a hydraulic fracture. Cooling of the primary fracture walls may produce enough secondary cracking (e.g. Barr, 1980) and overcome "scum" effects but overall volume cooling produces the best large-scale crack pattern. However, even for such a thermal process, two main wings will dominate (Section 3C). This bulk cooling does provide an effective means for initiating a hydrafrac without excess pressure but its main advantage may be the enhanced containment that it provides, through a reduction in reservoir confining stresses (Cleary, SPE 9260, 1980).

Injection of hot fluid or heat is also common, e.g. in steam drives and fire flooding for heavy oil extraction. This produces compression in the heated region but, for high enough temperatures, it may drive ahead of it an array of cracks which never fully close after the front sweeps by them: this mechanism may well explain the apparent success of recent field trials with a (downhole) rocket motor propelling superheated steam into some closely spaced reservoir wellbores -- which even gave rise to production enhancement in nearby wellbores and dramatic improvement all-round; many analogous schemes suggest themselves. Another major aspect of thermal processing is the phase-transformation and potentially very high pore-pressure, which can be induced by heat supply, thereby leading to the PPIC of Section 1A.1 ii) for instance. An example of this is conversion of oil shale kerogen to various gaseous and liquid hydrocarbons, which we are studying in the laboratory.

3). Chemical alteration of microstructure and induced stresses.

The industry is quite expert at finding new chemical treatments (e.g. acidisation, Williams, Gidley & Schechter, 1979) to clean up wellbores and enhance permeability: well-known chemical reactions of fluids with rock minerals govern this process. Less attention has been paid to two other aspects: the feasibility for the chemical treatment to drastically change the stress state in the reservoir (hence effects as in Cleary, SPE 9260, 1980), and the enormous potential for induction of high pore-pressures, e.g. great enough to maintain reaction products in the condensed state at reservoir temperatures (having the PPIC effect in Sec. 1A.1 ii)).

4). Other chemophysical processes and technology

Many other techniques have been suggested for cracking rock, varying from laser beams (which might sometime be considered for pre-notching the walls of a borehole) to flow of electricity (which has little effect except due to resistive heating) to radio-wave propagation (which can produce dielectric heating, hence 1A.2); these may merit more attention.

Among these various processes, many have played a role both in nature and in human technology: over-pressured reservoirs have become naturally fractured, thermal and shrinkage cracks are everywhere, and magmatic intrusions (dykes and sills) are fascinating examples of the earliest hydrafracs, fortunately usually stabilised by the cooling of molten rock as it approaches earth's surface (Delaney and Pollard, 1981).

1B. General Equations Governing Hydraulic Fracturing.

With reference to figure 1B.1 and the notation section at the end, the most general form of the equations to be satisfied by any hydrafrac model may be phrased as follows:

1). Mass conservation for each fluid component in the fracture

Using a superscript i for the i-th phase, we may write

$$\underset{\sim}{\nabla}^{s}\cdot(\rho\underset{\sim}{q}\delta)^{i} + \frac{\partial}{\partial t}(\rho\delta)^{i} + (\rho q_{L})^{i} = s^{i}, \quad q_{L} = \underset{\sim}{q}\cdot\underset{\sim}{n} \tag{1B.1}$$

where the sources s^i arise from changes of phase (i.e. saturation levels) between components of the frac-fluid. Clearly, the flow velocities $\underset{\sim}{q}^{i}$ depend dominantly on the crack opening $\delta = \underset{\sim}{\delta}\cdot\underset{\sim}{n}$, so we need to study the displacement discontinuity δ as follows.

2). Equation for opening of the fracture under excess pressure

For a fracture surface with normal $\underset{\sim}{n}$, tractions change by

$$\underset{\sim}{n}\cdot\underset{\sim}{\sigma}_{f} - \underset{\sim}{n}\cdot\underset{\sim}{\sigma}_{c} \equiv \underset{\sim}{n}(\underset{\sim}{x}_{o})\cdot\underset{\sim}{\sigma}(\underset{\sim}{x}_{o}, t), \quad \underset{\sim}{\sigma}_{c} \equiv \underset{\sim}{\sigma}_{T} + \underset{\sim}{\sigma}_{B} \tag{1B.2a}$$

due to the opening of the fracture and fluid flow, as given by

$$\underset{\sim}{\sigma}, \underset{\sim}{\sigma}_{B} = \int_{-\infty}^{t} d\tau \int_{S_c(\tau)} dS \left[\underset{\sim}{\Gamma}_{S}^{D}(\underset{\sim}{x}_{o}, \underset{\sim}{x}; t, \tau)\cdot\underset{\sim}{\nabla}^{s}[\underset{\sim}{n}\underset{\sim}{\delta}(\underset{\sim}{x},\tau)], \quad \underset{\sim}{\Gamma}_{S}^{F}(\underset{\sim}{x}_{o}, \underset{\sim}{x}; t, \tau)\cdot\underset{\sim}{n}q_{L}(\underset{\sim}{x},\tau) \right] \tag{1B.2b}$$

We have written the equation in a singular integral form, which is the most efficient way to solve for $\underset{\sim}{\delta}$ and the one we use in all our numerical schemes -- although we are studying hybridisation with various volume discretisation methods for describing severe heterogeneity/nonlinearity of material response. As well, we note that eqn. (1B.2) gives the change in stress $\underset{\sim}{\sigma}$ at any point in the region due to the discontinuity created by the fracture, once we know the basic

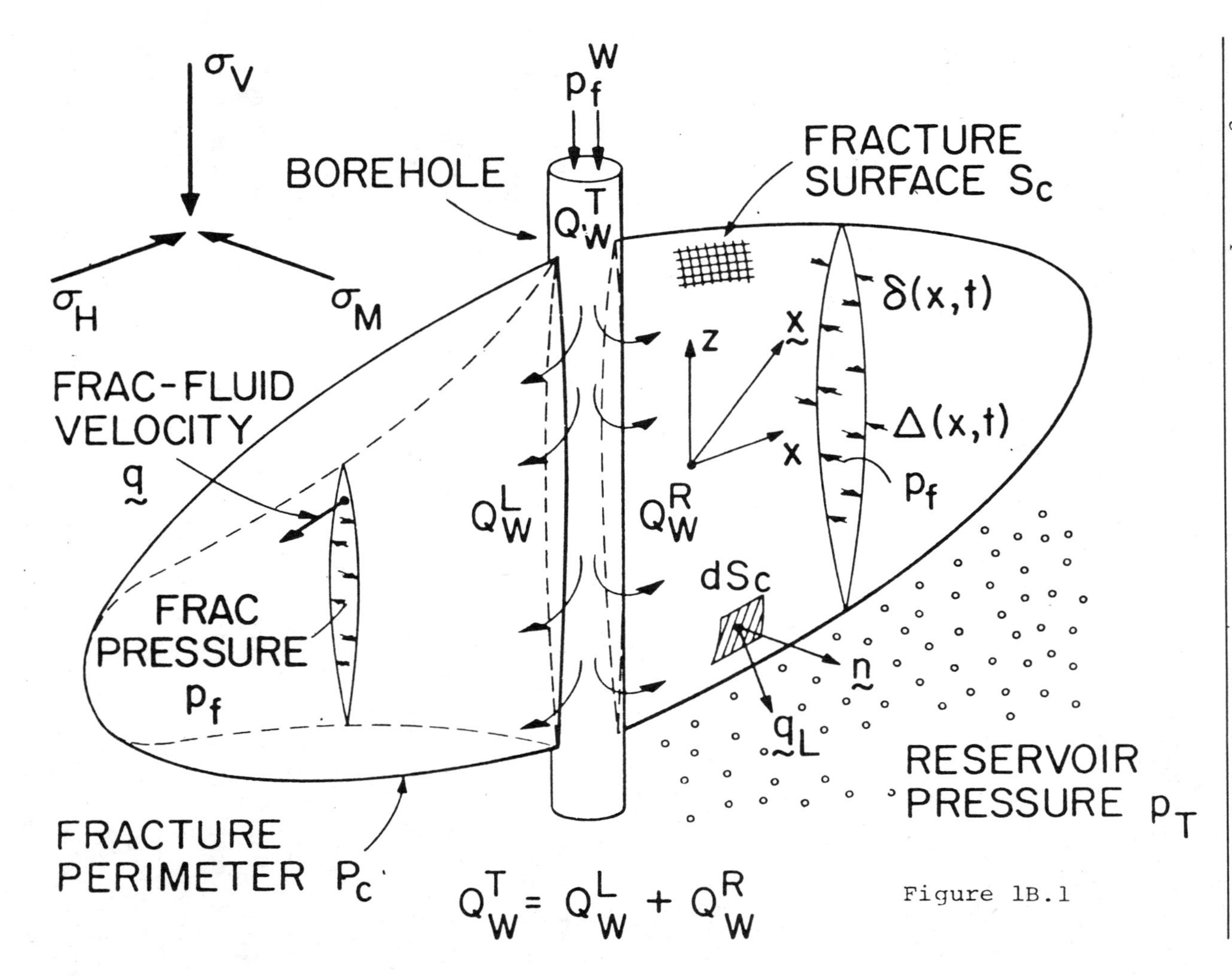
σ_V
p_f^W
BOREHOLE
Q_W^T
FRACTURE SURFACE Sc
σ_H
σ_M
δ(x,t)
z
$\tilde{x}$
x
Δ(x,t)
p_f
FRAC-FLUID VELOCITY
$\tilde{q}$
Q_W^L
Q_W^R
dSc
FRAC PRESSURE
p_f
$\tilde{n}$
$\tilde{q}_L$
RESERVOIR PRESSURE p_T
FRACTURE PERIMETER P_c
$Q_W^T = Q_W^L + Q_W^R$

Figure 1B.1

influence function $\underset{\sim}{\Gamma}$ for the region in question. This change in stress must be known or simultaneously determined on the surface of the fracture, hence the need to incorporate frac-fluid rheology (dictating fluid stress $\underset{\sim}{\sigma}_f$) and fluid loss to formation (q_L in eqn. (1B.1)); the latter affects confining stress $\underset{\sim}{\sigma}_c$ through the back-stress $\underset{\sim}{\sigma}_B$, which is superimposed on tectonic stress $\underset{\sim}{\sigma}_T$.

Fracture propagation is now determined by a condition such as energy balance at all points on the perimeter of the fracture; the latter can be translated to a critical stress intensity factor or fracture toughness, eqn. (2c) of Cleary, SPE 9259, 1980. An interesting result, proved in eqn. (7b) of SPE 9259, is that the energy required to break rock is typically negligible by comparison to that needed to open the crack by amount δ large enough to take the fluid and proppant pumped during large hydrafrac jobs. This has at least three important repercussions: the models can often employ zero toughness without appreciable loss of accuracy (hence reducing the amount of parametric study needed); the complex measurement of fracture toughness (e.g. Schmidt, 1975, Abou-Sayed, 1977), from expensive protected core, fortunately is often unnecessary; and we can design experiments in the laboratory which do not require any breaking or throwaway of specimens used, hence allowing facile repeatable verification of models (Sec. 4A).

3). Rheology of fracturing fluids in channel flow.

The fluids used in fracture treatments (gels, suspensions, foams etc.) can be quite complex in their flow behavior and the stress $\underset{\sim}{\sigma}_f$ can display the character of a viscoelastic material; we are studying the dominant characteristics of this history dependent behavior and finding efficient ways to describe the effects on fracture evolution, clean-up and reservoir production but the details will require much work. Meanwhile, we are working with a broad class of history-independent fluid behavior, encompassing all existing industrial frac-fluid models; this class is described by the form

$$R(q/\delta)\ \underset{\sim}{q}/q = -\ \delta^{U}\ \underset{\sim}{\nabla} p_f/\overline{n} \tag{1B.3a}$$

where R (e.g. for laminar flow) is the rheology function relating shear stress to shear-strain rate in a rheometric test, and the effective viscosity $\bar{\eta}$ incorporates a channel flow factor, arising from integration of the constitutive equation (in R) to get the shear-stress at the walls (which balances the gradient ∇p_f). A particularly useful specialization of R is the class of power-law relations, which give the channel-flow relation and a further specialisation to laminar flow gives

$$R(q/\delta) = (q/\delta)^m, \; U = 2n-2m+1. \tag{1B.3b}$$

$$\bar{\eta} \equiv 2K'(4+2/n)^n \text{ if } m = n, \tag{1B.3c}$$

where K' is consistency and n is the fluid power.

4). Fluid exchange with the formation.

The exchange of fluid with the reservoir, by loss of frac-fluid into the rock pores and by the important influx of porefluid into the low-pressure region near the perimeter, generally requires a reservoir flow simulation by volume discretisation; such simulators do exist (apparently all based on a fixed grid of nodal points) and most companies in industry will have access to one or more of these, with which our fracture geometry prediction schemes may be interfaced. However, we will be studying this aspect from two fresh points of view:

i) The amenability to volume discretisation with a mesh which moves along with the fracture shape and also takes advantage of the natural flow patterns around the fracture (streamlines, curvilinear coordinates, conformal mapping etc.); this should lead to great increases of efficiency in the simulation, both during stimulation and afterwards (in clean-up and production).

ii) The use of (singular) surface integral equations to describe the fluid exchange. A typical form of these would be

$$p_f - p_T = \int_{-\infty}^{t} d\tau \int_{S_c(\tau)} dS \left[\underset{\sim}{\Gamma}{}^F_p(\underset{\sim}{x}_o, \underset{\sim}{x}; t, \tau) . \underset{\sim}{n} q_L(\underset{\sim}{x}, \tau) + \underset{\sim}{\Gamma}{}^D_p(\underset{\sim}{x}_o, \underset{\sim}{x}; t, \tau) . \underset{\sim}{\nabla}[\underset{\approx}{n}\underset{\sim}{\delta}(\underset{\sim}{x}, \tau)] \right] \qquad (1B.4)$$

in which $\underset{\sim}{\Gamma}{}^F_p$ is the influence function for pore-pressure at $\underset{\sim}{x}_o$ due to a fluid source (or dipole) at x, and p_T is the reservoir pressure in the absence of flow q_L. We have included a stress induced pore-pressure term Γ^D due to crack opening $\underset{\sim}{\delta}$. Although normally restricted to linear and homogeneous reservoir conditions (on flowing fluids and permeability), eqn. (1B.4) can be used with adequate accuracy to describe the penetration of a frac-fluid (typically for small distances) into rock pores near the fracture surface, and will thus serve adequately during the fracturing treatment, after which i) or a conventional simulator may be used (if reservoir conditions are not sufficiently linear and homogeneous to employ (1B.4)). Again, hybridisation with reservoir volume discretisation (e.g. finite differences or finite elements) schemes may be employed, e.g. to capture heterogeneous and nonlinear effects.

5). <u>Heat transfer between fracture and reservoir.</u>

It is vitally important to incorporate the heat flow during a fracture treatment, e.g. because cold frac-fluid, heated up, can undergo severe changes in its rheology, and perhaps because secondary thermal cracking may be induced (e.g. Barr, 1980). Again various reservoir simulators are available to do this (based on a fixed mesh as in 4.) and our efforts are mainly an examination of i) moving grid dictated by evolving fracture shapes and ii) use of surface integral equations, exactly as in eqn. (1B.4), with p_f-p replaced by temperature $\theta_f - \theta$ and q_L replaced by "heat-loss" $h_L \equiv \underset{\sim}{h} . \underset{\sim}{n}$, namely

$$\theta_f - \theta = \int_0^t d\tau \int_{S_c(\tau)} dS \left[\underset{\sim}{\Gamma}_T^H (\underset{\sim}{x}_0, \underset{\sim}{x}; t, \tau) . \underset{\sim}{n} h_L (\underset{\sim}{x}, \tau) + \underset{\sim}{\Gamma}_T^Q (\underset{\sim}{x}_0, \underset{\sim}{x}; t, \tau; \theta) . \underset{\sim}{n} q_L (\underset{\sim}{x}, \tau) \right] \qquad (1B.5)$$

We note the very important convection term, based on a influence function $\underset{\sim}{\Gamma}_T^Q$ which we are presently developing, first for 2-D cross-sections. However, we should emphasize that a fairly simple one-dimensional conduction/convection (and surface transfer) calculation may prove adequate for many practical purposes.

6). <u>Transport and Deposition of Proppant by the Frac-Fluids.</u>

It is obviously essential to leave some propping agent in the fracture after excess pressures, which cause the opening and propagation, are removed to produce the reservoir. This proppant must be carried in by the fracturing fluids and may be regarded as part of these, simply contributing to their overall rheology. However, the proppant does segregate, due to differential pull of gravity and drag against fracture walls etc., and thus must be modelled as a separate phase in the fluid, governed by eqn. (1B.1) in particular. The simplest description (e.g. Daneshy, 1978) is to regard the proppant as moving along with the fluid laterally, so that a simultaneous solution of eqns. (1B.1-4) gives its lateral distribution directly, in terms of mass content per cross-section: then, simultaneously the mobile density ρ is allowed to change as the proppant settles onto an immobile sand-bed, which also affects the area of cross-section available for flow. The latter process is described by something like a Stokesian settlement relation, and extensive work has been done on the investigation of appropriate formulae for this (e.g. Clark and Quadir, 1981); the major novelty in our work may be in the incorporation of proppant as an impedance to vertical growth, in fracture closure by propagation (e.g. after shut-in) and in more general models of 2-D flow over a fracture surface.

1C. Models of fracture processes, existing/potential technology.

I. Modelling. The equations in Section 1B are presently being implemented in various special (approximate) forms by interested groups in industry, government laboratories and academia. Very few of these models (e.g. Settari and Cleary, 1982) actually solve all six equations together but, more importantly, most of them make assumptions which are valid only in special field circumstances and then give (at best) approximate estimates of the actual results. Foremost among these assumptions are the following:

i) The geometry of the fracture is prespecified; typically, height is assumed (or computed by consideration of some equilibrium height, without accounting for time needed to reach that height) and some characteristics of the shape are imposed, such as CGD versus PKN (Cleary, 1980, SPE 9259, fig. I.1).

ii) The storage of fluid in the fracture is poorly accounted. The flow rate is often assumed to be constant, which effectively means that storage term and loss term in eqn. (1B.1) are being neglected in the computation of pressure distribution (from eqn. (1B.3a)), although (1B.1) is still satisfied in a rough overall way for the whole fracture; otherwise storage is computed by some approximate/iterative technique.

iii) The opening profile of the fracture is assumed or estimated, using some approximate form of the pressure distribution (e.g. constant, or one of a class satisfying the critical conditions at the crack tip), which is not usually that consistent with flow in the fracture; as well, approximations to the opening equations (eqn. (1B.2)) are typically used (e.g. valid only for homogeneous isotropic reservoir).

iv) Fluid loss to the formation (Sec. 1B.4) is computed from some simplified model, such as one-dimensional flow normal to the fracture and approximate time integration with increasing area.

v) The back-stresses induced by changes of reservoir pressure and temperature (e.g. σ_B in eqn. (1B.2)) are neglected.

vi) Proppant segregation from fluid is assigned to gravity only.

vii) The fracture is not allowed to propagate during shut-in or flowback; the models thus require frac-fluid leak-off (and a longer shut-in time) for closure on proppant, which may physically be achieved quickly/effectively by fracture extension (hence width reduction at constant volume), with low loss fluids.

The other extreme would be a model which makes no such assumptions, solving exactly at each time step all of the equations in Section 1B (or even generalisations of these, for instance accounting for rock inelasticity in eqn. (1B.2) and history dependence of fluids in eqn. (1B.3)). Aspirations to such a model may be considered unwise on at least two counts: 1) it would be practically unnecessary and 2) it would be excessively cumbersome to run and interpret. [Indeed, worse may be said about models, including some of our own (Secs. 3A, 3B), which take extraordinary care to solve some of the equations, indeed only special cases of these (e.g. eqns. (1B.1, 2, 3)), while neglecting other first-order effects.] It is becoming clear (e.g. Sections 2, 3) that only judicious approximations will allow such general theoretical models to be employed toward practical design. While this may well be true of theory, we are now becoming hopeful that laboratory models (Section 4) will redeem the situation, a recourse which has surely typified development of most complex technologies over human history.

Actually, as well, it is possible to develop tractable theoretical models (Sections 2,3E), which remove many of the assumptions above and give a more realistic picture of the fracture evolution, without getting bogged down in forbidding numerical details and prohibitive computer expense. These are based on approximations, verified by more complex numerical simulation, which do not lose any of the character and also effectively retain the precision of the more general schemes (Sec. 3).

II. Field technology. The existing scope of field experience and techniques is quite vast and diverse; a review here would be impossible but the many recent overviews should be mentioned (e.g. Abou-Sayed et al., 1981, White and Daniel, 1981, Allen and Roberts, 1978). [Any novice can just riffle through Journal of Petroleum Technology, especially over the past dozen years since Howard and Fast, 1970;

and service companies provide extensive information in handbooks]. Primary attention has been paid to developing suitable fracturing fluids, with particular emphasis on damage to the formation, lossiness to reservoir rock, viscosity to achieve sufficient widths and ease of flow-back/clean-up; expense and toxicity to humans/materials are obvious constraints. Equally important has been work on proppants suitable for the rocks and depths: cheaper (but carefully screened) sand had served well until depths greater than 10,000 ft. became commonplace, requiring expensive particles (e.g. sintered bauxite) and other concerns such as embedment (e.g. Cooke, 1977). [It is possible that cheaper proppants will be developed as a result of research but another approach is to use channels or rock spall particles: these may be produced by suitable sequencing of frac-fluids and pressures -- e.g. by acidisation and/or pore-pressure-induced cracking].

With all this goes the very impressive field equipment Fig. 1C.1 and Journal of Petroleum Technology, August 1981, p. 1416), now often capable of pumping of order 30 barrels/min. for tens of hours at depths approaching 30,000 ft. Fracturing jobs have typically been conducted at a fixed flow rate Q, but staging is now occasionally being adopted to better control the pressures; how this should be done depends, of course, on the character of the growing fracture (e.g. increasing vs. decreasing Q are required to hold pressure constant, for the extremes of CGD vs. PKN). Decisions will eventually require something like the models of Section 2 to be run in real time at the wellhead, during a job. Indeed, there are a broad range of recommendations that follow from such modelling (a la Cleary SPE 9260, 1980), especially prior breakdown/treatment of the wellbore/reservoir and continuous monitoring of the fracture's progress during the operation; a little of this is coming into vogue (e.g. limited use of prior mini-frac and tiltmeter monitoring) but much research is needed.

Fig. 1C.1. Photograph of typical massive hydrafrac operation.

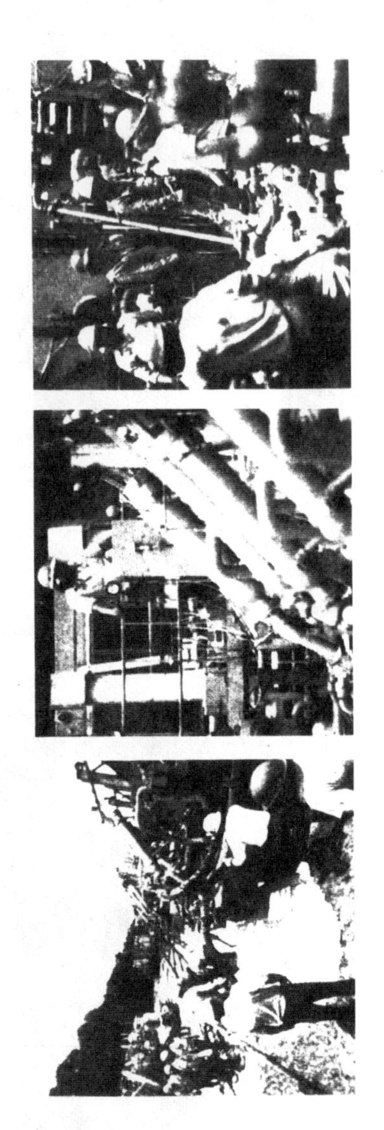

Fig. 1C.1. Photograph of typical massive hydrafrac operation.

2. FIRST-ORDER MODELS AND DESIGN OF HYDRAULIC FRACTURES

2A.1. Equations governing lumped P3DH-type models

With reference to fig. 2A.1, we assume that growth in one direction (say height H) is governed by a CGD-type model, while the other (say length L) is described by a PKN-type model; this is a reasonable description if, as we wish, the fracture is well contained but these interpretations, assumed henceforth, can be reversed (i.e. L interchanged with H) if the fracture is very poorly contained (H > L). The model seems to work quite well also in the middle ground of relatively equiaxed geometries (L ≃ H) but a better description (taking account of limited flow through perforations) may be circular growth (Cleary and Wong, 1982), which captures the higher pressures needed for flow.

Based on these fairly unrestrictive geometric assumptions, we have been able to derive some very simple formulae for the growth of H and L in time. Without any loss in generality for the moment, the height-growth expression can be written as

$$\dot{H} = \overline{\gamma}_2 H/\tau_c, \qquad (2A.1)$$

which may be applied to either wing (upper H_U or lower H_L, fig. 2A.1). Here the characteristic time τ_c has been obtained by a combination and non-dimensionalisation of the governing equations (3A-3C) for a hydraulic fracture (e.g. Cleary, 1980), leading to

$$\tau_c^m \equiv \overline{n}/(\overline{E}\hat{\sigma}^{2n+2-m} H^{2n-2m}) \qquad (2A.2a)$$

in which the elastic crack opening relation has appeared through

$$\Delta_E/H = \gamma_1\hat{\sigma} \equiv \gamma_1\sigma/\overline{E}, \; \sigma \equiv p_f-\sigma_c \qquad (2A.2b)$$

(Obviously, any power-law relation can also be used for crack-opening, still leading to formulae of the same type; we specialize for simplicity). Here the coefficient γ_1 is obtained by solving for crack

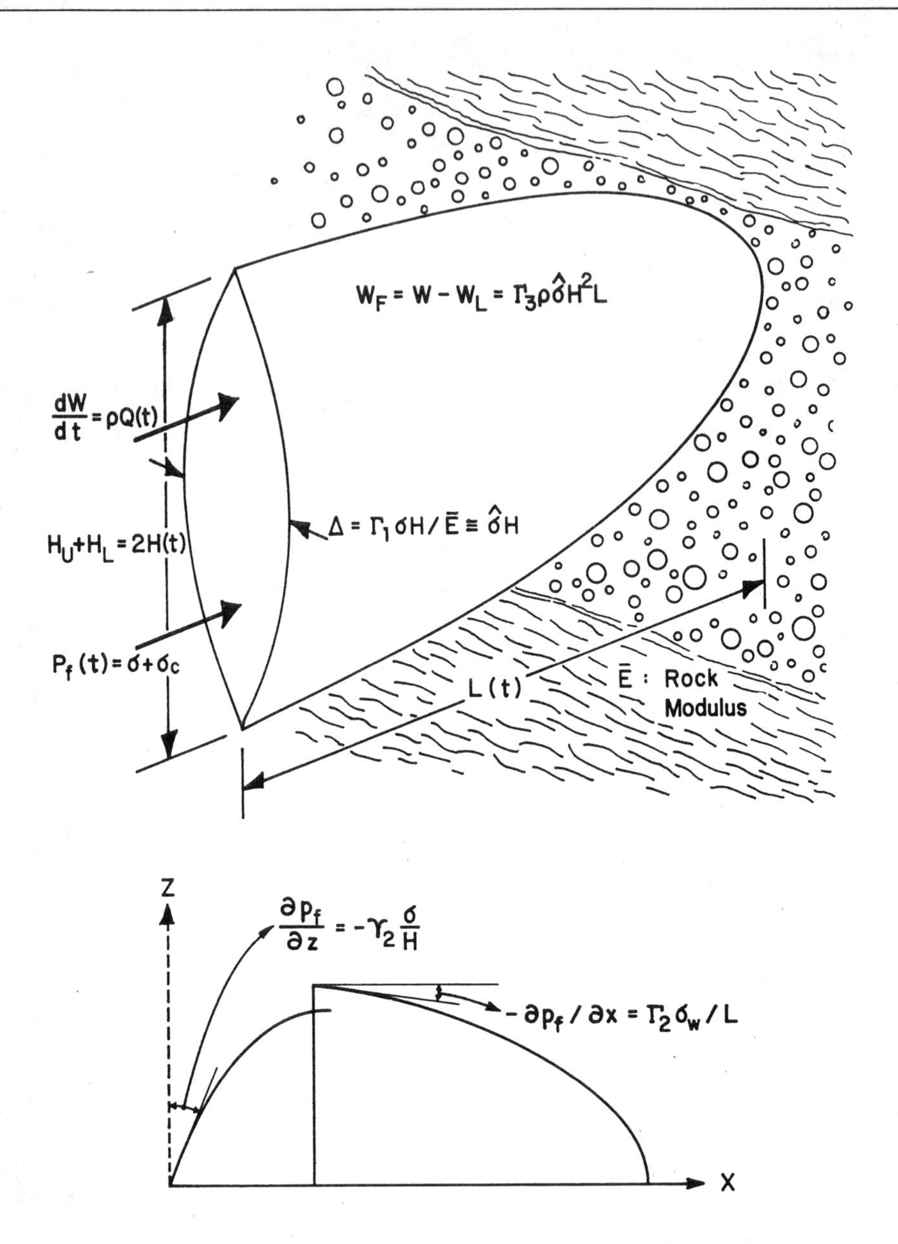

Fig. 2A.1. Basic geometry and parameters of lumped hydrafrac models.

opening (e.g. Cleary et al., 1981) with the relevant distribution of excess pressure $\sigma = p_f - \sigma_c$ in the fracture: clearly, it will vary somewhat in time (e.g. from $2/\pi$ to 1 in going from L=H to L>>H) but so will most other coefficients, especially γ_2 (discussed later, see figure 2A.1b).

Actually, the form in eqn. (2A.1) can be written to describe the growth of any fracture dimension, including the length. However, it is more appropriate to conduct a specific analysis for lateral flow, assuming only that width depends dominantly on height H (as in eqn. (2A.2b)), as detailed in Sec. 2B; the result is an important variation on eqn.(2A.1):

$$\dot{L}^m = \overline{\Gamma}_2^m H^{m+1}/L\tau_c^m \qquad (2A.3)$$

in which H/L clearly modifies the propagation rate. Eqns. (2A.1,3) are the most general forms of the equations governing CGD-type and PKN-type models, respectively: all other forms and solutions can be derived from them, and they are also remarkably simple (even by comparison to schemes presently used in the industry). Particular forms are given in SPE 9260 (Cleary, 1980).

Implementation of eqn. (2A.1,3) requires a knowledge of the coefficients $\overline{\gamma}_2$ and $\overline{\Gamma}_2$. These may be obtained numerically, e.g. as described in Sec. 2B for $\overline{\Gamma}_2$ and by quite accurate self-similar approximations for $\overline{\gamma}_2$ (Settari and Cleary, 1982). However, it is worth having expressions which relate those to another pair of physically transparent parameters, namely the slopes of the pressure distribution at the well-bore (x=0) and center of the fracture (z=0), Γ_2 and γ_2 respectively (fig. 2A.1b). Straightforward implementation of mass conservation for the whole fracture (e.g. using the equations in Section 2B) leads to an identity, which follows directly by combining eqns (2A.4c,d,2b):

$$\bar{\Gamma}_2^m = \frac{\Gamma_2\Gamma_4}{\Gamma_1\gamma_4}\left[\frac{1-\dot{W}_L/\dot{W}}{\Gamma_3+\Gamma_3\Lambda_L}\right]^m , \qquad (2A.4a)$$

The change of cross-sectional mass-content is parameterised by

$$\Lambda_L = \frac{L\ d(\Gamma_3\rho\hat{\sigma}H^2)/dt}{(dL/dt)\Gamma_3\rho\hat{\sigma}H^2} \qquad (2A.4b)$$

and this must be determined (using eqn.(2A.1)) at each instant in marching out eqn. (2A.3). Here Γ_3 is the (volume x density) factor for the overall fracture (fig. 2A.1), namely

$$W-W_L = 2\Gamma_3\rho\hat{\sigma}H^2L\ \Big|_0^t \qquad (2A.4c)$$

and the mass injection rate $\dot{W}$ is determined by the flow law

$$(\pm\dot{W}/2H\rho)^m = (\Gamma_4/\gamma_4\bar{\eta})\Delta^{2n+1}(\pm\Gamma_2\sigma/L),\ \Delta = \Delta_E + \Delta_A \qquad (2A.4d)$$

in which Δ_A is the anelastic component of displacement. The ratio Γ_4/γ_4 accounts for the difference in channel-flow factor for the crack profile in question (e.g. elliptical channel/parallel plates gives $12\pi/64$). The mass loss-rate W_L is to be obtained from a separate computation. Although the new coefficients, Γ_2 and Γ_3, are not constants, they are much more readily determinate than Γ_2, as in Section 2B, being relatively independent of fluid loss and pumping conditions.

By an entirely analogous argument for vertical flow along the cross-section defined by H, we may obtain the relation

$$\bar{\gamma}_2^m = \frac{\gamma_2}{\gamma_1}\left[\frac{1-\dot{W}_L/\dot{W}}{2\gamma_3+\gamma_3\Lambda_V}\right]^m \qquad (2A.5a)$$

with pressure alteration at the cross-section parameterized by

$$\Lambda_V \equiv \frac{H\ d(\gamma_3 \rho \hat{\sigma})/dt}{(dH/dt)\ \gamma_3 \rho \sigma} \tag{2A.5b}$$

This equivalency follows from mass-conservation relation for the cross-section and the vertical flow law, respectively given by

$$w - w_L = \gamma_3 \rho \hat{\sigma} H^2 \Big|_0^t \tag{2A.5c}$$

$$(\pm \dot{w}/\rho)^m = \Delta^{2n+1}(+\gamma_2 \sigma/H)/\sqrt{n} \tag{2A.5d}$$

in which Δ_E is given by eqn. (2A.2b) with $\Gamma_1 \equiv \gamma_1$ (having distinct symbols only to clearly identify their context).

The slope coefficient γ_2 depends primarily on contrasts in confining stress and moduli between strata in the reservoir and it may best be phrased for applications as follows:

$$\gamma_2 = \overset{\circ}{\gamma}_2 S(h);\ S(0)=1,\ S(\infty) = \gamma_2^{\infty}/\overset{\circ}{\gamma}_2;\ S=S_D S_S\ ;\ h=H/H_R \tag{2A.5e}$$

in which the shape of the slope function S depends on the details of the barriers impeding vertical growth: specifically, it has a different form for moduli vs. stress contrast, which requires the minimum level of rationalisation just provided -- i.e. different functions S_D and S_S for "deformability" and confining stress contrasts (e.g. as in fig. 2A.a). The values γ_2° and γ_2^{∞} are those for the limits of very small and very great heights respectively; these would typically be the values in a homogeneous region (γ_2°, fig. 2A.2b) and that in a stratified region (γ_2^{∞}). For stress and moduli contrast only, such limits may be related by an expression as simple as

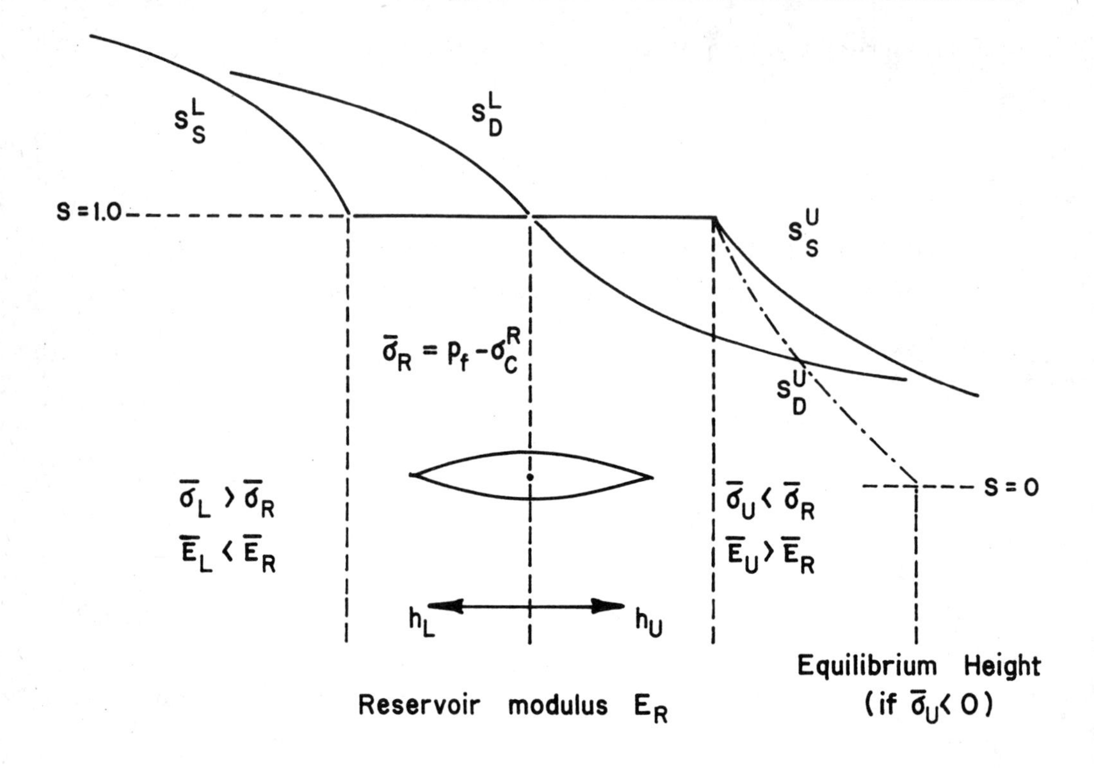

Fig. 2A.2. Schematic of cross-sectional modelling and results.

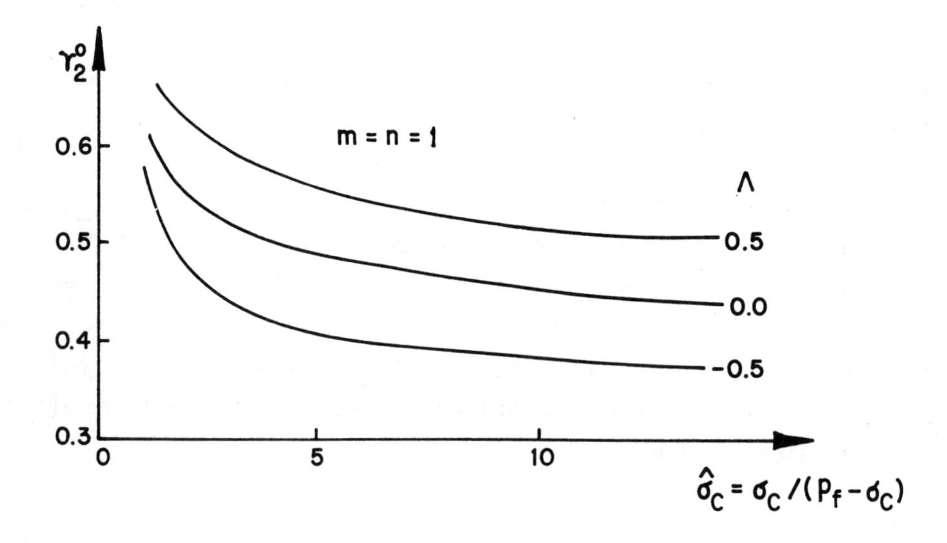

$$(\gamma_2^{\infty}/\gamma_2^{\circ})^m = (\sigma_A/\sigma_R)^{2n+2-m} \, (\bar{E}_R/\bar{E}_A)^{2n+1-m} \qquad (2A.5f)$$

which derives directly from the definition of characteristic time in eqn. (2A.2a). When the effective moduli of the adjacent strata are not the same, $\bar{E}_L \neq \bar{E}_U$, then the operative adjacent modulus $\bar{E}_A$ must be a suitable combination of $\bar{E}_L$ and $\bar{E}_U$ (which interchange in going from $\bar{E}_A{}^U$ to $\bar{E}_A{}^L$); this and other exact details of eqn. (2A.5e) can be worked out by using a self-similar model, for a cross-section like fig. 2A.2 (Cleary et al., 1981).

Reformulation for Specified Flow-Rate. To allow for the more typical field conditions of specified flow rate, we must transform from pressure to flow in eqns. (2A.1,3), which is achieved by use of eqn. (2A.4c), leading to

$$\frac{\dot{L}^m LH}{\bar{\Gamma}_2^m} = \frac{\bar{E}}{\bar{\eta}} \left[\frac{W-W_L}{2\Gamma_3 \rho HL} \right]^{2n+2-m} = \frac{\dot{H}^m H^2}{\bar{\gamma}_2^m} \qquad (2A.6)$$

Again, for any specified injected mass W, the evolution of L and H can be computed, simultaneously with loss W_L, by algebraic or numerical means, as follows.

2A.2. Algebraic solutions of lumped model equations.

Some broad features and specific algebraic formulae can be extracted, with which results of more general numerical solution schemes can be compared and verified. The most general result is obtained by direct comparison of eqns. (2A.1,3), namely

$$\bar{\gamma}_2 \frac{d}{dt} L^{1+1/m} = \bar{\Gamma}_2 \frac{d}{dt} H^{1+1/m} \qquad (2A.7a)$$

If the ratio of coefficients is time-independent, then we obtain

$$H^{1+1/m} - H_o^{1+1/m} = \bar{\gamma}_2 \left(L^{1+1/m} - L_o^{1+1/m}\right)/\bar{\Gamma}_2 \qquad (2A.7b)$$

Various special solutions have been worked out, based on eqn. (2A.7), by Cleary et al., 1981 and by Cleary, SPE 9259, 1980.

Solutions for Specified Flow. If we assume a time-independent non-zero ratio γ_2/Γ_2 for the moment, and suitable initial conditions, then eqn. (2A.7c) applies and may be used to eliminate H from eqn. (2A.6); an integration on time then allows us to write

$$\frac{mL^{\left(\frac{4n+6-m}{m}\right)}}{4n+6-m} = \int_o^t \frac{\bar{\Gamma}_2 \, dt}{(\bar{n}/\bar{E})^{1/m}} \left[\frac{\bar{\Gamma}_2}{\bar{\gamma}_2}\right]^{\frac{2n+3-m}{m+1}} \left[\frac{W-W_L}{2\rho\Gamma_3}\right]^{\frac{2n+2-m}{m}} \qquad (2A.8a)$$

from which H may be obtained directly in eqn. (2A.7c). Note that the interpretation of Γ_2 in eqn. (2A.4a) now facilitates the integration. A special case of eqn. (2A.8a) is that of constant pumping rate W: assuming constant values of parameters $\bar{\gamma}_2$, $\bar{\Gamma}_2$, Γ_3, Γ_4 and $\bar{n}/\bar{E}$,

$$L^{4n+6-m} = \Gamma_{WL} \frac{\bar{E}}{\bar{n}} \left(\frac{\rho}{W}\right)^m \left[\frac{W-W_L}{2\rho\Gamma_3}\right]^{2n+2} \qquad (2A.8b)$$

we obtain in which the length coefficient for specified flow is

$$\Gamma_{WL} = \frac{\Gamma_2\Gamma_4}{\Gamma_1\gamma_4} \left[\frac{8n+12-2m}{(1+\Lambda_L)(2n+2)}\right] \left[\frac{\bar{\Gamma}_2}{\bar{\bar{\gamma}}_2}\right]^{\frac{2n+3-m}{1+1/m}} \qquad (2A.8c)$$

It is now also interesting to work out the behavior of pressure (hence crack-opening) with time; use of eqn. (2A.4c), with (A.7c, 8b), leads to the result

$$\hat{\sigma}^{4n+6-m} = \Gamma_{WP} \left[\frac{\bar{\eta}\dot{W}^m}{\bar{E}\rho^m}\right]^3 \left[\frac{W-W_L}{2\rho\Gamma_3}\right]^{-2n-m} \qquad (2A.8d)$$

in which the pressure coefficient for specified flow is

$$\Gamma_{WP} = \left[\frac{(2n+2)(1+\Lambda_L)}{8n+12-2m}\right]^{3m} \frac{\Gamma_1^3\,\gamma_4^3}{\Gamma_2^3\,\Gamma_4^3} \left[\frac{\bar{\Gamma}_2}{\bar{\gamma}_2}\right]^{\frac{2n+3+m}{1+1/m}} \qquad (2A.8e)$$

There are two coefficients of cross-sectional volume variation to be determined in using the foregoing equations; these are defined in eqns. (2A.4b,5b) and can now be determined from/for the foregoing approximate solutions. Using eqns. (2A. 8b, 8d,7c), and temporarily neglecting density changes, we get

$$\Lambda_V \simeq \frac{Ld\hat{\sigma}/dt}{\hat{\sigma}dL/dt} = \frac{-2n-m}{2n+2} \qquad (2A.8f)$$

$$\Lambda_L \simeq \frac{L\ d(\hat{\sigma}L^2)/dt}{\sigma L^2(dL/dt)} = \frac{2n+4-m}{2n+2} \qquad (2A.8g)$$

so that the assumptions of constancy are justified. As well, note that this Λ_L simplifies eqns. (2A.8b,d), and (2A.9b,d) later, as another approach will show (in eqns. (2c.1b,c).

Note that pressure always drops with increasing total volume of fracture $(W-W_L)/\rho$, according to these special solutions; however, these presume initial conditions consistent with eqn. (2A.7c) and require constancy of most coefficients. Departure from these assumptions may lead to non-monotonic pressure variation with fracture volume, as the numerical results show in Sec. 2A.3. Still, the results provide central insights into anticipated behavior and they also permit easy scaling from normalised results, once obtained numerically for a range of parameters (e.g. with $W/\rho = 1$ for volume and $\bar{\eta}/\bar{E} = 1$ for time-scale) to the relevant dimensional results in the field.

There is also another (related) source of deviation from results in eqns. (2A.8): this occurs when neglible vertical growth develops $\overline{\gamma}_2 \to 0$. Actually, this is the limit where initial conditions consistent with eqn. (2A.7c) are never reached, and it gives us a clear idea of what happens when the intial height is greater than that required by (2A.7c). Here we simply regard H as independently determined and insert into eqn. (2A.6) to get the length

$$\frac{mL^{\frac{2n+3}{m}}}{2n+3} = \int_0^t \frac{\Gamma_2 \, dt}{(H\overline{\eta}/\overline{E})^{1/m}} \left[\frac{W-W_L}{2\Gamma_3 \rho H}\right]^{\frac{2n+2-m}{m}} \tag{2A.9a}$$

Again, we can perform the integration explicitly when $\dot{W}$ (and H) is constant, using eqn. (2A.4a), to get

$$L^{2n+3} = \Gamma^{o}_{WL} \frac{\overline{E}}{\overline{\eta}} \left(\frac{\rho}{\dot{W}}\right)^m \left[\frac{W-W_L}{2\rho\Gamma_3}\right]^{2n+2} H^{m-2n-3} \tag{2A.9b}$$

in which the length coefficient is now also different:

$$\Gamma^{o}_{WL} = \left[\frac{2n+3}{2n+2}\right] \frac{\Gamma_2\Gamma_4/\gamma_4}{\Gamma_1(1+\Lambda_L)m} \simeq \frac{\Gamma_2\Gamma_4}{\Gamma_1\gamma_4} \tag{2A.9c}$$

From eqn. (2A.4a), we can now obtain the behavior of pressure as a function of fracture volume, namely

$$\hat{\sigma}^{2n+3} = \frac{H^{-m-2n-3}}{\Gamma^{o}_{WL}} \frac{\overline{\eta}}{\overline{E}} \left[\frac{\dot{W}}{\rho}\right]^m \frac{W-W_L}{2\rho\Gamma_3} \tag{2A.9d}$$

This shows that pressure always rises with increasing fracture volume, a feature characteristic of the classical Perkins-Kern model; the associated value of the pressure coefficient is

$$\Lambda_L = L(d\hat{\sigma}/dt)/\hat{\sigma}(dL/dt) = 1/(2n+2) \qquad (2A.9e)$$

Obviously, conditions between that in eqn. (2A.7a) and H = constant will give amounts of pressure rise or drop, depending on the degree of vertical growth, $\overline{\gamma}_2 \neq 0$, as numerical results will show (Sec. 2A.3).

Solutions for Specified Pressure. These can be obtained directly by integrating eqns. (2A.1, 2) with constant characteristic time τ_c; for constant coefficients $\overline{\gamma}_2$ and $\overline{\Gamma}_2$, this gives

$$H = H_o \exp (\overline{\gamma}_2 t/\tau_c) \qquad (2A.10a)$$

from which L is given by eqn. (2A.7c) and total volume becomes

$$W - W_L = 2\Gamma_3 \rho\hat{\sigma} \left(\frac{\overline{\Gamma}_2}{\overline{\gamma}_2}\right)^{\frac{m}{m+1}} H_o^3 \exp (3\overline{\gamma}_2 t/\tau_c) \qquad (2A.10b)$$

This obviously increases exponentially in time and the limits of the pumps would quickly be reached; however, the result again presupposes constant coefficients, and appropriate initial conditions.

Again, an extreme counterexample is that of constant height (γ_2=0), when eqn. (2A.2) integrates directly to

$$(L/H)^{1+1/m} = \frac{m+1}{m} \int_0^t \overline{\Gamma}_2 \frac{dt}{\tau_c} \qquad (2A.11a)$$

and the resulting fracture volume grows (like length), as square-root of time for m=1,

$$W-W_L = 2\Gamma_3 \rho H^3 \hat{\sigma} \left[\overline{\Gamma}_2 \frac{m+1}{m\tau_c} t\right]^{\frac{m}{m+1}} \qquad (2A.11b)$$

so that the rate drops as inverse square-root of time for m=1. This square root behavior is interesting, since it is precisely that predicted (e.g. by a linear porous media theory) for fluid front penetration in channels with constant permeability although the present channel is actually being opened up by (and only exists because of) the driving pressure.

2A.3 Numerical Solutions of Lumped Model Equations

The equations (2A.1,3,6) can more generally, of course, be solved numerically for arbitrarily complicated initial conditions and time-dependence of the coefficients in $\overline{\gamma}_2$, $\overline{\Gamma}_2$ -- especially due to fluid loss W_L, which will not typically follow the same behavior as W and therefore will not allow homogeneous (e.g. power-law) solutions of the kind we often use (Cleary, 1980) for demonstration. Sample results are shown in fig. 2A.3; dominant effects of moduli and stress contrasts are clearly demonstrated.

2B. Summary of P3DH model equations and results.

For many practical reservoir conditions of fairly good containment (i.e., restricted height growth), the much-quoted outstanding problem of fracture height determination is solved correctly (for the first time, we believe) by the pseudo-three dimensional hydrafrac (P3DH) modelling schemes developed by Cleary (SPE 9259, 1980); as well, these are sufficiently simple, yet general, that realistic (quite complex) reservoir properties can be incorporated--a feature not soon promised in any of the more complex 3-D model developments that we know of (even including our own efficient version, Cleary et al. 1981). The basic assumption is that one-dimensional lateral flow can be coupled to a suitable two-dimensional model for crack-opening and vertical fluid flow at each cross- section, but that a fully 3-D grid is unnecessary, thus saving greatly on computation.

A suitable mesh spacing distribution is used, one which moves with the fracture (in contrast to Settari and Cleary, 1982); we also use a convenient expression for height growth at each cross-section (based on the experience with models for eqn. (2A.1)) and thus march out the shape of the fracture after each pressure solution is obtained. In

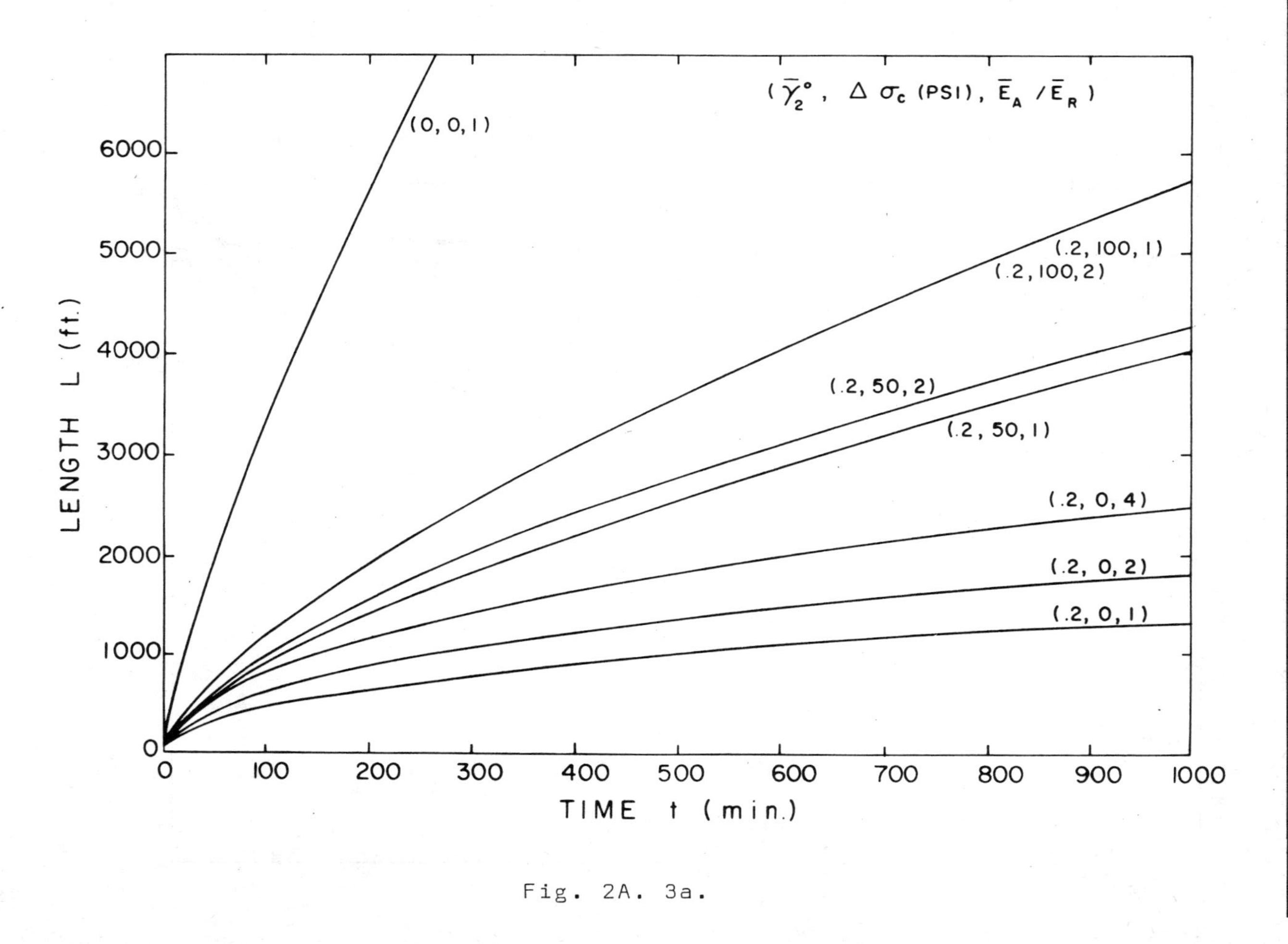
($\bar{\gamma}_2^\circ$, $\Delta\sigma_c$ (PSI), $\bar{E}_A/\bar{E}_R$)
(0, 0, 1)
(.2, 100, 1)
(.2, 100, 2)
(.2, 50, 2)
(.2, 50, 1)
(.2, 0, 4)
(.2, 0, 2)
(.2, 0, 1)
LENGTH L (ft.)
0
1000
2000
3000
4000
5000
6000
TIME t (min.)
0
100
200
300
400
500
600
700
800
900
1000

Fig. 2A. 3a.

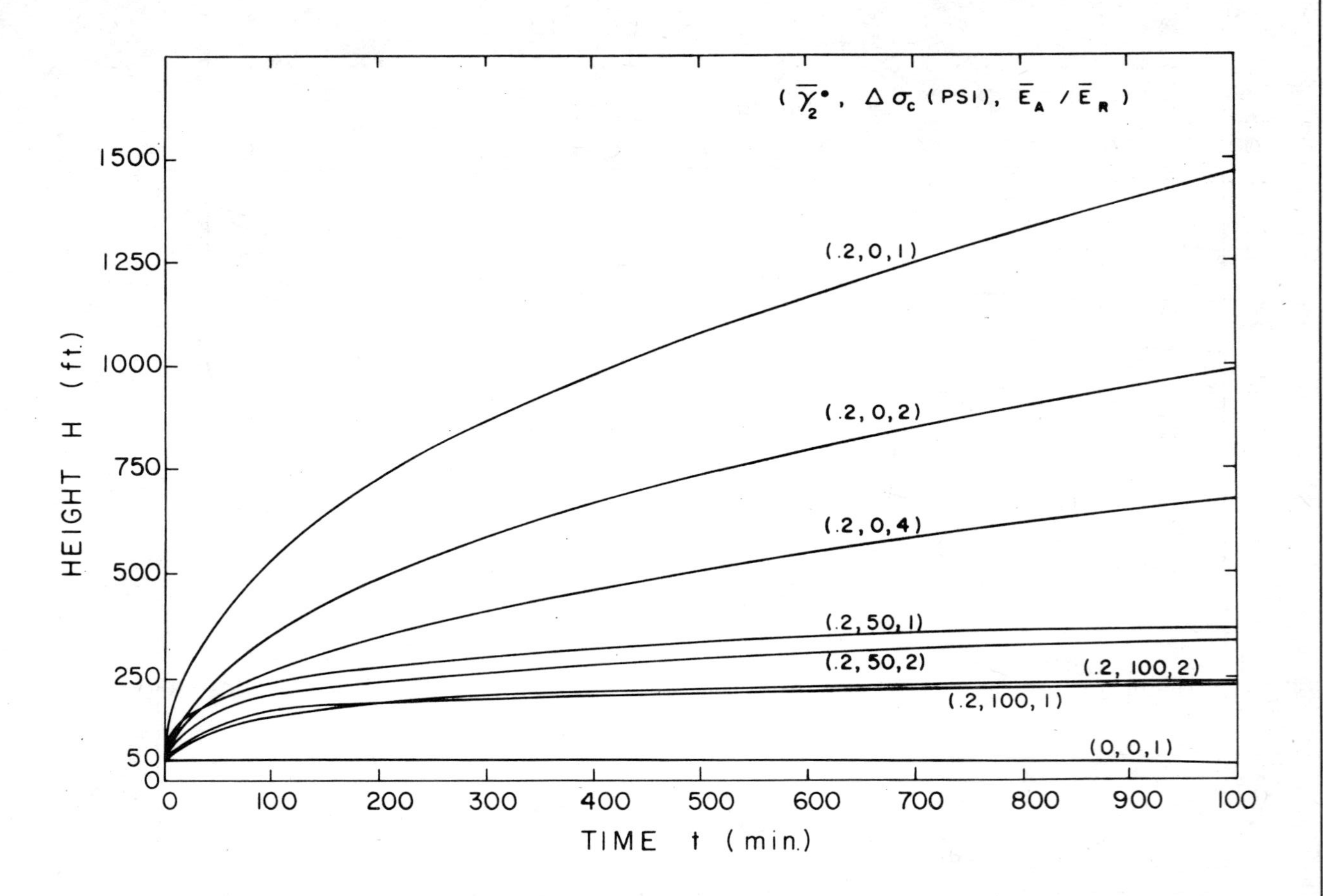

($\bar{\gamma}_2^{\circ}$, $\Delta\sigma_c$ (PSI), $\bar{E}_A$ / $\bar{E}_R$)
HEIGHT H (ft.)
1500
1250
1000
750
500
250
50
0
(.2, 0, 1)
(.2, 0, 2)
(.2, 0, 4)
(.2, 50, 1)
(.2, 50, 2)
(.2, 100, 2)
(.2, 100, 1)
(0, 0, 1)
0
100
200
300
400
500
600
700
800
900
100
TIME t (min.)

Fig. 2A. 3b.

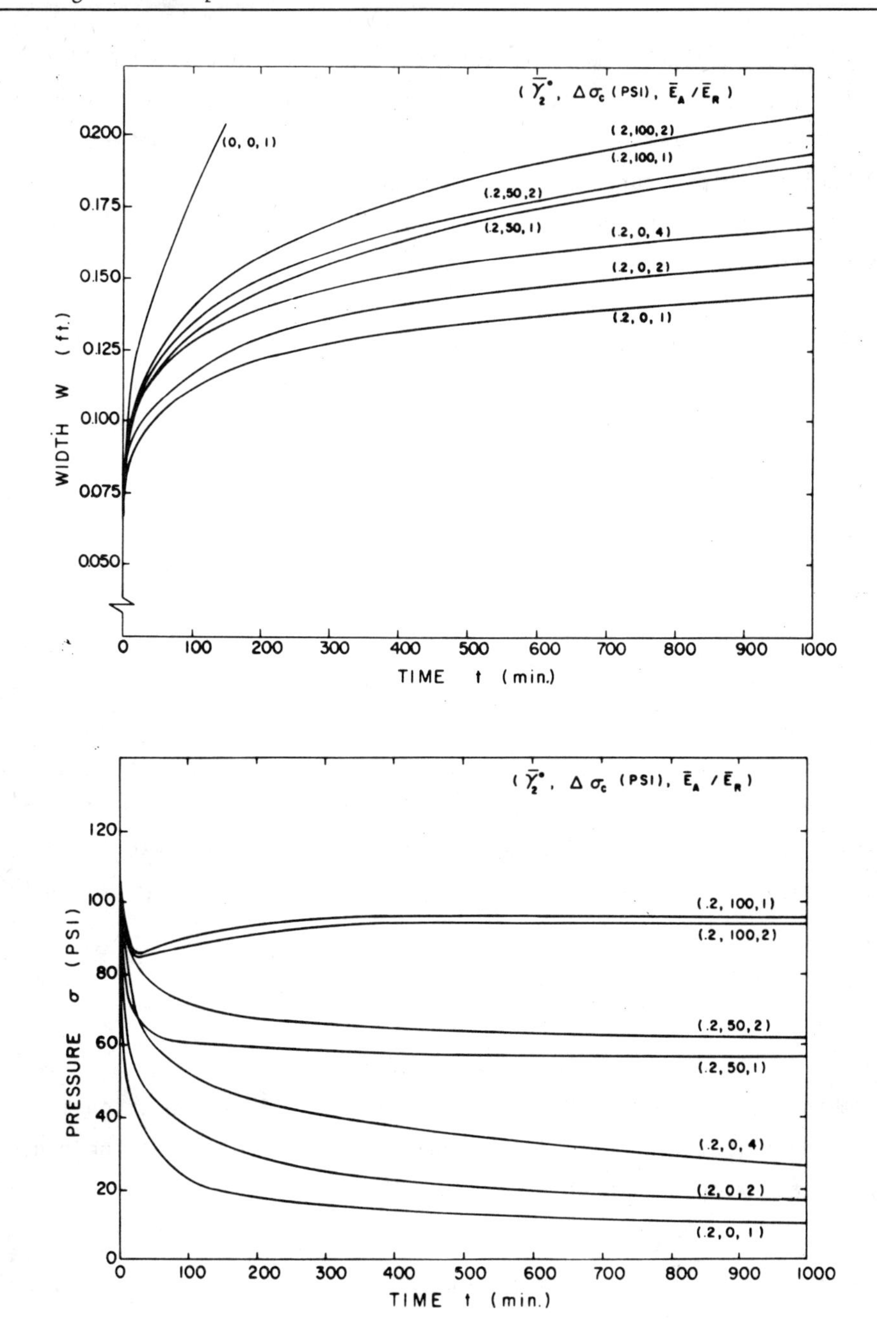

($\bar{\gamma}_2^\circ$, $\Delta\sigma_c$ (PSI), $\bar{E}_A/\bar{E}_R$)
(0, 0, 1)
(.2,100,2)
(.2,100,1)
(.2,50,2)
(.2,50,1)
(.2, 0, 4)
(.2, 0, 2)
(.2, 0, 1)
0.200
0.175
0.150
0.125
0.100
0.075
0.050
WIDTH W (ft.)
0
100
200
300
400
500
600
700
800
900
1000
TIME t (min.)
($\bar{\gamma}_2^\circ$, $\Delta\sigma_c$ (PSI), $\bar{E}_A/\bar{E}_R$)
(.2, 100,1)
(.2, 100,2)
(.2, 50, 2)
(.2, 50,1)
(.2, 0, 4)
(.2, 0, 2)
(.2, 0, 1)
120
100
80
60
40
20
0
PRESSURE σ (PSI)
0
100
200
300
400
500
600
700
800
900
1000
TIME t (min.)

Fig. 2A. 3c.

this way, the evolution of fracture shape can be traced and the (downhole) pressure history can be obtained for any specified injection history--and vice-versa; an interesting outcome is the potential for type-matching the pressure curve to determine how much vertical fracture growth is occurring during a stimulation operation. Such applications and many others have been described by Settari and Cleary (1982) and by Cleary et al (1981).

The basic assumptions and derivations have already been outlined in SPE 9259, so we merely summarize the governing equations, with reference to fig. 2B.1. Mass conservation is expressed as follows, for the net lateral fluid flow Q.

$$\frac{\partial}{\partial x}(\rho Q) + 2\rho H\overline{q}_L = -\frac{\partial}{\partial t}(2\gamma_3 \rho H \Delta) \tag{2B.1}$$

where $\overline{q}_L$ is the average fluid loss rate and γ_3 is the volume (shape) factor for the cross-section at x. The central crack opening Δ is still given by eqn. (2A.2b), where H and σ are now the height and excess pressure at any cross-section (at x). The system description is completed by the rheological equation for the frac-fluid in channel-flow:

$$\left(\pm\frac{Q}{2H}\right)^m = \mp\frac{\Gamma_4}{\gamma_4}\frac{\Delta^{2n+1}}{\overline{\eta}}\frac{\partial p_f}{\partial x} \tag{2B.2}$$

Combination of eqns. (2B.1,2; 2A.2b) gives a single equation in the pressure distribution, describing non-linear diffusion on the region (0,L), for any wing of length L:

$$\frac{\partial}{\partial x}\left[\pm\rho\overline{\Gamma}_5\left(\mp\frac{\partial f}{\partial x}\right)^{\frac{1}{m}}\right] + \rho\overline{H}q_L = -\frac{\partial}{\partial t}\left(\gamma_3\rho H^2\hat{\sigma}\right) \tag{2B.3a}$$

in which the flow is dominated by the new pressure function and transmissivity, defined by

$$\hat{f} = \hat{\sigma}^{2n+2} \text{ and } \overline{\Gamma}_5^m = \frac{\Gamma_4 \overline{E}}{\gamma_4 \bar{n}} \frac{H^{2n+1+m}}{(2n+2)\Gamma_1} \left[1 + \frac{\Delta_A}{\Delta_E}\right]^{2n+1} \tag{2B.3b}$$

However, completion of the description requires the determination of the boundaries $H(x,t)$ and $L(t)$, and this is where options begin to appear (Cleary et al., 1981).

The length growth requires a criterion, for what we have called the leading-edge (Fig. 2B.1), which is just the segment of the fracture near the front that cannot be completely described by eqns. (2B.1, 2, 3). We denote by $\hat{\sigma}_F$ the excess pressure required (at the front $x = L$) to drive this leading-edge at speeds compatible with the growth rate L deduced from solving (2B.1 2, 3) in the main body of the fracture; this rate can be derived from overall mass conservation for the fracture which may be manipulated into an equivalent expression

$$dL/dt = Q_F/2\gamma_3 H_F \Delta_F \; ; \quad \Delta_F = \hat{\sigma}_F H_F + \Delta_A^F \tag{2B.4a}$$

provided that eqn. (2B.3a) is satisfied at all points x. The leading-edge criterion itself has a similar character to the models which produced eqn. (2A.1), because it describes a segment of fracture (e.g. semi-circular) at the front, having width dominated by effective length (e.g. radius); it may also look like eqn. (2A.3), if we wish to lump part of the channel (described by eqns. (2B.1, 2, 3)) into it. The best way to write a propagation equation for it is as follows

$$dL/dt = \overline{\gamma}_2^F \, H_F/\tau_c^F \tag{2B.4b}$$

where τ_c^F has the definition in eqn. (2A.2a), with $\hat{\sigma} = \hat{\sigma}_F$. Combination of eqns. (2B.4 a, b) leads to the boundary condition

$$\left(\frac{Q_F/H_F^3}{2\gamma_3\bar{\gamma}_2^F}\right)^m = \frac{\bar{E}}{\bar{n}} \frac{\hat{\sigma}_F^{2n+2}}{H_F^{2m-2n}} \qquad (2B.4c)$$

Specification of Q_w or σ_w, or a combination, completes the model.

Now, it is possible to satisfy eqn. (2B.3a) accurately by using established numerical solution techniques for (nonlinear parabolic) equations which are now common practice in the business of reservoir engineering. However, these techniques are quite demanding computationally and have been developed only with fixed meshes, through which the fracture has to move; this means that detailed solution near the fracture front is not feasible, since a prohibitively fine mesh would be needed over the whole trajectory of the fracture (Settari and Cleary, 1982). A mesh which moves with the fracture avoids this need, except near the front; some sample solutions are given by Cleary et al., 1981. The resulting shapes look like those on the "self-determining height" side of fig. 2B.1; when these are unrealistic (e.g. for $L \sim H$), the shapes may be specified, e.g. as shown, also in fig. 2B.1.

2B.1. Reduction of P3DH model to Ordinary Differential Equation.

Our first approximation is simply that of employing the model described by eqns. (2A.5c, d) to determine the storage term in eqn. (2B.3a) at each cross-section, namely

$$\frac{\partial}{\partial t}(\gamma_3 \rho H^2 \hat{\sigma}) + \rho H \bar{q}_L = \bar{\gamma}_5 \rho H^{\frac{2n}{m}} \hat{f}^{\frac{1}{m}} \qquad (2B.5a)$$

in which the coefficient $\bar{\gamma}_5$ corresponds to $\bar{\gamma}_2$ in eqn. (2A.1),

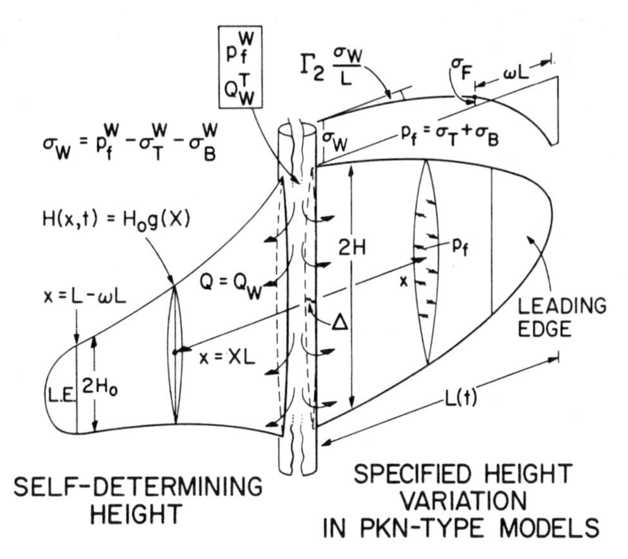

Fig. 2B.1a. Schematic of fracture shapes which can be described by P3DH models.

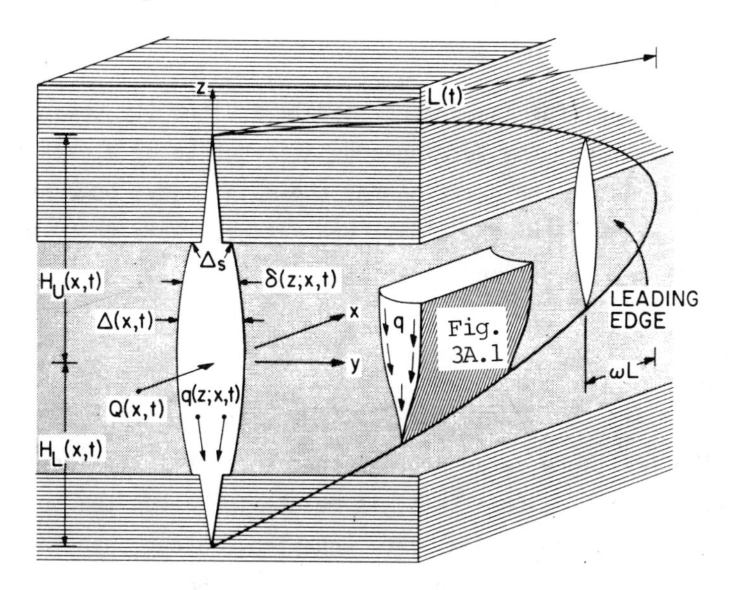

Fig. 2B.1b. Illustration of component concepts in pseudo-three-dimensional hydrafrac (P3DH) models.

$$\bar{\gamma}_5^m = \frac{\bar{\gamma}_2}{\gamma_1} \frac{\bar{E}}{\bar{n}} \qquad (2B.5b)$$

but does not need corrections for loss, pressure change Λ_v etc. (as in eqn. (2A.5a)).

We can now non-dimensionalise eqn. (2B.3a) to get

$$[\pm \Gamma_5(\mp g^{2n+1+m}\,\hat{f}')^{\frac{1}{m}}]' + \gamma_5[(L/H_R)^{m+1}\, g^{2n}\,\hat{f}]^{\frac{1}{m}} = -q_{LR} \qquad (2B.6a)$$

in which the primes stand for ordinary differentiation with respect to non-dimensional position X = x/L and we have introduced the following non-dimensional parameters

$$g \equiv \frac{H}{H_R} \; ; \; \frac{2n+2}{\Gamma_4/\gamma_4}\,\Gamma_5^m \equiv \frac{\rho^m}{\rho_w^m}\frac{\bar{E}}{\bar{E}_w}\frac{\bar{n}_w}{\bar{n}} \equiv \frac{\gamma_1}{\gamma_2}\,\bar{\gamma}_5^m \qquad (2B.6b)$$

The reference H_R can be (for the present model) arbitrarily chosen to be H_F, since this allows the leading-edge at front to be always of the same size in the view of the equation for the main body; however, the values of f are scaled to their wellbore values. Note that we have retained a loss term q_{LR}, defined by

$$q_{LR} \equiv (g\hat{f}_w\,\bar{q}_L\,\tau_c/H_R)(L/H_R)^{1+1/m} \qquad (2B.6c)$$

where τ_c is the (wellbore value of the) characteristic time defined in eqn. (2A.2a) with $H = H_R$; this term is meant to allow the option of lumping the losses into the lateral flow equation, rather than including them in the vertical propagation model of eqns. (2A.5c, d).

Our achievement is clearly an ordinary differential equation for the pressure function $\hat{f}$, to be solved on (0, 1), subject to wellbore boundary conditions on $\hat{f}\,(X=0)=\hat{f}_w$ or on the flow-rate

$$Q_W = Q(X=0) \;;\quad Q = \pm\, 2\Gamma_5(+\hat{f}'/L)^{\frac{1}{m}} \tag{2B.6d}$$

and subject to the (mixed) front condition in eqn. (2B.4d), with $Q_F = Q(X=1)$; we will actually also perform some numerical testing with specified $\hat{\sigma}_F$. The height-growth is then determined from eqn. (2A.1) or (2B.5a), while the length evolution is dictated by eqn. (2B.4b); numerical implementation is straightforward, but the resulting predictions are fairly unrealistic (Cleary et al., 1981). To make effective use of the concept in eqn (2B.5a), we need to incorporate it with the self-similar model described next.

2B.2. Self-Similar Approximations for Storage in Lateral Flow.

As we found in Sec. 2A, there is a difficulty in applying the combined vertical and lateral propagation models (2A.1, 3) to the limit of very little vertical growth; here the approximation in eqn. (2B.5a) begins to break down, since it is based on a model of continuing vertical flow while the dominant storage now arises from change in width (without appreciable flow). Thus, it is necessary to adopt a different approximation which adequately reflects the change in pressure (and also height) at each cross-section as the crack length increases. For this, we make an assumption that is absolutely precise only in the case of constant height and homogeneous growth in length (e.g. a power law resulting from a fluid loss of the same form as fracture volume): this approximation is simply that the storage can be computed from the present shape of the cross-sectional volume curve, as if that remained the same in going from time t to time t + Δt namely

$$\frac{\partial f}{\partial t} = -\,X\,\frac{\dot{L}}{L}f' + \frac{\dot{f}_W}{f_W}\,f \;;\quad f \equiv \frac{\gamma_3 \rho H^2 \hat{\sigma}}{(\gamma_3 \hat{\sigma}_W H_R^2\, \rho_W)} \tag{2B.8}$$

Here the normalization (in parentheses) is done after imposing the assumption in eqn. (2B.8). We call this a self-similar approximation, but we emphasise that it is applied only <u>instantaneously</u> and does not require the shape of f to remain the same over any finite period of time: our numerical solutions will actually show f varying in time.

Insertion of eqn. (2B.8) into eqn. (2B.3a), and suitable non-dimensionalisation as in eqn. (2B.6a), now leads to the ordinary differential equation

$$[\pm \Gamma_5(\mp g^{2n+m+1}\hat{f}')^{\frac{1}{m}}]'+\lambda_L[-Xf'+\Lambda_L f]=-q_{LR} \quad (2B.9a)$$

where q_{LR} has been defined in eqn. (2B.6c), $\Lambda_L \equiv \hat{\dot{f}}_w L/\hat{f}_w \dot{L}$ has the same interpretation as in eqn. (2A.4b) and the <u>eigenvalue</u> λ_L has the definition

$$\lambda_L^m \equiv \dot{L}^m \tau_c^m \hat{f}_w L/H_R^{m+1} \quad (2B.9b)$$

which is seen to correspond with the role of $\overline{\Gamma}_2$ in eqn. (2A.3), so that it provides special values of Γ_2 for the approximations in eqn. (2B.8) -- after correction by $\hat{f}_w (H_R/H)^{m+1}$ if necessary. Note that we can, again, scale the pressure function σ to its value at the wellbore (hence eliminating $\hat{f}_w$), and we choose $H=H_R$ in the definition of τ_c (eqn. (2A.2)).

Eqn. (2B.9a) could, of course, be solved by the matrix methods or shooting techniques, but a more effective procedure can be achieved by the following integration by parts and imposition of a consistency condition at x=L:

$$\lambda_L^m q^m(X) \equiv \lambda_L^m \left[Xf + \int_X^1 dX[(1+\Lambda_L)f + q_{LR}/\lambda_L] \right]^m$$

$$= \Gamma_5^m \, g^{2n+m+1}\hat{f}' \quad (2B.10a)$$

$$\equiv \hat{f}_w \, \tau_c^m \, (\rho Q/2\gamma_3^w \rho_w \hat{\sigma}_w H_R^2)^m \, L/H_R^{m+1} \tag{2B.10b}$$

The latter simply requires that eqn. (2B.9b) be satisfied with

$$\dot{L} = \lim_{X\to 1} [\rho Q/2\gamma_3^w \rho_w \hat{\sigma}_w H_R^2 f] = \lim_{X\to 1} [Q/2\gamma_3 \hat{\sigma} H^2] \tag{2B.10c}$$

which is exactly the condition employed in eqn. (2B.4b). Indeed, the function q simply represents the distribution of flow-rate Q along the fracture; it is clearly not constant, as often assumed by conventional models in the industry. Eqn. (2B.10b) also allows the use of a specified flow rate $Q_w = Q(X=0)$ for the determination of $\hat{\sigma}_w$, once λ_L has been obtained, as follows:

$$\hat{\sigma}_w^{2n+2} = (Q_w/2\gamma_3^w \rho_w H_w^2)^m \, L/\lambda_L^m \, H_R^{m+1} q^m(0) \tag{2B.10d}$$

To find λ_L, a further integration must be performed:

$$\hat{f} = \hat{f}_w - \lambda_L^m \int_0^X dX q^m(X)/\Gamma_5^m g^{2n+m+1} \tag{2B.11a}$$

The value of λ_L is now obtained by imposing boundary-condition at x=L,

$$\hat{f}_F = \hat{f}_w - \lambda_L^m \int_0^1 dX q^m(X)/\Gamma_5^m \, g^{2n+m+1} \tag{2B.11b}$$

for which we should employ the leading-edge condition in eqn. (2B.4c, 4d); the latter may be rephrased in a form suitable for direct application to eqn. (2B.11a):

$$\left(\frac{\hat{f}_F}{f_w}\right)^{\frac{2n+2-m}{2n+2}} = \left(\frac{\bar{\eta}\,\bar{E}_w}{\bar{\eta}_w \bar{E}}\right) \frac{H_F}{L} \left(\frac{\lambda_L}{\gamma_2^F}\right)^m \tag{2B.11c}$$

The latter expression crystallises the role of the leading edge: for any known ratio λ_L/γ_2^F (e.g. constant of order unity for a homogeneous medium), it requires a high front pressure only for short lengths L of order H_F, and viscosity degradation then plays an important role in the process, besides its presence in Γ^m of eqn. (2B.10). Eqn. (2B.11a) imposes a dominant control on the value of λ_L, since only major variations in the function q(X) will cause appreciable changes in λ_L; in a sense, this explains why non-ridiculous numbers can be obtained with models based on constant q (or any other approximation), but only when height is constant!

The foregoing derivations have now led to an integral equation formulation (eqn. (2B.11a)), which allows for a very effective and stable iterative numerical solution procedure.

2B.3. Hybridisation of Self-Similar and O.D.E. Models.

A study of results in Sections 2B.2, 3 shows that neither approximation for the storage terms (eqn. (2B.5a) or eqn. (2B.8)) can alone adequately describe the behavior observed: the self-similar assumption (eqn. (2B.8)) works best near the fracture front and/or when there is little height growth, whereas the vertical flow model of storage (eqn. (2B.5a)) has merits only where there is dominant height growth, and then only in the region of higher excess pressure back near the wellbore. Clearly, we should really have a combination of the two effects, reflecting the sharp rise in pressure near the front and the rapid height growth in the main body of the fracture; this is easily achieved by splitting the storage term into two parts as follows:

$$\frac{\partial}{\partial t}(\gamma_3\rho H^2\hat{\sigma})=\gamma_3 H^2\frac{\partial}{\partial t}(\rho\hat{\sigma})+\rho\hat{\sigma}\frac{\partial}{\partial t}(\gamma_3 H^2) \qquad (2B.12a)$$

We now use eqn. (2B.8) for the first term and a modification of eqn. (2B.5a) for the second term, with, $f\equiv\rho\hat{\sigma}/\rho_w\hat{\sigma}_w$ in which we replace

$$\frac{\partial}{\partial t}[\gamma_3\rho H^2\hat{\sigma}]=(1+\hat{\Lambda}_v)\rho\hat{\sigma}\frac{\partial}{\partial t}(\gamma_3 H^2) \tag{2B.12b}$$

Here the ratio of opening storage to vertical flow requirements is given by

$$\hat{\Lambda}_v \equiv \gamma_3 H^2\frac{\partial}{\partial t}(\rho\hat{\sigma})/\rho\hat{\sigma}\frac{\partial}{\partial t}(\gamma_3 H^2) \tag{2B.12c}$$

which can be computed directly at each time t and position x (from previous times or by iteration on the first solution, using eqns. (2B.8, 2A.1)). Combination of all the foregoing rationalisation now leads to a dimensionless equation very similar to both eqns. (2B. 6a, 9a), namely

$$[\pm\Gamma_5(\mp g^{2n+m+1}\hat{f}')^{\frac{1}{m}}]'=\lambda_L[Xf'-\hat{\Lambda}_L f]-\hat{\gamma}_5[(L/H_R)^{m+1}g^{2n}\hat{f}]^{\frac{1}{m}} \tag{2B.12d}$$
$$-q_{LR}\hat{\Lambda}_v/(1+\hat{\Lambda}_v)$$

in which the new coefficients are simply

$$\hat{\Lambda}_L\equiv\dot{f}_w L/f_w\dot{L},\quad f\equiv\rho\hat{\sigma}/\rho_w\hat{\sigma}_w,\hat{\gamma}_5\equiv\gamma_5/(1+\hat{\Lambda}_v). \tag{2B.12e}$$

Note now that the ambiquity, about how much loss to include in the lateral flow model, as against how much should be lumped into the vertical flow, has been resolved quite naturally by $\hat{\Lambda}_v$; our result is now a very appealing combination of all the identifiable dominant ingredients representable by a P3DH model. Again, we can solve by the methods of Sections 2B.2 or 2B.3, but the latter seems preferable since it so naturally incorporates the boundary-conditions at the front and so readily allows inclusion of the vertical growth component in the satisfying role of a fluid loss term like q_{LR}: indeed, eqns. (2B.10a, 11a) still apply directly now, provided we make the identification

$$q_{LR} \rightarrow q_{LR}\,\hat{\Lambda}_V/(1+\hat{\Lambda}_V)+\hat{\gamma}_5[(L/H_R)^{m+1}\,g^{2n}\,\bar{f}]^{\frac{1}{m}} \qquad (2B.12f)$$

The numerical procedure is therefore exactly the same as it is for the treatment of equation 2B.11a, requiring only an extra term in q_{LR}. Some sample results are shown, in Fig. 2B.2 for pressure change/ distribution, length growth, height growth distribution and flow rate Q_F at the front of the fracture -- using specified flow rate at the wellbore, as typical of field conditions.

2C. Hydrafrac designs based on lumped model solutions.

a. Algebraic Solutions. Expressions for fracture length have been derived for eqns. (2A.1,3), as in Cleary, SPE 9259, 1980 and Cleary et. al., 1981; such solutions are also provided in eqns. (2A.8,9). However, it may be seen that these solutions require iteration to find the appropriate fluid loss mass W_L, which depends on length and height; recognizing this, we have developed an alternative approach of finding the treatment time needed to produce the desired fracture dimensions, which effectively eliminates such tedious iteration. The method is simply to combine eqns. (2A.4c, d), with eqn. (2A.7) if necessary for H, to get the following natural sequence of solution steps:

i) A desired length of fracture L is chosen and a corresponding height H computed; if H is to be limited (e.g. avoiding break-out), then H may be chosen and L computed. This step requires use of the coefficients $\bar{\gamma}_2$ and $\bar{\Gamma}_2$ (eqns. (2A.4a, 5a)), which both actually involve the rate of fluid loss; it can be implemented precisely (using eqn. (2A.7a), as in but our purpose here is to provide simpler algebraic formulae. Thus, we employ eqn. (2A.7b) and the approximation

$$\frac{\bar{\gamma}_2}{\bar{\Gamma}_2} = \frac{\gamma_2\gamma_4}{\Gamma_2\Gamma_4}\left[\frac{\Gamma_3+\Gamma_3\Lambda_L}{2\gamma_3+\gamma_3\Lambda_V}\right]^m \qquad (2C.1a)$$

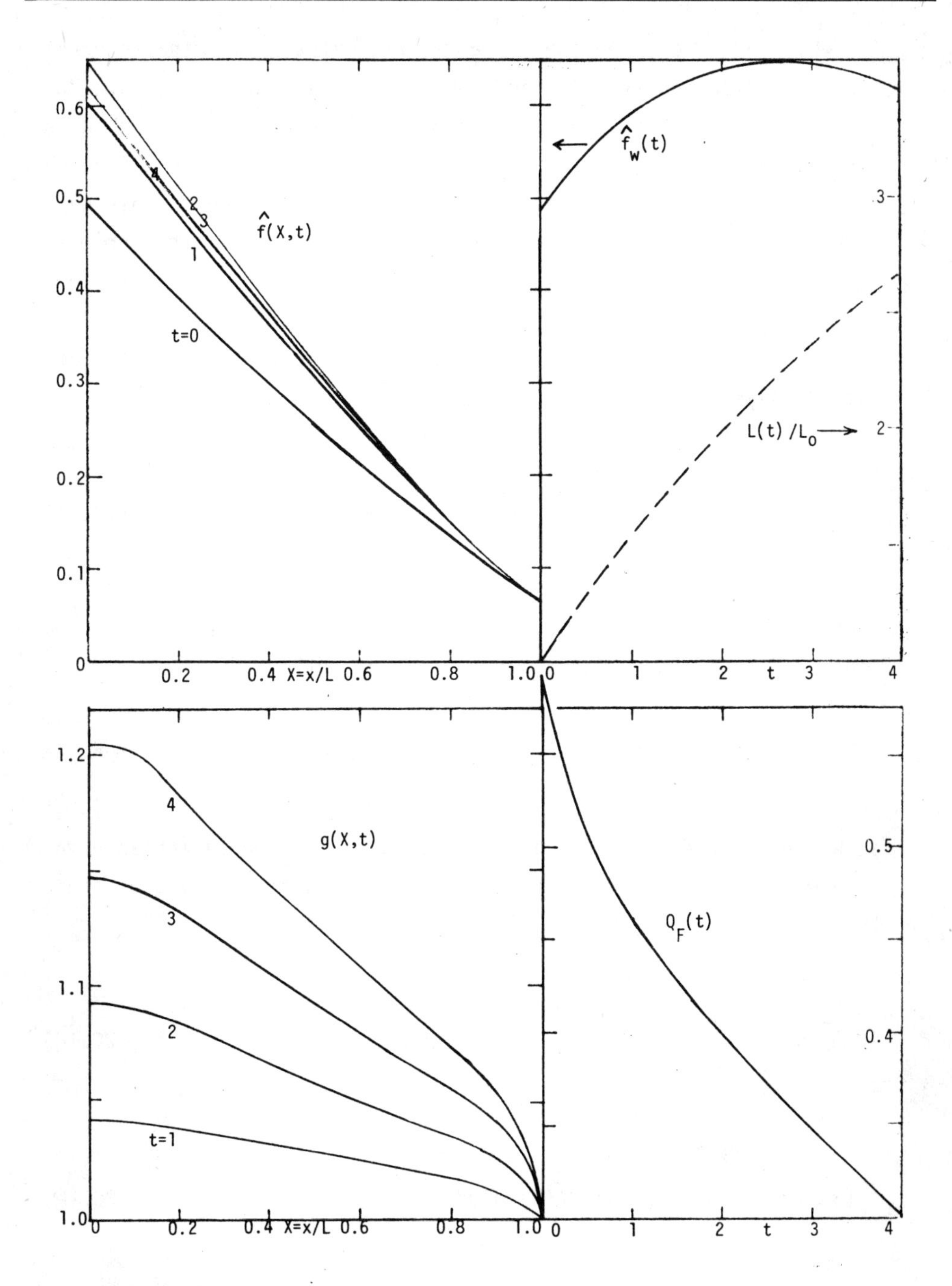

Fig. 2B.2. Sample results for hybridisation of self-similar and ODE models.

in which all of the coefficients have values determinable relatively independently of the pumping rates and loss rates. The primary variable dictating containment is the slope $\bar{\gamma}_2$, which is given by eqn. (2A.5e); clearly, this may be time-dependent through its dependence on H/H_R but we assume (for calculation in this part only) that γ^{∞} determines most of the growth. The results serve as a check on the numerical solutions, run with the same assumption, and the effects of varying γ_2 can then be demonstrated.

ii) The width of this fracture is obtained by combining eqn. (2A.2b) and (2A.4d) to get

$$\left[\frac{\Delta}{H_R}\right]^{2n+2} = \frac{\Gamma_1\gamma_4 L H^{1-m}\dot{W}^m H_R^{2m-2n}\,\bar{n}}{\Gamma_2\Gamma_4 H_R^{2-m}\,(2\rho H_R^3)^m\,\bar{E}} \qquad (2C.1b)$$

iii) The treatment time required to produce this geometry is now computed from eqn. (2A.4c), rewritten as follows

$$[W(t)-2\rho LH\bar{\bar{q}}_L(t)]t = 2\Gamma_3\rho LH\Delta \qquad (2C.1c)$$

in which the average injection rate is W(t), allowably variable in a known fashion with the time t -- as is the average speed of fluid loss q_L, defined by

$$2LH\bar{\bar{q}}_L(t)t = \int_0^t \bar{q}_L(\tau)A(\tau)d\tau\ ,\ A \equiv 2\bar{\Gamma}_3 LH \qquad (2C.1d)$$

$$\bar{q}_L(\tau)A(\tau) = \int_0^{\tau} q_L(\tau,s)[dA(s)/ds]ds \equiv \dot{W}_L(\tau) \qquad (2C.1e)$$

An interesting special case is the Carter square-root law for loss, arising from a one-dimensional model of flow through both walls of the fracture with a popular loss coefficient k_L^W:

$$q_L(\tau,s) = 2k_L^W/(\tau-s)^{\Phi} \ , \ \Phi \simeq 0.5 \qquad (2C.2a)$$

If we now assume a power-law growth of area with time, we can derive the following expression for loss:

$$A(t)=A_0 \ t^{\alpha} \rightarrow \bar{\bar{q}}_L = \left[\frac{1}{\Gamma_\alpha} \equiv \frac{2\alpha\Gamma(\Phi)\Gamma(\alpha)\bar{\Gamma}_3}{(1-\Phi+\alpha)\Gamma(\alpha+\Phi)}\right]\frac{k_L^W}{t^{\Phi}} \qquad (2C.2b)$$

in which the gamma functions in Γ_α have well-known formulae and tabulation (e.g. Abramowitz and Segun, 1965). Here the coefficient k^W should be essentially proportional to the pressure difference p_f-p_T (see eqn. (1B.4)), which drives the flow through the walls; the result from eqn. (2C.1b), is a simple quadratic equation for the treatment time

$$\left[\frac{\dot{\bar{W}}t}{2\rho LH}\right]^2 = \left[\left(\frac{k_L^W}{\Gamma_\alpha}\right)^2 t^{1-2\phi} + \frac{\dot{\bar{W}}\bar{\Delta}}{\rho LH}\right] t - \left(\bar{\Delta}\equiv\Gamma_3\Delta+\Delta_{sp}\right)^2 \qquad (2C.2c)$$

in which we have lumped the effective opening $\Gamma_3\Delta$ with a popular "spurt loss" Δ_{sp} which is regarded as a time-independent quantity (known for a given fluid and formation):

$$\bar{\Delta} \equiv \Gamma_3\Delta + \Delta_{sp} \qquad (2C.2d)$$

If is found that solutions of eqns. (2C.1,2) agree with those quite well of eqns. (2A.1,3), provided in figure 2A.3.

b. Use of numerical lumped solutions. To implement our more general numerical solutions, we need only insert the correct relationships between the physical dimensions involved and the specified wellbore variables, such as flow rate and pressure. We could use a standard set of units (e.g. ft-lb-sec or cgs) but we provide a broader applicability of each simulation performed if we refer all lengths to a suitable reservoir dimension (e.g. height of pay-zone H_R), all stresses to reservoir modulus $\bar{E}$, and all times to the dominant time-scale $(\bar{\eta}/\bar{E})^{1/m}$ at the wellbore. Thus, given any initial height H_o and length L_o, any injected volume W/ρ or rate $\dot{W}/\rho$ over a treatment time t, the program computes the following dimensionless (which we call "numerical") quantities, employing an arbitrary constant C to avoid very small or large numbers in the computation namely

$$H/H_R \to H, \; L/H_R \to L, \; \Delta \to C\Delta/H_R; \tag{2C.3a}$$

$$\sigma \equiv C\,\gamma_1(p_f^w - \sigma_c^w)/\bar{E}; \tag{2C.3b}$$

$$t^m\bar{E}/\bar{\eta}H_R{}^{2m-2n}\,C^{2n+2-m} \to \hat{t}^m; \tag{2C.3c}$$

$$C^{2n+2} = \rho H_R^3\,(\bar{E}_R/\bar{\eta})/\dot{W}; \tag{2C.3d}$$

$$H_R^{\frac{m-n}{m}}\,Ck_L^w/H_R\left((\bar{E}/\bar{\eta}C^{2n+2-m})^{1/m}\right)^{\phi} \to k_L^w \tag{2C.3e}$$

Because of their great generality, as against the specific results in Fig. 2A.3, we call these dimensionless results "Universal Charts"; sample versions are provided in Figs. 2C.1, showing the predominant effects of modulus and stress contrast between reservoir and adjacent strata. Corresponding plots of pressure behavior show a dramatic effect of these contrasts, which also dominate the shape of fracture growth: this leads to a major application of the models, beyond design, namely for monitoring of job performance and feedback control of pumping schedules through microcomputers mounted at the wellhead.

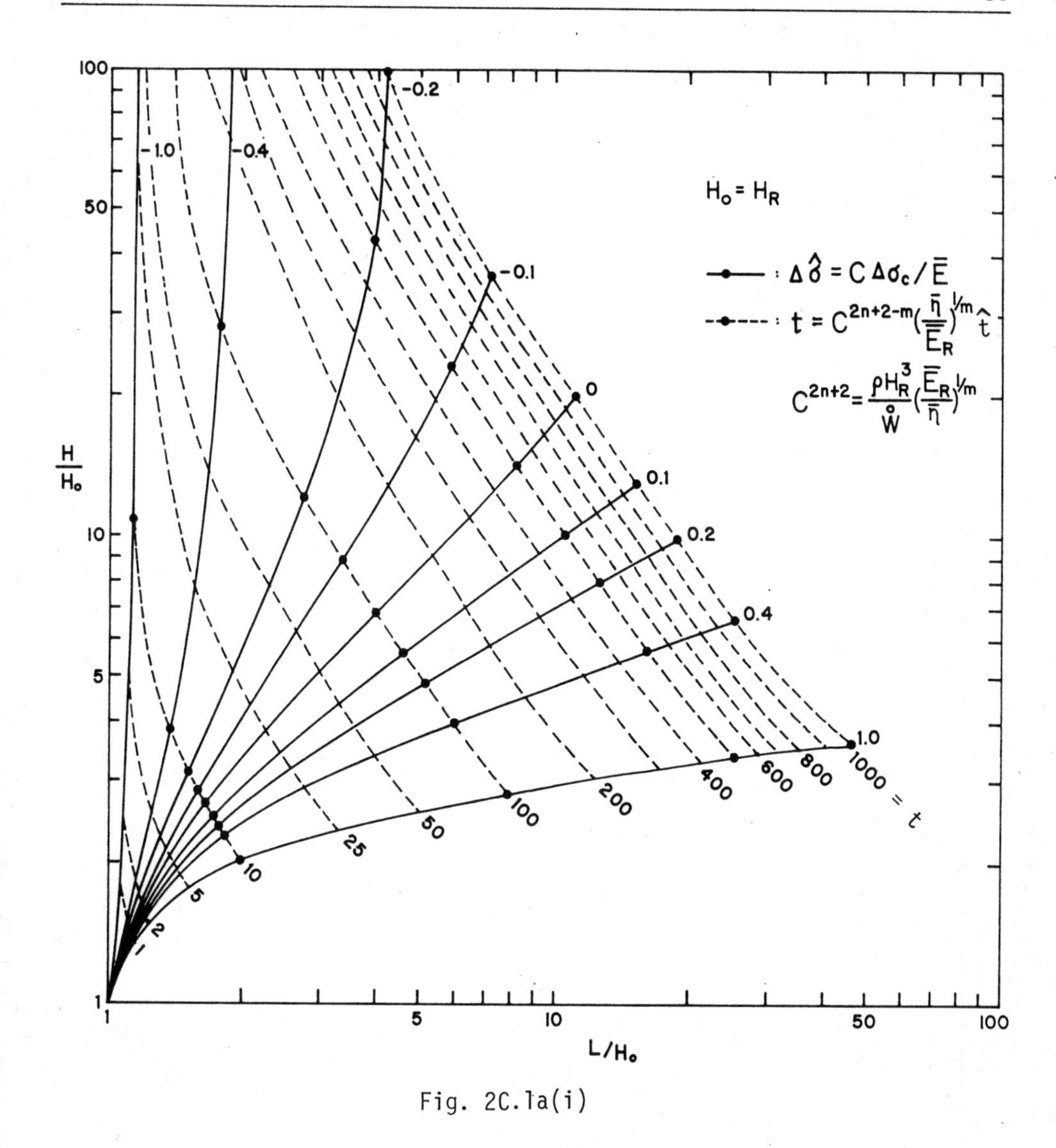

Fig. 2C.1a(i)

Fig. 2C.1. Universal plots for: (i) height and length growth, and (ii) pressure or width of hydrafracs with various barriers to growth: a) confining stress contrast $\Delta\sigma_c$ and b) modulus contrast $\bar{E}_A/\bar{E}_R$ between adjacent stratum and reservoir.

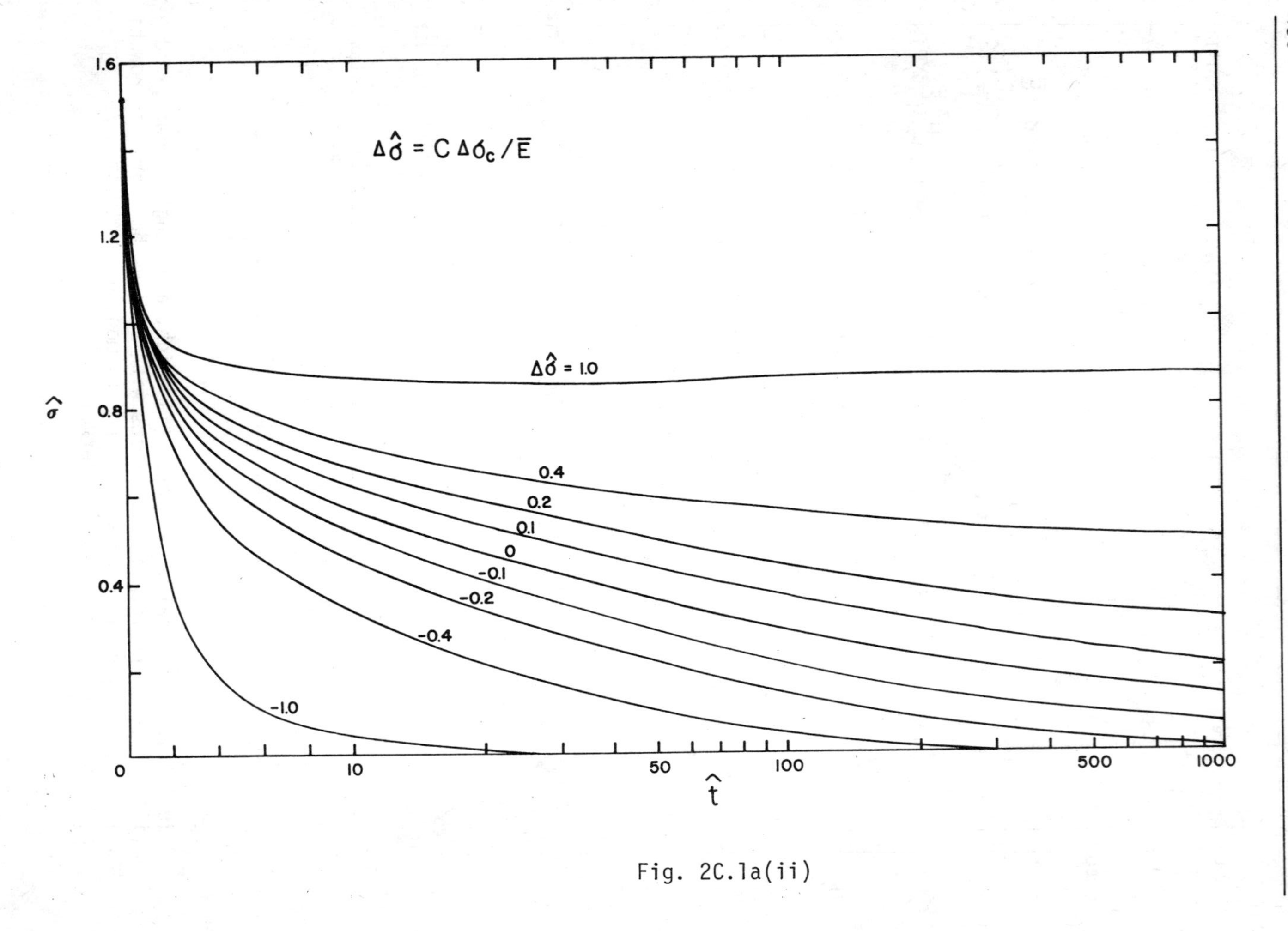

$\Delta\hat{\sigma} = C\,\Delta\sigma_c/\bar{E}$
$\Delta\hat{\sigma} = 1.0$
0.4
0.2
0.1
0
-0.1
-0.2
-0.4
-1.0
$\hat{\sigma}$
1.6
1.2
0.8
0.4
0
10
50
100
500
1000
$\hat{t}$

Fig. 2C.1a(ii)

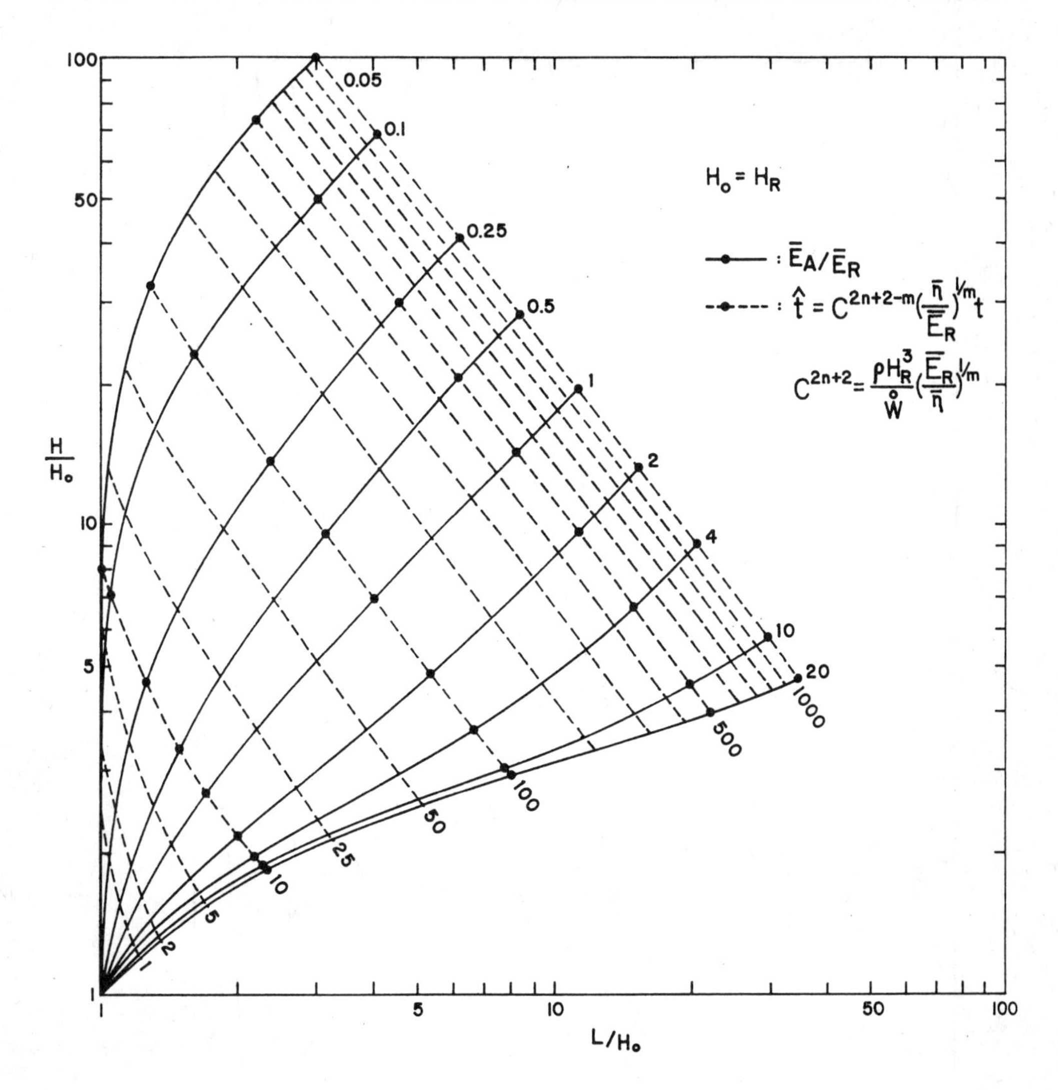
$H_o = H_R$
: $\bar{E}_A/\bar{E}_R$
: $\hat{t} = C^{2n+2-m}(\frac{\bar{\eta}}{\bar{E}_R})^{1/m}t$
$C^{2n+2} = \frac{\rho H_R^3}{\mathring{W}}(\frac{\bar{E}_R}{\bar{\eta}})^{1/m}$
0.05
0.1
0.25
0.5
1
2
4
10
20
1000
500
100
50
25
10
5
2
1
$\frac{H}{H_o}$
L/H_o
100
50
10
5
1
5
10
50
100

Fig. 2C.1b(i)

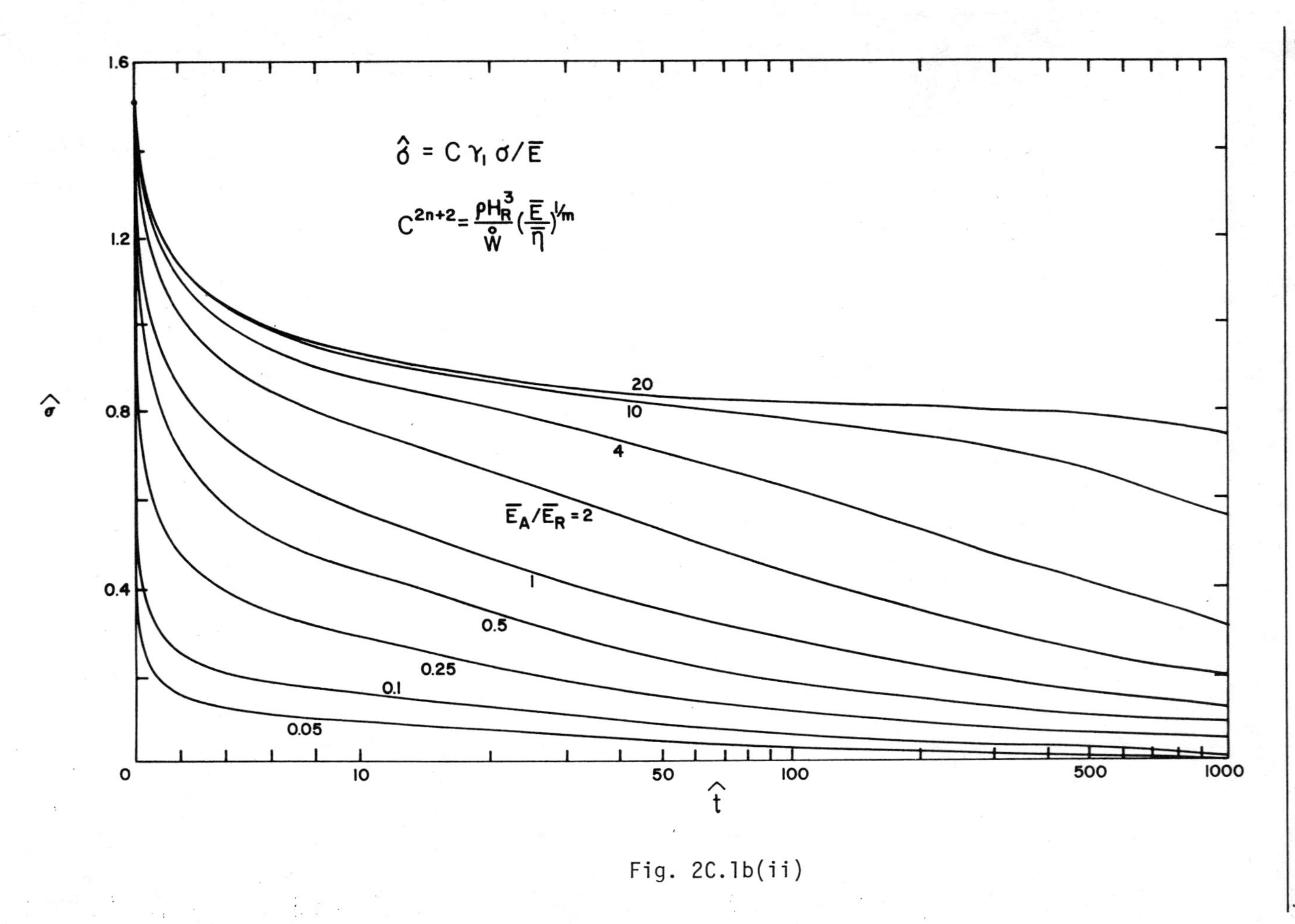

$\hat{\sigma} = C\gamma_1 \sigma/\overline{E}$
$C^{2n+2} = \frac{\rho H_R^3}{\mathring{W}}(\frac{\overline{E}}{\eta})^{1/m}$
$\overline{E}_A/\overline{E}_R = 2$
20
10
4
1
0.5
0.25
0.1
0.05
$\hat{\sigma}$
1.6
1.2
0.8
0.4
0
10
50
100
500
1000
$\hat{t}$

Fig. 2C.1b(ii)

3. DETAILED THEORETICAL MODELLING OF HYDRAFRAC.

There are a number of special tasks which required our attention, both in employing the approximate models of Chapter 2 and in adapting/extending them to the more complicated circumstances which may arise in field operations:

A. The need for baseline precise solutions of the equations governing any generic cross-section of a fracture: without these, we cannot make any estimates of how complex/expensive would be a full three-dimensional simulation or indeed of how good are the various approximations which have been used to model such problems. In particular, the conventional models with constant flow rate and, especially, our own claims of goodness for the self-similar approximation can be tested; such simpler models can then be employed judiciously as components of more realistic overall descriptions (e.g. for height growth in P3DH).

B. The exact solution for one geometry which can be precisely reproduced in the laboratory and allows very rigorous experimental testing of model predictions; such a circular crack model also has practical applications in the field, it allows further testing of self-similar approximations and it serves as a first check for more general 3D simulators.

C. The extension of our coupled fluid-flow and crack-opening solutions to multiple cracks, emanating from one or more (adjacent) wellbores: this problem has profound importance for our long-standing ideas of mechanical guidance and creation of fracture networks underground, among many other applications.

D. Analysis of mechanisms which could control the growth of fracture geometries in typical inhomogeneous reservoirs, especially the role of stratification interfaces and natural fractures; and the special problems of slippage, blunting reorientation and reinitiation near such geological features.

E. The development of a capability for fully 3-D simulation of hydrafrac growth, exploiting the experience gained and efficiencies developed with the various component studies in A-D.

3A. Complete Simulation for a Representative Cross-Section

We have performed the first complete analysis (e.g., Petersen, 1980, 1982), without approximations, for the simplest conceivable fracture geometry (Fig. 3A.1)--one actually assumed in various industrial models (see CGD of Fig. 1C.1). Our threefold purpose was: to fully characterise a generic vertical cross-section of P3DH (Sec. 2B); to evaluate the complexity of a fully 3-D analysis, without the benefit of judicious approximations; and to check the accuracy of various approximate models which have been employed by ourselves and others. In contrast to others available to us, we found that our own results, based on a weak assumption of self- simlarity (Cleary, Wong, Narendran and Settari, 1982) are very accurate, agreeing also with our experiments (Sec. 4A); this observation allows use of that simpler self-similar model to describe cross-sectional growth in P3DH and it also leads to a great simplification in our fully 3-D modelling (Sec. 3E), by means of a leading-edge concept.

For such a cross-sectional model, we have also proceeded to incorporate detailed fluid exchange and heat transfer around the fracture (Cleary, Slutsky, et al., 1982), providing the first efficient procedure for calculating also the stresses induced by these processes--which dramatically affect the fracture growth and the resulting production of reservoir fluids (also in the model).

3B. Development of a Reference Circular Hydrafrac Model

Again, along the lines of Section 3A, we developed the first complete analysis (Wong and Cleary, 1982) of this simplest realistic geometry (Fig. 3B.1), and demonstrated the accuracy of a model based on a weak but greatly simplifying assumption of instantaneous self-similarity between successive crack-opening profiles (Cleary and Wong, 1982); results contrast with others available but are generally in agreement with our own laboratory experiments (Sec. 4A). This justifies use of the more convenient self-similar model, which can now be embellished with reservoir fluid flow and heat transfer capabilities (a la Cleary and Slutsky, 1982, in Sec. 3A). The model

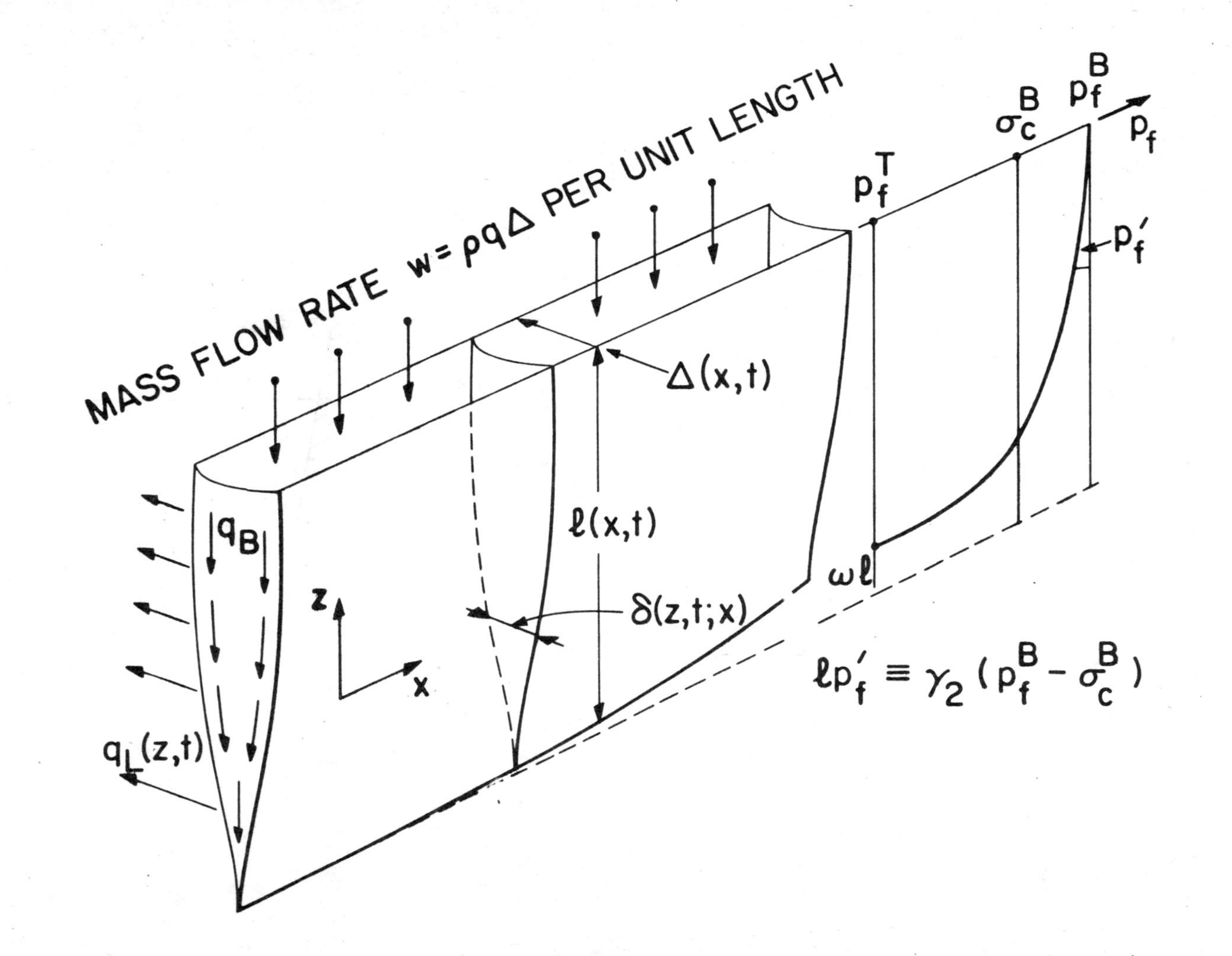

Fig. 3A.1. Schematic and details of typical hydrafrac cross-section.

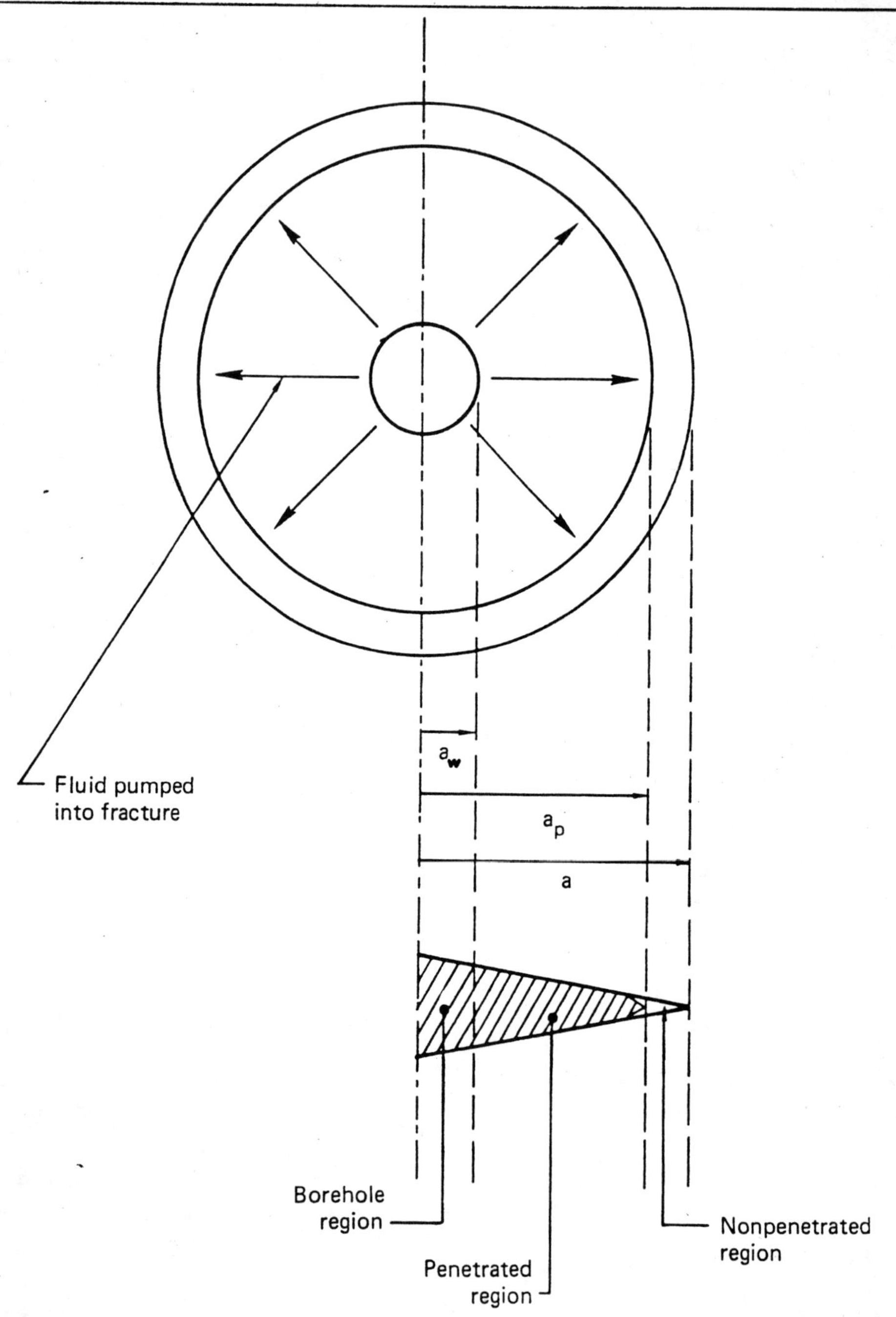

Fig. 3B.1. Sketch of circular hydrafrac model.

serves as a central test program for the full 3-D model (Sec. 3E) but it also stands alone as a useful tool for understanding horizontal fractures, created at shallow depths or along lateral discontinuities--for instance, in the context of tertiary recovery or steam flooding of heavy oil reservoirs; the model also applies to other situations, such as vertical fractures much higher than the perforation interval.

3C. Modelling of Multiple Fractures, Interaction with Reservoir Conditions of Stress, Pore-Pressure and Material Variations

We have performed the first analysis (Crichton, 1980; Narendran and Cleary, 1982) to describe both the mechanical interaction of two or more fractures growing in a reservoir and even the first to trace the re-orientation associated with initiation at an angle to the eventual preferred direction, dictated by tectonic stress orientation/ variation, stratification, inclusions, pore-pressure distribution etc. (Fig. 3C.1). Although apparently not seriously considered to date by the industry, this curving evolution of (multiple) fractures has dominant importance: for instance, in near-wellbore growth (including competing wings and the potential for shear-induced propping); in efforts to achieve horizontal fractures where vertical is preferred; in the design of fracture intersection schemes to kill blow-outs; and in the vital question of creating linked fracture networks (e.g., to form more stable and efficient plane sweeps rather than line drives from, typically five-spot, arrays of injection and production wells in tertiary flooding operations).

A comprehensive computer program has been developed (Narendran and Cleary, 1982). This not alone provides a capability to solve the vast range of associated elasticity problems (most previously inaccessible, but including a variety of established solutions for verification), but it also allows these fracture geometries to evolve in typical reservoir structures, driven by fluid flow from any number of wellbores. Sample configurations have been tested in the laboratory (Sec. 4B) and have verified the accuracy of the program.

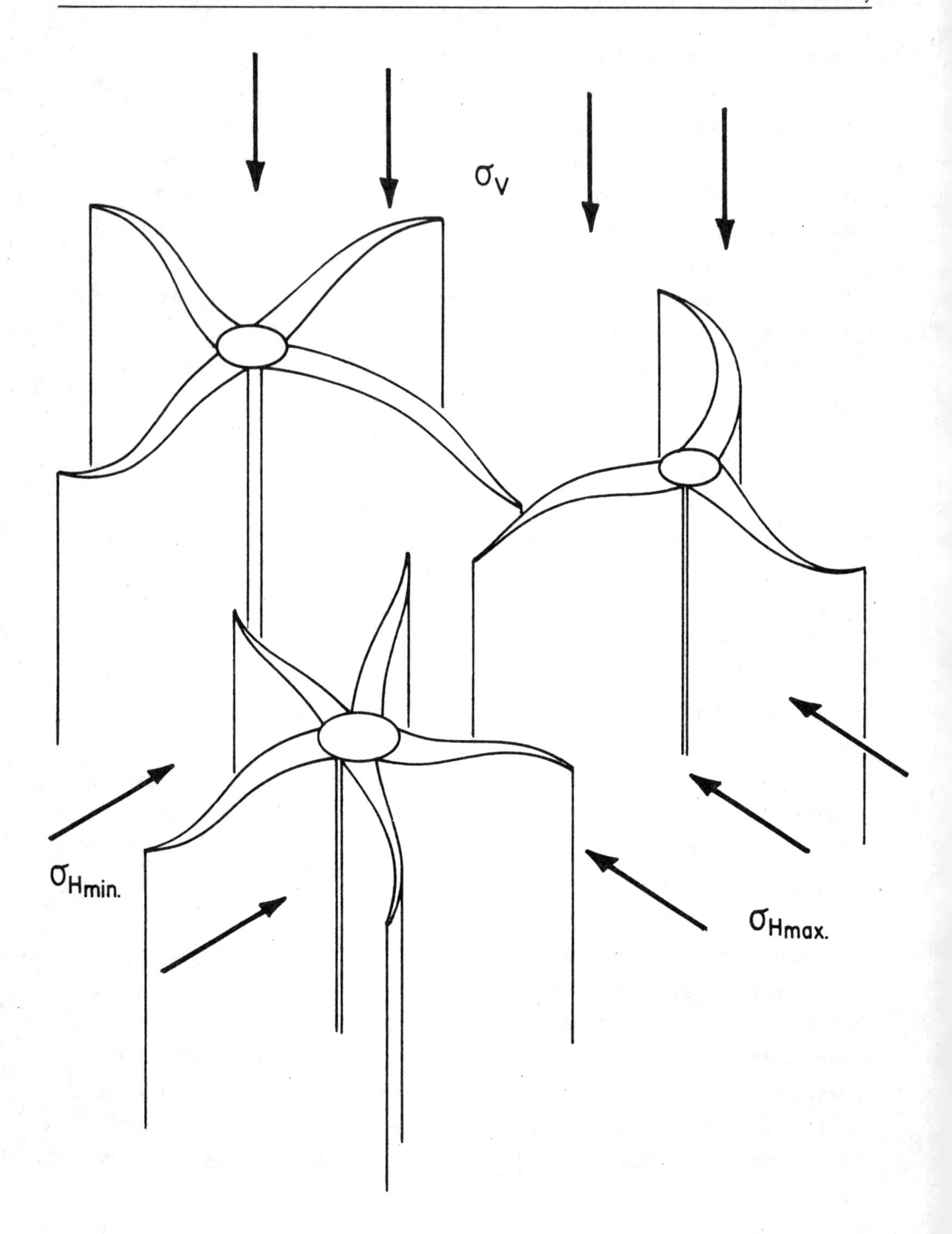

Fig. 3C.1. Illustration of multiple fracture/wellbore configuration.

3D. Fracture Impedance Mechanisms, Branching and Slipping

Although heuristically understood for some time (Cleary, 1978), we have only recently performed the first detailed analysis (Lam and Cleary, 1981) of some primary mechanisms which can dominate the growth of fractures in typical reservoir structures (Fig. 3D.1). A comprehensive computer program has been written to study the role of slippage and blunting at frictional interfaces (including cohesive boundaries between strata), and also to evaluate the potential for branching into pre-existing fractures. The results are quite complex but we have also managed to derive some very simple criteria (e.g., vertical minus horizontal stress greater than tensile strength, for reinitiation across interfaces). Good agreement with limited experimental results (Sec. 4B) has been obtained to date and field observations also support the models: major applications are the containment of vertical fractures within strata and the turning of fractures into natural discontinuities, which has real interest for generation of production-wise more effective but stress-wise unfavorable horizontal fractures at greater depths.

3E. Fully Three-Dimensional Simulation of Fracturing

We have developed a tractable formulation and numerical implementation of the full 3-D fracture problem (Cleary et al, 1979, Wong and Cleary, 1981; Kavvadas and Cleary, 1982). Based on our experience with components in the simpler models (Secs. 2A,2B,3A,3B), we have been able to render the analysis amenable to practical usage, a feature which seemed evasive in previous attempts at such modelling. The principal ingredients are an effective surface integral scheme based on local interpolation for crack opening, (and eventually for heat/mass transfer in the reservoir also), a finite element representation of fluid flow inside the fracture and a leading-edge model for the detailed mechanics of coupled opening and flow near the perimeter (Fig. 3E.1).

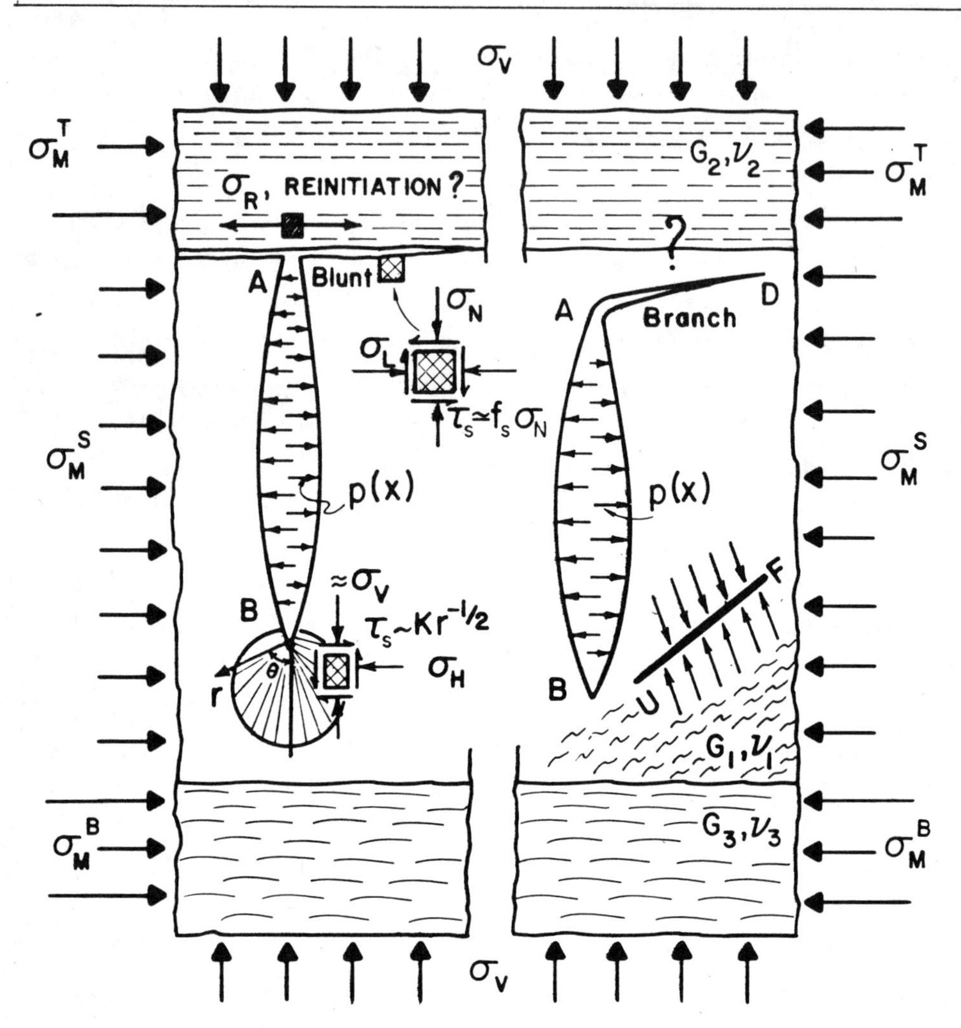

Fig. 3D.1. Sample of mechanisms controlling fracture growth. For reinitiation after interface slippage, a simple criterion is $\sigma_V - \sigma_M > \sigma_T$ (tensile strength).

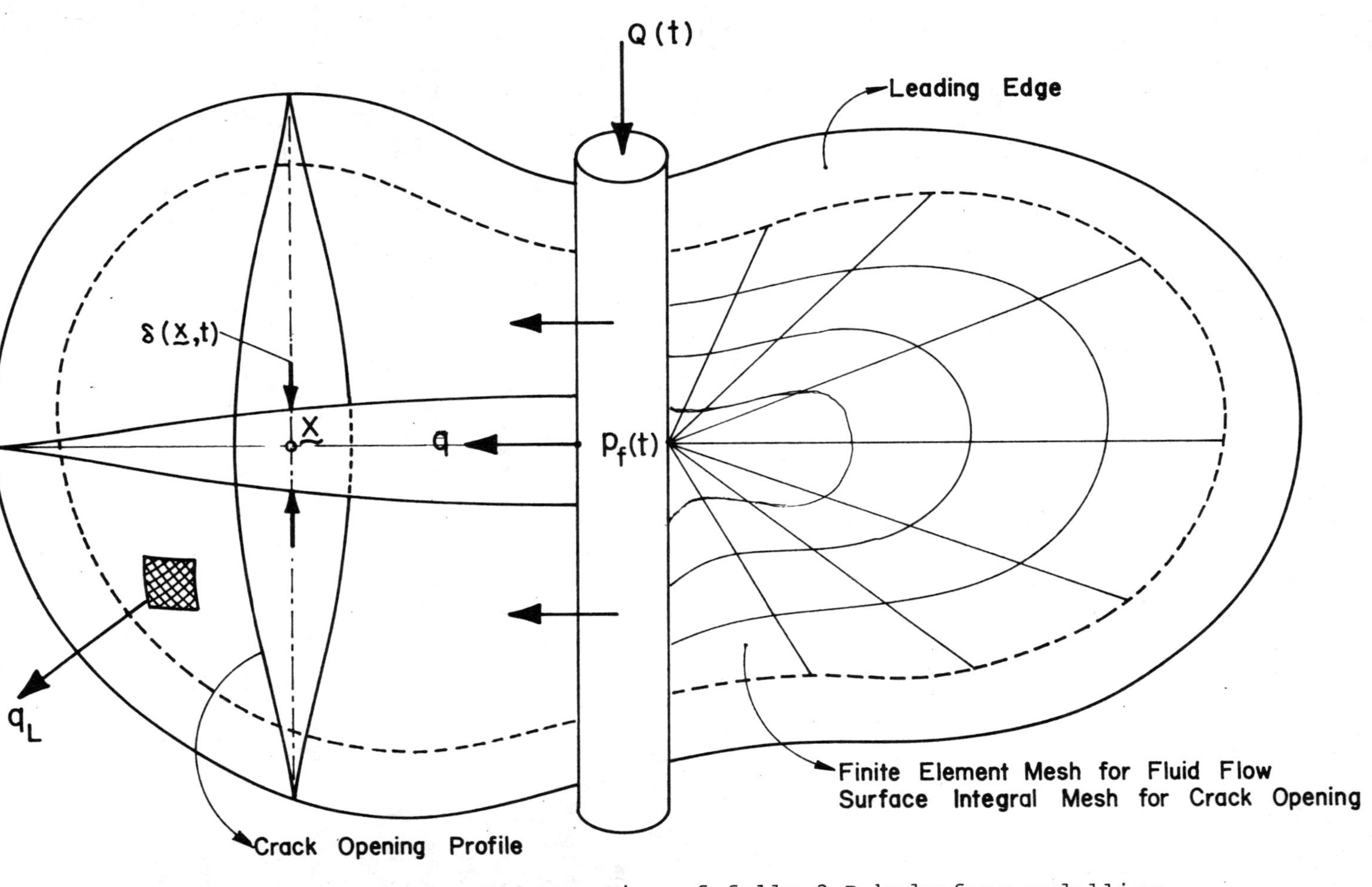

Fig. 3E.1. Illustration of fully 3-D hydrafrac modelling.

4. LABORATORY SIMULATION OF FRACTURING.

We have long maintained that much could be done in the laboratory to verify developing models and gain new insight on phenomenology, before field testing of the purely research kind (e.g., not tied to a well which would be drilled for production anyway) should be pursued at great expense. Our experimental work to date strongly supports that contention, both in the progress we've made and the failures we've had: we believe we now have the potential to simulate field conditions up to a high level of complexity, with precise control of all the parameters, but we have also learned how many things can go wrong in a nominally perfect arrangement of wellbore, perforated casing and prefractured walls-- most of which we have eventually remedied, owing to the accessibility of the "test location." Progress has been made on the following primary projects: A. Construction of general-purpose simulator for highly repeatable observation of hydrafrac growth at low expense of time and materials. We achieve this by interface separation, a novel concept in the present context; this achievement is based on scaling of calculations for field conditions, which show that fracture toughness most often plays a negligible role in the overall energy balance, and hence can be neglected. B. Examination of crack evolution, interaction and re-orientation in typical reservoir structures with biased stress fields, where details of material decohesion do play an important role in deciding direction of growth. C. Development of the capabilities for facile data acquisition and control in each of the relevant experiments, mainly employing new effective software and combining a number of purchased or constructed hardware components. D. Testing of various schemes for monitoring the growth of cracks, using acoustic or deformation or electrical signals. E. Investigation of the rheological behavior of various fluids to be used both in our lab simulators and in the field; one final goal may be to design new fluids with suitable mechanical properties.

4A. Interface Separation Apparatus, DISLASH

This Desktop Interface Separation Laboratory Apparatus for Simulation of Hydrafrac (Fig. 4A.1) is the first of a kind which allows repeatable verification of the role played by various parameters in the evolution of hydraulic fractures. It is based on our theoretical demonstration (Cleary, SPE 9259, 1980) that actual fracture energy typically plays a negligible role in the overall balance of work done against confining stresses during hydraulic injection to achieve reasonable crack openings; this may be verified eventually by re-opening of created fractures. Thus, we can avoid breaking samples and throwing them away, provided we are interested mainly in fracture growth along a predetermined surface (normally dictated by the minimum confining stress). The result is a dramatic simplification in the process of reliable laboratory simulation of hydrafrac. The apparatus has the following principal features:

- it is simple and low cost (running off any laboratory air pressure supply), encouraging duplication
- tests are rapidly doable (roughly 20 per day),
- it is convenient in size (actually portable),
- it is easily modified to accommodate new parameter studies,
- instrumentation can be simple or complex, as desired (Sec. 4C), analogous to field requirements.

As well, it is an excellent tool for research, teaching and even design:

- the fracture growth is clearly visible for demonstration and monitoring (e.g., by laser),
- components are very well characterised,
- it is easily instrumented, and
- a wide range of design conditions can be investigated, without major modifications

Indeed, the apparatus can accommodate many complexities (e.g., fluid rheology, boundaries) more readily than computer simulators--

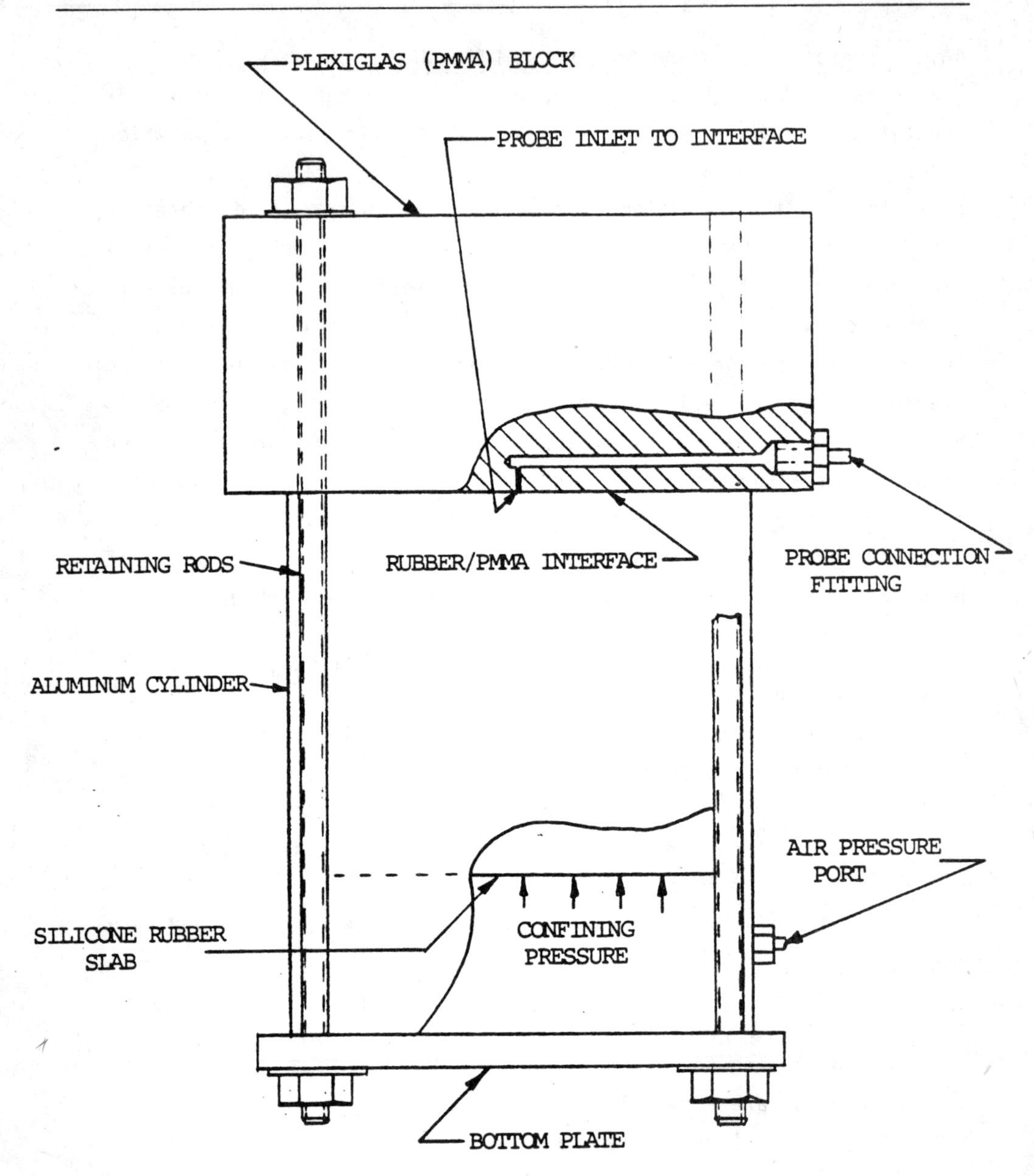

Fig. 4A.1a. Desktop Interface Separation Laboratory Apparatus for Simulation of Hydrafrac (DISLASH): outline of apparatus.

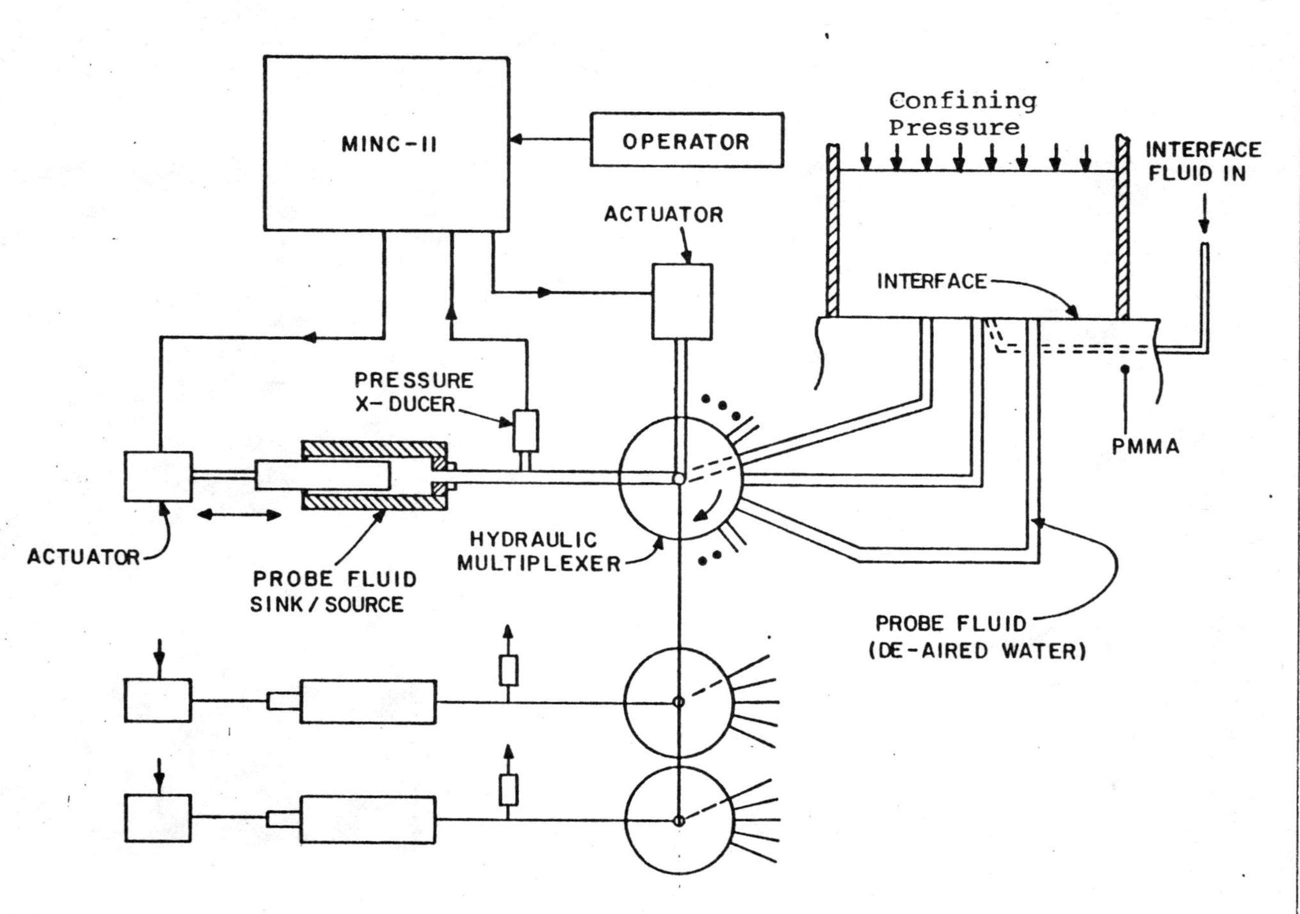

Fig. 4A.1b. Desktop Interface Separation Laboratory Apparatus for Simulation of Hydrafrac (DISLASH): pressure probe system.

Components of DISLASH; Mylar-covered cylinder and cylinder extension in foreground; PMMA blocks with borehole/slit for Penny/CGDD simulation.

Fig. 4A.1c.

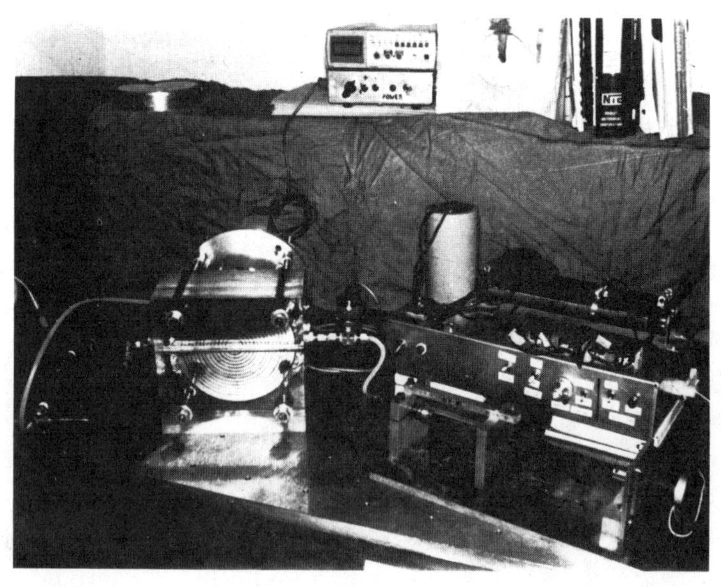

DISLASH set up to run, tipped onto front for photograph. To its right are motorized syringe and stepping motor power supply. Above is transducer power supply and voltmeter readout.

which it can also be used to test, a major purpose of its development.

We have invested a great deal of time in the development of this novel apparatus (Papadopoulos, Summa and Cleary, 1982): although the basic concepts and equipment were established quickly, there is an extraordinary amount of potential for apparatus deviation from the ideal model conditions--a lesson to be well learned for application of these models to underground operations.

Primary difficulties have been the following:

- The control of temperature, which primarily generates thermal stress changes on the interface, hence changing growth patterns and rates; however, we must note that our constrained apparatus (a potential thermometer) is much more sensitive than rock structures in this respect.

- Adhesion at the interface; this can be non-repeatable and non-negligible, so we have reduced it to a minimum; but we will later be interested in using it to capture the role of rock fracture toughness/energy.

- Inhomogeneity of the fluid injected, especially with dissolved solids (such as otherwise Newtonian sugar solutions) and viscoelastic clumping instabilities.

- Solid material departures from ideal behavior--especially chemical reactivity with injected fluid and creep.

Although all of these effects are very interesting in themselves, they need to be eliminated for comparison with simpler computer model results and then isolated for separate study (sometimes more readily, indeed, than they can be added to the theoretical simulators).

To date, we have used DISLASH to verify some essential components of the basic theory, such as the following:

- the validity of a theoretical characteristic time,

- the secondary role of confining stress, for a given excess pressure.

- the numerical results for some simple geometries (e.g., circular, CGD and PK plane fractures).

Although data-taking on these aspects is not complete, we are working on a number of other features, such as:

- confining stress variation on the interface
- modulus contrast between strata,
- permeable model material to allow fluid loss and to investigate induced backstreses, and
- deliberate thermal effects on confining stress and fluid rheology, hence on growth shapes of the fracture.

When the potential of the apparatus is fully exploited, we should have the capability to perform a very broad range of simulation in an attractive convincing way, which may lead to proper understanding of hydraulic fracturing, in all its complexity.

4B. Apparatus for full 3-D Fracture Growth and Interaction

Although DISLASH will be able to show the effects of many diverse factors on MHF (Sec. 4A), by its very nature it precludes study of fracture turning and branching ("reorientation"), except perhaps into specifically designed cuts in the model material. This aspect is of interest for a great number of reasons:

- If hydraulic fractures generally turn towards each other and link up, the resulting massive fracture could be practically used, e.g., in tertiary oil recovery or proposed solution mining schemes;
- Such knowledge would also let us predict how a fracture behaves near a free surface or a formation inhomogeneity (e.g., stratification interface or sand lense);
- We could determine how fractures evolve after initiation from a borehole, e.g., at an unfavorable angle to the tectonic stress field;
- We could predict growth in varying stress (and moduli) fields, e.g., caused by drawdown of pressure or fluid/heat injection in nearby wellbores;
- Finally, the topic actually deserves fundamental study, as

it is by no means generally understood, and many experiments (with inherently unstable growth) are currently ongoing, with various resulting claims about growth laws.

We are completing development on an apparatus with which we can address the many questions. Originally, it was designed for qualitative insight into the behavior of MHF: with it we can see the interaction of arbitrarily placed hydraulic fractures with each other, or biased stress field, or a free surface. However, since the fractures grow stably and continually (in contrast to any others that we are aware of) the system promises to be useful for studying fracture turning at a number of levels, especially if we can find or develop materials with various scales of heterogeneity (e.g., flaws).

The setup is shown in Fig. 4B.1 (Papadopoulos, Narendran and Cleary, 1982). We use cement specimens for their unusually low toughness, low cost, high strength, and easy preparation in any size. Acetate wings, representing initial fractures (and joined to fluid reservoirs) are cast into the specimen in any desired position and orientation. The specimen is placed in a large pressure vessel (which supplies the confining stress necessary for stable growth), and an axial hydraulic actuator provides a stress bias. Finally, the fractures are propagated by pressurising them internally to pressures higher than that in the vessel. The whole process is run by the laboratory micro-computer (Sec. 4C).

The particular values of this technique are several:

- it is unique in providing stable fracture growth;
- it is very versatile in allowing placement of internal features with any orientation and geometry;
- we can, both phenomenologically and quantitatively study factors affecting the propagation direction of fractures viz., we can begin to characterise the effect of material and earth stresses (pre-existing or induced) in various turning/branching configurations; and
- we can develop the capability to allow better representation of reservoir rock properties.

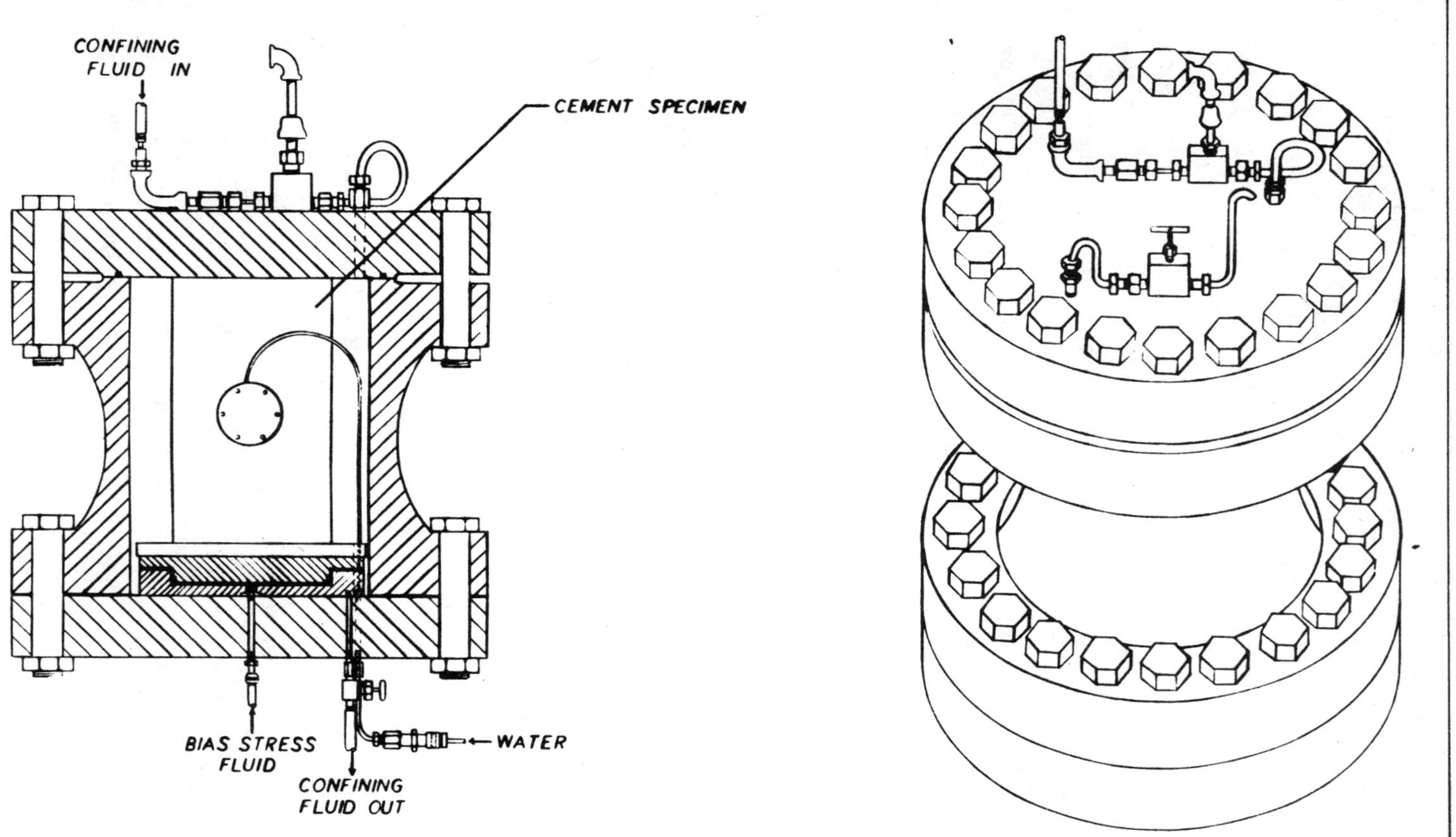

Fig. 4B.1. Schematic of experimental set-up: cement block, in pressure vessel.

To date, we have performed a large number of tests with the apparatus, developing confidence in it, extracting much new phenomenology and verifying the predictions of the corresponding theoretical models (Sec. 3C). Principal among these experiments have been the following (see samples in Fig. 4B.1b):

i) Single cracks in homogenous specimens under hydrostatic pressure. These grow straight ahead, as expected, except for turning induced as they approach specimen boundaries--which is also in agreement with model predictions;

ii) Two cracks in homogeneous specimens under hydrostatic pressure. These show distinct strong interaction, as they curve to embrace and link up in configurations typical of adjacent wellbore fracturing in the field;

iii) Single cracks in a biased (i.e., nonisotropic or nonhydrostatic) confining stress field, at various initial orientations. These turn dramatically toward the maximum stress direction, with a rapidity which increases as a strong function of bias stress divided by excess pressure, again as predicted;

iv) Two cracks in a biased stress field. The resulting growth is an interesting combination of (ii) and (iii), with competing effects in the middle region.

A large number of repeatibility and parameter-study tests have been run for the foregoing configurations, but we have also performed a number of exploratory tests for the following situations (which we plan to examine further):

v) Growth of a single fracture near an inclusion. Here we observed such phenomena as breakthrough over one section of the fracture while another segment managed to curve around the (stiffer) inclusion as it approached;

vi) Propagation of a hydraulic fracture across another fracture, precreated but closed by confining pressures great enough (in agreement with our predictions) to allow frictional grab and reinitiation across the interface.

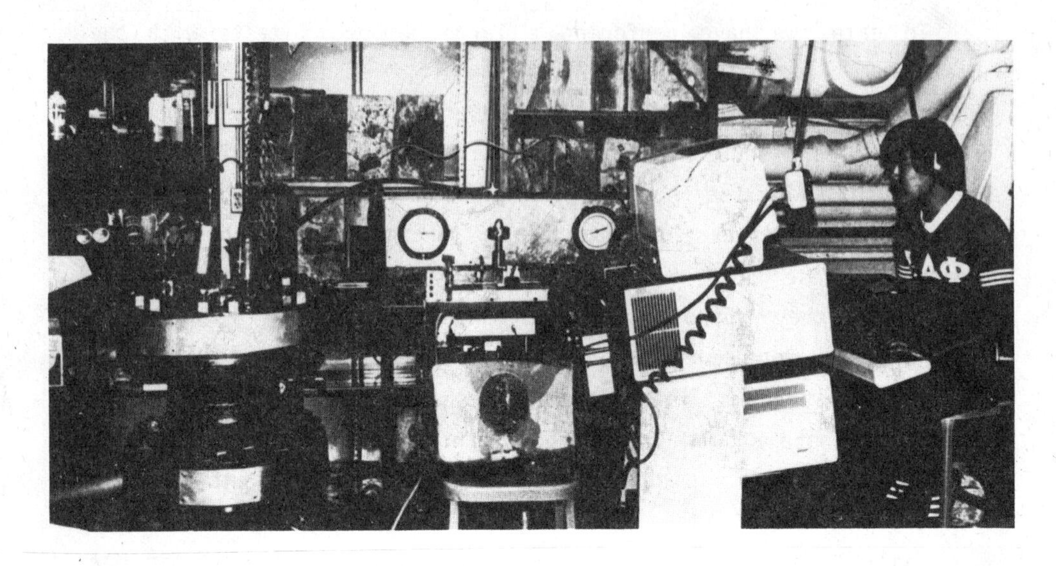

Fig. 4B.1b. Photograph of apparatus for crack interaction experiments (above) and typical sample before/after hydraulic fracturing (below).

As well, we plan to conduct extensive tests in investigating the following additional phenomena:

vii) Growth of a fracture in a layer of (e.g., reservoir) material different from surroundings, showing the role of modulus, permeability and stress contrast;

viii) The role of pore-pressure distribution in fracture growth, especially the phenomenon of fracture attraction by drawing down wellbores (e.g., in killing blowout wells);

ix) Growth of fractures from locally admissible to globally-preferred shapes (e.g., from favorable dished geometries to reverse curvature under biased stress, which we've already seen in our experiments).

4C. Development of Data Acquisition and Control Systems

An important capability necessary to any modern experimental laboratory is an automated data acquisition system. Such a system permits a large amount of data to be taken in a very short time interval and also frees the experiment operator from the tedious job of recording the experiment variables as well as eliminating the unreliability of human observation. Also important, particularly in areas involving fluids, is the capability to accurately control specific variables during an experiment. Keeping these variables constant or changing them with prescribed time histories allows the remaining variables to be more readily analyzed.

The most accurate method of controlling variables such as fluid flow and pressure, particularly if these are changing with time, involves the use of a servo-control system. The simplest servo-control system consists of a transducer to measure the variable, an actuator to change the variable and a "black box" that senses the difference between the actual value of the variable and its desired value, amplifies this difference and uses the amplified result to drive the actuator. The decision making or difference measurement part of the system may be implemented in either hardware or software, the latter being more flexible. Common to both areas of data acquisition and servo-control is the laboratory computer. Its high speed storage capabilities make it ideal for collecting data and its computational capabilities can be used for data manipulation and servo-control. Fig. 4C.1 shows the general data acquisition and control configuration. Details of this system and applications are provided by Morris and Cleary (1982), who give references to our other work on various component development.

4D. Monitoring of Fracture Growth in the Laboratory

The objective of this project is to develop a system to monitor crack propagation in laboratory simulations of underground fracturing. Without the ability to monitor the crack growth with time, the numerical models cannot be adequately correlated with the simulations, and even the physical lab process cannot be fully described. A crack monitoring system would play a key role in the further development and verification of numerical models of hydraulic fracturing. As well, the techniques developed in the laboratory should have potential for scaling-up to reservoir conditions. This would yield a much needed tool for the evaluation and improvement of hydraulic fracturing technology in the field.

A laboratory crack monitoring system must meet several requirements.

- it must detect cracks in cement specimens at depths of order 10 cm.,
- it must be able to monitor crack depth, shape, size, and orientation as well as crack propagation rate,
- a computer graphics system giving three-dimensional pictures from crack monitoring data would be desirable, and
- one monitoring system must be able to operate in the apparatus described in Sec. 4B.

In principle, any method used in the broad area of non- destructive testing for detecting flaws or cracks in solids could be adapted to this application. However, we had to undertake

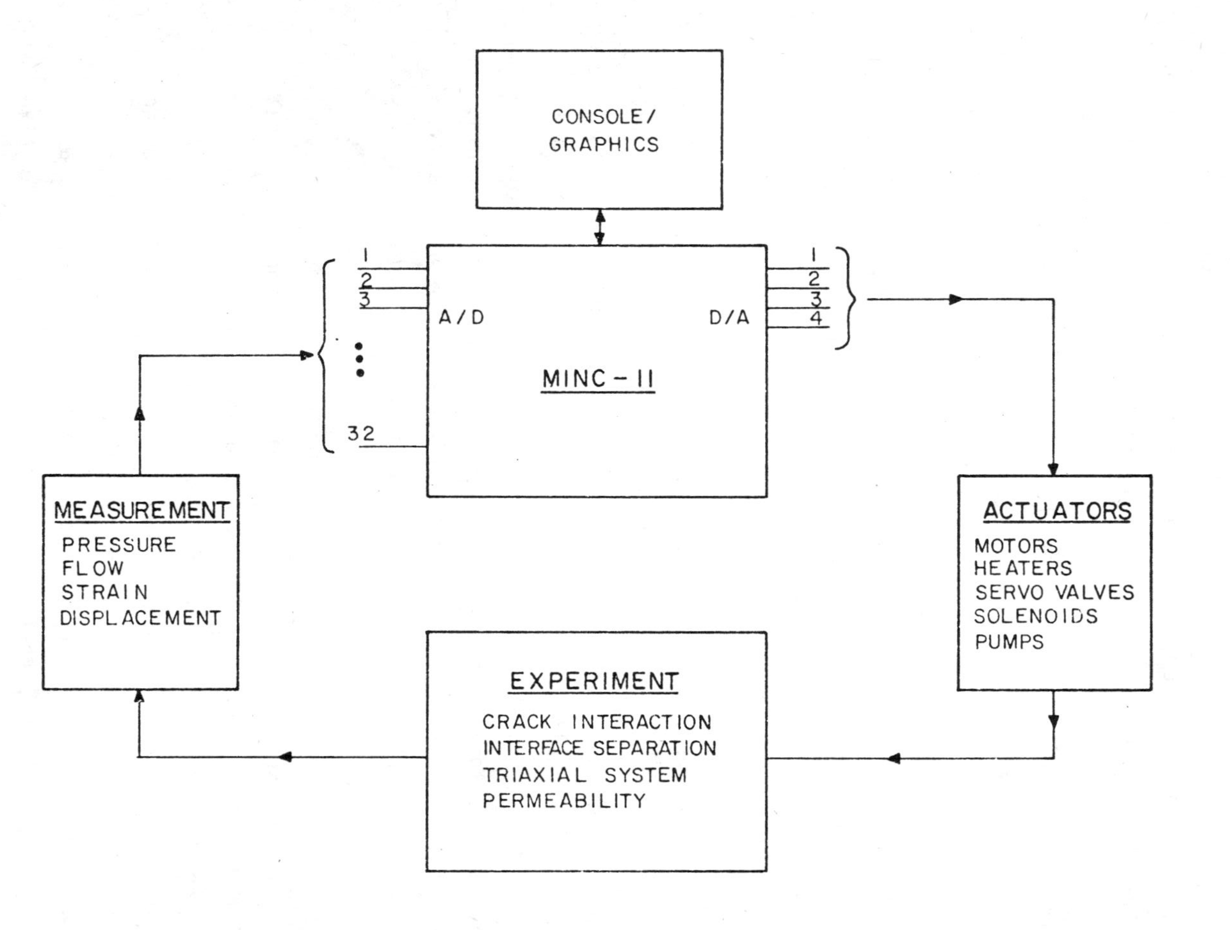

Fig. 4C.1. Overview of laboratory data acquisition and control system.

considerable work in order to develop a method which would apply to our situation.

We focused particularly on ultrasonics and did extensive testing on wave propagation in cement specimens with various geometries. We designed a workable scanning system (like that used in biomedical engineering) but the cost discouraged construction. Meanwhile, we developed a simple marker system (based on pressure pulsing at regular intervals) to delineate successive crack perimeters; this definitely rendered the ultrasonic scanning system unattractive for immediate laboratory development but we have pursued similar studies for field applications (aimed toward seismics and potential logging tools). In a corresponding effort, using the optical (vs. sonic) transparency of DISLASH, we have been developing a laser scanning technique for mapping the geometry evolution of interface separation.

4E. Some Studies of Fluid Rheology

Hydraulic fracturing has been successful without a complete knowledge of the fluid characteristics in the fracture but some fascinating modifications are possible to improve performance (e.g., Cleary, SPE 9260, 1980). Fracturing fluids can be various polymer-water solutions, gels and oil emulsions, foams etc., which are usually non- Newtonian in behavior. Shear thinning or thickening, history dependence, and normal stress differences are non-Newtonian and viscoelastic fluid characteristics, which the fracturing fluids may exhibit. A feature of dominant importance typical of all fluids is a change in viscosity with temperature; much less pressure-sensitivity is expected. This project aims to combine rheological theory and experimental data to develop a model which will predict fracturing fluid flow behavior under practical conditions. Prospective fluids for injection into DISLASH (Sec. 4A) are being tested and classified so that the growth data can be interpreted correctly. To date, a number of rheometers have been employed and a special reciprocating pipe-flow rheometer has been built to give long-term data on history, temperature and pressure effects.

5. LABORATORY TESTING OF MATERIAL RESPONSE

The simulators we develop, in theory or laboratory, will require the input of parameters which adequately describe the rock behavior underground. As well, there are still many mechanisms that we have postulated, which have to be thoroughly investigated in the laboratory. Both aspects require the availability of test systems which can reproduce reservoir conditions with significant accuracy. Thus we have pursued two major equipment building projects:

A. A pressure vessel and peripherals capable of generating and sustaining high temperatures typical of oil/gas/geothermal reservoirs and pyrolysis processes (e.g., shale oil and tar sands); and

B. A novel self-contained large triaxial cell (Like A.,requiring no test frame), capable of providing highest anticipated reservoir stresses and allowing general flow conditions through samples (including deformation, rupture and hydrafrac).

An excellent example of one major application for both is the novel concept of pore-pressure-induced-cracking (PPIC); samples will crack extensively (due to an effective stress law for strength) if they are subjected to high pore-pressures and low confining stress, a natural phenomenon in overpressured pressured (hence fractured) reservoirs. We have achieved this state in a variety of ways:

- by the simplest artifact of dropping a sample in the vessel, thoroughly soaking the sample, with pressure greater than the "tensile strength," using various flow configurations, and then suddenly dropping the pressure;
- by heating samples with volatile phases in their pore-space, without allowing time for release.

There are numerous other means which we are exploring, such as injection of chemicals suitable for reactions with pore or matrix material. The concept seems to work very well: we have found extensive fracture networks in samples subjected to PPIC by the above procedures (e.g., Fitzpatrick and Cleary, 1982).

5A. High Temperature Triaxial Test System

The purpose of this work is to develop a system from which to gain information on the characteristics of reservoir rocks at elevated temperatures and pressures, particularly information on the potential for inducing vapor pressures in the porespace of the rock. Pressures can be created by the vaporisation of water and other porespace fluids as well as by the decomposition of porespace material or minerals in the rock. It is possible that induced pressures would be sufficient to form enough cracks to significantly improve resource extraction in reservoir fracturing. In order to gain representative data on inhomogeneous materials, it is necessary to test large samples. A sample four inches in diameter and four inches long was considered to be a good compromise between the desire for large samples and the constraints of equipment size.

A high temperature triaxial test system has been developed for the purpose of simulating in-situ reservoir conditions (Fig. 5A.1, Heintz, Barr and Cleary, 1982). Maximum working temperatures and pressures are currently 600 degrees F and 10,000 psi, respectively; both of these limits are to be increased in the future, with the implementation of new equipment and the use of a new confining fluid. The system is designed for testing "large" samples (10 cm. = 4 in. in diameter and length), an important feature when working with inhomogeneous materials. The system also has the tested capability to measure pore pressures with the option to control the release of any produced by preactions of various species in the porespace. Possible applications of this system are as follows:

- conventional extraction from oil and gas reservoirs,
- wet geothermal energy sources,
- in-situ retorting of oil shale,
- underground gasification of coal,
- chemical reactions in the pore-space,
- heavy oil and tar sands.

Fig. 5A.1. Photograph of 69MPa pressure vessel and some sample experiments: pore-pressure-induced-cracking (PPIC) and multiple fractures from jacketed borehole.

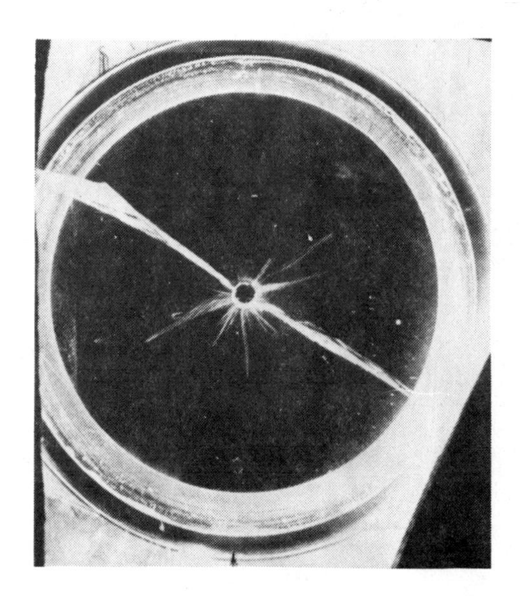

5B. Permeability, PPIC, and High Pressure Triaxial Test System

The major objective is measurement of permeability and study of PPIC on test specimens (nominalply between 10 microdarcy and 100 millidarcy) that are under either isotropic or anisotropic stress conditions. Since the permeability data must relate to field stress conditions a pressure vessel for confining the samples is required. The system must include a stable pore fluid supply pressure for flow through the samples. It soon became apparent that a properly designed permeability system also could be used to do various other types of experiments, including conventional triaxial testing, cracking, burst tests, and hydraulic fracture growth/ interaction.

The nominal upper limit for isotropic stress applied to either the sample or the pore fluid is 345 MPa (50,000 psi) which is also the maximum compressive deviatoric stress. The maximum extensional deviator stress is 69 MPa (10,000 psi). The maximum space to be available inside the triaxial cell is 17.8 cm. (7 inches) diameter by 41.9 cm. (16.5 inches) high, although the maximum sample size to be used will be 10.2 cm. (4 inches) diameter by 20.3 cm. (8 inches) long. Both inflow and outflow will be measured during permeation at flow rates between 1 and .001 cc./sec. The fluid flow subsystem will be capable of handling a wide variety of fluids; temperature limit for system operation should be the temperature stability limit of the fluids. Data will be collected remotely to ensure adequate safety standards.

The overall basic design for a functioning triaxial/permeability system is given by Maley, Martin and Cleary, 1982. The various interdependent components of the system are:

1) permeability measuring subsystem;
2) triaxial test cell (pressure vessel);
3) hydraulic subsystem (pumps and exterior plumbing);
4) transducers for measuring pressure, load, deformation, flow, and temperature; and
5) safety arrangements (specifically, a "strong room").

LIST OF REFERENCES

Abe, H., Mura, T., and L. M. Keer, "Growth Rate of a Penny-Shaped Crack in Hydraulic Fracturing of Rocks," Jour. Geophys. Res., 81 (29), 5335-5340, 1976.

Abramowitz, M. and I. A. Stegun, Handbook of Mathematical Functions, Dover, New York, 1965.

Abou-Sayed, A.S., "Fracture Toughness K of Triaxially-loaded Limestones," pp. 2A3, 1-8 in Proc. 18th Symposium on Rock Mechanics, June, 1977.

Abou-Sayed, A. S. Jones, A. H., and E. R. Simonson, "On the Stimulation of Geothermal Reservoir by Downward Hydraulic Fracturing, "Paper No. 77-Pet-81 presented at ASME Energy Technology Conference and Exhibit, Houston, Sept. 1977.

Abou-Sayed, A. S., Brechtel, C. E., and R. J. Clifton, "In-Situ Stress Determination by Hydrofracturing -- a Fracture Mechanics Approach," Jour. Geophys. Res., 83, B6, 2851-2862, 1978.

Abou-Sayed, A.S., Ahmed, U. and A. Jones, "Systematic approach to massive hydraulic fracturing treatment design", Paper No. SPE/DOE 9877, presented at the Symposium on Low Permeability Gas Reservoirs, Denver, May 1981.

Advani, S. H., "Finite element model simulations associated with hydraulic fracturing," Paper No. SPE/DOE, 8941 (presented at Unconventional Gas Recovery Symposium, Pittsburgh), May 1980.

Agarwal, R.G., Carter, R.D., and C. B. Pollock, "Evaluation and Performance Prediction of Low-permeability Gas Wells Stimulated by Massive Hydraulic Fracturing," Jour. Pet. Technology, 31, 3, 363-372, 1979.

Allen, T.O. and Roberts, A.P., "Production Operations", 2 Volumes, pub. by Oil and Gas Consultants, Inc., Tulsa, 1978.

Atkan, T. and S.M. Faroug, Ali, "Finite Element Analysis of Temperature and Thermal Stresses Induced by Hot Water Injection," Soc. Pet. Eng. Journal, 457-469, 1978.

Barr, D.T.,"Thermal Cracking in Nonporous Geothermal Reservoirs",

S.M. thesis at MIT, May, 1980

Batchelor, A.S., "The creation of hot dry rock systems by combined explosive and hydraulic fracturing", Report of Geothermal Energy Project, Cambourne School of Mines, Cornwall, U.K., presented at Intl. Conf. on Geothermal Energy, Florence, Italy, May 1982.

Berry, P., Cataldi, C., and M.E. Dantini, "Geothermal Stimulation with Chemical Explosives," ASME Paper No. 78-Pet. 67, 1978.

Christianovich, S. A., and Yu. P. Zheltov, "Formation of Vertical Fractures by Means of Highly Viscous Fluid," Proc. 4th World Pet. Cong., Vol. 2, pp. 579-586, 1955.

Cinco-Ley, H., F. Samaniego-V, and Dominuez, A. N.: "Transient Pressure analysis for fractured wells", pp. 1749-1766, Jour. of Pet. Tech., Sept. 1981.

Clark, P. E., "Prop Transport in Vertical Fractures", Paper No. SPE 10261, 1981.

Cleary, M. P., "Continuously-Distributed Dislocation Model for Shear-Bands in Softening Materials," Int. Jour. Num. Methods in Eng., 10, 679-702, 1976. Cleary, M.P., "Primary Factors Governing Hydraulic Fractures in Heterogeneous Stratified Porous Formations", Paper No. 78-Pet-47 presented at Energy and Technology Conference of the Petroleum Division, ASME, Houston, Nov. 1978.

Cleary, M. P., "Rate and Structure Sensitivity in Hydraulic Fracturing of Fluid-Saturated Porous Formations," pps. 127-142 , in 20th U.S. Symp. on Rock Mech., pub.by SPE-AIME and University of Texas at Austin, Texas June 1979.

Cleary, M. P., "Comprehensive Design Formulae for Hydraulic Fracturing", SPE 9259, and "Mechanisms & Procedures for Producing Favourable Shapes of Hydraulic Fractures," SPE 9260, Papers presented at Fall Annual Meeting of SPE-AIME, Sept. 1980.

Cleary, M.P., et at. "Theoretical and Laboratory Simulation of underground Fracturing Operations", First Annual Report of the MIT UFRAC Project, Resource Extraction Laboratory, MIT, Sept. 1981.

Cleary, M. P., "Mechanisms and Procedures for Producing Favourable

Shapes of Hydraulic Fractures," paper no. SPE 9260, presented at Fall Annual Meeting, SPE-AIME, 1980

Cleary, M. P., et al., Quarterly Reports to Lawrence Livermore Laboratory, published in "LLL Gas Simulation Program," UCRL-50036, available from NTIS, 1978 - 1981

Cleary, M.P., S. K. Wong, V. M. Narendran and A. S. Settari, "General Solutions for Unsymmetric Unidirectional Hydraulic Fracture Propagation in Reservoir Structures," in Reports of Research in Mechanics and Materials, Dept. of Mech. Eng., MIT REL-82-10, Oct.. 1982.

Cleary, M. P., S. Slutsky et al., "Surface Integral Models for Fluid Exchange and Induced Stresses around Fractures in Underground Reservoirs," in Reports of Research in Mechanics and Materials, Dept. of Mech. Eng., MIT REL-82-9, Oct. 1982

Cleary, M. P., "Theory and Application of Hydraulic Fracturing Technology," First Japan-U.S. Seminar on Hydraulic Fracturing and Geothermal Energy, Tokyo, Nov. 1982.

Cleary, M. P., "Applications of Rock Fracture Mechanics," Symposium on Fracture Mechanics," Sendai, Japan, Nov. 1982

Cleary, M. P., and S. K. Wong, "Self-Similar and Spatially-Lumped Models for Tracing the Evolution of Circular Hydraulic Fractures," to be published in Jour. Applied Mechanics, 1963.

Clifton, R. J. and A.S. Abou-Sayed, "On the computation of the three-dimensional geometry of hydraulic fractures", SPE 7943, presented at SPE/DOE Tight Gas Symposium, 1979.

Clifton, R. J. and A. S. Abou-Sayed, "A Variational Approach to the prediction of the three-dimensional geometry of hydraulic fractures", SPE/DOE 9879, 1981 SPE/DOE Symposium, Denver, 1981.

Cooke, C., "Fracturing with high-strength proppant", Jour. Petroleum Tech., 1222-1226, 1977.

Crichton, A., "Crack Interaction in Hydraulic Fracturing of Cement Blocks", B.S. Thesis, MIT, 1980

Daneshy, A. A., "Opening of a Pressurized Fracture in an Elastic

Medium," Paper No. 7616, presented at 22nd Annual Tech. Meeting of CIM Petroleum Society, June, 1971.

Daneshy, A. A., "A Study of Inclined Hydraulic Fractures," Soc. Petroleum Eng. Journal, 61-68, April, 1973.

Daneshy, A. A., "Experimental Investigation of Hydraulic Fracturing Through Perforations," Jour. Pet. Tech., Transactions of SPE, u(255); 1201-1206, Oct., 1973.

Daneshy, A. A., "On the Design of Vertical Hydraulic Fractures," Jour. Petroleum Technology, pps. 83-93, 1973.

Daneshy, A. A., "Hydraulic Fracture Propagation in Layered Formations," Soc. Pet. Eng. Jour., 18, 1, 33-41, 1978.

Daneshy, A. A., " Numerical solution of sand transport in Hydraulic fracturing", Paper No. SPE 5636, 1975.

Delaney, P. T. and D. D. Pollard, "Deformation of host rocks and flow of magma during growth of minette dikes and breccia-bearing intrusions near Ship Rock, New Mexico", U.S.G.S. Prof. paper No. 1202, U.S. Govt. Printing Office, Wash. D.C. 20402, 1981.

Delort, F., and F. Supiot, "Nuclear Stimulation of Oil Reservoirs," pps. 649-661 in Vol. 1 of Proc. Symp. Eng. with Nuclear Explosives, pub. by Am. Nuc. Soc., 1970.

Dobkins, T. A., "Methods to better determine hydraulic fracture height", paper no. SPE 8403, 1979.

Evans, K.F., Holzhausen, G.R. and M.D. Wood, "The mapping of nitrogen gas induced hydraulic fractures in Devonian shale by observation of the associated surface deformation", Paper No. SPE/DOE 8933, presented at Unconventional Gas Recovery Symposium, Pittsburgh, May 1980.

Geertsma, J. and Haafkens, "A Comparison of the theories for predicting width and extent of vertical hydraulically-induced fractures," J. Energy Resources Technology, 101, 8-19, March 1969.

Geertsma, J. and F. de Klerk, "A Rapid Method of Predicting Width and Extent of Hydraulically-Induced Fractures," Jour. of Petroleum Technology, p. 157, Dec. 1969.

Greenfield, R.J., Schuck, L.Z., and T. W. Keech, "Hydraulic Fracture Mapping Using Electrical Potential Measurements," In Situ, 1 (2), pp. 146-149, 1977.

Haimson, B. C. and C. Fairhurst, "In-situ Stress Determination at Great Depth by Means of Hydraulic Fracturing," in "Rock Mechanics-Theory and Practice" (ed. W. H. Somerton), pub. by Soc. Mining Engineers (AIMMPE Inc.), pp. 559-584, 1970.

Haimson, B. C., "The State of Stress in the Earth's Crust," Rev. Geophys. Space Physics 13 (3), 350-352, 1975.

Heintz, J. A., D. T. Barr and M. P. Cleary, "Development of a High Temperature Triaxial Testing System and Measurement of Thermally-Induced Pore-Fluid Pressures in Oil Shale," in Reports of Research in Mechanics and Materials, Dept. of Mech. Eng., MIT, REL-82-3, Jan. 1982.

Howard, G. C. and C. R. Fast, "Hydraulic Fracturing", Soc. Pet. Eng. Monograph, 1970.

Kavvadas, M., K. Y. Lam and M. P. Cleary, "A fully 3-D Numerical Model of Hydraulic Fracturing," in Reports of Research in Mechanics and Materials, Dept. of Mech. Eng., MIT, REL-82-12, expected Nov. 1982.

Keck, L. J., and Schuster, C. L., "Shallow Formation Hydrofracture Mapping Experiment," Trans. ASME, J. Pres. Ves. Tech., Feb., 1978.

Kehle, R. O., "The Determination of Tectonic Stresses Through Analysis of Hydraulic Well Fracturing," Jour. Geophys. Res., 69, 259-273, 1964.

Komar, C. A., Shuck, L. Z., Overbey, W. K., and T. O. Anderson, "Delineating a Subsurface Fracture System in a Petroleum Reservoir - An Experiment," J. Pet Tech., pp. 531-537, May, 1973, (see also JPT, pp. 951-952, Aug., 1975).

Lam, K. Y. and M. P. Cleary, "General Branching and Frictional Slippage at Crack Tips with Applications to Hydraulic Fracturing," in Reports of Research in Mechanics and Materials, Dept. of Mech. Eng., MIT, REL-82-1, Jan.. 1982.

Lam, K.Y., "Numerical Analysis of Crack-Branching and Slippage at a Frictional Interface", S.M. Thesis at MIT, January 1982.

Laspe, C., "How to predict explosive stimulation results," pps. 68-76 in Petroleum Engineering, May 1971

Lewis, J.B., "New Uses of Existing Technology for Controlling Blowouts: Chronology of a Blowout Offshore Louisiana," in Jour. Pet. Technology, 30, 10, 1473-1480, 1978.

Maley, K. R., R. T. Martin and M. P. Cleary, "Design of a Triaxial Test System to Measure Permeability and Deformation in Reservoir Rocks," in Reports of Research in Mechanics and Materials, Dept. of Mech. Eng., MIT, REL-82-4, JAN. 1982

McKee, C.R., Hanson, M.E., and R.W. Terhune, "Permeability from single and Multiple Detontations of Explosive Charges," Lawrence Livermore Labs. Report No. UCRL-78207, May, 1976.

Medlin, W. L. and L. Masse, "Laboratory Investigation of Fracture Initiation Pressure and Orientation," paper No. SPE 6087 of AIMMPE, presented at 51st Annual Fall Conference, Oct. 1976.

Morita, N., Gray, K.E. and C.M. Kim, "Stress-state, porosity, permeability and breakdown pressure around a borehole during fluid injection", pps. 192-197 in Proc. 22nd U.S. Symp. in Rock Mech., MIT, June 1981.

Morris, K. M. and M. P. Cleary, "Servo-Controlled System for Fluid Injection in a Laboratory Simulator of Hydraulic Fracturing," in Reports of Research in Mechanics and Materials, Dept. of Mech. Eng., MIT REL-82-7, Sept. 1982.

Narendran, V. M. and Cleary, M. P., "Elastostatic Interaction of Multiple Arbitrarily Shaped Cracks in Plane Inhomogeneous Regions", Dept. of Mech. Eng., Report REL-8-6, 1982.

Narendran, V. M. and M. P. Cleary, "Analysis of Growth and Interaction of Multiple Hydraulic Fractures," in Reports of Research in Mechanics and Materials, Dept. of Mech., MIT, REL-82-11, Oct.. 1982

Nolte, K. G., "Determination of fracture parameters from fracturing

pressure decline," Paper No. SPE 8341, 1979.
Nolte, K. G. and M. B. Smith, "Interpretation of fracturing pressures," Paper No. SPE 8297, 1979.
Nordgren, R. P., "Propagation of a Vertical Hydraulic Fracture," Jour. of Soc. of Petroleum Engineers, p. 306, 1972.
Papadopoulos, J. M., Cleary, M. P., et al. "Laboratory Simulation of Hydraulic Fracturing," Reports of Resource Extraction Laboratory, M.I.T., 1980, 1981, 1982.
Papadopoulos, J. M., V. M. Narendran and M. P. Cleary, "Laboratory Studies of (Multiple) Hydraulic Fracture Growth and Interaction," in Reports of Research in Mechanics and Materials, Dept. of Mech. Eng., MIT, REL-82-11, Oct. 1982.
Papadopoulos, J. M., B. A. McDonough and M. P. Cleary, "Laboratory Simulation of Hydraulic Fracturing," in Reports of Research in Mechanics and Materials, Dept. of Mech. Eng., MIT, REL-82-2, Jan. 1982
Papadopoulos, J. M., D. Summa and M. P. Cleary, "Development and Result of a Desktop Laboratory Apparatus for Simulation of Hydrafrac (DISLASH)," in Reports of Research in Mechanics and Materials, Dept. of Mech. Eng., MIT, REL-82-14, OCT. 1982.
Perkins, T. K. and L. R. Kern, "Widths of Hydraulic Fractures," Journal of Petroleum Technology, p. 937, 1961.
Petersen, D. R. and M. P. Cleary, "Unsteady Cross-Section Model of Hydraulic Fracture Growth with an Overview of its Applications," in Reports of Research in Mechanics and Materials, Dept. of Mech. Eng., MIT, REL-82-8, OCT, 1982.
Pierce, Aaron E., Vela, Saul, and Koonce, K. T., "Determination of the Compass Orietnation and Length of Hydraulic Fractures by Pulse Testing," SPE 5132, SPE Annual Fall Meeting, 1974
RPower, D. V., Schuster, L., Hay, R. , and Twombley, J., "Detection of Hydraulic Fracture Orientation and Dimensions in Cased Wells," SPE 5626, SPE Annual Fall Meeting, 1975
Pugh, T. D., McDaniel, B. W. and R. L. Seglem, "A new fracturing

technique for Dean Sand," J. Pet. Tech., pp. 167-172, Feb. 1978.

Salz, L.B., "Relationship between fracture propagation pressure and pore-pressure", Paper No. SPE 6870, Fall Conference, 1977.

Schmidt, R. A., "Fracture Toughness Testing of Rock," Closed Loop (Magazine of Mechanical Testing), 5, 2, 1-12, Nov. 1975.

Schmidt, R.A., Boade, R.R. and R.C. Bass, "A New Perspective on Well-Shooting -- the Behavior of Contained Explosions and Deflagrations", Paper No. SPE 8346, Fall Annual Meeting, 1979

Settari, A. and M. P. Cleary, "Three-dimensional Simulation of Hydraulic Fracturing", Paper No. SPE 10504, presented at Symposium on Numerical Simulation, New Orleans, Feb. 1982, to be published in SPE Jour.., 1983.

Settari, A. and M. P. Cleary, "Development and Testing of a Pseudo-Three-Dimensional Model of Hydraulic Fracture Geometry (P3DH)," paper no. SPE 10505, presented at Symposium on Numerical Simulation, New Orleans, Feb. 1982, to be published in SPE Jour., 1983.

Simonson, E. R., Abou-Sayed, A. S., and R. J. Clifton, "Containment of Massive Hydraulic Fractures," Soc. Pet. Eng. Jour., 18, 1, 27-32, 1978.

Sinclair, A. R., "Heat Transfer effects in deep well fracturing", Jour. Pet. Tech., 1484-1492, Dec. 1971.

Smith, M. B., Holman, G. B. Fast, C. R., and Covlin, R. J., "The Azimuth of Deep, Penetrating Fractures in the Wattenberg Field, J. Pet. Tech., Feb. 1978

Swolfs, H. S. and C. E. Brechtel, "The Direct Measurement of Lon Term Stress Variations in Rock," proc. 18th Symposium on Rock Mech pub. by Colorado School of Mines Press, Golden, Colo. 80401, 4C5-1 to 4C5-3, June 1977.

Tester, J. W., Potter, R. M. and R. L. Bivins, "Interwell tr analyses of hydraulically fractured granite geothermal reservo Paper No. SPE 8270, Sept. 1979.

Voegele, M.D. and A.H. Jones, "A wireline hydraulic fracturing for the determination of in-situ stress contrasts", Paper No. S

DOE 8937, presented at Unconventional Gas Recovery Symposium, Pittsburgh, May 1980.

Warpinski, N.R., Schmidt, R.A., and Northrop, D.A., "In-situ stresses: the predominant influence on hydraulic fracture containment", Jour. Pet. Tech. 653-664, March 1982.

White, J.L. and E.F. Daniel, "Key Factors in MHF Design", Jour. Pet. Tech. 33, 8, 1501-1512, August 1981.

Williams, B.B., Gidley, J.L. and R.S. Schechter, "Acidising Fundamentals", SPE Monograph, 1979.

Wong, S. K. and M. P. Cleary, "Numerical Analysis of Axisymmetric and Other Crack Problems Related to Hydraulic Fracturing," in Reports of Research in Mechanics and Materials, Dept. of Mech. Eng., MIT, REL-81-4, June 1981

Young, C., "Evaluation of stimulation technologies in the Eastern Gas Shales Project," pps. G5/1-16, Proc. DOE Symposium on Enhanced Oil Recovery, Vol. 2, 1978.

Zoback, M. D. and D. D. Pollard, "Hydraulic Fracture Propagation and the Interpretation of Pressure-Time Records for in-situ Stress Determinations," pp. 14-27 in Vol. 1 of 19th U.S. Symposium on Rock Mechanics, Mackay School of Mines, 1978

Zoback, M. D., Rummell, F., Jung, R., and C. B. Raleigh, "Laboratory Hydraulic Fracturing Experiments in Intact and Pre-fractured Rock," Int. Jour. Rock Mech., 14, 49-58, 1978.

ACKNOWLEDGEMENTS

The author wishes to thank particularly K.Y. Lam, Jim Martinez and Mark Mear for their help in preparation of this manuscript. The overall work described has been supported mainly by the author's MIT UFRAC Project, sponsored by sixteen energy companies, by the National Science Foundation, by a Soderberg career development chair and by various internal sources at MIT.

AUTHOR INDEX

Abel J.F. 203
Abou-Sayed A.S. 385,401
Abramowitz M. 433
Achenbach J.D. 229,245,279
Addy S.K. 230
Ahmed U. 401
Alani G. 70
Allen T.O. 401
Alm O. 121
Anderson T.O. 389
Andrews W.R. 125
Ash R.L. 321
ASTM 6,69,75,258
Atchison T.C. 321
Atkinson B.K. 119,276
Atluri S.N. 160
Awaji H. 70

Baker B.R. 274
Barker D.B. 321,344
Barker L.M. 20,80
Barr D.T. 390,464
Barsoum R.S. 164
Bartelds G. 119
Barton C.R. 90,92
Bass K.C. 353,390
Beech J.F. 98
Begley J.A. 4
Benzley S.E. 26
Berkvist H. 278
Berlie J. 85
Bhandari S. 321
Bieniawski Z.T. 158,307
Biot M.A. 230
Birch F. 22
Blackburn W.S. 119
Blandford G. 172,188
Bligh T.P. 324
Boade R.K. 353,390
Bombolakis E.G. 158
Bowie O.L. 92,304
Brace W.F. 22
Bradley C. 264
Brillhart L.V. 216
Broberg B.K. 274
Broek D. 6
Brown Jr. W.F. 14,22,172
Bubsey R.T. 113
Budiansky B. 65,127
Burger C.P. 216
Bush A.J. 98

Carlsson J. 57
Catalano D. 188
Chan S.F. 163
Cherepanov G.P. 11
Chona R. 263
Chong K.P. 90
Christianovich S.A. 385
Christie D.G. 216
Clark G.B. 31
Clark P.E. 399
Cleary M.P. 46,80,92,278,383
Clifton R.J. 385
Cook N.G.W. 4,31
Cooke C. 402
Cooper G.A. 85
Cooper P.W. 354
Corten H.T. 23,65
Costin L.S. 82,89,144
Cotterell B. 158
Coulomb C.A. 4

Coursen D.L. 321
Covlin R.J. 389
Cranz C. 215
Crichton A. 389,443
Cruse T.A. 171

Dally J.W. 210,226,253,259,301
314,321,371
Daneshy A.A. 385,399
Daniel E.F. 385,401
DeBremaecker J.C. 371
de Klerk, F. 385
de Koning, A.U. 119
Delale F. 92
Delaney P.T. 393
deLorenzi H.G. 165
Der V.K. 268
Dey S. 230
Dick E. 199
Dobkins T.A. 389
Douglass P.M. 90
Durelli A. 254
Duvall W.I. 321

Erdogan F. 92,176,347
Eshelby J.D. 57,277
Etheridge M.J. 314
Evans K.F. 389
Ewing J.M. 230
Ewing W.M. 379

Fairhurst C. 321
Fast C.R. 389,401
Finnie I. 85
Föppl L. 210
Forootan-Rad P. 72
Fossum A.F. 288
Foster C.L. 303
Fourney W.L. 115,209,290,301
Freese C.E. 92,165
Freund L.B. 274,284
Friedman M. 72,133

Gallagher R.H. 196
Gangal M.D. 196
Gariepy S. 90
Geertsma J. 385
German M.D. 165
Gidley J.L. 390
Goodier J.N. 324
Gray K.E. 388
Greenberg D.P. 196
Greenfield R.J. 389
Griffith A.A. 4,10,72,151
Gunsallus K.L. 98
Gupta I.N. 358,371

Haber R.B. 196
Hagen T.N. 321
Hahn G.T. 13,74
Haimson B.C. 387
Handin J. 72
Hangen J.A. 90,93
Hardy M.P. 13,31
Hawkes I. 90
Hay R. 389
Heintz J.A. 464
Henry J.P. 90
Heuze F.E. 193
Hills E.S. 22
Hoagland R.G. 13,74
Hoek E. 158,192
Holloway D.C. 115,218,268,344,372
Howard G.C. 401
Holzhausen G.R. 389
Hübner H. 121
Huddle C.W. 17
Hussain M.A. 181
Hutchinson J.W. 127

Ingraffea A.R. 14,66,70,75,98,151
Irwin G.R. 3,8,74,258,260,304,315

Jaeger J.C. 4,7,31
James L.A. 22
Jardetzky W.S. 379
Johansson C.H. 37,321,367
Johnson A.M. 90
Johnson J.N. 20
Jones A.H. 388,401
Jones M.H. 14
Jordan W.B. 163
Jung R. 388
Just G.D. 321

Kaufman S.G. 14
Kavvadas M. 445
Keck L.J. 389

Keech T.W. 389
Kerkhof F. 216
Kern L.R. 385
Kihlstrom B. 303,321
Kim C.M. 388
Kim K. 77
Kisslinger C. 358,371
Kleinlein W.F. 121
Knowles J.K. 57
Kobayashi A.S. 160,222,264
Kobayashi T. 115,268
Kolosov G.W. 6
Kolsky H. 216,229
Komar C.A. 389
Koonce K.T. 389
Kordisch H. 182
Kuske A. 253
Kutter H.K. 304,321

Ladegaard-Peterson A. 302
Lajtai E.Z. 196
Lajtai V.N. 196
Lam K.Y. 445
Landes J.D. 4,131
Langefors U. 303,321
Lehnigk S.H. 287
Lewis III, D. 371
Liebowitz H. 91
Liggett J.A. 172,189
Lindqvist P.A. 133
Lundborg N. 37,321
Lutz T.J. 14,75

Maley K.R. 466
Manu C. 167
Martin R.T. 466
Masse L. 388
Maue A.W. 274
Maxwell J.C. 210
McCabe D.E. 131
McClintock F.A. 14
McDaniel B.W. 385
Medlin W.L. 388
Mellor M 90
Merkle J. 23,85
Miller B.L. 92
Mills W.J. 22
Mindess S. 14
Moavenzadeh F. 72

Mönch E. 210
Morita N. 388
Morris K.M. 460
Mubeen A. 75
Munari A.C. 90
Munz D. 77,113
Muskhelishvili N.I. 6,37

Nadeau J.S. 14
Nakagaki M. 160
Narendran V.M. 46,440
Nelson F.G. 14
Nelson P. 98
Nordgren R.P. 385
Northrop D.A. 354,387

Olofsson T. 98,133
Ouchterlony F. 4,35,69,90,304,326
Overbey W.K. 389
Ozdemir L. 199

Pacquet J. 90
Papadopoulos J.M. 389,453
Paris P.C. 6,18,85,125
Parks D.M. 160
Paul B. 196
Pekeris C.L. 250,374
Peng S. 90
Perkins T.K. 385
Perrucchio R. 203
Persson P.A. 37,321,367
Peter K. 199
Peterson D.R. 440
Pierce A.E. 389
Pierce W.S. 114
Pollard D.D. 393
Porter D.D. 321
Post D. 216,258
Power D.V. 389
Press F. 379
Pu S.L. 181
Pugh T.D. 385

Raleigh C.B. 388
Reinhardt H.W. 216
Rice J.R. 65,82
Riley W.F. 210,253
Rinehart J.S. 231
Roberts A.P. 401

Robertson G. 253
Rose L.R.F. 280
Rosenfield A.R. 13,74
Rossmanith H.P. 13,209
Rudnicki J.W. 4
Rummel F. 119,388

Sanford R.J. 259
Saouma V.E. 66,188
Sato S. 70
Schardin H. 215
Schechter R.S. 393
Schilling P.E. 14
Schmidt D.W. 125
Schmidt R.A. 13,18,26,75,90,133
352,387,390
Schuck L.Z. 389
Schulman M.A. 196
Schuster C.L. 389
Schwalbe K.H. 127
Seglem R.L. 385
Settari A. 400,416
Setz W. 127
Shannon Jr. J.L. 113
Shephard M.S. 196
Shih C.F. 165
Shukla A. 222,254,290
Sih G.C. 6,91,176,347
Simonson E.R. 385
Singh B. 350
Slutsky S. 440
Smith C.W. 264
Smith D.G. 290
Smith J.W. 90
Smith M.B. 389
Snyder L. 199
Sommer E. 182
Srawley J.E. 22,109,113,172
Starfield A.M. 321
Stecher F.P. 358
Stegun I.A. 433
Sternberg E. 57
Steverding B. 287
Stillborg B. 133
Summa D. 389,453
Sun Z. 105,119
Swan G. 98,118,121
Swan P. 31
Switchenko P.M. 90

Tada H. 8
Tancrez J.P. 90
Thau S.A. 226,253
Timoshenko S. 324
Tolikas P.K. 279
Tracey D.M. 163,165
Tresca H. 16
Tseng A.A. 66,70
Tuba I.S. 163
Turner C.E. 110
Tuzi Z. 213
Twombley J. 389

Uenishi K. 90
Underwood J. 181

Voegele M.D. 388
Voight B. 90
von Mises, R. 13

Walling H.C. 354
Wan F.D. 199
Wanhill R.J.H. 119
Wang F.D. 31
Wang S.S. 65
Warpinski N.R. 354,387
Wells A.A. 3,216,258
Westergaard H.M. 6,262
White J.L. 385,401
Wijk G. 119
Wilkening W.W. 93
Williams B.B. 393
Williams J.A. 22
Wilson R.B. 171
Wilson W.K. 163
Winter R.B. 119
Wong S.K. 403,440
Wood M.D. 389
Woods R. 371

Yau J.F. 65
Yoffe E. 272
Young C. 390

Zheltov Yu.P. 385
Zoback M.D. 388

SUBJECT INDEX

Air-filled borehole 364
Aluminum 11,231
Angle-notched plate 156
Anisotropy 25,90
Automatic remeshing 191
Apparent fracture toughness 14
Approximate fracture toughness 79
Arrester 26

Back stress 400
Barrier branching 330
Basalt 306
Bench blasting 36,342,373
Bend specimen, 3P- , 19,71
Blasting 31,301
- configuration 31
Blunting 84
Borehole 32,302,349
- crushing 362
- , jacketed 465
- pressure 313,366
- stresses 325
Boulder splitting 36
Boundary condition 234
Boundary element method 170
Branching, crack 223,330,445
Burial depth 379
Buried explosion 250,323,373

Camera, high speed 215
Cavity, underground 195
Channel flow 396
CENRBB-specimen 110
Circular crack 8,439
Cohesive forces 6,83
Cohesive joint 234
Compression 16,155
Complex analysis 36
Compliance 100
- method 123
Computer graphics 154
- program 154,190,383
Confining stress 17
Control, fracture 301,310
- of crack initiation 303
Core specimen 22,98,114
Corrosion 17
Crack 6,31,69,151,209,229,253,271, 301,321,341,353,371,383
- branching 223,330,445
- closure 77
- curving 154
- extension 286
- growth, subcritical 17
- increment length 183
- initiation 154,178,281,458
- mouth opening displacement C(M)OD 22,78
- , multiple 153
- opening 359
- pressurization 355,457
- propagation, dynamic 267,311
propagation, mixed-mode 267
- resistance 114
- system 32
- tip opening displacement C(T)OD 10
- tip zone 12
- trajectory 155
- tunneling 119

CR-39 158
CT-specimen 22,71,266

Data Analysis 256
DCB-specimen 267
Deflagration, propellant 363
Delay, detonation 349
Deposition of proppant 399
Detachment 249
Detonation pulse 244
Diffraction, wave 224,241,284,290
Displacement control 82
Dissimilar media 245
Divider 26,89
Double compliance method 125
Dynamic holography 217,301,372
- photoelasticity 213,301

Edge-cracked specimen 19,71
Ekeberg marble 102,126,135
Elastodynamics 209
Energy rate balance 71
Energy release rate 31,69,271
Epoxy 11,231,321,341,353
Expansion force 49
Explosive 32,301,391
Exposure 214

Failure curve 122
- mechanics 3,222
Fatigue 17
Finite elements 162
Flaw 307,321,341
- initiation 332
- , natural 307
Flow-back 401
Fluid pressure 391
Frac-fluid 399
Fracture control 301
- creation in rock 391
- criteria 10,181
- height 405
- , joint-initiated 344
- mechanics 3,271,301,383
- modelling 159
- propagation 151,189,271 301
- , stem-induced 361

Fracture toughness 11,74,306
- - , apparent 14
- - testing 19,69
- , underground 391
Fragmentation 31,321
Fragment size 349
Framing rate 214
Friction 7
Fringe pattern 253,376

Gabbro 7
Gas flow 354
- penetration 45
- pressure 317
- volume 40
- well stimulation 353
Geothermal energy 152
Glass 158,231
Granite 7,90,136,157,231,306
- , Bohus 136
- , Chelmsford 73,90
- , Westerly 11,25,76,87,90,306
Griffith criterion 10,72
Grooving 307

Heat transfer 398
Holography, dynamic 217,301,372
Homalite-100 315
Hybrid method 160
Hybridisation 428
Hydrafrac model 406
Hydraulic fracturing 35,383
Hydrostatic compression 16,25
Hysteresis 132

Image 214
Impact breaking 36
Impedance, fracture 445
- , wave 229
Indiana limestone 11,21,73,76,90,156 306
Initiation, crack growth 83,154,178
Interaction, crack-wave 222,271,290
- , fluid-solid 387
Interactive computer graphics 155
Interface 234,295
- separation apparatus 449
Interferometry 218

Internal pressure 35
Isochromatic fringe 254
Isoparametric element 163
Isotropy, transverse 88

J_{Ic} 21,85
J-Integral 21,57,82
- resistance 82
Joint 234,295,342
- -initiated fracture 344
Jump, crack 279

K_{Ic} 19,69,74
K_J 140
K_Q 75
K_R-curve 126,135
K-calibration 20

Laboratory simulation 448
Lateral flow 425
Layer detachment 249
Layered rock 241,321,341
Limestone 11,21,73,76,90,156,306
Linear-elastic fracture mechanics LEFM 3,57,153
Loading rate 361
Loose joint 238

Marble 7,90,102
MCT-specimen 266
Micro-cracking 12,79
Mixed-mode 10,153,174,259
Model scale blast 33,301
Modes of fracture 5
Momentum transfer 232
Monitoring of crack growth 460
Multiple cracks 439

Nonlinearity 132,383
Notched beam 22,69
- borehole 302

Oblique wave incidence 282
Oil shale 89,94
Over-burden pressure 6

Path-independent integral 57
Pattern, crack 327
Permeability 466
Photoelastic fringe pattern 253,290
Photoelasticity 210
- , dynamic 213
Photography, high speed 213
Plane problem 37
Plastic zone 13
Plasticity adjustment 79
Plate wave 240
PMMA 33,158
Polariscope 212
Pore pressure 391
- - induced cracking (PPIC) 465
Pre-cracking 23,74
-stress 237
Pressurized crack 42,355,457
- borehole 42,301,353,383

Quarter-point singular element 164
Quartzite 306

Radial crack system 41,303,327
Rayleigh-wave 225,241,373
R-curve 72,84,115
Reflection 236
Re-initiation 446
Remeshing technique 154
Reservoir 383
Resolution 214
Rheology, fluid 396,448,462
Rock 7,31,69,231
- mechanics 152
- splitting 35,63

Sandstone 7,90,306
Secant offset procedure 108
Secondary crack 156
Segregation, proppant 400
Shale 89,92,94
Shear layer cracking 298
Shock wave 367
Short rod specimen 80,114
Side grooving 308
Simulation of fracturing 445
Single-edge-crack-round-bar-bending (SECRBB)specimen 98
Singularity element 163
Slate 90

Slipping 445
Small scale yielding 15
Specific work of fracture 70
Specimen geometry 70
- requirements 75
Stability criterion 52
Star crack 34,41
Steel 11,231
Stem-induced fracture 361
Stimulation, gas well 353
Straight-through-crack assumption STCA 113
Strain energy density 179
- - release rate 5,53
Stress, confining 17
- corrosion cracking 17
- intensity factor 5,106,159
- - - , dynamic 274,316
- - resistance 142
- wave 220,282
Subcritical crack growth 17
- failure cycle 131
Superposition 234
Surface wave 225,241,373
Synchronisation 214
System of cracks 31

Temperatur, high 464
Tensile strength 11
Tension test 22
Thermal stress 392
Trachyte 7
Transverse, short 26
Tunnel boring 36
- crack 8,246
Tunneling of cracks 119

Underground cavity 195
Uniform crack growth 54
Unloading 126

Vibration, ground 371
Viscoelasticity 396

Wave, cylindrical 232,257
- diffraction 224
- , nonplanar 243
- , plane 229,257
- propagation 220,229
- reflection 236
- , spherical 232,257
Wedge, cracked 59
Wellbore 354
Westerly granite 11

Yield criterion 13
Yielding, small scale 15
Yield strength 69